SO-BKP-413

probability;
decision;
statistics

james v. bradley
Professor of Psychology
New Mexico State University

probability;
decision;
statistics

prentice-hall, inc.
Englewood Cliffs, New Jersey

Library of Congress Cataloging in Publication Data

BRADLEY, JAMES VANDIVER,
    Probability, decision, statistics.

    Includes index.
    1. Probabilities. 2. Statistical decision.
3. Mathematical statistics. I. Title.
QA273.B847   519   75-1083
ISBN 0-13-711556-3

© 1976 by Prentice-Hall, Inc.
Englewood Cliffs, New Jersey

10   9   8   7   6   5   4   3   2   1

Printed in the United States of America

PRENTICE-HALL INTERNATIONAL, INC., *London*
PRENTICE-HALL OF AUSTRALIA, PTY. LTD., *Sydney*
PRENTICE-HALL OF CANADA, LTD., *Toronto*
PRENTICE-HALL OF INDIA PRIVATE LIMITED, *New Delhi*
PRENTICE-HALL OF JAPAN, INC., *Tokyo*
PRENTICE-HALL OF SOUTHEAST ASIA (PTE.) LTD., *Singapore*

contents

preface   xvii

1   introduction   *1*

2   description   *3*

2–1   Describing Entire Sets   4
    2-1-1 Accounting separately for every value, 4
    2-1-2 Accounting for intervals of values, 6
    2-1-3 Describing sets of two-valued units, 10

2–2   Summary Indices   12
    2-2-1 Indices of location, 12
    2-2-2 Indices of dispersion, 13
    2-2-3 Appropriateness of summary indices, 15

2–3   Populations and Samples   16

2–4   Problems   18

3   possibilities   *20*

3–1   Permutations   20
    *3-1-1 Number of permutations of n things
    taken r at a time, 21*

3-1-2 Number of permutations of n things, 22

*3-1-3 Number of distinguishable permutations of n things
some of which are alike, 24*

*3-1-4 Permutations in experimental design: Counterbalancing of
spurious sequential effects, 27*

3–2 Combinations 29

*3-2-1 Number of combinations of n things
· taken r at a time, 30*

3-2-2 Pascal's triangle, 32

*3-2-3 Relationship between combinations and
distinguishable permutations, 32*

3–3 Groupings
(Allocations Of Things To Containers) 33

*3-3-1 Number of allocations of n things
to k unique categories,
each category receiving a specified number of things, 33*

*3-3-2 Number of divisions of n things
into k groups of specified sizes, 36*

*3-3-3 Total number of allocations of n things
to k distinguishable containers, 38*

3–4 Sampling Distributions 39

3-4-1 The set of all possible samples, 40

3-4-2 The set of means of all possible samples, 41

3-4-3 Sampling distributions, 42

3–5 Problems 47

4 randomness *50*

4–1 Equal Likelihood,
Its Implications And Consequences 50

4–2 Quantification of Likelihood 55

4–3 Randomness 56

4-3-1 Random selection, 56

4-3-2 Random sampling, 57

*4-3-3 Likelihood that a sample statistic will have
a certain value under random sampling, 57*

4–4 Attaining Randomness 58

4–5 Prevalence And Consequences of Nonrandomness
(Both Innocuous And Drastic) 61

4–6 Random Variables 64

4–7 Randomness As A Prototype 65

4–8 Problems 66

5   probability   *68*

5–1   Chance   68

5–2   The Concept of Probability   69
  5-2-1 Vernacular probability, 69
  5-2-2 The reference class of similar incidents, 70

5–3   Definitions of Probability   73
  5-3-1 Radically objective probability, 73
  5-3-2 Objective probability, 74
  5-3-3 Subjective probability, 74

5–4   Methods of Obtaining Numerical Probabilities   75
  5-4-1 The deductive method, 75
  5-4-2 The past relative frequency method, 81
  5-4-3 The subjective estimation method, 82
  5-4-4 Critique, 83

5–5   Problems   89

6   general probability laws   *91*

6–1   The Urn Model   91

6–2   The Range Of Proportions
       Or Of Probabilities   95

6–3   The Proportion Of Units Having A Certain Characteristic
       Or The Probability That An Incident
       Will Have It   96

6–4   The Proportion Of Units Having The Complementary
       Characteristic Or The Probability That An Incident
       Will Have It   96

6–5   The Proportion Of Units
       Having Either Or Both Of Two Characteristics
       Or The Probability That An Incident Will Do So   97

6–6   The Proportion Of Units
       Having One Or More Of Several Characteristics
       Or The Probability That An Incident Will Do So   99

6–7   Proportions Within A Subgroup,
       Or Conditional Probabilities   102

6–8   Bayes' Formula   104

6–9   The Proportion Of Units
Having Both Of Two Characteristics
Or The Probability That An Incident Will Do So   109

6–10 The Proportion of Units
Having All Of Several Characteristics
Or The Probability That An incident Will Do So   111

6–11 Computational Strategems   114

6–12 Problems   116

7   specific probability laws   *121*

7–1   Family Of Laws To Be Considered
And Circumstances Under Which They Apply   121

7–2   The Hypergeometric Probability Law   123

7–3   The Multivariate Hypergeometric Probability Law   126

7–4   The Binomial Probability Law   130

7–5   The Multinomial Probability Law   133

7–6   Relationships Between Preceding Specific Laws;
And Some Special Features   136

7–7   The Negative Binomial Probability Law   139

7–8   The Geometric Probability Law   141

7–9   The Poisson Approximation
To The Binomial Probability Law   142

7–10 The Poisson Probability Law   146

7–11 Computational Strategems   148
7-11-1 Point probabilities, 149
7-11-2 Recursion formulas, 150
7-11-3 Cumulative probabilities, 151
7-11-4 Tables of probabilities, 154

7–12 Problems   155

8   games   *161*

8–1   Value   161
8-1-1 Objective value, 162
8-1-2 Expected value, 162
*8-1-3 Hazards in using expected value
as a criterion for action, 163*

8-1-4 *Utility or subjective value, 165*
8-1-5 *Measuring utility, 166*
8-1-6 *Expected utility, 167*

8–2 Two-Person Zero-Sum Games 168

8-2-1 *General theory, 168*
8-2-2 *Two-action games, 170*
8-2-3 *Mixed strategies, 173*
8-2-4 *Multi-action games, 175*
8-2-5 *Summary and critique, 175*

8–3 Problems 177

9 decisions *181*

9–1 Decision-Making Under "States Of Nature" 181

9-1-1 *The game against nature, 181*
9-1-2 *Degrees of knowledge
about the state of nature, 182*
9-1-3 *Criteria and methods for choosing actions, 183*

9–2 The Maximum Utility Criterion 184

9–3 The Maximum Expected Utility Criterion 185

9-3-1 *Method, 185*
9-3-2 *Examples, 185*
9-3-3 *Use of sample information to revise the $P(\theta_j)s$, 188*
9-3-4 *The value of the sample, 190*
9-3-5 *Taking the cost of the sample into account, 192*
9-3-6 *Use of subjective probabilities, 193*
9-3-7 *Critique, 195*

9–4 The Maximum Utility Under Most Probable State
Criterion 195

9-4-1 *Method, 195*
9-4-2 *Example, 196*
9-4-3 *Use of sample information, 197*
9-4-4 *Critique, 197*

9–5 The Maximum Likelihood Criterion 198

9-5-1 *Method, 198*
9-5-2 *Rationale, 198*
9-5-3 *Example, 199*
9-5-4 *Critique, 200*

9–6 The Maximin Expected Utility Criterion 201

9-6-1 *Method, 201*
9-6-2 *Rationale, 201*
9-6-3 *Example, 202*
9-6-4 *Critique, 203*

9–7   The Maximin Criterion   206

    9-7-1 Method, 206
    9-7-2 Example, 206
    9-7-3 Critique, 207

9–8   Problems   208

10   decisions about the state of nature   *213*

10–1   Expedient Announcements Of The State Nature   213

10–2   Estimating The State Of Nature   215

    10-2-1 Estimation, 215
    10-2-2 The most probable state estimate, 215
    10-2-3 The conditionally most probable state estimate, 216
    10-2-4 The maximum likelihood estimate, 221

10–3   Choosing Between Conflicting Hypotheses   222

    10-3-1 Rationale, 222
    10-3-2 Choosing accurately: the maximum likelihood estimate, 224
    10-3-3 Choosing cautiously: the likelihood ratio test, 228

10–4   Problems   234

**11   hypothesis testing—
the neyman–pearson procedure   *240***

11–1   Case Where There Are
      Only Two Possible Values Of θ 241

    11-1-1 Illustration, 241
    11-1-2 Generalization, 246

11–2   Case Where There May Be Any Number
      Of Possible Values Of θ—Left-Tail Test   248

11–3   Case Where There May Be Any Number
      Of Possible Values Of θ—Right-Tail Test   252

11–4   Case Where There May Be Any Number
      Of Possible Values Of θ—
      The Conventional Two-Tail Test   254

    11-4-1 Illustration, 254
    11-4-2 Critique, 255

11–5   Possible Varieties of Cases   259

11–6   Summary of Neyman-Pearson
      Hypothesis Testing Procedures   260

11–7   Problems   264

12 generally applicable tests and estimates *271*

12–1   Binomial Tests   272

> 12-1-1 Test for the median, 272
12-1-2 Test for the median difference, 274
12-1-3 Test for trend, 276
12-1-4 Test for concentration, 277

12–2   Hypergeometric Tests   278

12-2-1 Test for dependence, 278

12–3   Rank Tests   281

12-3-1 Treatment populations, 281
12-3-2 Special features of the tests, 282
12-3-3 Hotelling and Pabst's test for correlation, 284
12-3-4 Wilcoxon's signed-rank test
   for differences in treatment effects, 286
12-3-5 Wilcoxon's rank-sum test
   for nonidentical population distributions
   (or unequal treatment effects), 289
12-3-6 Friedman's test
   for nonidentical treatment effects, 293
12-3-7 The Kruskal-Wallis test
   for nonidentical population distributions, 298
12-3-8 Tests of main effects and interactions
   when there are two causal variables, 302
12-3-9 Treatment of tied values, 307

12–4   Estimation Of Population Indices   311

12-4-1 Estimation versus hypothesis testing, 311
12-4-2 Interval estimation, 312
12-4-3 Confidence intervals
   for population percentiles, 313
12-4-4 Point estimation, 316

12–5   Problems   318

13 expected values and moments *329*

13–1   Laws Of Expectation   329

13-1-1 Laws involving a single variable
   or function of it, 329
13-1-2 Laws involving several basic variables, 332

13–2   Moments   335

13-2-1 General case, 335
13-2-2 Moments about the origin, 335
13-2-3 Central moments, 335
13-2-4 Standardized values, 336
13-2-5 Standardized central moments, 337
13-2-6 Example, 338

13–3   Laws Of Variance   339

13–4   Expected Value And Variance Of The Sample Mean   342
    13-4-1 Formulas, 342
    13-4-2 Proof of formulas,
        in case where population is undepletable, 343
    13-4-3 Proof of formulas
        in case where population is depletable, 344

13–5   Expected Values And Variances
      Of Other General Sample Statistics   346
    13-5-1 Expected value of the sample variance, 347
    13-5-2 Expected value and variance
        of the difference between sample means, 348
    13-5-3 Expected value and variance
        of the sum of sample observations, 348

13–6   Expected Values And Variances
      Of Sample Statistics
      Following Specific Probability Laws   349
    13-6-1 Formulas, 349
    13-6-2 Proofs of formulas, 351

13–7   Mean-Unbiased Estimators   354

13–8   Problems   356

14   the distribution of the sample mean   *360*

14–1   The Exact Distribution of $\bar{X}$ 360
    14-1-1 Case where sampled population is dichotomous, 360
    14-1-2 General case where sampled population
        is discretely distributed, 361
    14-1-3 Critique, 363
    14-1-4 Information about $\bar{X}$ when population
        distribution is incompletely known, 363

14–2   Mean And Variance Of The Distribution Of $\bar{X}$
      And Behavior Of Var($\bar{X}$) As *n* Increases   364

14–3   Tchebycheff Inequalities   365
    14-3-1 Tchebycheff's inequality, 365
    14-3-2 Complementary and alternative forms
        of Tchebycheff's inequality, 368
    14-3-3 Limitations of Tchebycheff's inequality, 369
    14-3-4 Tchebycheff's inequality for the sample mean, 370
    14-3-5 Tchebycheff's inequality for the sample proportion
        and other sample statistics, 372
    14-3-6 Critique, 375

14–4   The Law Of Large Numbers   376

14-4-1 The general case, 376
14-4-2 Bernoulli's theorem, 378
14-4-3 Illustrations, 379

14–5   The Central Limit Effect   382

14-5-1 The normal distribution, 382
14-5-2 The central limit theorem, 386
14-5-3 The central limit effect, 388

14–6   Problems   433

15   normal-approximation statistics   *437*

STATISTICAL METHODS BASED UPON
THE NORMAL APPROXIMATION
TO THE DISTRIBUTION OF $\bar{X}$

15–1   The One-Sample Z Statistic   437

15-1-1 Confidence intervals for μ,
    based on the Z statistic, 438
15-1-2 Test of an hypothesized value for μ,
    using the one-sample Z statistic, 441
15-1-3 The difference-score Z statistic, 443

15–2   The Two-Sample Z Statistic   447

15-2-1 Confidence intervals for $\mu_X - \mu_Y$
    based on the two-sample Z statistic, 448
15-2-2 Test of an hypothesized value of $\mu_X - \mu_Y$,
    using the two-sample Z statistic, 450

15–3   Critique of Z Statistics   452

15–4 $\hat{Z}$ Statistics   453

15-4-1 Rationale, 453
15-4-2 The one-sample $\hat{Z}$ statistic, 457
15-4-3 The difference-score $\hat{Z}$ statistic, 459
15-4-4 The two-sample $\hat{Z}$ statistic, 460

15–5 t Statistics   461

15-5-1 The t distributions, 461
15-5-2 The one-sample t statistic, 462
15-5-3 The difference-score t statistic, 466
15-5-4 The two-sample t statistic, 468

15–6   General Critique   478

15–7   The Normal Approximation
       To The Distribution of Successes   479

15-7-1 The normal approximation
    to the binomial distribution, 479

15-7-2 *The normal approximation*
       to the Poisson distribution, 482
15-7-3 *The normal approximation*
       to the hypergeometric distribution, 482

15–8   Chi-Square Statistics   483

15-8-1 *The chi-square distributions, 483*
15-8-2 *The chi-square approximation*
       to the multinomial, 484
15-8-3 *The chi-square test*
       of hypothesized population proportions, 487
15-8-4 *The chi-square test of independence, 489*
15-8-5 *The chi-square test of goodness-of-fit, 493*

15–9   Problems   496

appendix a   an algebraic proof of the central
             limit theorem   *504*

appendix b   tables

       Table 1   Random Digits   514

       Table 2   Squares of Integers   515

       Table 3   Binomial Probabilities   518

       Table 4   Poisson Probabilities   532

       Table 5   Critical Values for Binomial Tests
                 for Median, Median Difference, or Trend   541

       Table 6   Critical Values for Fisher's Hypergeometric
                 Test for Dependence   551

       Table 7   Critical Values for Hotelling
                 and Pabst's Test for Correlation   558

       Table 8   Critical Values
                 for Wilcoxon's Signed-Rank Test   559

       Table 9   Critical Values
                 for Wilcoxon's Rank-Sum Test   561

       Table 10 Critical Values for Friedman Test,   565

       Table 11 Cumulative Probabilities
                 for a Standardized Normal Variable ϕ 566

       Table 12 Values of t with v Degrees
                 of Freedom Having Certain
                 Cumulative Probabilities   568

Table 13 Values of Chi-Square with v Degrees
of Freedom Having Certain
Cumulative Probabilities   569

**appendix c   answers to odd numbered
problems** *570*

index *579*

preface

Probability, decision-theory, and statistical methods all present mathematical models for the solution of real-world problems. In order to use these models properly, the practitioner must understand both the mathematical and the empirical conditions qualifying their application. However, textooks in this field often tend to be either ritualistic "cookbooks" devoid of mathematical rationale, or mathematical treatises so cryptic that the implicit logical rationale may escape even the mathematically sophisticated reader. This is unfortunate for *the fundamentals of probability, decision-theory, and statistics can be thoroughly derived, developed, and explained using only high school algebra and a modicum of common-sense logic.* This has been done in this book—largely by confining the *initial* development of a concept to statistics that are discretely distributed.

Most textbooks define probability in a way that pleases mathematicians but confuses laymen. What is confusing is the omission of the relationship between the mathematical abstraction and the empirical (scientific) reality. In an effort to make the concept of probability experimentally meaningful to the practitioner, two entire chapters have been devoted to developing and explaining the empirical meaning of the related concepts of randomness and probability.

No effort has been made to pretend that mathematically desirable "assumptions" will necessarily be satisfied in practice. On the contrary, the impossibility of *exactly* satisfying some assumptions (such as normality) and the unlikelihood—at least in certain areas of application—of satisfying others (such as random sampling) has been stressed. And considerable effort has been made to give the reader an inkling of what types of assumption-violating situations are relatively innocuous, in a practical sense, and what types are dangerous. This has necessitated considerable theoretical exposition and empirical illustration, especially in the last chapters. However, it is believed that it is better to face the issue squarely than to convey the impression that one can "assume" the required conditions into existence, as is so often done.

Problems have been selected for their appropriateness, which means that they have come from many diverse areas of application. The computational steps for derivations of formulas and solutions of illustrative problems have generally been given in enough detail so that the reader can follow them, without pencil and paper, using only mental arithmetic.

The book takes up in sequence the topics of Descriptive Statistics (Chapter 2), Probability (Chapters 3–7, and possibly 13), Decision-Theory (Chapters 8–10), and Inferential Statistics (Chapters 10–15), each topic being logically developed from its predecessors. The first eleven chapters are easy as are the first and last portions of Chapter 12, which concern statistical tests and estimates based on the binomial distribution. This portion of the book covers the most important *concepts*, although not all of the important *topics*, and it alone could provide the material for an elementary course. The interior of Chapter 12 covers distribution-free tests based on ranks. Chapter 13 concerns moments. And the last two chapters are concerned primarily with the distribution of the sample mean and methods related to it. Virtually all of the logical and mathematical rationale is given for the material in the first 13 chapters. For the last two chapters much of the logical and mathematical rationale is given, but the normal, t, and χ^2, distributions are not mathematically derived. Therefore, the highest level of mathematics used in this book is high school algebra.

The writer is indebted to the following persons and institutions for tables or other materials reproduced in the Appendices or elsewhere: The American Cyanamid Company (Lederle Laboratories Division), The American Society for Quality Control, *Annals of Mathematical Statistics*, the Editor and Trustees of *Biometrika*, R. Davids, W. Eastman, D. J. Finney, The Florida State University, M. Friedman, The General Electric Company (Defense Systems Department), G. J. Glasser, Dr. and Mrs. D. P. Hunt, *Journal of the American Statistical Association, Journal of Quality Technology*, S. K. Katti, W. J. MacKinnon, the Macmillon Company (and the Free Press of Glencoe), New Mexico State University (Statistical Laboratory), The Rand Corporation, W. Teichner, the Van Nostrand Reinhold Company, L. R. Verdooren, R. Wilcox, F. Wilcoxon, and R. F. Winter.

J. V. B.

1

introduction

It is a bemusing paradox that we live in a world that appears to be both stable and erratic. The stock market is an example. There is enough regularity in its gyrations to entice investors and enough irregularity to ruin them. The paradox can be resolved if we assume that Nature is completely lawful but that man's knowledge of these laws is imperfect. This position has its roots in Neo-Laplacian determinism, which asserts that an omniscient demon who knew everything about the universe (and its laws) at a given instant of time could predict everything about it at any future instant, and in Descartes' philosophical conclusion that the only thing about which man could be absolutely certain was his own existence—and of that only when he thought about it. An alternative viewpoint attributes some of the apparent chaos to Nature itself rather than to man's ignorance of its laws. However, partisans of both schools and indeed all sophisticated scientists and philosophers would agree that (a) Nature is largely lawful, (b) man is partially ignorant of those laws, and (c) there are few if any natural phenomena that are not accompanied by unaccountable fluctuations, i.e., fluctuations whose specific cause cannot (at present) be pinpointed. In the well-developed physical sciences most of these unaccountable fluctuations tend to be minor; however, precisely because the physical sciences are well developed, even these minor fluctuations are important. In the behavioral sciences, unaccountable fluctuations typically represent a large part of the behavior of the phenomenon under investigation.

Indeed, fluctuation is so much the rule that it is difficult to think of an example of a true physical constant. One might suppose that the weight of a cubic centimeter of water should be a constant. But on further reflection one realizes that he must introduce endless qualifications: the water must be distilled, the temperature, elevation, barometric pressure, latitude, and position of the moon are all relevant and must be specified—that is, "held constant" with all of the difficulties entailed in achieving *those* "constancies"! Yet, even if we were to omit all of these qualifications, we would achieve a sort of rough constancy in the sense that the major factor influencing the weight of water, under ordinary circumstances, is its volume. Thus virtually everything is to some (often large) extent constant (the sun comes up every morning) and to some extent variable (the time of sunrise varies). Probability and statistics are sometimes called the study of variability. They are concerned with unaccounted-for variability, whether that variability is unaccounted for because its sources are unknown or simply because it is too troublesome or expensive to try to take them into account.

Despite ubiquitous variability in his information about the world, man must still make decisions. But since his knowledge about the relevant facts and relationships is incomplete, he is often in doubt as to the appropriate decision and must therefore guess in the face of this uncertainty. Guesses however, can be blind, uninformed, and unintelligent, or they can be shrewd and sagacious, based upon an optimally weighted appraisal of all relevant known facts and relationships. One of the earliest books on probability and statistics was called *The Art of Guessing*. It was well titled. We guess when we are uncertain. And probability and statistics are simply systematic, commonsensical, mathematically aided, methods of guessing intelligently when required to make a decision under conditions of uncertainty.

2

description

Statistics has two functions. One is to describe. The other is to serve as a basis for inference. Probability has only one function, the latter. That which is described, or about which inferences are drawn, is some particular measurable characteristic of the units comprising a set. The units may be objects, persons, things, or events (such as trials). The measurable characteristic is called a unit's **value** and it may be simply the name of a category to which the unit belongs (such as "defective" or "middle-class"), a frequency count (such as number of errors committed during a trial or number of children in a family), or position on a quantitative scale of measurement (such as the weight of a rat or the time required to complete a trial). A measurement actually taken on the characteristic in question is called an ***observation***. (Due to imprecision of measurement an observation of a unit's value is not necessarily the same as the value, especially when the value is a position on a continuous quantitative scale. However, the term "value" is often used to mean simply a reasonably accurate actual or potential observation.)

At this point let us briefly introduce some symbolic notation. Let N be the number of units in the set. Consider the N units to be numbered from 1 to N in ascending order of value so that the 1st unit has the smallest value and the Nth unit the largest. Now let u_j be the value of the jth unit. Thus we have $u_1 \leq u_2 \leq u_3 \ldots \leq u_{N-2} \leq u_{N-1} \leq u_N$. The $\leq$ sign acknowledges the fact that some units may have the same value. Now let the number of *different* values possessed by the N units be k and imagine these k different values to be arranged in order of increasing size. Let x_i be the ith such value in order of increasing size so that $x_1 < x_2 < x_3 \ldots < x_{k-2} < x_{k-1} < x_k$. Now let

3

f_i be the *number* of units and $p_i = f_i/N$ be the *proportion* of units that have the value x_i. Finally, let F_i be the number of units, and $P_i = F_i/N$ be the proportion of units, that have values less than or equal to x_i.

2-1   Describing Entire Sets

2-1-1   Accounting Separately for Every Value

In order to describe a set of units completely, one would have to list every individual unit together with its value. However, one is seldom interested in the individual identities of the units, and it is usually sufficient simply to know the frequencies with which the various values occur in the set. A wholesale clothing manufacturer, for example, might like to know how many men there are in the United States having each of the various chest sizes, without being interested in also obtaining their names and addresses. Such cases are common; hence one of the most useful ways of describing a set of units is simply to report for each value x_i the number f_i of units in the set that have that value. This information can be conveyed by either a table or a graph. In either case the result is known as a *frequency distribution.*

Often one is not interested in absolute frequencies so much as in relative frequencies, i.e., in the *proportion* of units in the set having each value. (Surely this would be true of the clothing manufacturer.) In such cases, instead of telling for each x_i the corresponding f_i as one would for a frequency distribution, one tells for each x_i the corresponding p_i, that is, the corresponding f_i/N. A table or graph that does this is called a *relative frequency distribution.*

Sometimes one is interested not so much in the number f_i of units having the value x_i as in the number of units having a value that is equal to or less than x_i. For example, a coach would probably be more interested in knowing how many men on his team could chin themselves only nine times or less than in knowing how many reached their limit at exactly nine chins. We can accommodate such cases by means of a *cumulative frequency distribution* (or *cumulative relative frequency distribution*) which tells for each specificable value the number (or proportion) of units in the set having values less than or equal to it. Thus, to obtain a cumulative frequency distribution we tell for each x_i the corresponding value of $F_i = f_1 + f_2 + \cdots + f_i$ and to give a cumulative relative frequency distribution we tell for each x_i the corresponding value of P_i, i.e., of F_i/N. Again, these distributions can be presented in either tabular or graphic form.

To illustrate these various modes of description, suppose that the set of units with which we are concerned is the set of words comprising Lincoln's Gettysburg Address and that the characteristic we wish to describe is length-of-word whose value is simply the number of letters in the word. All four types of distribution of word-lengths are presented in Table 2-1. The upper portion of Figure 2-1 serves as a graph for both the frequency distribution and relative frequency distribution, the left vertical scale applying for the former and the right vertical scale for the latter. The lower portion of Figure 2-1 serves in the same way as a graph for both the cumulative frequency distribution and cumulative relative frequency distribution. In the upper graph the frequency (or relative frequency) of occurrence of a given word-length is represented by the length, or height, of the vertical line erected above that word-length. In the lower graph the cumulative frequency (or cumulative relative frequency) of any word-length is represented by the height of the point lying directly above it on the step-shaped "curve," or, when there is more than one such point, by the highest of these points.

These four modes of description work well when the number of different values is both (a) small to moderate and (b) considerably smaller than the number of units, so that the frequency of occurrence of some values must be considerably greater than 1. These conditions are often met when it is only possible for values to be whole numbers (integers) or some constant multiple thereof, as in Lincoln's word-lengths, or when values are categories. (Figure

Table 2-1

Distribution of Word-Length in Lincoln's Gettysburg Address

Length of word (Number of letters contained)	Frequency of words of that length	Relative frequency of words of that length	Cumulative frequency of words of that length or less	Cumulative relative frequency of words of that length or less
x_i	f_i	f_i/N (or ρ_i)	F_i	F_i/N (or P_i)
1	7	.025830	7	.025830
2	50	.184502	57	.210332
3	60	.221402	117	.431734
4	58	.214022	175	.645756
5	34	.125461	209	.771218
6	24	.088561	233	.859779
7	15	.055351	248	.915129
8	6	.022140	254	.937269
9	10	.036900	264	.974170
10	4	.014760	268	.988930
11	3	.011070	271	1.000000

$\Sigma = 271 = N$

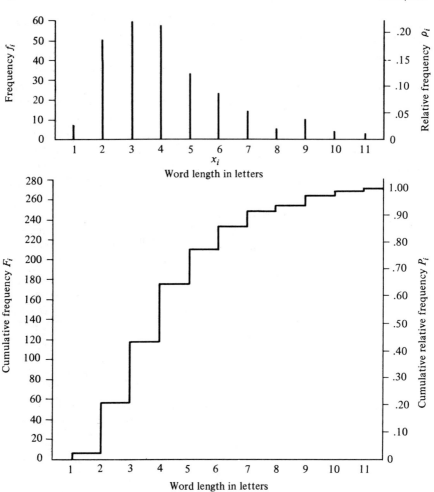

FIG. 2-1. Distribution of length of word among the 271 words of Lincoln's Gettysburg Address.

2-2 shows the frequency and relative frequency distributions for a case where units have categorial values; since the values are not ordinally related, there is little point in graphing cumulative distributions.)

2-1-2 Accounting for Intervals of Values

When nearly every unit has a value different from all the rest, the above methods of description offer little economy over the tedium of listing every individual unit together with its value. Furthermore, the graphs of the frequency or relative frequency distributions would be composed of vertical

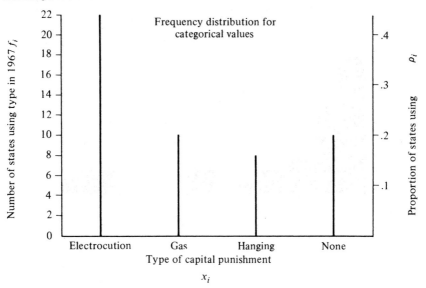

FIG. 2-2. Frequency distribution for categorical values.

lines all of which tended to have the same height, so that height-of-line would become virtually devoid of information. But even if the f_i's vary sufficiently, if the number of different x_i's is quite large, it may be impossible to represent each of them with a distinguishable line so that it becomes necessary to resort to other methods.

Consider, for example, the case where one wished to tell how many of the countries of Europe had what populations, without actually listing all of them. As one might except, no two countries have exactly the same population. Therefore, in order to obtain something analogous to a frequency distribution with frequencies greater than 1, it is necessary to tell how many countries have populations falling within entire regions of values. This is done in Figure 2-3. Since an entire region of values on the horizontal axis is being represented, frequency is represented by the *area* of a rectangle whose base is the interval of values represented. If one chooses equal intervals of values (as one ordinarily should), the bases of the rectangles are all equal and their *heights* therefore are proportional to the number of units in the set having values that fall in the interval represented. This type of graph is known as a **histogram**, and it may be used to represent either frequencies or relative frequencies. The fact that every country has a different population rendered the frequency distribution a rather ineffective method of economical description and forced us to replace it with a histogram. However, it does not prevent us from using the exact cumulative frequency distribution as shown in Figure 2-3. In this respect, and in several others, cumulative distributions are a superior method of describing data. However, there are other cases in

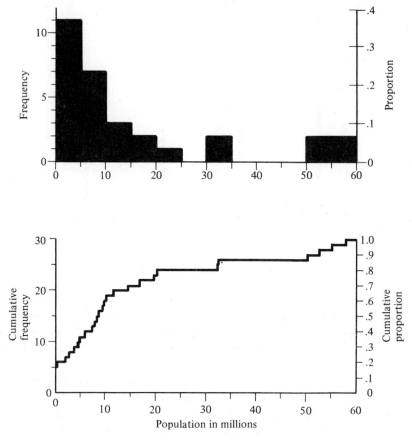

Fig. 2-3. Histogram and cumulative step-graph showing distribution of populations of entirely European countries.

which we do not know the exact value of every unit, but only the interval into which it falls. For example, the time-scores for a set of trials may each be measured only to the nearest hundredth of a second. In that case we might take .01 sec as the width of our histogram interval so that if x_i were our nominal measurement in hundredths of a second, the base of the rectangle representing it would extend from $x_i - .005$ to $x_i + .005$, its height being f_i. However, the graph giving cumulative frequencies would be steplike, rising by an increment of f_i at $x_i + .005$, i.e., at the *upper endpoint* of the interval—since we do not know *where* the f_i values that occurred within the interval from $x_i - .005$ to $x_i + .005$ are actually located within the interval but only that they are in the interval and are $\leq x_i + .005$. The graph therefore would be exact only at the interval boundries.

When the number of units in the set is infinite but the number of different values they take is finite, the f_i's (or at least some of them) are infinite, so meaningful frequency distributions cannot be tabulated or plotted. However, one can still tabulate or plot relative frequency distributions and cumulative relative frequency distributions since relative frequencies are *proportions*, which are finite. And if the number of different values is small, we could plot these distributions exactly, accounting individually for every different value. However, if the units in the infinite set have an infinite number of different values (as they will if the value is a point on a continuum such as time or length), the set of units could be represented by a histogram with equal bar widths. Since the units take on an infinite number of different values along a continuum, the histogram bars could be as thin as a fine pencil line and there would still be few if any gaps between them. Therefore, the sides of adjacent histogram bars would almost always touch and the tops of the bars would describe a smooth curve. The smooth curve, then, gives the same information as the histogram and can therefore replace it. The smooth curve is called a **density function**. (See Figure 2-4.) The histogram represented the relative frequency of values falling within the base interval of a histogram bar by the area of the bar (the area, therefore, being regarded as "equal to" the relative frequency). Analogously, the density function represents the relative frequency of values falling within *any* interval by the area under the portion of the curve above that interval. The total area enclosed within all of the histogram bars is the sum of the relative frequencies for all the x_i's, $\sum_{i=1}^{k} f_i/N$ $= N/N = 1$. So the area under a density function is 1. And since there are an infinite number of different x-values under the density function, the area representing any one of them is $1/\infty = 0$. Since area represents relative frequency, the relative frequency of occurrence of any single x-value under a density function is therefore zero. However, the height of the density function curve above a given x-value may be regarded as representing the "density" of actual unit values in the immediate vicinity of the given x-value.

The proportion of units having values equal to or less than x is simply the proportion of the total area under the density function curve that lies above values less than or equal to x. A **cumulative proportion graph** (see Figure 2-4) plots these proportions, represented on the vertical axis, against the corresponding values of x, represented on the horizontal axis. Thus cumulative proportions can be either read directly off of a cumulative proportion graph or obtained by measuring the area under a density function to the left of x.

Sometimes when N is not infinite one would like to have an idea of what the density function would look like if N were to become infinite. This can sometimes be accomplished by proceeding just as if one were going to construct a histogram, but instead of actually drawing the histogram bars, each histogram bar (or interval that would have contained a bar if there were any observations in it) is replaced by a single point—the midpoint of the top of the bar (or interval). These points are then connected by straight lines. The

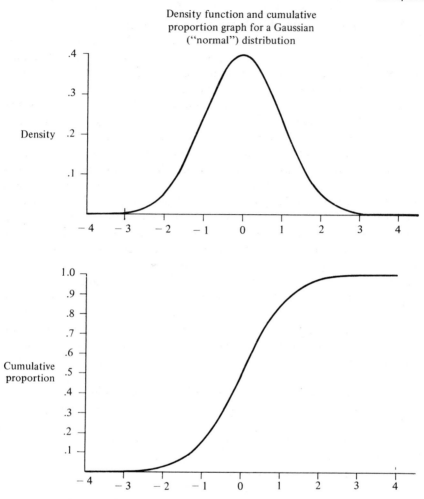

FIG. 2-4. Density function and cumulative proportion graph for a Gaussian ("normal") distribution.

result is called a *frequency* (or *relative frequency*) *polygon*. (See Figure 2-5.) Notice, however, that if N is large enough and interval-width is small enough, the histogram itself would be a fairly accurate picture of the density function, whereas if N is small or if interval-width is large, even the frequency polygon may present a misleading picture.

2-1-3 Describing Sets of Two-Valued Units

So far we have discussed the situation in which each unit in the set has only a single value in which we are interested. Sometimes, however, every

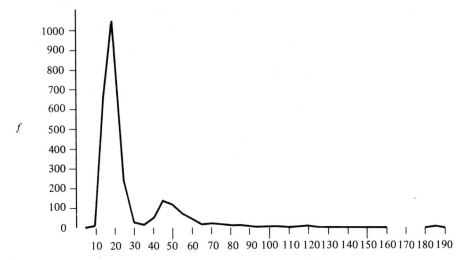

FIG. 2-5. Frequency polygon for a distribution of time scores measured to the nearest twentieth of a second.

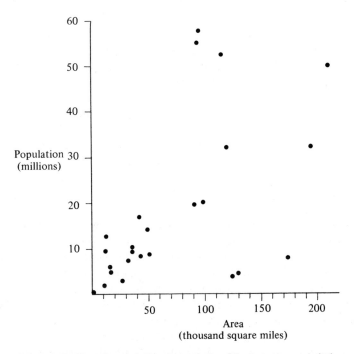

FIG. 2-6. Scatter diagram showing relationship between population and area of entirely European countries.

unit has both an X-value and a Y-value and we are interested in the relationship between them. If N is not too large and if units having almost exactly the same pair of X and Y values are rare, we can show the relationship quite effectively by plotting each unit as a point on a graph whose horizonal axis represents X-values and whose vertical axis represents Y-values. Such a plot is called a **scatter diagram**. Figure 2-6 presents a scatter diagram showing that, in general, European countries having large populations tend also to have large areas, as one might expect.

2-2   Summary Indices

The methods we have discussed so far tell us how many, or what proportion, of the units in the set have each of the values (or approximate values) we choose to consider. Generally we do not wish our information to be any more fine-grained than that, so, in a sense, these methods yield complete information, providing a total description of the set. However, tables and graphs that present entire distributions are often bulky and tedious to prepare or even to read, whereas economy of space and effort is often a powerful consideration in publishing, or otherwise publicizing, information. Therefore, a strong motivation exists to represent an entire distribution, or at least some special characteristic of it, by means of a single, summary, numerical index or perhaps by a pair of such indices. Furthermore, even when one presents the entire distribution, such indices have the advantage of pointing up explicitly certain characteristics of the distribution that may not be readily apparent in the diffuse form in which it is tabled or graphed. There are several such indices and they fall into two major categories: **indices of location** that attempt to indicate where some namable part of the distribution is on the horizontal axis and **indices of dispersion** that attempt to tell how spread out the distribution is according to some criterion.

2-2-1   Indices of Location

Perhaps the crudest index of location is the **mode**. It is the value that occurs with the greatest frequency, i.e., that is possessed by the greatest number of units in the set. Thus the mode is the value associated with the largest of the f_i. In the distribution of word-lengths in Lincoln's Gettysburg Address the largest f_i is 60 which is the frequency of occurrence of words three letters in length, so the mode is 3. Two or more values may be tied for the greatest frequency in which case the distribution has two or more modes and is called *bimodal* or *multimodal*, respectively.

A considerably more refined index of location is the **median**. Consider the N units arranged in a line in order of increasing value. The median is the value of the middle unit, if N is odd, and is the midpoint between the values of the two midmost units when N is even. Thus if the values of the units arranged in order of increasing size are $u_1, u_2, u_3, \ldots, u_{N-2}, u_{N-1}, u_N$, the median is $u_{(N+1)/2}$ if N is odd, and is $(u_{N/2} + u_{(N+2)/2})/2$ if N is even. The median of the 271 word-lengths in Lincoln's Gettysburg Address is $u_{(N+1)/2} = u_{136}$ which is 4, since every unit value from u_{118} to u_{175} is 4.

The median is a special case of a general class of indices of location called **percentiles**. The 100 Pth percentile is that value of a unit in the set such that at least a proportion P of the units in the set have either the same or a smaller value *and* at least a proportion 1-P of them have values that are the same or larger—unless there are two such values in which case the midpoint between them is, by convention, taken to be the 100 Pth percentile. (If a proportion $> P$ of the units have values $\leq x_i$ and a proportion $< P$ have values $< x_i$, the matter is simple: x_i is the 100 Pth percentile). The 30th percentile of the distribution of word-length in Lincoln's Gettysburg Address is 3 since the percentage of words having lengths ≤ 3 is ≥ 30 and the percentage having lengths ≥ 3 is ≥ 70 (or since more than 30% have values ≤ 3 and less than 30% have values < 3). The 25th, 50th, and 75th percentiles are also known as the 1st, 2nd, and 3rd **quartiles** since at least one, two, and three quarters of the units have values less than or equal to them. The 50th percentile is still better known as the median. The 1st, 2nd, and 3rd quartiles for word-lengths in Lincoln's Gettysburg Address are 3, 4, and 5. Percentiles can be read directly off of cumulative relative frequency graphs. If we want the 100 Pth percentile, we simply draw a horizontal line through the cumulative relative frequency of P until the line crosses the step-graph. We then draw a vertical line through the point (or midpoint) of intersection until it touches the horizontal axis of the graph. The point at which it does is the 100 Pth percentile.

Perhaps the best known index of location is the **mean**. This is simply the average of the N unit values. It can be calculated in any of several equivalent ways:

$$\text{Mean} = \frac{\sum_{j=1}^{N} u_j}{N} = \frac{\sum_{i=1}^{k} f_i x_i}{N} = \sum_{i=1}^{k} p_i x_i \qquad (2\text{-}1)$$

The mean word-length in Lincoln's Gettysburg Address is

$$\frac{\sum_{i=1}^{11} f_i x_i}{N} = \frac{1,149}{271} = 4.23985$$

2-2-2 Indices of Dispersion

The simplest index of dispersion is the **range**. If H is the highest and L the lowest value in a set of values, then the range of that set of values is $w = H - L$. The range of word-lengths in Lincoln's Gettysburg Address is $11 - 1 = 10$.

A somewhat more sophisticated index of spread is the ***interquartile range***. If Q_1 and Q_3 are the 1st and 3rd quartiles, respectively, then the interquartile range is their difference in value $Q_3 - Q_1$. The ***semi-interquartile range*** is $(Q_3 - Q_1)/2$, half of the interquartile range. The interquartile range for the distribution of word-lengths in Lincoln's Gettysburg Address is $5 - 3 = 2$ and the semi-interquartile range is 1.

The interquartile range is a special case of ***interpercentile range***. The interpercentile range between the $100 P_1$th and $100 P_2$th percentiles is simply the positive difference between their values. The 5th and 95th percentiles for word-lengths in Lincoln's Gettysburg Address are 2 and 9, respectively, so the interpercentile range between the 5th and 95th percentiles is $9 - 2 = 7$. If one uses a percentile, such as the median, as an index of location, then an interpercentile range is a particularly appropriate index of dispersion. Since there is a choice of percentile endpoints, one should, of course, exercise discretion in making that choice. The proper choice in particular cases will depend on one's objectives.

Perhaps the most common indices of dispersion are the ***variance*** and its square root, the ***standard deviation***. If we call the difference between a unit's value and the mean a "deviation," the variance is the average squared deviation. Thus, if we let $\bar{u}$ stand for the mean of the N unit values,

$$\text{Variance} = \frac{\sum_{j=1}^{N} (u_j - \bar{u})^2}{N} = \frac{\sum_{i=1}^{k} f_i(x_i - \bar{u})^2}{N} \tag{2-2}$$

A more convenient formula for calculation can be derived by squaring the difference in the numerator, applying the summation individually to each of the three resulting terms, and making use of the facts that

$$\sum_{j=1}^{N} \bar{u}^2 = N\bar{u}^2$$

and that since

$$\bar{u} = \frac{\sum_{j=1}^{N} u_j}{N}$$

then

$$\sum_{j=1}^{N} u_j = N\bar{u}$$

$$\text{Variance} = \frac{\sum_{j=1}^{N} (u_j - \bar{u})^2}{N} = \frac{\sum_{j=1}^{N} (u_j^2 - 2u_j\bar{u} + \bar{u}^2)}{N}$$

$$= \frac{\sum_{j=1}^{N} u_j^2 - 2\bar{u} \sum_{j=1}^{N} u_j + \sum_{j=1}^{N} \bar{u}^2}{N}$$

$$= \frac{\sum_{j=1}^{N} u_j^2 - 2\bar{u}(N\bar{u}) + N\bar{u}^2}{N} = \frac{\sum_{j=1}^{N} u_j^2 - N\bar{u}^2}{N}$$

$$= \frac{\sum_{j=1}^{N} u_j^2}{N} - \bar{u}^2 = \frac{\sum_{i=1}^{k} f_i x_i^2}{N} - \bar{u}^2 \tag{2-3}$$

The standard deviation is given by any of the formulas

$$SD = \sqrt{\frac{\sum_{j=1}^{N}(u_j - \bar{u})^2}{N}} = \sqrt{\frac{\sum_{i=1}^{k} f_i(x_i - \bar{u})^2}{N}}$$

$$= \sqrt{\frac{\sum_{j=1}^{N} u_j^2}{N} - \bar{u}^2} = \sqrt{\frac{\sum_{i=1}^{k} f_i x_i^2}{N} - \bar{u}^2}$$

(2-4)

If the values of the units are numbers of inches, then the variance will be some number of *square* inches, whereas the standard deviation will be some number of inches, i.e., the standard deviation will be expressed in the same units of measurement as were the original values. This is an advantage for the standard deviation that is not shared by the variance. However, as we shall see later, the variance is superior in almost all other respects, especially in mathematical tractability. (Indeed, it is precisely this mathematical tractability that motivates us to go to all the trouble of squaring deviations, taking their average, and then the square root of that average, rather than simply using the average absolute deviation from the mean as our index of dispersion; the latter is not mathematically tractable.) The variance of lengths of words in Lincoln's Gettysburg Address is

$$\frac{\sum_{i=1}^{11} f_i x_i^2}{N} - \bar{u}^2 = \frac{6{,}081}{271} - (4.23985)^2 = 22.43911 - 17.97633 = 4.46278$$

and the standard deviation is its positive square root 2.1125.

2-2-3 Appropriateness of Summary Indices

A unit's value may be the *amount of some standard quantity* (such as pounds, or letters in the case of Lincoln's word-lengths) that it possesses, a *categorial label implying a position in some meaningful order* (such as one of the labels, "idiot," "imbecile," "moron," "subnormal," "normal," "bright-normal," "superior," and "genius," or one of the labels, "first quality," "second quality", etc.), or a *categorical label having no quantitative or ordinal implications* of interest to the statistician (such as one of the labels, "dry goods," "hardware," "comestibles," etc., or "dachshund," "boxer," "collie," "chow," "fox terrier," etc.). The mean, variance, and standard deviation are meaningful indices in the first case, but not in the second or third. The range and indices based upon percentiles (e.g., the median, the quartiles, the inter-quartile or interpercentile range) are meaningful in the first and second cases, but not in the third. The mode is a meaningful index in all three cases.

Sometimes a summary index provides all that one needs or cares to know about the values of the units in a set. If units are votes and their values are the names of candidates in a plurality-rule election, the mode identifies the winner. Certain graduate schools accept all applicants who score above a certain percentile and none of the applicants who score below it in the distribution of scores for all students taking the Graduate Record Examination; in such cases the only information needed to determine admission, in addition to the applicant's score, is the critical percentile. If median and mode happen to coincide in a distribution, their common value provides all the information one needs if he is seeking only the value most *typical* of the units. Finally, a rancher might be quite content to know only the mean weight $\bar{u}$ of his beef cattle and the number N of such cattle that he owns, since calculating $N\bar{u}$ gives him their total weight from which he can calculate their total worth in dollars.

Sometimes a pair of indices, one for location, the other for dispersion, provides all the information really needed. For example, a 300-pound man considering a career as a jockey might well be able to arrive at an intelligent decision if he knows only the median and an interpercentile range (say, between 1st and 99th percentiles) for the distribution of weights of present jockeys. The complete frequency distribution of weights would be almost superfluous.

It cannot be emphasized too strongly, however, that whether or not a single index or pair of indices will suffice depends on the needs and objectives of the user of the information. Often they do not suffice, but are resorted to for reasons of economy. But worst of all are the all-too-frequent cases in which they are insufficient, and not necessitated by economic considerations, but are resorted to out of pure institutional inertia because they appear to be "traditional." If the information in a complete frequency distribution is useful and if it is not prohibitively expensive or tedious to present it, then it is highly desirable to do so. This is the case because the information contained in a pair of indices does not imply any particular pattern for the distribution. An enormous variety of vastly differing distributions can have the same mean and variance or the same median and interquartile range.

2-3   Populations and Samples

The methods we have described so far apply to the distribution of any set of N units. We must now distinguish between two types of set. A *population* is the *entire* set of units that meet some qualifying definition or have some defining characteristic, such as the set of all stars, or the set of all stars constituting the big dipper; the set of all U.S. military weapons, or the set of all Civil War Cannons decorating courthouse lawns; the set of all college students, or the set of all right-handed coeds with 20-20 vision enrolled in Psychology 314 at Purdue. Clearly it is the definition that determines what is the entire set and therefore the population. A *sample* is simply a selection of units from a population, i.e., a set of units drawn from a population, or the set of observations made upon those units. It is a subset of the set of units that comprises the population (or the observations made upon the subset's units).

The units comprising a population can be either *actual*, such as the items produced yesterday by a certain factory (or the subjects available to take part in an experiment) or *potential*, such as the items that will be produced tomorrow or for the rest of eternity (or the potential trials associated with a set of potential time scores). If the units are *actual*, the population must be *finite* and its distribution must be *discrete*. That is, N, the number of different units, will be finite; and k, the number of different unit values (which cannot exceed N) will also be finite; and, as a consequence of the latter, there will be gaps between different, but adjacent, values that occur in the population, i.e., the array of actual values that occur in the population will be discontinuous or discrete. If the units comprising the population are *potential*, the population can be either *finite*, such as next year's graduating class, or *infinite*, such as all future college graduates. If it is *finite*, it will also be *discrete*. But if it is *infinite*, it can be either *discrete*, such as the population of eternal future tosses of a coin whose outcomes, i.e., values, can be only "heads" or "tails," or *continuous*, such as the population of heights of all future adult members of the human race (since, presumably, every height within a continuum containing an infinite number of possible heights would be represented in the population).

A sample can be either actual, in the sense that it presently exists, or potential, in the sense, generally, that it has not yet been drawn but definitely will or could be. Actual samples are always finite and therefore discrete.

It should be clear that potential populations, at least, are often highly idealized. An infinite population, of course, can never become actual, and, in that sense, is not even truly potential. Furthermore, while *values* such as lengths or times can be continuously distributed, *observations* are doomed to be discretely distributed simply because it is impossible to measure with precision to an infinite number of decimal places. Idealizations are common in science and are necessary to its progress and to the efficient solution of problems. However, it is important to keep idealizations in their proper perspective as conveniences rather than regarding them as the distilled essence of Truth.

We may regard a sample as a set of n observations made upon the values of the n units drawn from a population. (So the *observations* are the units comprising the sample set.) And we may think of the n units as having been removed one at a time in a sequence of n individual withdrawals of a single unit from the population. If the n units are permanently withdrawn from the population, or if they are returned to it only after all n sample observations have been made, we say that we have sampled **without replacements**. On the other hand, if after each unit is removed, we observe its value, record the observation, and then return the unit to the population before removing the next unit, we say that we have sampled *with replacements*. When the sampled population is finite, the two methods lead to different results. For example, if a sample of two observations is drawn without replacements from a population consisting of $N = 2$ items, whose values are x and y, the sample must consist of one x and one y; whereas if sampling had been with replacements, the sample could have consisted of two x's, or of two y's, as well as of an x and a y. (When the population is finite and sample units are drawn in sequence and without replacements, each of the units comprising the sample is drawn from a slightly different population—the original population altered by different amounts of depletion.)

Since it is usually important to differentiate between populations and samples, different symbols are used in the two cases to describe essentially the same characteristic of the units in the set. When dealing with populations, it is conventional to denote the number of units by N, the mean by μ, variance by σ^2, and standard deviation by σ. When dealing with samples, it is conventional to let n represent the number of units in the sample and to let $\bar{x}$ represent their mean value. What we have previously called a summary index is called a *statistic* when the set involved is a sample, but it is often loosely referred to as a *parameter* when the set involved is a population.

2-4 Problems

1. Tabulate and plot on graph paper the frequency distribution, relative frequency distribution, cumulative frequency distribution, and cumulative relative frequency distribution for word-lengths in the following quotation from John B. Watson's *Behaviorism* (treating hyphenated words as two words and "I'll" as one):

> *Give me a dozen healthy infants, well-formed, and my own specified world to bring them up in and I'll guarantee to take any one at random and train him to become any type of specialist I might select—doctor, lawyer, artist, merchant-chief and, yes, even beggar-man and thief, regardless of his talents, penchants, tendencies, abilities, vocations, and race of his ancestors.*

Give the numerical value of the following summary indices: (a) mode, (b) median, (c) mean, (d) 10th percentile, (e) 90th percentile, (f) range, (g) interpercentile range between 10th and 90th percentile, (h) interquartile range, (i) semi-interquartile range, (j) variance, (k) standard deviation.

2. In the 1970 volume of *Econometrica* there were 58 articles, having the following lengths in number of pages:
17, 21, 11, 16, 7, 20, 4, 21, 10, 25, 12, 6, 26, 12, 6, 20, 5, 20, 5, 17, 13, 13, 7, 14, 10, 6, 17, 12, 8, 19, 4, 15, 5, 9, 8, 17, 10, 16, 9, 17, 15, 21, 16, 37, 5, 16, 22, 8, 30, 12, 18, 25, 15, 25, 9, 24, 18, 14. Tabulate the frequency distribution of article lengths, and give the mode, median, mean, 5th percentile, 95th percentile, range, interquartile range, 5th to 95th interpercentile range, variance, and standard deviation.

3. For each of the following distributions, state which type of graph would be most appropriate, a frequency distribution graph giving f_i for each x_i, a histogram, or a frequency polygon:
   (a) A distribution of 30 error scores taking integral values in the range from 0 to 8.
   (b) A distribution of 10,000 error scores taking integral values in the range from 0 to 8.
   (c) A distribution of 200 time scores ranging from .5 to 1 sec and measured to the nearest hundredth of a second.
   (d) A distribution of 100,000 time scores ranging from .5 to 1 sec and measured to the nearest tenth of a second.
   (e) A distribution of 1,000,000 time scores ranging from .5 to 1 sec and measured to the nearest thousandth of a second.

4. What summary index of location and what summary index of dispersion would be most appropriate in each of the following situations:
   (a) A sociologist has the distribution of family incomes in Saudi Arabia and wishes to show what is typical.
   (b) A sports writer has a distribution of weights of linesmen on a football team and wants to indicate the power of the team.
   (c) A psycholinguist has a distribution of lengths of the 10,000 most commonly used words and wants to show what length is typical.

3

possibilities

In this chapter we shall be concerned with the variety of possibilities of composition and arrangement of one or more subsets of units drawn from a set of n units.

3-1 Permutations

Suppose that four of the five members of a family line up shoulder to shoulder for a group photograph while the fifth member takes the picture. The particular four family members in the picture, arranged in the particular order shown, constitute a *permutation*. A **permutation** is the arrangement resulting when r units are removed from a set of n units and assigned to r fixed positions or places in such a way that each place is occupied by one unit. It may be regarded as an assignment of things to places where each place receives one and only one thing. Thus, a permutation is an ordered subset (i.e., arrangement into an order relationship) of some or all of the objects of a set. The order can be spatial (pattern) or temporal (sequence). Always in the latter case and often in the former the arrangement is linear. The n units may all be distinguishable from one another, or certain subsets of them may be composed of units that are mutually indistinguishable. For example, two of the five candidates for the family photograph might have been identically dressed identical twins.

Two permutations are different if they do not contain exactly the same r units arranged in exactly the same pattern. Two permutations are *distin-*

guishably different only if you can tell them apart, i.e., only if they differ in a perceptible way. Thus, if the family consists of a father (F), mother (M), son (S), elder daughter (D), and younger daughter (d), the permutation $DMFd$ is, of course, different both from the permutation $DMSd$ and from the permutation $dMFD$. However, if the two daughters are identically dressed identical twins, the first and third permutations above are not *distinguishably* different. This can easily be demonstrated by substituting T for both D and d, in which case both the first and third permutations become $TMFT$ and therefore indistinguishable.

3-1-1 Number of Permutations of *n* Things Taken *r* at a Time

It is often desirable to know how many different permutations are possible under certain specified conditions. Consider first the general case in which the subset may be any portion of the set, i.e., where $r \leq n$, so that it may consist of only some of the units in the set. In that event, permutations may differ because of differences in the identities of the units comprising the subset as well as because of differences in their arrangement. In this case, the number of different ordered subsets of r things that might possibly result if one drew r things from a set of n things and assigned them to r fixed positions is called the *number of permutations of* n *things taken* r *at a time* and is represented by the symbol $_nP_r$. **The number of permutations of n things taken r at a time is**

$$_nP_r = \frac{n!}{(n-r)!} \tag{3-1}$$

Proof: The number of permutations of n things taken r at a time is simply the number of different possible assignments of r of n things to r fixed positions, or, roughly, the number of different ways of choosing r of n things and assigning them to r positions so that each position is occupied by one thing. Any one of the n things can be chosen to occupy the first position; and, no matter which one is, any one of the $n - 1$ remaining things can be assigned to the second position; and, no matter which one is, any one of the remaining $n - 2$ things can be assigned to the third position; until, finally, any one of the remaining $n - (r - 1)$ things can be assigned to the rth position. Thus, there are n different ways of filling the first position. For *each* of these ways there are $n - 1$ different ways of filling the second position. So there are $n(n - 1)$ different ways of filling the first two positions. For *each* of these $n(n - 1)$ different ways there are $n - 2$ different ways of filling the third position. So there are $n(n - 1)(n - 2)$ different ways of filling the first three positions. Continuing in this fashion, there are

$$n(n - 1)(n - 2) \cdots (n - r + 2)(n - r + 1) = \frac{n!}{(n-r)!}$$

different ways of filling all r different positions, each of which represents a different assignment—a different ordered subset.

Example: The number of permutations of 5 things taken 2 at a time is

$$_5P_2 = \frac{5!}{(5-2)!} = \frac{5!}{3!} = \frac{(5)(4)(3)(2)(1)}{(3)(2)(1)} = (5)(4) = 20$$

Denoting the 5 things as A, B, C, D, E, and letting the 2 positions be adjacent positions on a horizontal line, the 20 permutations are as shown below:

AB	BA	CA	DA	EA
AC	BC	CB	DB	EB
AD	BD	CD	DC	EC
AE	BE	CE	DE	ED

Illustrative Problems: If the outcome of a horse race is a list identifying the winner, the horse finishing second, and the horse coming in third, and if ties are impossible, how many different possible outcomes are there for a race in which 8 horses compete?

ANSWER: The problem amounts to asking in how many different ways an unspecified 3 of 8 horses may be assigned to 3 positions. So the answer is

$$_8P_3 = \frac{8!}{(8-3)!} = \frac{8!}{5!} = \frac{(8)(7)(6)(5)(4)(3)(2)(1)}{(5)(4)(3)(2)(1)} = (8)(7)(6) = 336$$

The chairman of a 5-man committee seats himself at the head of a 10-chair conference table. In how many different ways may the remaining 4 committeemen seat themselves at the table?

ANSWER: The question amounts to asking in how many different ways 4 of 9 chairs can be assigned to 4 men. It really makes no difference whether we assign things (people) to fixed places or fixed places to things, so long as the larger number is n and the smaller is r. (The first man can select any one of 9 chairs, the second any one of 8, etc.) Therefore, the answer is

$$_9P_4 = \frac{9!}{(9-4)!} = \frac{9!}{5!} = (9)(8)(7)(6) = 3,024$$

3-1-2   Number of Permutations of n Things

Now consider the case in which the ordered subset consists of all units in the set so that $r = n$. In this case, since every subset contains the same units, the number of different permutations is simply the number of different possible arrangements of n different things within n fixed places or positions, under the restriction that each place must contain exactly one thing. We shall call this simply the *number of permutations of* n *things.* **The number of permutations of n things is**

$$_nP_n = n! \tag{3-2}$$

Proof: The number of permutations of n things is simply the number of different possible assignments of n things to n different fixed positions so that each position is occupied. It is a special case of the number of permutations of n things taken r at a time, where $r = n$. So substituting n for r in the formula

$$_nP_r = \frac{n!}{(n-r)!}$$

we obtain

$$_nP_n = \frac{n!}{(n-n)!} = \frac{n!}{0!} = \frac{n!}{1} = n!$$

Alternatively, following the same reasoning as used in the derivation of $_nP_r$, any one of the n things can be chosen to occupy the first position, any one of the remaining $n-1$ things can be picked for the second position, etc., etc. So the number of different ways of filling all n positions is

$$n(n-1)(n-2) \cdots (3)(2)(1) = n!$$

Example: The number of permutations of 4 things is $4! = (4)(3)(2)(1) = 24$. Denoting the 4 things as A, B, C, and D, and letting the 4 positions be four equally spaced points on a horizontal line, the 24 permutations are shown below:

ABCD	*BACD*	*CABD*	*DABC*
ABDC	*BADC*	*CADB*	*DACB*
ACBD	*BCAD*	*CBAD*	*DBAC*
ACDB	*BCDA*	*CBDA*	*DBCA*
ADBC	*BDAC*	*CDAB*	*DCAB*
ADCB	*BDCA*	*CDBA*	*DCBA*

Illustrative Problems: A motorist removes the wheels from his car in order to rotate the tires. He forgets which position each wheel comes from. In how many different ways may he assign the four wheels to the four positions?

ANSWER: $_4P_4 = 4! = (4)(3)(2)(1) = 24$.

An experiment involves six treatments, all of which are administered to every subject. In order to balance out possible sequential effects (such as learning or fatigue), the experimenter wants each subject to receive the treatments in a different order and wants to use as many subjects as there are different possible orders of treatment-administration. How many subjects must he use?

ANSWER: The question amounts to asking how many different possible assignments there are of six different treatments to six different positions in sequence. Therefore, the answer is

$$_6P_6 = 6! = (6)(5)(4)(3)(2)(1) = 720.$$

3-1-3 Number of Distinguishable Permutations of *n* Things Some of Which Are Alike

Now consider the case in which the subset consists of all the units in the set so that $r = n$, but the units are not all distinguishable, r_1 of the units belonging to Type 1, r_2 to Type 2, etc., where units of one type are distinguishable from units of a different type but not from each other. In that event, if there are k different types so that

$$\textstyle\sum_{i=1}^{k} r_i = r = n$$

then *the number of distinguishable permutations is*

$$\frac{n!}{r_1! r_2! \cdots r_k!} = \frac{n!}{\prod_{i=1}^{k} r_i!} \tag{3-3}$$

Proof: Consider any particular one of the distinguishable permutations of n things. The r_1 like things of the first type can be permuted among themselves in $r_1!$ ways without changing the appearance of the overall permutation. For each of these $r_1!$ ways, the r_2 like things of the second type can be permuted among themselves in $r_2!$ ways without changing the appearance of the overall permutation. . . . And the r_k things of the kth type can be rearranged within the r_k fixed positions they now occupy in $r_k!$ ways without altering the apparent pattern of arrangement. So for *each distinguishable* permutation of n things, there are

$$r_1! r_2! \cdots r_k! = \textstyle\prod_{i=1}^{k} r_i!$$

different rearrangements of its units that leave its overall appearance unchanged; that is, for *each distinguishable* permutation there are $\prod_{i=1}^{k} r_i!$ *in*distinguishable permutations. Therefore, the total number $n!$ of possible permutations of all varieties (distinguishable or otherwise) must be the product of the number D of distinguishable permutations and the number $\prod_{i=1}^{k} r_i!$ of indistinguishable permutations that can be formed or generated from each distinguishable one. Thus,

$$n! = D \textstyle\prod_{i=1}^{k} r_i!$$

and

$$D = \frac{n!}{\prod_{i=1}^{k} r_i!}$$

The rationale of the above proof can be illustrated as follows: Suppose that we have $n = 4$ units identified as A_1, A_2, B_3, and B_4, where the four different subscripts identify the four different units, but where A_1 and A_2 are regarded as indistinguishable from each other as are B_3 and B_4, although A's are distinguishable from B's. There are $n! = 4! = 24$ different permutations of the four units within four equally spaced positions on a horizontal line, as shown below. But in any distinguishable permutation we can interchange A's in 2! ways and B's in 2! ways, making 2! times 2! or 4 ways in all, without interchanging any A's with B's, and therefore without altering the distinguishable permutation with which we started. Therefore, if we ignore subscripts, there are only

$$\frac{n!}{r_1! r_2!} = \frac{4!}{2! 2!} = 6$$

distinguishably different permutations. Each of the six blocks below contains the four indistinguishable permutations of a different distinguishable one.

$A_1A_2B_3B_4$	$A_1B_3A_2B_4$	$A_1B_3B_4A_2$	$B_3A_1A_2B_4$	$B_3A_1B_4A_2$	$B_3B_4A_1A_2$
$A_1A_2B_4B_3$	$A_1B_4A_2B_3$	$A_1B_4B_3A_2$	$B_4A_1A_2B_3$	$B_4A_1B_3A_2$	$B_4B_3A_1A_2$
$A_2A_1B_3B_4$	$A_2B_3A_1B_4$	$A_2B_3B_4A_1$	$B_3A_2A_1B_4$	$B_3A_2B_4A_1$	$B_3B_4A_2A_1$
$A_2A_1B_4B_3$	$A_2B_4A_1B_3$	$A_2B_4B_3A_1$	$B_4A_2A_1B_3$	$B_4A_2B_3A_1$	$B_4B_3A_2A_1$

Example: The number of distinguishable permutations of 5 things of which 3 are indistinguishably alike and of one kind (circle), 1 is of a second kind (square), and 1 is of a third kind (triangle), where kinds are distinguishable, is

$$\frac{5!}{3! 1! 1!} = (5)(4) = 20$$

These 20 distinguishable permutations are shown below for the case in which the places to which the things are assigned are equally spaced positions on a horizontal line.

The number of distinguishable permutations of n things of which r_1 are of type 1, r_2 are of type 2, . . . , and r_k are of type k, where things of one type are distinguishable only from things of a different type, is sometimes represented by the symbol

$$\begin{bmatrix} n \\ r_1, r_2, \ldots, r_k \end{bmatrix}$$

Since $n!$ increases rapidly with increasing values of n, it is sometimes convenient to use Stirling's approximation for $n!$ in preference to calculating it directly. In its crudest, but simplest, form **Stirling's approximation** is

$$n! \cong \left(\frac{n}{e}\right)^n \sqrt{2\pi n} \tag{3-4}$$

where e is the base of natural logarithms, 2.71828 . . . . The extent of error hazarded by its use can be judged from the inequality

$$\left(\frac{n}{e}\right)^n \sqrt{2\pi n} < n! < \left(\frac{n}{e}\right)^n \sqrt{2\pi n}\left[1 + \frac{1}{12n - 1}\right] \tag{3-5}$$

Illustrative Problems: Pool balls numbered 1 to 6 are arranged side by side on a rack. If the balls numbered 1 to 3 are painted red (thereby obscuring their numbers and rendering them indistinguishable except by position), those numbered 4 and 5 are painted white, and ball number 6 is painted blue, how many different permutations are there of (a) the six balls on the rack, (b) the red balls within the three positions they now occupy, (c) the two white balls within their present positions, (d) the blue ball within its present position? (e) How many distinguishable permutations are there of the six colored balls?

ANSWERS: (a) $6! = 720$, (b) $3! = 6$, (c) $2! = 2$, (d) $1! = 1$,

(e) $\dfrac{6!}{3!\,2!\,1!} = \dfrac{720}{6 \times 2 \times 1} = 60$

An experiment to test hearing involves twelve presentations of a "middle C" tone—three presentations at the same inaudible volume, four at a constant volume that is just barely perceptible to a person with normal hearing, and five at a fixed volume that is clearly perceptible to anyone without a hearing defect. In order to prevent fakery, it is deemed desirable to vary the order of presentation of the twelve tones to different patients. How many different orders of presentation are possible?

ANSWER: Since the only thing that distinguishes two tones of the same volume is the order in which they are presented (and since you *cannot* interchange the second tone with the fifth tone, presenting the second one fifth and the fifth one second), the question really asks for the number of distinguishable permutations of twelve things of which three are of one indistinguishable kind, four of another, and five of a third, where kinds are distinguishable from each other. So the answer is

$$\frac{n!}{r_1!\,r_2!\,r_3!} = \frac{12!}{3!\,4!\,5!} = \frac{12 \times 11 \times 10 \times 9 \times 8 \times 7 \times 6}{3 \times 2 \times 1 \times 4 \times 3 \times 2 \times 1}$$

$$= 11 \times 10 \times 9 \times 4 \times 7 = 27{,}720$$

3-1-4 Permutations in Experimental Design:
Counterbalancing of Spurious Sequential Effects

A common technique in experimentation is to test each subject under every treatment, thereby letting each subject serve as his own control. The potential flaw in this procedure is that a subject's reaction to a given treatment may be influenced by the general or particular effects of preceding treatments. For example, in many psychological experiments, the greater the number of preceding treatments the greater is the opportunity for the subject to learn, become bored, or become fatigued; and certain *particular* preceding treatments may be more conducive to this than others or may have idiosyncratic "transfer" effects. Therefore, it is desirable to distribute these spurious sequential effects to the various treatments as nearly equally as possible.

The nearest that one could come to this would be to run equal groups of subjects under each of the different possible permutations of treatments within sequential positions. Thus, if there were three treatments T_1, T_2, and T_3, equal numbers of subjects would be given the treatments in each of the six different possible sequences of treatment presentation: $T_1T_2T_3$, $T_1T_3T_2$, $T_2T_1T_3$, $T_2T_3T_1$, $T_3T_1T_2$, $T_3T_2T_1$.

This is usually feasible if there are no more than four treatments since $4! = 24$ is not an excessive number of subjects. But as the number of treatments increases, the number of treatment-permutations increases very rapidly, and few experimenters could afford to run the number of subjects called for when the number of treatments is as large as six.

When the number of treatments is too large to permit all permutations to be used, one may settle for a partial balance using only those permutations that, as a group, provide the most crucial of the balancing features desired. Often a good partial balance can be obtained with a surprisingly small number of permutations.

One very common type of experiment is that in which several variables are being investigated and treatments consist of the simultaneous administration of one "level" (i.e., fixed value) of each variable. In that event, one can achieve a reasonable partial balance by concentrating upon the complete balancing of *levels* rather than treatments. For example, if each treatment consisted of the pairing of one of three noise levels (designated I, II, III) with one of two illumination levels (designated 1, 2), in an experiment measuring the effects of these variables upon reading speed, and if one's interest were

Table 3-1

Permutation-of-Levels Counterbalancing Scheme

Sequence of treatment presentation is left to right.
Each pairing of a Roman with an Arabic number is
a treatment. Subject groups G_i's contain equal
numbers of subjects.

	I		II		III	
G_1	1	2	1	2	1	2
G_2	2	1	2	1	2	1

	I		III		II	
G_3	1	2	1	2	1	2
G_4	2	1	2	1	2	1

	II		I		III	
G_5	1	2	1	2	1	2
G_6	.2	1	2	1	2	1

	II		III		I	
G_7	1	2	1	2	1	2
G_8	2	1	2	1	2	1

	III		I		II	
G_9	1	2	1	2	1	2
G_{10}	2	1	2	1	2	1

	III		II		I	
G_{11}	1	2	1	2	1	2
G_{12}	2	1	2	1	2	1

focused primarily upon the noise variable, Table 3-1 would give the sequence of treatment presentation. The number of equal-sized subject groups required by this type of scheme is $\prod_i L_i!$ where L_i is the number of levels of the ith variable. In the example given, the number of groups required is $(3!)(2!) = 12$, so as few as 12 subjects could be used in contrast to a minimum of $6! = 720$ if all treatment permutations had been used. This type of scheme presents each *treatment* (e.g., combination of a level of Variable 1 with a level of Variable 2) equally *often* in each possible sequential position and completely *permutes* the presentation of *levels*. It is particularly appropriate when one's major interest is focused upon a particular variable, which may then be made the superordinate variable.

In other cases, when it is impractical to use all possible permutations of treatments and no single variable is of primary interest, a third type of scheme is available which has several desirable features. It requires only as many subject groups as there are treatments. Each treatment is presented exactly once at each of the possible points in sequence. And each treatment is immediately preceded (and immediately followed) exactly once by each of the other treatments. Therefore, both sequential position and preceding treatment are completely counterbalanced. The possible influence of more remote preceding treatments, however, is not balanced out, and the scheme requires

Table 3-2

Scheme Counterbalancing Both Sequential Position and
Preceding Condition for Six Treatments (Numbered 1 to 6)

Equal-sized group of subjects	Sequential position in which treatment is to be presented					
	First	Second	Third	Fourth	Fifth	Sixth
G_1	1	6	2	5	3	4
G_2	2	1	3	6	4	5
G_3	3	2	4	1	5	6
G_4	4	3	5	2	6	1
G_5	5	4	6	3	1	2
G_6	6	5	1	4	2	3

that the number of treatments be an even number. The scheme is given in Table 3-2 for the case of six treatments. It is easy to obtain such a table for any even number T of treatments. In the cells of the first row of a table, with T rows and T columns, one enters the sequence of integers $1, T, 2, T - 1, 3, T - 2, \ldots, T/2, (T/2) + 1$. Then in each column, starting with the integer in the top row, successive integers are written in successive rows, except that the integer T is followed by the integer 1.

3-2   Combinations

Suppose that a cat has a litter of seven kittens and that five of them are picked to be given away. The particular five so identified constitutes a *combination*. A **combination** is a subset of some or all of the units in a set, where the order of arrangement of the units contained in the subset is immaterial. Thus, a combination is defined only by the individual identities of the units of which it is comprised, and not by any order relationships among them. While a permutation is an ordered subset, a combination is simply a subset. Thus if four of the five members of a family pose for a group photograph while the fifth member takes the picture, the four constitute a combination, *irrespective* of the order in which they align themselves for the picture. In this book we shall generally assume that each of the original n units is identifiable from all the rest, even though some units may have the same value (just as George is identifiable from Mike even though they have the same IQ and are therefore indistinguishable in IQ score).

Two subsets represent the same combination only if they contain the same number of units and if every unit contained in one is also contained in the other. Thus, two combinations are different if one contains one or more units not contained in the other. So if the parent set consists of the units A, B, C, and D, the two combinations ABC and ABD are different.

3-2-1   Number of Combinations of *n* Things Taken *r* at a Time

The "number of combinations of n things taken r at a time" is simply the number of different unordered subsets of r things that might possibly result if one drew r things from a set of n things. It is represented by the symbol

$$\binom{n}{r}$$

The ***number of combinations of*** n ***things taken*** r ***at a time is***

$$\binom{n}{r} = \frac{n!}{r!(n-r)!} \tag{3-6}$$

Proof: The number of combinations of n things taken r at a time

$$\binom{n}{r}$$

is the number of different possible *unordered* subsets of r units that might result when r units are taken from n units. The r units in each of them can be ordered, or permuted, in $r!$ different ways. So

$$r!\binom{n}{r}$$

must be the number of different possible *ordered* subsets that might result when r units are taken from the original set of n objects and assigned to r fixed positions. But we already know (see Formula 3-1) that the latter is

$$_nP_r = \frac{n!}{(n-r)!}$$

so

$$r!\binom{n}{r} = \frac{n!}{(n-r)!}$$

and dividing both sides by $r!$ we obtain

$$\binom{n}{r} = \frac{n!}{r!(n-r)!}$$

Example: The number of different possible combinations of five things taken three at a time is

$$\binom{5}{3} = \frac{5!}{3!(5-3)!} = \frac{5!}{3!2!} = \frac{(5)(4)(3)(2)(1)}{(3)(2)(1)(2)(1)} = \frac{(5)(4)}{2} = 10$$

If the five things are A, B, C, D, and E, the ten different possible combinations of three things are:

ABC	*ADE*
ABD	*BCD*
ABE	*BCE*
ACD	*BDE*
ACE	*CDE*

Notice that the number of combinations of n things taken r at a time is the same as the number of combinations of n things taken $n - r$ at a time:

$$\binom{n}{r} = \frac{n!}{r!(n-r)!}$$

$$\binom{n}{n-r} = \frac{n!}{(n-r)![n-(n-r)]!} = \frac{n!}{(n-r)!r!} = \binom{n}{r}$$

This should not be surprising. When we "take" a combination of r things from n things, we "leave" a combination of $n - r$ things. So the number of different combinations "leavable" should be the same as the number of different combinations "takable." This should give us some insight into a couple of other matters. When we "take" a combination of n things, we "leave" a combination of 0 things. The number of combinations of n things taken n at a time is obviously 1. Therefore, the number of combinations of n things taken (or "left") zero at a time must also be 1. And since

$$\binom{n}{n} = \frac{n!}{n!(n-n)!} = \frac{n!}{n!0!} = \frac{1}{0!}$$

which must equal 1, then 0! must equal 1.

Illustrative Problems: In how many different ways could a 3-man committee be formed from the personnel in a 10-man department?

ANSWER:

$$\binom{10}{3} = \frac{10!}{3!7!} = \frac{10 \times 9 \times 8}{3 \times 2 \times 1} = 10 \times 3 \times 4 = 120$$

From a "subject pool" of 12 laboratory rats, 4 rats are to be withdrawn for use in an experiment. How many possibilities are there for the composition of the experimental group?

ANSWER: The question amounts to asking the number of combinations of 12 things taken 4 at a time. Therefore the answer is,

$$\binom{12}{4} = \frac{12!}{4!8!} = \frac{12 \times 11 \times 10 \times 9}{4 \times 3 \times 2 \times 1} = 11 \times 5 \times 9 = 495$$

3-2-2 Pascal's Triangle

If one compiles a table with values of n as row headings and values of r as column headings and giving the corresponding values of

$$\binom{n}{r}$$

as cell entries, it turns out that each cell entry is the sum of the cell entry directly above it and the cell entry just to the left of the latter (if there is one). By making use of this fact, it is easy to extend the table without excessive labor. The relationship can be expressed as a formula, known as **Pascal's Rule,**

$$\binom{n}{r} = \binom{n-1}{r} + \binom{n-1}{r-1} \tag{3-7}$$

and it is easily proved by substituting the factorial expressions involved and then obtaining a common denominator:

$$\frac{n!}{r!(n-r)!} = \frac{(n-1)!}{r!(n-r-1)!} + \frac{(n-1)!}{(r-1)!(n-r)!}$$

$$= \frac{(n-1)!(n-r) + (n-1)!r}{r!(n-r)!}$$

$$= \frac{(n-1)!n}{r!(n-r)!} = \frac{n!}{r!(n-r)!}$$

The table is known as **Pascal's Triangle**. Table 3-3 gives Pascal's Triangle up to $n = 10$.

3-2-3 Relationship Between Combinations and Distinguishable Permutations

The number of combinations of n things taken r at a time $n!/[r!(n-r)!]$ is the same as the number of distinguishable permutations of n things, of which r are of one type and $n - r$ of another (and only units of different types are distinguishable). The reason that the two have the same formula is that the indistinguishable units of a single type may be regarded as an unordered subset (since the relative order of mutually indistinguishable units contributes nothing to distinguishability and therefore doesn't matter)—that is, as a combination. Or taking a more analytic approach, the reason for the equivalence may be seen as follows. Imagine the n positions that will be occupied by the n things of both types to be numbered from 1 to n. For every distinguishable permutation of the n things within the n positions, the numbers of the positions occupied by the r things of the first type will constitute a different combination of r of the n numbers. (Contrariwise, each different combination of r of the n numbers will identify a set of positions for type 1 occupancy that will result in a different distinguishable permutation.) So for every combination there is a different distinguishable permutation (and vice

Table 3-3

Pascal's Triangle

	r	0	1	2	3	4	5	6	7	8	9	10
	0	1										
	1	1	1									
	2	1	2	1								
	3	1	3	3	1							
	4	1	4	6	4	1						
n	5	1	5	10	10	5	1					
	6	1	6	15	20	15	6	1				
	7	1	7	21	35	35	21	7	1			
	8	1	8	28	56	70	56	28	8	1		
	9	1	9	36	84	126	126	84	36	9	1	
	10	1	10	45	120	210	252	210	120	45	10	1

versa). And, as a result, the number of different possible combinations is the same as the number of different possible distinguishable permutations.

3-3 Groupings (Allocations of Things to Containers)

Each of the topics we have discussed so far may be regarded as a special case of a more general topic: the number of possible allocations of r of n things to k containers. Permutations present a special case in which "containers" are distinguishably different places each of which must receive exactly one thing, so that $r = k$. Combinations represent the case in which there is just one container ("distinguishable" by virtue of its uniqueness) so that $k = 1$, and all n things are mutually distinguishable. Groupings, the topic of this section, represents still another special case in which all n things are allocated to containers, so that $r = n$, and all n things are mutually distinguishable. Groupings differ from permutations in that more than one thing may be allocated to a container. They differ from combinations in that there may be more than one container and (in some cases to be discussed) some containers may be indistinguishable from others.

3-3-1 Number of Allocations of *n* Things to *k* Unique Categories, Each Category Receiving a Specified Number of Things

Suppose that the manager of a 5-man sales force wants to know the number of different possible assignments of personnel to sales regions in which two men are assigned to the eastern region, two to the western region, and one to the central region. His problem is a specific case of the general problem of the number of different possible allocations of n distinguishable things to k uniquely predesignated categories, where each category receives a specified number of things. And that, in turn, may be conceptualized as the problem of the number of different possible allocations of n mutually distinguishable things to k mutually distinguishable containers, where each container is to receive a specified number of things.

The number of different possible allocations of n *mutually distinguishable things to* k *mutually distinguishable containers where the number of things to be assigned to each container is specified in advance* (r_1 things to go to the first container, r_2 to the second, . . . , r_i to the *i*th, . . . , and r_k to the *k*th, where $\sum_{i=1}^{k} r_i = n$) is

$$\frac{n!}{r_1! r_2! \cdots r_i! \cdots r_k!} \quad \text{or} \quad \frac{n!}{\prod_{i=1}^{k} r_i!} \tag{3-8}$$

Proof: If the containers are filled in sequence, any one of the

$$\binom{n}{r_1}$$

different possible combinations of r_1 things that can be obtained from n things could become the contents of the first container. For each of these

$$\binom{n}{r_1}$$

subsets of r_1 units, there are

$$\binom{n - r_1}{r_2}$$

different possible combinations of r_2 things that can be obtained from the remaining $n - r_1$ things and that could therefore become the contents of the second container. So there are

$$\binom{n}{r_1}\binom{n - r_1}{r_2}$$

different possible pairs of subsets that could be the respective contents of the first two containers. For each of these

$$\binom{n}{r_1}\binom{n - r_1}{r_2}$$

pairs, there are

$$\binom{n - r_1 - r_2}{r_3}$$

different possible combinations of r_3 things that can be obtained from the remaining $n - r_1 - r_2$ things and that could become the contents of the third container. Continuing in this fashion, the total possible number of complete allocations in which all n things are assigned is

$$\binom{n}{r_1}\binom{n-r_1}{r_2}\binom{n-r_1-r_2}{r_3}\cdots\binom{n-\sum_{i=1}^{k-1}r_i}{r_k}$$

which equals

$$\left[\frac{n!}{r_1!(n-r_1)!}\right]\left[\frac{(n-r_1)!}{r_2!(n-r_1-r_2)!}\right]\left[\frac{(n-r_1-r_2)!}{r_3!(n-r_1-r_2-r_3)!}\right]\cdots$$
$$\left[\frac{(n-\sum_{i=1}^{k-1}r_i)!}{r_k!0!}\right]$$

One of the factorials in the denominator of each term inside square brackets equals, and therefore cancels out, the numerator of the following term. And since the last term can be rewritten in the form $r_k!/[r_k!0!]$ or simply $r_k!/r_k!$, the product of all k terms is $n!/[r_1!r_2!\cdots r_k!]$ or $n!/\prod_{i=1}^{k}r_i!$. To illustrate, in the case where $n=100$, $r_1=40$, $r_2=25$, $r_3=18$, $r_4=10$, $r_5=7$, substituting the appropriate values for the terms in square brackets we obtain

$$\left[\frac{100!}{40!60!}\right]\left[\frac{60!}{25!35!}\right]\left[\frac{35!}{18!17!}\right]\left[\frac{17!}{10!7!}\right]\left[\frac{7!}{7!0!}\right]=\frac{100!}{40!25!18!10!7!}$$

Example: The number of different possible allocations of 4 distinguishable things A, B, C, D to 3 containers so that container I receives 2 things and containers II and III each receive 1 thing is $4!/[2!1!1!]=12$. The 12 different allocations of distinguishable things to distinguishable containers are shown below:

I	II	III		I	II	III
AB	C	D		AB	D	C
AC	B	D		AC	D	B
AD	B	C		AD	C	B
BC	A	D		BC	D	A
BD	A	C		BD	C	A
CD	A	B		CD	B	A

The formula for number of allocations is identical to that for the number of distinguishable permutations. This is not surprising since we may regard the ith container as having r_i indistinguishable "places" for the r_i things it is to receive. The number of allocations is simply the number of distinguishable assignments of n distinguishable things to n places located in k containers where within-container places are indistinguishable from each other. But it is equally valid to regard these assignments as assignments of n places, certain subsets of which are indistinguishable, to n distinguishable things; it makes no difference whether things are assigned to places or places to things. And if we adopt the latter point of view, the distinguishable-permutation formula is obviously appropriate.

Illustrative Problems: The maze-running ability of rats is to be tested under three different drug conditions: caffeine, alcohol, and nothing (control). Ten rats are available of which three are to be assigned to the caffeine group, three to the alcohol treatment, and four to the control condition. How many different possible allocations are there of the ten distinguishable rats to the three distinguishable conditions?

ANSWER:

$$\frac{10!}{3!\,3!\,4!} = 4{,}200.$$

Ten men are to be divided into two 5-man basketball teams, one to be called the white sox, the other the red sox. How many different possible allocations are there of men to teams.

ANSWER:

$$\frac{10!}{5!\,5!} = 252$$

3-3-2 Number of Divisions of *n* Things into *k* Groups of Specified Sizes

Suppose that the manager of the 5-man sales force were interested merely in the number of different possible divisions of the 5-man force into two 2-man teams and one 1-man team, to be available for assigment anywhere. This is a different problem from that of allocations. The assignment of Hank and Harry to the eastern region, Pete and Paul to the western region, and Bill to the central region is one allocation. And the assignment of Pete and Paul to the eastern region, Hank and Harry to the western, and Bill to the central is a different allocation. But both groupings correspond to just one division of the five men into two groups of two men each and one group composed of one man.

If *n* things are to be divided into *k* groups of specified sizes, we may regard the intended sizes of the groups as categories to which the *n* things are to be allocated in specified numbers. However, if several groups have the same intended size there is no way of distinguishing between those groups prior to the actual allocation. The problem therefore may be conceptualized as that of the number of distinguishable allocations of *n* distinguishable things to *k* containers where each container is identified solely by the number of things it is to receive; all containers to receive the same specified number of things are therefore indistinguishable from each other.

The number of different possible divisions of n mutually distinguishable things into k groups of specified sizes $r_1, r_2, \ldots, r_k$ is

$$\frac{n!}{\prod_{i=1}^{k} r_i! \prod_{j=1}^{n} G_j!} \tag{3-9}$$

where G_j is the number of groups to be of size j (i.e., the number of groups to contain exactly j things or the number of r_i's having the value j).

Proof: We may regard a division as a distinguishable assignment of n distinguishably different things to k containers labeled "container to receive r_1 things," "container to receive r_2 things," ..., "container to receive r_k things," where each container receives the number of things its label calls for and containers to receive the same specified number of things are indistinguishable. If each container is to receive a different number of things, then the k containers are all mutually distinguishable and the number of distinguishable assignments was shown in the previous section to be $n!/\prod_{i=1}^{k} r_i!$. So when the group sizes are all different, the number of divisions is the same as the number of allocations.

Now consider the general case where some containers may receive the same specified number of things and therefore be indistinguishable from each other. The entire contents of containers receiving the same number of things can be interchanged, or permuted, without producing any distinguishable change. Therefore, for any distinguishable assignment, the entire contents of the G_1 containers containing only one thing can be permuted in $G_1!$ ways without changing the distinguishable assignment. For each of these ways, the contents of the G_2 containers containing exactly two things can be permuted in $G_2!$ ways without altering the distinguishable assignment, etc. So, in all, there are $G_1! \, G_2! \cdots G_n!$ or $\prod_{j=1}^{n} G_j!$ "indistinguishably" different assignments for every *distinguishably* different assignment. (Remember that $0!$ and $1!$ each equal 1 so that $\prod_{j=1}^{n} G_j!$ will equal 1 unless some groups contain the same number of things.) The number of *different* assignments is the same as the number of *distinguishably different* assignments when containers are all mutually distinguishable, which we already know to be $n!/\prod_{i=1}^{k} r_i!$. Now the number of distinguishable assignments times the number of different indistinguishable assignments per distinguishable one equals the number of assignments of all types. Therefore, letting D stand for the number of distinguishable assignments, and therefore for the number of divisions,

$$D \prod_{j=1}^{n} G_j! = \frac{n!}{\prod_{i=1}^{k} r_i!} \quad \text{and} \quad D = \frac{n!}{\prod_{i=1}^{k} r_i! \prod_{j=1}^{n} G_j!}$$

Example: The number of different possible divisions of four distinguishable things A, B, C, D into three groups so that one group contains two things and each of two groups contain one thing is

$$\frac{4!}{(2!\,1!\,1!)(2!\,1!\,0!\,0!)} = 6$$

The six different possible divisions are shown below:

Group Containing Two Things	One Group Containing One Thing	Other Group Containing One Thing
AB	C	D
AC	B	D
AD	B	C
BC	A	D
BD	A	C
CD	A	B

Illustrative Problems: Ten distinguishable rats are to be divided into three groups containing 4, 3, and 3 rats, respectively, for use in an experiment. How many different divisions into groups of the sizes named are possible?

ANSWER:

$$\frac{10!}{(4!\,3!\,3!)[0!\,0!\,2!\,1!\,(0!)^6]} = \frac{10!}{4!\,3!\,3!\,2!} = 2{,}100$$

Ten men are to be divided into two 5-man basketball teams. How many such divisions are possible?

ANSWER:

$$\frac{10!}{(5!\,5!)[(0!)^4\,2!\,(0!)^5]} = \frac{10!}{5!\,5!\,2!} = 126$$

3-3-3 Total Number of Allocations of *n* Things to *k* Distinguishable Containers

Sometimes we are interested simply in the number of different possible assignments of things to categories where any category may receive any number of things including none. If there is no restriction on the number of units that can occupy a container or on the number of containers that must be occupied, then *the number of different possible allocations of* n *mutually distinguishable things to* k *mutually distinguishable containers is*

$$k^n \qquad\qquad\qquad\qquad (3\text{-}10)$$

Proof: The first thing can be dropped into any one of the k different containers and therefore can be assigned in any of k different ways. For each of these ways, the second thing can also be dropped into any one of the k different containers and therefore can be assigned in k different ways, etc.,

etc. So the total number of different ways of assigning all n things is $k(k)(k) \cdots (k) = k^n$.

Example: $n = 2$ distinguishable units can occupy $k = 3$ distinguishable containers in $k^n = 3^2 = 9$ different ways. Thus, a husband H and wife W can be distributed within the three rooms of a house in the following ways:

HW		
	HW	
		HW
H	W	
W	H	
H		W
W		H
	H	W
	W	H

Illustrative Problems: How many different possible allocations are there of five business letters to three baskets labeled "in," "out," and "hold"?

ANSWER: $3^5 = 243$.

The same two dignitaries have been invited to parties held by four rival embassies on the same night. If each dignitary attends exactly one party, how many different possible allocations are there of dignitaries to parties?

ANSWER: $4^2 = 16$.

3-4 Sampling Distributions

If a sample of n units is to be drawn from a population of N units, there are many different possible outcomes. That is, there are many different possible ordered subsets of n units that could result from drawing the n units, in sequence, from the N units and, therefore, that could become the sample. So if some sample statistic, such as the mean or median, were calculated from the values of the n units comprising each of these possible samples, the value of that statistic would generally tend to vary considerably from one sample to another. The relative frequency distribution of the values taken by the statistic in all possible samples of n units is the subject of this section.

3-4-1   The Set of All Possible Samples

A sample of n units could be drawn from a population of N units in many different ways. If sampling is without replacements, the first unit drawn could be any one of the N units in the population, the second could be any one of the $N - 1$ units remaining after the first has been withdrawn, . . . , and the nth could be any one of the $N - (n - 1)$ or $N - n + 1$ units remaining after the first $n - 1$ sample units have been withdrawn. For each of the N different ways of obtaining the first sample unit there are $N - 1$ different ways of obtaining the second, so there are $N(N - 1)$ different ways of obtaining the first pair, etc., and ultimately $N(N - 1) \cdots (N - n + 1) = N!/(N - n)!$ different ways of obtaining the entire sample. And, to each of these $N!/(N - n)!$ different ways there corresponds a different ordered subset of n units.

If sampling is with replacements, the first unit drawn could be any one of the N units, and (because of replacement) so could the second, . . . , and so could the nth; so the number of different ways of obtaining the entire sample is $N(N) \cdots (N) = N^n$. And, to each of these N^n different ways there corresponds a different ordered subset of n units.

We shall define samples drawn in different ways to be different samples (so samples containing the same units drawn in different orders are different samples). Hence, the set of all possible samples of n units that could be drawn from a population of N units contains $N!/(N - n)!$ different samples if sampling is without replacements or N^n different samples if sampling is with replacements. The set of all possible samples of two units that could be drawn from a population of three units (A, B, and C) is given in Table 3-4 for both sampling with and without replacements.

Table 3-4

Set of All Possible Samples of Two Units Drawn
From a Population of Three Units

First Unit Drawn	Second Unit Drawn		First Unit Drawn	Second Unit Drawn
Samples drawn with replacements			*Samples drawn without replacements*	
A	A		A	B
A	B		A	C
A	C		B	A
B	A		B	C
B	B		C	A
B	C		C	B
C	A			
C	B			
C	C			

3-4-2 The Set of Means of All Possible Samples

Now suppose that unit A's value is 1, unit B's is 1, and unit C's is 7. For every different possible sample we can list the included units, their values, and some sample statistic calculated from those values, such as the mean. This is done in Table 3-5.

Table 3-5

Table 3-4 Augmented by Values of Units and Means of Samples

Samples drawn with replacements					Samples drawn without replacements				
First unit drawn		Second unit drawn		Mean	First unit drawn		Second unit drawn		Mean
Identity	Value	Identity	Value		Identity	Value	Identity	Value	
A	1	A	1	1	A	1	B	1	1
A	1	B	1	1	A	1	C	7	4
A	1	C	7	4	B	1	A	1	1
B	1	A	1	1	B	1	C	7	4
B	1	B	1	1	C	7	A	1	4
B	1	C	7	4	C	7	B	1	4
C	7	A	1	4					
C	7	B	1	4					
C	7	C	7	7					

For every different possible sample we now have a mean, and we can prepare a frequency distribution and a relative frequency distribution for those means, as shown in Table 3-6. The type of relative frequency distribution given in Table 3-6 is an example of a *sampling distribution*.

Table 3-6

Frequency and Relative Frequency Distribution
of Means in Table 3-5

Samples drawn with replacements			Samples drawn without replacements		
Value of mean $\bar{x}$	Number of means f	Proportion of means p	Value of mean $\bar{x}$	Number of means f	Proportion of means p
1	4	4/9	1	2	2/6
4	4	4/9	4	4	4/6
7	1	1/9			

3-4-3 Sampling Distributions

A *sampling distribution* is the relative frequency (or sometimes frequency) distribution for the set of values of a sample statistic obtained when that statistic is calculated for every possible sample of a given size that could be drawn in a specified manner (with or without replacements) from a given population. Thus we may regard all possible samples as comprising the units of a "population of samples," the values of the units being the calculated values of the sample statistic, and their relative frequency distribution being the sampling distribution.

The set of all possible samples is likely to be quite large. Therefore, it is of some interest to know what shortcuts can be taken in obtaining a sampling distribution. The value of a sample statistic generally depends only on the identities of the units comprising the sample and not on the order in which they were drawn. In such cases, instead of listing the samples corresponding to all different distinguishable orders of the same set of units, we can list a single order and consider that sample to have occurred as many times as there are distinguishable permutations of its units (*not* values). Thus, for a sample of three units, instead of listing all six permutations of the units A, B, C, we list ABC once and give it (and the value of the sample statistic calculated from it) a frequency of 6. Likewise, instead of listing the samples AAB, ABA, and BAA, we list only AAB and give it a frequency of 3. When sampling is without replacements, the set of all possible samples consists of each of the $n!$ possible permutations of each of the

$$\binom{N}{n}$$

possible combinations of units. Therefore, since each combination occurs the same number $n!$ of times, one can obtain a sampling distribution (in *relative* frequency form) simply by calculating the sample statistic for each of the

$$\binom{N}{n}$$

different combinations of units and compiling a relative frequency distribution for its values. (Indeed, even the *frequency* distribution in such cases is regarded as the equivalent of a sampling distribution and is often referred to as a sampling distribution.) This frame of mind, however, can quickly get one into trouble when sampling is with replacements.

Anything that can be calculated numerically from the sample is a statistic and can have a sampling distribution.

In order to illustrate the above points, we shall obtain the sampling distributions of the mean, median, and range for samples of 3 units drawn (both with and without replacements) from a population consisting of the units A, B, C, D, and E, whose values are respectively 1, 1, 7, 19, and 37. This is done below in two stages. The first step is taken in Table 3-7 which lists for each different set of 3 sample units the number d of distinguishable permutations of the units (*not* values) comprising the sample, the sample mean $\bar{x}$, the sample median m, and the sample range w. The first 10 rows apply to the case of sampling without replacements, whereas all 35 rows apply to the case of sampling with replacements.

Using the information in this table, we may now obtain the frequency distribution of each sample statistic by simply listing the *different* values taken by the statistic, and for each such value, listing the sum of the d-entries

Table 3-7

Illustration of Shortcut Method of Obtaining
Sampling Distributions—First Stage

Units	Values	d	$\bar{x}$	m	w
A B C	1, 1, 7	6	3	1	6
A B D	1, 1, 19	6	7	1	18
A B E	1, 1, 37	6	13	1	36
A C D	1, 7, 19	6	9	7	18
A C E	1, 7, 37	6	15	7	36
A D E	1, 19, 37	6	19	19	36
B C D	1, 7, 19	6	9	7	18
B C E	1, 7, 37	6	15	7	36
B D E	1, 19, 37	6	19	19	36
C D E	7, 19, 37	6	21	19	30
A A B	1, 1, 1	3	1	1	0
A A C	1, 1, 7	3	3	1	6
A A D	1, 1, 19	3	7	1	18
A A E	1, 1, 37	3	13	1	36
B B A	1, 1, 1	3	1	1	0
B B C	1, 1, 7	3	3	1	6
B B D	1, 1, 19	3	7	1	18
B B E	1, 1, 37	3	13	1	36
C C A	7, 7, 1	3	5	7	6
C C B	7, 7, 1	3	5	7	6
C C D	7, 7, 19	3	11	7	12
C C E	7, 7, 37	3	17	7	30
D D A	19, 19, 1	3	13	19	18
D D B	19, 19, 1	3	13	19	18
D D C	19, 19, 7	3	15	19	12
D D E	19, 19, 37	3	25	19	18
E E A	37, 37, 1	3	25	37	36
E E B	37, 37, 1	3	25	37	36
E E C	37, 37, 7	3	27	37	30
E E D	37, 37, 19	3	31	37	18
A A A	1, 1, 1	1	1	1	0
B B B	1, 1, 1	1	1	1	0
C C C	7, 7, 7	1	7	7	0
D D D	19, 19, 19	1	19	19	0
E E E	37, 37, 37	1	37	37	0

corresponding to that value. The sum of these d's is the number of distinguishably ordered subsets of units whose unit-values produce the statistic
-value in question; it is therefore the frequency f with which the statistic takes
that value in the set of all possible samples. Dividing f by the number of
possible samples ($N!/(N - n)! = 5!/2! = 60$ without replacements or
$N^n = 5^3 = 125$ with replacements) yields p, the relative frequency of the
statistic-value in question. The set of statistic-values and their corresponding
relative frequencies is the sampling distribution for the statistic in question.
Table 3-8 gives the results of this procedure applied to Table 3-7, producing a
sampling distribution for each of the three statistics under each of the two
methods of sampling.

Table 3-8

Illustration of Shortcut Method of Obtaining
Sampling Distributions—Second Stage

Sampling distributions of $\overline{x}$, m, and w for sampling without replacements

$\overline{x}$	f	p	m	f	p	w	f	p
3	6	6/60	1	18	18/60	6	6	6/60
7	6	6/60	7	24	24/60	18	18	18/60
9	12	12/60	19	18	18/60	30	6	6/60
13	6	6/60		60		36	30	30/60
15	12	12/60					60	
19	12	12/60						
21	6	6/60						
	60							

Sampling distributions of $\overline{x}$, m, and w for sampling with replacements

$\overline{x}$	f	p	m	f	p	w	f	p
1	8	8/125	1	44	44/125	0	11	11/125
3	12	12/125	7	37	37/125	6	18	18/125
5	6	6/125	19	31	31/125	12	6	6/125
7	13	13/125	37	13	13/125	18	36	36/125
9	12	12/125		125		30	12	12/125
11	3	3/125				36	42	42/125
13	18	18/125					125	
15	15	15/125						
17	3	3/125						
19	13	13/125						
21	6	6/125						
25	9	9/125						
27	3	3/125						
31	3	3/125						
37	1	1/125						
	125							

Shortcuts are also possible when many units have the same value. Suppose that we wished to obtain the sampling distribution of some statistic based on a sample of two units drawn from a population consisting of the units A, B, C, D, E, F, G, H, I, J, whose values are respectively 0, 0, 0, 0, 0, 0, 0, 0, 0, 1. If sampling is without replacements, the first sample unit drawn can be any of ten units and the second can be any of the remaining nine, so there are $10 \times 9 = 90$ different possible samples. But the first and second sample unit respectively can have only three different possible ordered pairs of values 0, 0; 0, 1; and 1, 0. Now consider the number of different ways in which these pairs of values could be obtained. First consider the pair 0, 0. Any one of nine population units whose value is zero could be drawn as the first sample unit, and any one of the remaining eight could be drawn as the second. So the pair 0, 0 can be obtained in $9 \times 8 = 72$ different possible ways. The pairs 0, 1 and 1, 0 can each be obtained in nine different ways since any one of nine population units can become the sample unit whose value is zero, but only unit J can become the sample unit whose value is 1.

If sampling is with replacements, there are $10^2 = 100$ different possible samples, whose units must have one of four different possible ordered pairs of values, 0, 0; 0, 1; 1, 0; 1, 1. Irrespective of its position in the sample, each sample unit can be any one of the ten population units; so there are nine ways that it can have a value of 0 and one way that it can have a value of 1. Hence, there are $9 \times 9 = 81$ different ways of obtaining 0, 0; there are $9 \times 1 = 9$ ways of obtaining 0, 1; $1 \times 9 = 9$ ways of getting 1, 0; and $1 \times 1 = 1$ way of obtaining the sample observations 1, 1. Thus, for the two different methods of sampling we have the situation shown in Table 3-9.

Now suppose that the population, instead of consisting of ten units nine of which are 0's and one of which is a 1, consisted of a very large number N of

Table 3-9

Frequency Distribution for Samples of Two Observations
When Population Contains Nine Zeros and a One

Sampling without replacements			Sampling with replacements		
First observation	Second observation	f	First observation	Second observation	f
0	0	72	0	0	81
0	1	9	0	1	9
1	0	9	1	0	9
		$90 = 10 \times 9$	1	1	1
					$100 = 10^2$

units 90% of which were 0's and 10% of which were 1's. The set of all possible samples of two units would consist of $N(N-1) \cong N^2$ samples if sampling is without replacements and of N^2 samples if sampling is with replacements. In the latter case, the N^2 possible samples are obtained by letting each of the N different population units be the "first sample unit" N times and for each such block of N listings of the same population unit as first sample unit, letting each of the N different population units be the "second sample unit" once. It follows from this procedure that in the set of all possible samples, 90% of the first sample units are 0's and 10% are 1's, and that 90% of the second sample units are 0's and 10% are 1's both when the first sample unit is a 0 and when it is a 1. Therefore, in the set of all possible samples the proportion of the time that the sample 0, 0 occurs is $.9 \times .9 = .81$, the sample 0, 1 occurs $.9 \times .1 = .09$ of the time, 1, 0 occurs $.1 \times .9 = .09$ of the time, and 1, 1 occurs $.1 \times .1 = .01$ of the time. If sampling is without replacements, the number of possibilities for the second sample unit is reduced by one unit and so is the number of units having the same value as the first unit, so the proportions are altered accordingly. Both situations are summarized in Table 3-10.

Table 3-10

Relative Frequency Distribution for Samples of Two Observations When Population Contains Many Units, 90% of Which Have a Value of Zero and 10% of Which Have a Value of One

Sample		Proportion of samples whose units have these values if samples are drawn	
First observation	Second observation	With replacements	Without replacements
0	0	$.9 \times .9 = .81$	$.9 \times \frac{.9N-1}{N-1} \simeq .81$
0	1	$.9 \times .1 = .09$	$.9 \times \frac{.1N}{N-1} \simeq .09$
1	0	$.1 \times .9 = .09$	$.1 \times \frac{.9N}{N-1} \simeq .09$
1	1	$.1 \times .1 = .01$	$.1 \times \frac{.1N-1}{N-1} \simeq .01$

It should be clear that if the number n of units in the sample is finite and the number N of units in the population is infinite, the proportions calculated under the two sampling methods become equal in all cases. And even if N is finite, if $n \ll N$, i.e., if n is much smaller than N, the proportions calculated under the two sampling methods will generally be approximately the same.

Thus, if n is very small and N is either very small or represents a very large number of units that take a very small number of different values, it is fairly easy to obtain the relative frequency distribution of the sets of n ordered sample values within the set of all possible samples. And from this it is easy to obtain the sampling distribution of any sample statistic for which one is desired.

3-5 Problems

1. How many permutations are there of the letters in the word MISSIS-SIPPI? How many permutations are there of the letters in the word MISSISSIPPI that leave the word unchanged, correctly spelling the name of the river? How many distinguishable permutations are there of the letters in the word MISSISSIPPI?

2. How many different possible allocations are there of 11 mutually distinguishable things to 4 predesignated categories so that the first, second, third, and fourth categories receive 1, 4, 4, and 2 things, respectively?

3. How many different possible divisions are there of 11 mutually distinguishable things into 4 piles, one containing 1 thing, two each containing 4 things, and one containing 2 things?

4. The owner of a combination padlock remembers everything required to open the lock except the sequence in which the three numbers are to be used. What is the maximum number of sequences he may have to try in order to open the lock?

5. From ten subjects a group of five is to be selected. How many differing sets of five people could become the selected group? How many differing sets of five people could become the remaining group? Five of ten subjects are to be assigned to Treatment A, the remaining five to Treatment B. How many differing pairs of groups identified with treatments are possible? Ten subjects are to be broken into two groups of five subjects each. In how many different possible ways could this be done?

6. How many different possible peck-orders are there in a flock of eight chickens if no chicken acknowledges any peers?

7. How many different possible allocations are there of four differently colored envelopes to (a) ten differently labeled mail slots, (b) two differently labeled mail slots?

8. A marriage broker has 12 female and 3 male clients. How many different possible solutions are there in which the 3 males are paired with 3 females?

9. An experiment is going to investigate 4 major conditions, each of which has 3 subconditions, making 12 "treatments" in all. It is relatively easy to change the apparatus so as to change from one subcondition to another within a single major condition, but it is relatively difficult to change to a subcondition within a different major condition. Therefore, the 3 subconditions within each major condition must be presented consecutively. How many different sequences of presentation of the 12 subconditions are there which meet this restriction?

10. A family contains five boys and three girls. In how many different boy-girl sequences (e.g., *BBGBGBBG*) could the children of such a family have been born?

11. In how many different possible sequences might an unspecified six of nine subjects show up for an experiment?

12. A telegraphic code identifies each letter of the alphabet by a series of elements, each element being either a dot or a dash. If n, the number of elements in a series, is a constant, how large must n be to provide a different series of elements for each of the 26 letters? What is the answer if n is not a constant and different letters can be represented by series of different lengths?

13. In how many different ways could the seven members of a family line up so that the three male members are all together on one side and the four female members are all together on the other side?

14. Four players are seated at a bridge table. How many permutations are there of players within the four seats in which the positions of the players *relative to each other* are different? Why?

15. Stocks are represented on a ticker tape by symbols composed of a series of capital letters. If a symbol can consist of any number of letters up to and including three, how many different symbols can be formed if a given letter (a) can appear only once in a single symbol, (b) can be used repeatedly in a single symbol?

16. A multiple-personality psychotic has four different personalities A, B, C, D. How many different sequences of four personalities are possible if the only restriction is that a personality cannot be immediately followed by itself?

17. A social psychologist studying intrafamily alliances defines a coalition as an alliance between at least two but less than all of the members of a family. Coalitions of the same size differ if they do not include exactly the same persons. How many different coalition patterns are possible in a family of five consisting of a Father, Mother, Oldest, Second, and Youngest child?

18. A message is sent via a ten-word telegram. If each word used has exactly one additional synonym, how many ten-word messages are there that would say essentially the same thing?

19. A committee consisting of a chairman, a vice chairman, a secretary, and two nonofficeholding members is to be formed from the personnel of an eight-man department. How many distinguishably different possibilities are there for the composition of the committee?

20. What is the answer to the above problem if it is stipulated that the committee must exclude the head of the department and must include Smith, the newest member of the department?

21. An ice cream parlor offers 32 different kinds of ice cream. Two triple-dip ice cream cones are of the same type only if they contain the same varieties of ice cream located in corresponding layers. How many different types of triple-dip cones can be made if (a) each cone contains three different varieties, (b) each triple-dip cone can contain any number of varieties up to three, (c) each triple-dip cone contains only two different varieties?

22. A population consists of five units having the values 9, 0, −3, 0, 0. For samples of size $n = 3$, obtain the sampling distributions of the mean, median, and range when sampling (a) without replacements, (b) with replacements.

4

randomness

4-1 Equal Likelihood, Its Implications and Consequences

The term "equally likely" is a primitive concept; we cannot break it down into something more elementary, and its definition must rest largely in the intuition. It is a fortunate characteristic of such terms that "everyone" knows what they mean—at least in a vague primitive sense. But we need more than that. And although we cannot analyze the meaning of "equally likely" through definition, we *can* refine our understanding of it by determining what it does not mean and by exploring its implications and consequences.

Buridan's ass starved to death because it found itself at exact dead center between two bales of hay, was attracted with exactly equal valence in opposite directions, and was therefore unable to move in either. But if we toss a fair coin, it cannot come up half a head and half a tail—it *will* come up either heads or tails; and if it comes up heads, what can we mean if we say that tails had been equally likely? We do not mean that the coin's trajectory did not obey the laws of physics. (Given the initial conditions, the outcome "heads" was a certainty. The uncertainty resided in our failure to control all of the relevant factors.) Nor can we mean that in an even number of tosses heads and tails would come up equally often. No matter how many times we tossed a fair coin, it would be possible for it to come up heads every time. Therefore, we cannot define equal likelihood in terms of the actual outcome of a finite number of tosses. And if we defined it in terms of an infinite number of tosses, we would be basing a definition upon the occurrence of an impossibility.

However, *whatever* "equally likely" means, it surely implies that heads and tails should be given equal weight in our considerations of the possible results of *n* tosses. And *that* implies that every *possible* pattern of outcomes in a sequence of tosses is exactly as likely as every other. Thus if *H* and *T* are equally likely, then in two tosses of the coin the sequences *HH, HT, TH, TT* are also equally likely; in three tosses the sequences *HHH, HHT, HTH, THH, HTT, THT, TTH, TTT* are all equally likely; and in four tosses the sequences *HHHH, HHHT, HHTH, HTHH, THHH, HHTT, HTHT, HTTH, THHT, THTH, TTHH, HTTT, THTT, TTHT, TTTH, TTTT* are equally likely. We notice that when the number of tosses *n* is 2, there are four possible sequences: one consisting of 2 *T*'s, one of 2 *H*'s, and two of an *H* and a *T*. So in 25% of the sequences the proportion of heads P_H is 0, in 50% P_H is .5, and in 25% P_H is 1. Likewise, when $n = 3$, P_H is 0 in 12.5% of the eight possible sequences, .3333 in 37.5%, .6667 in 37.5%, and 1.0000 in 12.5%. And when $n = 4$, P_H is 0 in 6.25%, .25 in 25%, .5 in 37.5%, .75 in 25%, and 1.0000 in 6.25%. Continuing in this fashion, we obtain the information presented in Table 4-1.

There are two important features to be noted in Table 4-1. First, although it is possible for heads to come up exactly half the time in an even number of tosses, the proportion of possible sequences in which it does so decreases steadily as the even number of tosses increases. So we cannot say that if *H* and *T* are equally likely they should come up exactly the same number of times in the long run. The second feature is far more important. As *n* increases, smaller and smaller proportions of sequences have extreme proportions of heads and (although there are some striking exceptions at small values of *n*) larger and larger proportions of sequences have a proportion of heads that is in the *neighborhood* of .5. Indeed, if ϵ is an infinitesimal value greater than zero, there is some huge value of *n* for which it is true that 99.999% of the possible sequences contain a proportion of heads lying somewhere in the region between $.5 - \epsilon$ and $.5 + \epsilon$. (This will be proved in Chapter 14.)

Suppose that we apply the same reasoning to the throw of a fair die, where there are six different possible equally likely outcomes. If we throw the die once, the possible outcomes are 1, 2, 3, 4, 5, 6. If we throw it twice, there are $6^2 = 36$ different possible equally likely sequences of outcomes, shown in Table 4-2. And, in general, if we throw the die *n* times, there are 6^n different possible equally likely sequences of outcomes.

Now if we continue to write down the 6^n different possible sequences for increasing values of *n*, we shall find that the information given in the left half of Table 4-3 is correct. As *n* increases (above a small value) a larger and larger proportion of the 6^n different possible sequences contain a proportion of 1's in the neighborhood of $1/6 = .1667$. And when *n* becomes huge, the proportion of 1's in a sequence becomes almost exactly 1/6 for almost 100% of the 6^n different possible sequences.

So to generalize, if an event has *N* possible outcomes all of which are

Table 4-1

Proportion of Sequences, Among All Possible Sequences of n Tosses of a Fair Coin, in Which the Proportion of Heads in the Sequence is P_H

P_H	1	2	3	4	5	10	20	50	100	500	1,000
0-.0499	.5	.25	.125	.0625	.03125	.00098	.00000	.00000	.00000	.00000	.00000
.05-.0999						.00977	.00002	.00000	.00000	.00000	.00000
.10-.1499							.00018	.00000	.00000	.00000	.00000
.15-.1999							.00109	.00000	.00000	.00000	.00000
.20-.2499				.25	.15625	.04395	.00462	.00015	.00000	.00000	.00000
.25-.2999							.01479	.00115	.00002	.00000	.00000
.30-.3499			.375			.11719	.03696	.01512	.00088	.00000	.00000
.35-.3999							.07393	.04304	.01671	.00000	.00000
.40-.4499					.3125	.20508	.12013	.18048	.11803	.01123	.00070
.45-.4999							.16018	.20392	.32458	.47094	.48669
Exactly .5	*.5*	*.5*	*.375*	*.375*	*.24609*	*.24609*	*.17620*	*.11228*	*.07958*	*.03566*	*.02522*
.5001-.55							.16018	.20392	.32458	.47094	.48669
.5501-.60					.3125	.20508	.12013	.18048	.11803	.01123	.00070
.6001-.65							.07393	.04304	.01671	.00000	.00000
.6501-.70			.375			.11719	.03696	.01512	.00088	.00000	.00000
.7001-.75				.25			.01479	.00115	.00002	.00000	.00000
.7501-.80					.15625	.04395	.00462	.00015	.00000	.00000	.00000
.8001-.85							.00109	.00000	.00000	.00000	.00000
.8501-.90						.00977	.00018	.00000	.00000	.00000	.00000
.9001-.95							.00002	.00000	.00000	.00000	.00000
.9501-1.00	.5	.25	.125	.0625	.03125	.00098	.00000	.00000	.00000	.00000	.00000
.40-.60	*0*	*.5*	*0*	*.375*	*.625*	*.65625*	*.73682*	*.88108*	*.96480*	*1.00000*	*1.00000*
.45-.55	*0*	*.5*	*0*	*.375*	*0*	*.24609*	*.49656*	*.52012*	*.72875*	*.97754*	*.99860*

Table 4-2

Possible Sequences of Outcomes in Two Throws of a Fair Die

First throw	Second throw	First throw	Second throw	First throw	Second throw
1	1	3	1	5	1
1	2	3	2	5	2
1	3	3	3	5	3
1	4	3	4	5	4
1	5	3	5	5	5
1	6	3	6	5	6
2	1	4	1	6	1
2	2	4	2	6	2
2	3	4	3	6	3
2	4	4	4	6	4
2	5	4	5	6	5
2	6	4	6	6	6

Number of 1's in sequence	Number of sequences containing that number of 1's	Proportion of 1's in sequence	Proportion of sequences containing that proportion of 1's
0	25	0	25/36 = .6944
1	10	.50	10/36 = .2778
2	1	1.00	1/36 = .0278

equally likely, then in n occurrences of the event there are N^n different possible sequences of outcomes each of which is equally likely. And if n is large enough, then an overwhelmingly large (e.g., .99999) proportion of sequences will have the following characteristic: the proportion of times that a specified one of the N possible outcomes occurs within the sequence will be a proportion differing negligibly from $1/N$ (i.e., lying in the region from $1/N - \epsilon$ to $1/N + \epsilon$, where ϵ is a negligible quantity).

To summarize the entire argument, *if* an event E has N different possible outcomes one of which is e, and all of which are equally likely, then if the event occurs n times in succession, there are N^n different possible sequences of outcomes, and it follows that each sequence must be equally likely if the N possible outcomes are equally likely. Let P_e be the proportion of times the outcome e occurs in a single sequence. Then when n exceeds some very large number (many many times greater than N) the proportion of sequences in which P_e is *almost* exactly $1/N$ becomes *almost* as large as 1.00. And since each of the N^n sequences is equally likely, what occurs in a proportion of them almost equal to 1.00 must be overwhelmingly likely, indeed almost

Table 4-3

Proportion of Sequences, Among All Possible Sequences of n Throws of a Fair Die, in Which a Certain Outcome Constitutes a Certain Proportion of the Sequence

Proportion of digits in sequence that are 1's — **Proportion of sequences of n throws that contain a proportion P_1 of 1's**

P_1	1	2	3	10	100	1,000
0-.0499	.833	.694	.579	.162	.000	.000
.05-.0999					.021	.000
.10-.1499				.323	.266	.071
.15-.1999					*.493*	*.926*
.20-.2499				.291	.198	.003
.25-.2999					.021	.000
.30-.3499			*.347*	*.155*	*.001*	*.000*
.35-.3999					.000	.000
.40-.4499				.054	.000	.000
.45-.4999					.000	.000
Exactly .5		*.278*		*.013*	*.000*	*.000*
.5001-.55					.000	.000
.5501-.60				.002	.000	.000
.6001-.65					.000	.000
.6501-.70			.069	.000	.000	.000
.7001-.75					.000	.000
.7501-.80				.000	.000	.000
.8001-.85					.000	.000
.8501-.90				.000	.000	.000
.9001-.95					.000	.000
.9501-1.00	.167	.028	.005	.000	.000	.000

Proportion of digits in sequence that are either 1's or 2's — **Proportion of sequences of n throws that contain a proportion $P_{1,2}$ of digits that are either 1's or 2's**

$P_{1,2}$	1	2	3	10	100	1,000
0-.0499	.667	.444	.296	.017	.000	.000
.05-.0999					.000	.000
.10-.1499				.087	.000	.000
.15-.1999					*.001*	*.000*
.20-.2499				.195	.027	.000
.25-.2999					.181	.011
.30-.3499			*.444*	*.260*	*.393*	*.850*
.35-.3999					.301	.139
.40-.4499				.228	.087	.000
.45-.4999					.010	.000
Exactly .5		*.444*		*.137*	*.000*	*.000*
.5001-.55					.000	.000
.5501-.60				.057	.000	.000
.6001-.65					.000	.000
.6501-.70			.222	.016	.000	.000
.7001-.75					.000	.000
.7501-.80				.003	.000	.000
.8001-.85					.000	.000
.8501-.90				.000	.000	.000
.9001-.95					.000	.000
.9501-1.00	.333	.111	.037	.000	.000	.000

certain. Therefore, the proportion of times the event E has outcome e in a *single* sequence involving a *huge* number of trials must be *almost* certain to be *almost* exactly $1/N$. So, if outcomes are equally likely, then in a huge number of trials the various equally likely outcomes should almost always occur with about equal relative frequencies.

4-2 Quantification of Likelihood

If a thing is impossible (such as a die coming up heads or 17) or happens a zero proportion of the time, it is natural to say that it has a likelihood of zero. If it is certain (such as a die tossed on to a flat surface coming up *something*) or happens a proportion 1.00 of the time, it is customary to say that it has a likelihood of 1.00. By convention, then, likelihood is expressed or measured on a scale running from zero to 1.00. If an event having N equally likely, mutually exclusive, and exhaustive outcomes is certain to occur, then its likelihood of occurrence is 1.00. But its occurrence must produce one and only one of N equally likely outcomes. Therefore, it is natural to divide the total likelihood of 1 equally among the N possibilities and say that each outcome has a likelihood of occurrence of $1/N$. This seems doubly appropriate since in almost all of the N^n different possible sequences of n outcomes the proportion of the time that a given outcome appears is almost exactly $1/N$ when n is huge. Therefore, if an event has N equally likely, mutually exclusive and exhaustive outcomes, we shall say that each different possible outcome has a likelihood of $1/N$.

So far we have considered our N different possible equally likely outcomes as "atomic" in the sense that they are incapable of being subdivided into more elementary outcomes. Suppose now that we combine several of these atomic outcomes into a "molecular" outcome which is considered to have occurred if any one of the atomic outcomes of which it is composed occurs.

Specifically, suppose that we take a fair die and paint its "1" and "2" sides red, its "3" and "4" sides white, and its "5" and "6" sides blue. We have now reduced our original six different possible equally likely outcomes to three different possible outcomes: "die comes up red," "die comes up white," "die comes up blue." Each of the three "color" categories is composed of the same number of "side" categories and each side contributes to one and only one color. Therefore, it seems reasonable to regard the three color outcomes as equally likely. If they are, then when n is very large nearly all of the 3^n different possible sequences of n color outcomes should contain about $1/3$ "red" outcomes. Or, equivalently, nearly all of the 6^n different possible sequences of n "side" outcomes (for large n) should contain digit outcomes about 1/3 of which are either a "1" or a "2." The right half of Table 4-3 shows that this is indeed the case. So we should be safe in saying that the likelihood of the molecular outcome "red" is 1/3.

But suppose that we had painted the "1" and "2" sides red and left the "3," "4," "5," and "6" sides as they were, unpainted. The right half of Table 4-3 shows us that the likelihood of "red" is still 1/3 and the left half convinces us that the likelihood of each of the four numbered sides is 1/6. Therefore, the likelihood of 1/3 for "red" obviously does not depend on the *molecular* outcome "red" being equally-likely to all other outcomes. Indeed, it can be shown that if "1" and "2" are painted red, "3," "4," and "5" are painted white, and "6" is painted blue, in practically all of the 3^n different possible sequences of outcomes, when n is huge, the proportions of "reds," "whites," and "blues" are almost exactly 1/3, 1/2, and 1/6, respectively.

We can therefore state the following principle: *If an event has N different possible, equally likely, mutually exclusive, and exhaustive, "atomic" outcomes, each of these outcomes has a likelihood of* 1/N. *And if a "molecular" outcome is regarded as having occurred if any one of K different atomic outcomes (of which it is "composed") occurs, then the likelihood of occurrence of the molecular outcome is* K/N.

There is, of course, no necessity for alternative atomic outcomes to be equally likely. If we toss a cone, for example, there are two atomic outcomes; "comes to rest on base" and "comes to rest on side"; and we would expect the latter to be "more likely" than the former. So the above principle is contingent upon a condition whose validity cannot simply be taken for granted. *If* the conditions are met, *then* the principle gives us the likelihood.

4-3 Randomness

4-3-1 Random Selection

A ***random selection*** of one unit from a set of N units is a selection made *in such a way* that each of the N units is *equally likely* to be the unit selected. Thus, it is the *process* by which the selection is made that is random, not the outcome of the selection. A selection process that does *not* give each of the N units an equal chance to be selected, i.e., a nonrandom selection, is called a ***biased selection***.

From our previous discussion of equal likelihood, it follows that a random selection if repeated (with replacements) a very large number n of times would tend (i.e., would be almost certain) to select each unit essentially the same proportion $1/N$ of the time. For if n is huge, almost 100% of the N^n possible sequences of outcomes would contain essentially equal proportions of the

N different outcomes, and the actually obtained sequence is almost certain to be one of the sequences in that almost exhaustive class of sequences. But although the sequences in that class contain nearly equal proportions of each of the N outcomes, every one of those sequences represents a different sequential *pattern* of arrangement, and each is equally likely to be the actually obtained pattern. Thus, while it is fairly safe to predict that if n is huge each of the N outcomes will occur about the same proportion of the time, the *sequential pattern* of outcomes is highly *un*predictable. Thus, in 1,000 tosses of a fair coin, it is fairly safe to predict that heads will come up about half the time. But a sequence in which the first 500 outcomes are H and the last 500 are T is just as likely as one in which H and T alternate $HTHTHT \cdots$, and indeed, just as likely as any one of the possible sequences that contain 500 H's and 500 T's. *Thus, unpredictability of sequential pattern of outcomes is just as much a characteristic of random selection as is predictability of long-run proportions.*

4-3-2   Random Sampling

A **random sample** of n units drawn from a set of N units is one drawn *in such a way* that every *possible* set of n units is *equally likely* to be drawn. So if sampling is with replacements, each of the N^n different possible samples is equally likely to be drawn. And if sampling is without replacements, each of the $N!/(N - n)!$ different possible samples enjoys an equal chance of being drawn. If we think of the n sample units as being drawn one at a time in sequence (and replaced before the next draw if sampling with replacements), the following definition holds regardless of whether sampling is with or without replacements: *A random sample is one drawn in such a way that at each draw every unit remaining in the population is equally likely to be drawn.* Thus, *a random sample is one composed of units each of which was drawn by a random selection from the units remaining in the population.* A nonrandom sample is called a **biased sample** because some sets of n units were more likely than others to be drawn.

4-3-3   Likelihood That a Sample Statistic Will Have a Certain Value under Random Sampling

Now suppose that it is intended to draw a random sample of n units from a population of N units and then calculate some specified sample statistic S from the n units actually drawn. We know from the preceding chapter that there are N^n or $N!/(N - n)!$ different possible samples that might be drawn, depending on whether sampling is with or without replacements. Let K stand for the number of different possible samples. Since sampling is random, we know that each of these K possible samples is equally likely. Suppose now that the sample statistic S is calculated for each of these K possible samples and that the number of samples in which S has the value s is k. Then since each possible sample is equally likely, samples in which $S = s$ must be k times as likely as a single possible sample. But the likelihood of a single possible sample is $1/K$, so the likelihood of a sample in which $S = s$ must be k/K. But k/K is simply the *proportion* of the sampling distribution of S in which $S = s$. Therefore, *if sampling is random, the likelihood of drawing a sample in which* S $=$ s *is simply the proportion of the sampling distribution of* S *in which it has the value* s.

4-4   Attaining Randomness

While it is fairly easy to define a random selection of a unit from N units as one in which each of the N units has equal opportunity to be the one selected, it is difficult if not impossible to specify explicitly the conditions that render opportunity equal. A wealth of experimental evidence suggests that it is impossible for a human being to make himself a random selector simply by deciding to be one. Various mechanical processes are an improvement over willed human" randomness." For example, if a unit is to be drawn randomly from a population of 50 units, one can associate each unit with one of the cards in a bridge deck discarding the two extra unassociated cards, then shuffle the cards thoroughly, and finally draw a card or "cut the deck" to identify the unit to be drawn. But experiments show that two or three shuffles do not really result in a thorough mix of the cards, especially if the cards are sticky to different degrees. Nor are draws or cuts equally likely to come from all parts of the deck. This, of course, is a crude and primitive mechanical procedure. Some fiendishly sophisticated mechanical (or other physical, such as electronic) procedures have been devised for the purpose of creating equal opportunity of selection, but curiously enough, despite their cunning they have failed to pass moderately sophisticated tests of randomness, and often the precise source of bias has been identified. One of the most nearly random processes devised so far is that by which the RAND Corporation created their series of one million random digits—a series of a million samples of one unit drawn randomly and with replacements from a population of ten units whose values are 0, 1, 2, 3, 4, 5, 6, 7, 8, 9, respectively, and recorded in the order in which drawn. (A portion of the RAND table is given in Table B-1 of Appendix B.) One can regard the entries as successive samples of one digit drawn from a population consisting of the ten different digits from 0 to 9 or as successive samples of d digits drawn from a population of 10^d integers consisting of every possible d-digit integer (where if $d = 3$, the first such integer is 000, the second is 001, etc., and the last is 999). Thus, if one wishes to simulate random sampling from a population of $N = 127$ units, he need only number the units arbitrarily from 1 to 127, then, starting anywhere, examine successive blocks of three digits in the RAND table, drawing the corresponding unit (if there is one) from the population until he has a sample of the required size. Thus, if the first five successive blocks of three digits in the RAND table were (a) 541, (b) 065, (c) 155, (d) 122, (e) 065, he would (a) do nothing, (b) draw unit number 65, (c) do nothing, (d) draw unit number 122, (e) do nothing if sampling without replacements or draw unit number 65 again if sampling with replacements, continuing of course in the same fashion until he had the sample of the required size. If one follows this regimen, his sample will be essentially random. But if he starts at the same point in the table every time he wishes to draw a random sample, his *set* of samples will be biased, and indeed even his *second* sample will be biased since its units were predetermined and therefore favored with better than equal opportunity to be drawn.

One can avoid such difficulties and create his own sequence of (at least roughly) random digits by successively performing an operation consisting of tossing a (fair) coin and throwing a (fair) die. If the die comes up 6, it is not counted and the die must be rethrown until it comes up a number other than 6. If the coin comes up heads, one's random digit is the number that comes up on the die. If the coin comes up tails, one's random digit is the number showing on the die plus 5, unless the die comes up 5 in which case one's random digit is zero. If one requires a random number containing d digits, he simply performs the operation d times.

To give a concrete example of random selection, suppose that in a club containing 16 members one member must be selected to perform a distasteful task for which no one volunteers. It is desired that the selection be made in a way that is *fair* to all. Each member's name is written on a different four of the 64 squares of a checkerboard. A few drops of rain are just starting to fall. The checkerboard is laid flat in an open field, with the understanding that the first square of the checkerboard to be struck by a drop of rain will identify the club member who must perform the distasteful task. Such a selection would be essentially random.

Now suppose that, whether intentionally or inadvertently, Brown's name was written on only three squares and Smith's name was written on five, the other members names being written on four squares each. Such a selection is biased. And this is the case even if the first drop of rain falls on one of the four squares labeled Jones, since it is the *process* of selection rather than the *outcome* that is referred to by the terms random and biased. The first *procedure* was fair, the second unfair, even though Jones may have been selected in both cases. (The second procedure may be regarded as a random selection of checkerboard square but as a biased selection of club member.) If the selecting mechanism truly had exactly equal valence for each of two or more objects during a single selection, it might be able to select none of them or perhaps all of them, but it certainly could not discriminate by selecting only one of them. The first raindrop cannot make its initial contact with *all* of the squares of the checkerboard. And the one square that it does strike is determined by its trajectory, which obeys the laws of physics. The randomness in selection-by-raindrop lies in the multiplicity and volatility of the factors determining its flight, not in any *absence* of determining factors. In any single instance, a random selection is just as determined as a biased selection. It differs from a biased selection only in that the determining factors are ephemeral and inaccessible to us. Selection-by-raindrop is random and therefore fair in the sense that we have no reliable clue as to which square will be the one selected, but we have every reason to believe that in the long run each square would tend to be selected about the same number of times.

It follows from the preceding definitions that we can never be absolutely certain that a selective process is random, or that it is biased, on the basis of any finite sequence of actual selections. Since the sequence of units to be chosen by a random selection is unpredictable, it *can* turn out to be *any* sequence that is possible; and since under random selection a given unit's relative frequency must be $1/N$ only in the infinitely long run, it *can* be *anything* from 0 to 1 in the finite run of whatever length. And even a sequence that "looks" as if it were the result of a random process could have resulted from a slightly, or erratically, biased process.

Since both randomness and bias are characteristics of a *process*, rather than of a finite number of its selections, we must examine the *mechanism* governing the process before we can feel entirely secure in judging that the process is random or that it is biased (rather than simply behaving atypically). If we can identify factors in the selective mechanism that we know, on a priori grounds, would tend to favor certain units, we can be reasonably sure that the process is biased. And if a sequence of units actually selected by the process tends to confirm our judgment, we can be virtually certain of bias. If we make the most thorough search for biasing factors in the selective mechanism of which we are capable and find none, *and* if a long sequence of actually selected units is convincingly haphazard (as determined, perhaps, by statistical test), we may be reasonably sure that the process is approximately random (but we can never be as sure of exact randomness as we can be of appreciable bias). Thus, if we examine a die and find that it is loaded with a lead weight embedded close to the side opposite the 5 and if in sixty throws the die comes up 5 twenty-seven times, we may safely conclude that the die is a biased selector of the integers from 1 to 6; but if we find no biasing factors and in sixty throws each of the six faces comes up at least seven but no more than thirteen times (and in a haphazard order), we may tentatively conclude that the die is a random selector.

4-5 Prevalence and Consequences of Nonrandomness
(Both Innocuous and Drastic)

Truly random sampling appears to be an unattainable ideal. We have already pointed out that although approximately random selective devices are easy enough to come by, man's best efforts to devise a "perfect" random selector have invariably been thwarted. Yet practical and economic considerations are even more responsible for appreciable degrees of nonrandomness than are the physical difficulties encountered in constructing the perfect selective mechanism. This is especially the case when the population of units to be sampled is very large and widely dispersed. It may be, and often is, extremely difficult simply to locate and identify the units comprising the population. For example, the census bureau invariably misses an appreciable portion of the population when taking the census, some of whom are virtually impossible to find, such as derelicts sleeping in culverts, and others are missed through carelessness, which, itself, is virtually ineradicable in such a large project. Obviously, units that have not been detected and listed cannot enjoy equal likelihood of being subsequently selected from a list of detectees. A much larger source of nonrandomness however is the economically motivated tendency to sample only from the accessible units at hand rather than randomly from the entire population, some of whose units may be located in remote regions. For example, a well-known IQ test, intended to be applicable to American adults in general, was standardized on a sample of adults living in the greater New York area. The reason was simply that it is cheaper for a team of testers to conduct their testing within commuting distance of their home office than to travel such circuits as Butte, Montana; Loco Hills, New Mexico; Troy, South Carolina; etc., as they would have had to do if they had taken a random sample. A smaller-scale, but more widespread, example of the same thing is the tendency of academically based psychologists to draw their samples from students in sophomore psychology classes, experiment on them, and then generalize their results to the much broader, and unrandomly sampled, population of adult Americans.

A nonrandom sample need not be objectionable; it depends on the purpose for which the sample is to be used. A sample obtained by drawing every hundredth name in a telephone directory might provide a typical group of telephone owners which might be all that is desired by the sampler. Such "systematic" sampling schemes have the advantage of being easy and economical and of producing samples that often behave in much the same way as random samples, so that at a considerably smaller cost one can in a sense obtain a "close approximation" to a random sample. Usually, however, when we speak of a biased sample, the word "bias" carries objectionable

connotations. Thus, if the reddest, ripest, largest, most nearly flawless strawberries are deliberately set aside and packed last into a carton of strawberries,
we say that the top layer of strawberries is a biased sample of the entire set
of strawberries in the carton and if we are consumers rather than merchants
we probably view the practice as objectionable.

A random sample from one population may be a biased sample from
another, larger, population. Thus, a random sample of American women
would be a biased sample of American adults. If the sampler drew a very
large random sample of women, measured or observed some characteristic
X of each of them, compiled a relative frequency distribution of X and then
used it as an estimate of the relative frequency distribution of X in the population of American *adults,* he would have committed a technical error having
negligible consequences if X is IQ, but having disasterous consequences if
X is weight. Thus, the practical consequences of bias depend on the extent to
which the favored units differ, in the measured characteristic, from the
unfavored units. And this, of course, is why systematic sampling is often
innocuous and why other types of bias do not inevitably thwart one's purpose.
For example, imagine a loaded die that comes up 1 one-twelfth of the time,
3 two-twelfths of the time, and 5 three-twelfths of the time. We may therefore regard a throw of the die as a selection of one of six unequally likely
units identified (i.e., labeled) as 1, 2, 3, 4, 5, 6. But if the units 1, 3, and 5 all
have the same *value,* O standing for "odd," and the units 2, 4, and 6 all have
the same value, E standing for "even," then O occurs six-twelfths or one-half
the time and so does E just as they would if each face came up one-sixth of
the time, i.e., the *values* O and E are equally likely although the *units* 1, 2,
3, 4, 5, 6 are not. Clearly, we have a nonrandom selection of units. But if
we were only interested in their *values* and recorded only O or E for a sequence
of throws, the sequence of O's and E's would be just as random as the sequence of H's and T's for tosses of a fair coin. Therefore, if, as is usually the
case, we regard *values* rather than *units* as our basic data, and if our sampling
procedure tends to sample every different value a proportion of the time equal
to the proportion of units in the population having that value, then even if
our sampling of units is biased, we may have a sampling technique that is
effectively random for our purposes.

By far the safest thing to do, however, is to take a random sample from
the population in which one is interested, since it is not always obvious what
influence a particular bias may have. A notorious example of this hazard is
the poll taken by the *Literary Digest* in an effort to predict the outcome of
the 1936 presidential election.

In 1920, 1924, 1928, and 1932 the *Literary Digest* had conducted polls
prior to the presidential election in an attempt to identify the winner. In all
cases they had succeeded, and in 1932 they forecast the popular vote correctly

to within 1%. As a result, by 1936 the poll had acquired so much prestige and was regarded as so authoritative that the *Digest* exuded confidence:

> *While Chairmen Farley and Hamilton noisily claim "at least forty-two States," and while the man in the street sighs "I wish I knew," The Digest's smooth-running machine moves with the swift precision of thirty years' experience to reduce guess-work to hard facts. . . . Once again The Digest was asking more than ten million voters—one out of four, representing every county in the United States—to settle November's election in October. . . . The Poll represents thirty years' constant evolution and perfection. Based on the "commercial sampling" methods used for more than a century by publishing houses to push book sales, the present mailing list is drawn from every telephone book in the United States, from the rosters of clubs and associations, from city directories, lists of registered voters, classified mail-order and occupational data. . . . The master-list represents every vocation, every voting age, every religion and every nationality extraction in the country. It is constantly revolving, so that a certain percentage of names change with each poll. . . . Thus an amazing machine goes into action. When it has finished its work, not only Farley and Hamilton, but millions of other guessers, will know how close they have come to the mark.*

Of the ten million ballots sent out by the *Digest*, over 2,350,000 were returned. Table 4-4 gives the results (for the two major candidates only) of the poll and of the election.

Table 4-4

Literary Digest Poll Votes and Actual Ballot
Votes in 1936 Election

	Candidate	Popular votes		States carried	Electoral votes
		Number	Percentage		
Digest Poll	Roosevelt	972,897	42.9	16	161
	Landon	1,293,669	57.1	32	370
Actual election	Roosevelt	27,751,597	62.5	46	523
	Landon	16,679,583	37.5	2	8

The disasterous *Literary Digest* poll is a celebrated example of the perils of biased sampling. The *Digest wanted* its sample to be representative of the electorate. But its mailing list, made up largely of telephone owners, club members and buyers-by-mail, tended to exclude those who had been most impoverished by the Great Depression. Perhaps more important is the fact that of the ballots mailed out only about a fourth were returned, and people who will answer and return a mailed questionnaire differ in many ways from those who will not. In discussing a variety of polls taken on the 1936 election, Katz and Cantril[1] remark.

> As a rule less than one-fifth of the mailed ballots are returned and these tend to come from selected groups. People with intense opinions (reformers, arch-conservatives, radicals) are more likely to return ballots than those who are luke-warm or undecided; more highly educated and economically secure persons take a greater interest in the ballots and feel more free to answer them. The American Institute found that the largest response (about 40 percent) came from people listed in Who's Who. Eighteen percent of the people in telephone lists, 15 percent of registered voters in poor areas, and 11 percent of people on relief returned their ballots. Men are more likely to reply than women.

So those who survived the Great Depression *well* were more likely to be polled and more likely to respond than those who faired badly. And the former were much more likely to be disenchanted with Roosevelt's social-welfare reforms than were the latter. The accuracy of the *Digest* poll in preceding elections is probably attributable to the fact that there was no radical difference in voting pattern between the affluent and the destitute. By 1936, however, the "haves" and "have-nots" had become politically polarized, the have-nots perceiving their interests to be served by the Democratic Party. Since the *Digest's* poll did not sample the have-nots in proportion to their numbers, the result was a disasterious undersampling of Democratic votes. There are two important lessons to be learned here: (a) *Taking a huge sample does not insure that it will be representative, if it is nonrandom.* (b) *The fact that a biased sampling technique has produced adequately representative samples in the past does not guarantee that it will continue to do so in the future.*

4-6 Random Variables

Some quantities, like π, are absolutely constant, but they tend to be found in mathematics rather than in empirical science. Others, like the number of fingers on a human hand (or heads on a human body) are only "typically" constant, admitting of occasional exceptions. Still other quantities are sys-

[1]D. Katz and H. Cantril, "Public Opinion Polls," *Sociometry*, Vol. 1, Nos. 1&2 (1937), 155–79.

tematically, periodically, or predictably varying variables, such as the position of the second hand on a watch—knowing its position at one instant, one knows where it will be 5 seconds later and how it will get there. Finally, some variables take on their various values in an apparently unsystematic sequence that reveals no pattern or periodicity and therefore defies prediction. If such a variable takes on each of its various values a definite fixed proportion of the time, it is called a ***random variable***.

Suppose that X is a random variable that takes on k different possible values $x_1, x_2, \ldots, x_i, \ldots, x_k$ proportions of the time $p_1, p_2, \ldots, p_i, \ldots, p_k$ respectively (so $\sum_{i=1}^{k} p_i = 1$). We shall say that X has "occurred" when it takes on a new value (which may be the same value, i.e., a repetition of the value that it had immediately before). Then we may regard an occurrence of X as a random selection from a population whose units are potential occurrences, a proportion p_i of which have the value x_i.

Random variables are not nearly so rare as (humanly produced) random selections. In making a random selection from a presently existing population of units there is great difficulty in making sure that the process would select each unit equally often in the long run, and we have seen that a biased selective process can exclude a unit or sample it only a fraction of an equal proportion of the time. But the occurrences of a natural variable are future happenings that are equally likely by definition; how can an occurrence not occur, or occur only partially, or occur more than once? (It may help here to think of the potential occurrences as forming an infinite population that is sampled without replacements.) Thus, future occurrences automatically qualify as equally likely in the sense that they *will* occur equally often (namely, once).

However, we are not guaranteed that the *sequence* of future occurrences will be random. Actually, it makes little sense to speak of the randomness of a sequence of units that are never replaced in their population and that therefore never appear more than once in the sequence. What we really require of the variable is that its sequence of *values* have the sort of pattern that would be obtained by random sampling. And *this* is *not* guaranteed by the situation. Therefore, while "truly" random variables are not as rare as truly random selectors, they are rare.

4-7 Randomness as a Prototype

It is commonplace in science to postulate idealized conditions that are never exactly met in practice but which seem to represent a variety of real-world conditions better than any *particular* one of them would. We cannot fault Euclid for basing Greek geometry upon dimensionless points and perfectly straight lines; what diameter should his points have had, and what particular crooked line should he have used?

Likewise, truly random sampling is extremely rare, if not totally absent, in the real world. Yet, though it is a perhaps unattainable ideal, it is more representative of a *variety* of actual real-world sampling processes (that are intended to approximate randomness) than any particular one of these actual processes would be. While truly random sampling is rare, haphazard sampling is common, and so is sampling from which all factors *known* to have a *large* biasing effect have been removed or otherwise canceled out. Therefore, we may regard random sampling as a prototype to which many sampling processes correspond roughly but to which hardly any correspond exactly.

4-8 Problems

1. Would the color of a randomly selected barn be a randomly selected color? Why?

2. If one wished a random selection from inebriates at a party, would it be appropriate to use the first partier to have too much to drink? Why?

3. Which of the following would be essentially random variables: (a) the initial depth of coffee in your morning coffee cup, (b) sex of children born (in sequence) at a certain hospital, (c) the number of times a cigarette lighter must be flipped before it lights, (d) the times at which a traffic light is red?

4. Would the *Literary Digest* poll have been unobjectionable if it had sampled *randomly* from among telephone owners, club members, buyers-by-mail, etc? Why?

5. Rank the following procedures from least to most vulnerable to the charge of bias: (a) to assess the likelihood of plague, a rodent count is taken at a cat shelter; (b) the average height of the University of Minnesota basketball team is used as an estimate of the average height of American male adults; (c) the average IQ of all U.S. military personnel is used as an estimate of the average IQ of all U.S. adults; (d) the average age of people the third letter of whose surname is a vowel is used as an estimate of the average age of people in general.

6. In a draft lottery, half of the first 50 numbers drawn are paired with birth dates in February and March. Those affected claim that the selections could not have been random. The draft officials claim that they *were* random. Who is right?

7. Explain the difference between a representative sample and a random sample. Are all random samples representative? If not, which ones are not? Must a representative sample be random? What makes it representative?

5

probability

5-1 Chance

The concept of probability and the associated concept of randomness are fraught with philosophical difficulties and defy completely satisfactory definition. Yet, paradoxically, their meaning is fairly easily grasped on an intuitive basis. Much of the difficulty with these terms stems from viewing "chance" as an attribute of the external world rather than as an attribute of our state of *knowledge about* the external world.

In studying the behavior of almost any scientific phenomenon, we find that its behavior appears to be governed by (i.e., related to) the net effect of a multiplicity of influencing factors (many of which are interacting with each other). These factors fall roughly into two classes, those that individually are capable of exerting large or moderate influence upon the phenomenon, and those whose individual influence is small both in an absolute and in a relative sense. Generally, there are few factors of the first type and many of the second. Because of their relatively large influence, factors of the first type are usually the first to be discovered. The next to be discovered are generally those factors of the second type whose influence is largest, leaving undiscovered a host of factors each of whose individual influence is negligible. As a science advances, what is negligible at one stage must be taken account of at a more sophisticated stage, so as the science develops more and more previously negligible factors become appreciable and are discovered. However, the reservoir of "negligible" factors appears to be virtually inexhaustible so that at *any* stage of scientific progress there remains a legion of factors no one of which has much individual influence but whose effect in concert may be appreciable. So past experience suggests that there are always a multitude of unknown factors which, in concert, may appreciably influence the observed phenomenon.

Since these factors are unknown, they are also uncontrolled. And because they are uncontrolled, some will be present, others absent, some will be augmented, others stifled; different subsets of these factors will be called into play on different occasions, and their interactions with one another will therefore differ. Therefore, their net influence will often vary unpredictably (since they are unknown), and, as a result, the observed phenomenon will vary from one observation to the next even though the relevant conditions of observation appear to have remained the same. Because of this the variation will appear to be inherent in the observed phenomenon rather than merely being a natural consequence of variation in as-yet-undiscovered causal factors. Such variation is usually attributed to "chance."

5-2   The Concept of Probability

5-2-1   Vernacular Probability

The word probability is not just an esoteric technical term; it belongs to the layman's vocabulary. When a layman says, "It will probably rain tonight," he means that he is not absolutely certain of rain tonight but that he expects it. He may be more explicit and say, "The chances are 2 to 1 that it will rain tonight," in which case he means that his expectation of rain is double his expectation of nonrain. And presumably his expectation is based upon the relative frequency with which he has observed the general class of atmospheric and other relevant conditions now present to have been followed by rain. But even if his prognostication is based only upon a hunch, the odds that he quotes are likely to be based, more or less, upon the relative frequency with which his hunches have proved right in the past.

Another type of probability statement common among laymen is a statement such as, "Had Socrates shown a more conciliatory attitude at his trial, the odds are 2 to 1 that he would have been acquitted." What can this mean? The trial of Socrates is not a repeatable incident, but then neither is tonight's rain. Both are unique members of a *class* of incidents having only certain gross characteristics in common. One cannot really *expect* an outcome of an incident that never happened, but he can *believe* that the outcome would have resulted if the incident had taken place. What may be meant is that in the class of incidents composed of actual trials made under circumstances grossly similar to that of Socrates, but in which the defendent's attitude was conciliatory, the relative frequency of acquittal is believed to be double that of conviction.

These two types of situation are quite different and there are many other intermediate types between them. Yet, in all cases, the term "probability" as used by the layman appears to refer (or at least can be interpreted as referring) to a *degree* of *belief* or *expectation* that an incident would have a certain outcome (or that an object would be found to have a certain trait) based upon the *relative frequency* with which that outcome (or trait) occurs, or is thought to occur, *within a class* of similar, but of course not identical, incidents (or objects).

5-2-2 The Reference Class of Similar Incidents

The layman's implicit definition of probability is quite close to the technical definition. Certain points, however, require clarification. By "similar incidents" we mean incidents having certain common characteristics. These common characteristics identify the reference class and therefore define the situation to which the probability applies. Since they qualify the probability they must be clearly specified. We cannot take as the reference class of similar incidents those having *all* of the precise characteristics of the incident in question because (a) broadly conceived, every incident has a virtually infinite number of characteristics, many of which are unknown or at least unthought of, and (b) if we specify *all* of the characteristics of *any*thing (its position in time and space included), what we have specified becomes unique and belongs to a class consisting only of itself. Nor can we generally take the reference class to be incidents having all of the *relevant* characteristics of the incident in question. Typically, there are many relevant factors of which we are unaware and we are seldom able to specify the exact values (characteristics) of those variables we actually know to be influential.

Yet in order for relative frequency to be a valid index of degree of belief, the reference class of similar incidents must have all of the characteristics that qualify one's degree of belief about the outcome. Therefore, the incidents comprising the reference class must have (at least) all of the characteristics that meet all three of the following criteria: (a) one *knows* that the characteristic is *relevant* to the outcome in question and has some inkling of what its influence might be; (b) one is able to *specify* the characteristic with precision (by which we mean merely to emphasize that the characteristic must be a constant, such as "80" or "red," rather than a variable, such as "age" or "color"); (c) one knows that the incident in question will *have* the characteristic (one may know this because the incident is *defined*, in part, by the characteristic, or because one intends to *control* or hold constant a certain variable, so as to produce the characteristic, at the time the incident takes place). Thus, if we want the probability that a man who is having his 80th birthday will live to see his 81st, the incident in which we are interested is the arrival of an 81st birthday and those comprising the reference class must have the characteristics that the person involved is a male and that he was alive at 80 since both are relevant to the outcome "alive at 81." But, if we know that the particular man in question has cancer, we must surely include "person involved has cancer" as a characteristic of the incidents in the reference class if we wish the relative frequency with which the incidents of the reference class have the outcome "alive at 81" to represent our degree of belief in the outcome.

An important special case is that in which it is known that outcomes tend to follow a known temporal pattern. We may consider the incident in which one is interested as a particular occurrence of a variable. However, we cannot use *all* occurrences of the variable as the reference class if the variable has a known pattern of variation and we know *where* in that sequence our future incident is to occur. Thus, we cannot use the proportion of bills among all letters delivered as the probability that the next letter to appear in our mailbox on the second day of the month will be a bill. Rather, the reference class must be all letters delivered on the second of the month. So we must not be able (even partially) to predict the behavior of the variable comprising the reference class of events on the basis of any periodicity or systematic sequentiality in its variations. This requirement is met if the variable whose occurrences comprise the reference class is *random*, or if it is nonrandom but we are unaware of it, or if it is nonrandom and we are aware of it but unable to anticipate how that nonrandomness might influence the relative frequency of the outcome in question within the reference class of occurrences.

In order that our basis of reference be undistorted, as we have said, the reference class of similar incidents must include *all* incidents having those exactly specifiable characteristics known to be relevant. The incident in question must be a member of the class of similar incidents, and the incident itself (or, at least, the knowledge of its outcome) lies in the future. Therefore, the class of *all similar* incidents must also include all similar *future* incidents. (And, unless one of the stated characteristics of the incidents in the reference class is that they occur within a specified time period, the class of *all* similar *future* incidents must be an *infinite* class.) It must therefore be largely (or entirely) a *potential* rather than an *actual* population of incidents. And the relative frequency with which the outcome in question occurs within this class must be *conceptual* rather than immediately measurable. This makes the *direct measurement* of probability impossible, although it does not prevent us from *defining* it to have a certain value under specified conditions, nor does it prevent us from *inferring* or *estimating* its value from indirect evidence or from measurements of *past* relative frequencies. The problem of estimating probabilities will be dealt with later. At present we are concerned primarily with the concept.

We can illustrate much of what we have said about chance, probability, and the reference class of similar incidents as follows. The throw of a die takes place under conditions that may be divided into three categories. First is a very small set of conditions that are both *known* to be relevant to the outcome and *specified* as qualifying the probability. Such conditions might be (a) that the die is a cube constructed of homogeneous material, (b) that the table must be flat and the backstop perpendicular, (c) that the throw must be made with a reasonable degree of insouciance and abandon (e.g., the die must leave the hand at least 6 in. above the table and bounce against the backstop before touching the table). Second is a much larger set of conditions that are *relevant*, whether known to be or not, but which are left *un*specified and *uncontrolled*, and which therefore *vary*. Such conditions might be (a) the exact orientation of the die at the moment it left the hand, (b) its exact height above the table at that moment, (c) its exact distance from the backstop, (d) its exact linear and angular velocities and accelerations, (e) its exact coefficients of friction with the backstop and with the table, (f) the exact time of day and year (i.e., the position from which the sun exerts its gravitational pull), (g) the exact position of the moon, (h) the exact altitude and latitude of the table, (i) the locus and velocity of any air movements above the table. Undoubtedly, there are many more conditions having some real influence upon the outcome. Third is a virtually infinite set of *irrelevant* conditions accompanying the incident and varying widely in the degree to which they appear to be associated with it. Such conditions might be (a) that the die is red, (b) that the thrower's social security number is 114-25-6517, (c) that at the moment of throw a vulture is circling over a dead snake at a certain spot in Afghanistan, (d) that the trading of 1,000 shares of AT&T has just appeared on the tape at the New York Stock Exchange, (e) a Swedish accountant is considering how to get rid of his wife.

If we could hold all conditions in all three categories constant during successive throws, the die would always come up the same number. But to demand this is to demand that everything in the universe be the same at each throw, and we cannot duplicate instances of the entire universe. The conditions present at a given instant of the universe can occur only once, whereas probability requires repeatability, at least of general circumstances. However, if we held all conditions in the first two classes constant and allowed those in the third class to vary, we would have repeatability. Indeed, since the third class is an infinite one, there would be an infinite number of ways of repeating the constant conditions of the first two classes. If we did this, the die would still come up the same way on successive occasions, since all *relevant* conditions would be the same. Now if we allow conditions in both the second and third classes to vary at will, and hold constant only those of the first class, it follows that (a) the throw of the die belongs to an infinite class—that is, there are an infinite number of combinations of conditions under which it could be done, (b) on successive throws the outcome will tend to vary since many of the relevant variables are uncontrolled.

We see therefore that *chance is simply uncertainty or variability of outcome due to the net effect of relevant but uncontrolled, and therefore varying, factors or characteristics.* The reference class of similar incidents are those having certain specified characteristics (i.e., constant values) that we *know* will be *present* in the incident in question, and preferably, that we *know* to be *relevant* to the outcome. The probability of the outcome is the relative frequency with which incidents in the reference class have that outcome.

5-3 Definitions of Probability

There are many definitions of probability. Many of them are operational definitions which tell you how to obtain the numerical value of a probability. Some of these will be dealt with later in a section on methods of obtaining and estimating probabilities, which we feel is a better place for them. At present we are concerned only with defining the *concept* of probability. There are three major conceptual definitions of probability, differing mainly in the role played by belief—whether or not it is involved at all and whose belief or what belief it is.

5-3-1 Radically Objective Probability

The most austerely objective definition ignores belief altogether, defining probability as a relative frequency within whatever reference class one chooses to specify, irrespective of whether or not that reference class is appropriate to one's state of knowledge. Thus, what we shall call **radically objective probability** is defined as follows. *The probability that a (partially) defined incident will have a stated outcome is the relative frequency with which that outcome occurs in the (potential) class of all incidents that have the defining characteristics.* Thus, the probability that an incident defined only as having characteristics C_1, C_2, and C_3 will have outcome K (actually a fourth characteristic) is the relative frequency with which outcome K occurs within the (probably infinite) class of all incidents having characteristics C_1, C_2, and C_3. This definition defines probability objectively and unambiguously. However, it is estranged from degree of belief and is therefore seldom what we really *mean* by "probability."

5-3-2   Objective Probability

In order for probability, as defined above, to correspond to an appropriate degree of belief, the defining characteristics must include all characteristics known or suspected to be relevant and of whose influence one has at least *some* inkling (such as whether, if included, it tends to increase relative frequency or decrease it). If we make this modification, we have a definition which forces relative frequency to correspond to an appropriate, rational degree of belief. Thus, what we shall call ***objective probability*** is defined as follows. *The probability that a given incident will have a stated outcome is the relative frequency with which that outcome occurs in a specified reference class of similar incidents; the latter consists of all incidents having all of the characteristics of the given incident that are known or suspected to influence the outcome in an at least vaguely foreseeable way.* This definition is a special case of the first definition and is therefore *objective*. It does not define probability in terms of degree of belief (which is not even mentioned in the definition). Rather it equates probability with an objective relative frequency but does so under circumstances which render that relative frequency an appropriate rational degree of belief. And that degree of belief is not *chosen* by particular individuals; it is implicitly *prescribed* by the definition. It is a *normative* degree of belief—a constant degree of belief that *everybody* (who possesses exactly the same information) "*ought* to have" but that is not necessarily characteristic of *any*body. Yet, although this definition is independent of a person's credulity, it is not independent of the amount of relevant information he possesses (any more than the first definition is independent of the amount of information he specifies). This definition is the one whose rationale was presented earlier (in Section 5-2-2). If we had to present a single definition of probability, this would be it. And when we refer simply to "probability" or "objective probability," this will be the definition we mean to imply.

5-3-3   Subjective Probability

The preceding definition "standardizes" probability by equating it to the relative frequency with which the outcome in question actually occurs within the specified reference class of similar incidents. This standardized probability serves as a sort of impersonal, "recommended" degree of belief, which is likewise standardized and therefore objective. If, in the definition of objective probability we change "occurs" to "is *thought* to occur," we lose the standardization and have a definition for a rational ***subjective probability*** where the probability is an index of the degree of belief of the particular, rational person doing the thinking. However, this makes the subjective probability simply a guessed-at objective probability, and therefore an estimate of the latter. Therefore, although permissible, it is not necessary to regard it as a unique definition of probability. We shall treat it as a method of estimating objective probability.

Despite their differences these three definitions of probability have much in common. In all cases probability is a *proportion* of incidents having a stated outcome within a specified reference class of incidents having certain well-defined characteristics. It can therefore be regarded as the proportion of units having a stated value within a well-defined population of units. Therefore, whatever laws hold for proportions of units in populations should be extendable to probabilities of any of the three types mentioned. And this means that we need not be concerned with which type of probability we are dealing with so long as we are only *manipulating* them mathematically. We need only concern ourselves with type of probability when we are considering the appropriate approach to a problem or interpreting the meaning of our final numerical results. The computations will be the same for all three types, being independent of the interpretation placed upon the proportions being manipulated.

5-4   Methods of Obtaining Numerical Probabilities

We have defined probability conceptually, but the definitions do not tell us how to obtain the number representing a probability in a particular case. We have said that the reference class of all similar incidents lies largely in the future. It therefore is potential, rather than actual, and, since it contains *all* similar *future* incidents, it tends to be an *infinite* population of incidents. This, of course, makes it impossible to *measure* the relative frequency of incidents having the outcome in question. However, although we cannot measure it directly, we can often infer it by other means. There are three major methods used to obtain numerical probability values:

1. deduction from an initial premise of equal likelihood
2. objective estimation—based on relative frequency of outcome in past occurrences
3. subjective estimation—based on the most educated guess one can make

5-4-1   The Deductive Method

In the chapter on randomness we distinguished between "atomic" and "molecular" outcomes. An atomic outcome is incapable of being subdivided into more elementary outcomes; an example would be a 1 coming up when tossing a single die. A molecular outcome is merely a common characteristic of several atomic outcomes. The molecular outcome is regarded as occurring if any one of the atomic outcomes of which it is composed occurs. Thus, the molecular outcome "even number" occurs if a die comes up either 2 or 4 or 6. Now consider an event having N different, possible, *equally likely*, mutually exclusive, and exhaustive atomic outcomes of which N_v belong to a common category which conveys upon them the value v—so that having value v is a molecular outcome. Sections 4-1 and 4-2 of the chapter on randomness imply that in an infinite sequence of occurrences of this event the proportion of occurrences that would have a particular atomic outcome would be $1/N$ (or, at least, would differ from it by only an infinitesimal amount) and the proportion of occurrences that would have the value v would be N_v/N.

The infinite sequence of occurrences is obviously the reference class of similar incidents. Therefore, the probability that the event will have a particular atomic outcome is $1/N$ since that is the proportion of times that it occurs in the reference class of similar incidents. And the probability that it will have the molecular outcome "value is v" is N_v/N for the same reason. Hence, we can obtain the *exact* value of the atomic probability if we know N, and of the molecular probability if we know N and N_v. Since they are properties of the event, rather than of the infinite sequence of occurrences, we often do know them.

The principle difficulty with probabilities obtained in this way is that they are, in effect, deduced from an initial premise that atomic outcomes are equally likely. The validity of the deduction depends upon that of the premise, just as it does in any application of deductive logic. But how does one know that the premise of equal likelihood is valid? There are essentially three approaches: (a) One can simply assume it, in which case the validity of the obtained numerical probabilities must be regarded as contingent upon the unsubstantiated truth of the assumption. This approach has little appeal outside of theoretical model building. (b) If the outcome in question is (or can be regarded as) a characteristic of a unit or sample of units actually drawn by us from a pre-existing population of real units, then we can be sure that the premise is (at least approximately) true by using a reliably random selector to identify the units to be drawn. This sounds a bit circular since a random selector is one that gives every unit an equal likelihood of being drawn. However, we are talking about a method of selection that has been *shown* in the past to draw each unit about the same proportion of the time and in an unanticipatable sequence. So when we speak of a random selector we are not talking about an idealized abstraction but rather about an existing and available method whose validity (at least to a close approximation) has been pragmatically established. (c) A final method is to infer equal likelihood as a self-evident consequence of the structural "symmetry" of the situation. In this method, under certain situations "equal likelihood" is concluded because it seems intuitively obvious. This is not the same thing as taking it for granted, because it is not intuitively obvious in all situations.

The second method—forcing equal likelihood by imposing randomness of selection, or knowing of equal likelihood through knowledge of randomness—is the most satisfactory. Indeed, it is the model we shall use in developing laws of probability. Its subcategories are discussed in the following three sections. Following that we shall discuss the third method—the use of symmetries.

5-4-1-1 PROBABILITY UNDER RANDOM SELECTION. Consider the random selection of a unit from a population containing N units, N_v of which have the value v. By the definition of randomness, each of the N units is equally likely to be selected. And the N units are obviously mutually exclusive and exhaustive atomic outcomes of the random selection. From our previous discussion it follows, therefore, that the probability that the selected unit will have value v is N_v/N. So if we know the numerical values of N_v and N and know that selection will be random, we can calculate the exact numerical probability. The premise of equal likelihood has been replaced by the stipulation that selection will be random. However, there are some fairly reliable methods of attaining randomness of selection, at least to a good approximation, so the premise of equal likelihood or the substitute premise of randomness need not be taken on pure faith. Indeed, if one uses a well-checked-out random selector (such as the proper use of the RAND tables), the premise of equal likelihood may generally be regarded as satisfied for practical purposes.

5-4-1-2 PROBABILITY UNDER RANDOM SAMPLING. Consider a random sample of n units drawn from a population of N units. By the definition of a random sample every possible sample of n units is equally likely. Therefore, we may regard the random sample as a random selection from a population consisting of all possible samples. Now suppose that there are K different possible samples in this population of all possible samples of size n. Suppose that some sample statistic S had been calculated for each of the K samples and that in k of them S had the value s. Then it follows from the previous section on random selection that the probability that S will have the value s is k/K. But this is simply the proportion of the (raw, uncondensed) sampling distribution of the statistic S that has the value s. Therefore, if we know the sampling distribution of S and know that sampling will be random, we can calculate the exact probability that the statistic S will have the value s.

5-4-1-3 PROBABILITY FOR A RANDOM VARIABLE. The occurrence of a random variable a proportion p of whose occurrences have an outcome of value v is analogous to a random selection from a population of N units a proportion $\rho = N_v/N$ of which have the value v. Therefore, the probability that the outcome of an occurrence will have the value v is ρ. So if we know that a variable is random and know the proportion of times it takes a certain value, we know the exact probability that it will take that value.

5-4-1-4 PROBABILITY AND LIKELIHOOD. In the chapter on randomness we dealt with the quantification of likelihood, and with likelihood under random sampling, arriving at essentially the same fractions obtained in the three preceding sections. This is entirely appropriate, since what we mean by probability is the same as what we mean by likelihood, and since in both cases the result was derived from an initial premise of equally likely atomic outcomes. We have arbitrarily used the term likelihood in referring to an undefined intuitive concept and the term probability in referring to a more formally defined (and elaborately discussed) concept. Actually, the two terms are essentially synonymous in common parlance.

5-4-1-5 PROBABILITY UNDER SYMMETRY. Atomic outcomes may generally be regarded as the ways in which the event can occur. Sometimes these ways entail common structural characteristics that appear, to the naive intuition, to entitle them to equal likelihood. Thus, a die is a symmetric structure; it has six square faces each having equal area and forming equal angles with adjacent faces; and most dice appear to be constructed of homogeneous material. Therefore, the "symmetry" of the situation is conducive to the conclusion that each side of the die is an equally likely outcome of tossing the die. Similarly, a roulette wheel is essentially a disc whose circumference is divided into equal arcs. From inspection of the symmetric structure of this situation, one is tempted to conclude that if the wheel is spun on its axis, each of its equal arcs is equally likely to come to rest in front of a fixed pointer.

One can restate deductive probability as follows. If an event can occur in W different, possible, equally likely, mutually exclusive, and exhaustive ways of which w are regarded as favorable outcomes, then the probability of a favorable outcome is w/W. For many years this was the classical definition of probability, and the premise of equal likelihood was regarded as validated if the "ways" possessed symmetries that made equal likelihood self-evident or intuitively obvious. Probabilities established or defined in this way were known as *a priori* or *mathematical* probabilities.

Some of the perils and shortcomings of the a priori method are obvious. The symmetry of dice and of roulette wheels does not prevent either of them from being "loaded" either intentionally or inadvertently. We pointed out earlier that it is virtually impossible to devise a perfect random selector. By the same token it is virtually impossible to devise a perfect die whose sides are equally likely or a perfect roulette wheel whose segments are equally probable outcomes. Playing cards that are of symmetrically equal size and weight tend to become differentially sticky. The friction between axle and bearing of a roulette wheel may not be the same for every possible position of the wheel. The corners of dice may become differentially worn or chipped. There are all kinds of actual or potential biases lurking in even the most symmetric-appearing structures. Therefore, one should be extremely cautious in inferring equal likelihood from apparent symmetry. Furthermore, even if after careful consideration (and empirical test if possible) the inferrence seems reasonable, it would be wise to regard *exactly* equal likelihood as an unattainable ideal and to conclude only that the method will yield a reasonably close estimate of the true probability. Even so, there is a definite element of risk. There is never any guarantee that the symmetric characteristics one views will insure equal likelihood. There may be some very influential asymmetries that are out of sight.

Other pitfalls in applying this method are less obvious but far more devastating. One notorious source of error is the incorrect determination of the total number of different possible outcomes. For example, consider a game in which a coin is tossed twice. If it comes up heads on either toss, the first player wins; if it does not, the second player wins. The mathematician D'Alembert considered that there were three possible outcomes: head on first toss, so first player wins and game is over; tail on first toss and head on second, so first player wins; tail on both tosses so second player wins. D'Alembert concluded that two out of three possible outcomes favored the first player so that his probability of winning was 2/3. Actually, there are four possibilities *HH*, *HT*, *TH*, and *TT* of which three are favorable to the first player, so his probability of winning is 3/4. A similar error would be to consider that there were three possible outcomes of two tosses: two heads, a head and a tail, and two tails. The error lies in considering every possible *combination* as equally likely even though some combinations may be obtained in more equally likely ways than others, e.g., the combination "a head and a tail" can be obtained in two different sequences *HT* and *TH*. Thus, in applying this method, a possible outcome must be elementary in the "atomic" sense—it must not be decomposable into a variety of still more elementary "possible outcomes." So the possible outcomes in tossing two dice are not the 11 different possible sums of the two numbers showing, but rather the 36 different possible outcomes obtainable by assigning integers from 1 to 6 to x and y in the statement: "number showing on first die is x; number showing on second die is y."

During the early development of probability theory it was sometimes thought that the number of different possibilities were simply the ones that one took into account and that if he were equally ignorant as to which one would occur, then they were equally likely. Thus, suppose that an urn contains balls that are known to be identical except for color, about which we have no information whatsoever. A blindfolded man is going to reach into the urn and draw out a ball. What is the probability that the ball will be red? In the eighteenth century one might have reasoned as follows. "There are two possibilities (a) the ball is red, (b) the ball is not red, and since I am totally ignorant of the true situation, they are equally likely, so the probability that the ball will be red is 1/2." On the other hand, one might have reasoned that there are eight possibilities, namely that the ball is (a) red, (b) orange, (c) yellow, (d) green, (e) blue, (f) chartreuse, (g) silver, and (h) other, and that my total ignorance renders them equally likely so that the probability that the ball will be red is 1/8. The principle fallacy consists in mistaking conceptual outcome categories for actual (physically existing) outcome alternatives; the various colors of the balls actually in the urn are the possible outcome-values, not the colors that "come to mind." The second fallacy consists in concluding that *values* rather than units are the possibilites; the third consists in assuming that one's ignorance of the size of a category renders them equal in size and therefore in likelihood. Thus, if one knows only that the urn contains balls that are either red, white, or blue, his ignorance of how many there are of each color does not justify the conclusion that 1/3 of the balls are of each color. We can make this same type of error in choosing between scientific hypotheses in areas about which we are totally ignorant. Was Freud right or wrong? Distributing our total ignorance equally between these two possibilities, we conclude that the probability that Freud was right is 1/2. Who was right, Freud, Jung, Adler, or none of them? With four possibilities made equally likely by our total ignorance of psychiatry, the probability that Freud was right shrinks to 1/4.

Errors of this type are not always as obvious as these examples would suggest. Sometimes they are quite subtle. Following Keynes, consider the following situation. We know that a certain object has a weight of exactly 1 lb. We are totally ignorant of its volume except that we know it lies somewhere in the region from 1 to 3 cubic inches. Our total ignorance moves us to regard it as equally likely that the volume is between 1 and 2 cubic inches as that it is between 2 and 3. The density of the object is its weight divided by its volume, so its density must lie somewhere in the region from 1 to 1/3 lb per cubic inch. Since our ignorance of volume implies a corresponding ignorance of its density, we regard it as equally likely that the density lies between 1/3 and 2/3 as that it lies in the region from 2/3 to 1. But if the density is in the region from 2/3 to 1, dividing weight by density to obtain volume we find that the corresponding volumes must be in the region from 1 to $1\frac{1}{2}$ cubic inches, rather than in the region from 1 to 2 as previously concluded. Thus, taking equal numerical intervals of density and volume as equally likely gives us mutually inconsistent results. (When weight is 1, density is simply the reciprocal of volume; although the range of the integers from 1 to 9 is divided in half by the number 5, the range of their reciprocals is not divided in half by 1/5 although there *are* the same number of reciprocals above 1/5 as there are below it. There is no guarantee that equal intervals necessarily contain equal numbers of possibilities, although it is sometimes a reasonable assumption.)

The a priori method has many pitfalls, but it has also been much maligned. True, it is rendered impotent by a loaded die, and is not impervious to errors of logic or to gaucheries in cataloging possibilities or appraising their relative likelihoods. However, there is a fairly large class of situations in which the most elementary and atomistic possibilities *are* essentially equally likely. In such cases if one is careful to deal only with these primordial and indivisible possibilities (or with nonoverlapping groups containing exactly equal numbers of them), the method is not only excellent but optimal.

5-4-2   The Past Relative Frequency Method

The most obvious method of obtaining the probability of an event having a certain outcome is simply to observe a large number of occurrences of similar incidents and let the proportion of times the particular outcome occurred represent its probability. Thus, if one wished to estimate the probability that a U.S. president would be assassinated, he might take as a gross approximation (there are too few presidents for much precision) the fraction whose denominator is the total number of past presidents (so the common characteristic is "elected president and assumed office") and whose numerator is the number of past presidents that were assassinated; so the probability of assassination is equated with the proportion of presidents assassinated. Or if a farmer wished to estimate the probability that it would rain at harvest time, he could take as his probability the proportion of times that it has done so since weather records were initiated. These two examples illustrate some of the difficulties encountered in the frequency method. First, there is no guarantee that the future will be similar to the past, e.g., that the conditions conducive to rain at harvest time, or presidential assassination in 1865, will enjoy exactly the same likelihood of occurrence a hundred years later— weather patterns are surely changing with time and one would suspect the same of "assassination patterns." Thus, one is not certain that an incident ("fate of president") to occur in the coming year belongs to the same class of incidents as those occurring in previous years. Indeed, he may have good reasons to believe otherwise, in which case "future" is a characteristic known or suspected to be relevant and should therefore be a defining characteristic of the reference class of similar events (unless one is dealing with radically objective probability). This, of course, excludes all past incidents from the reference class and therefore rules out the frequency method, although one could still use the method (based on past incidents) to obtain a gross approximation of unknown degree of grossness. Second, the number of obtainable past occurrences of the event may be too small to provide a very reliable estimate. There have been less than two-score presidents, and in many parts of the country weather records have been kept for less than 100 years; if one tossed a fair coin 100 times, it might easily come up heads 55 times rather than 50, yielding an error of 10 percent in estimated probability. The relative frequency of occurrence in a large but finite number of trials is generally a good estimate but not always since (although sequences with about equal numbers of heads and tails are overwhelmingly likely) it is *possible* for a fair coin to come up heads every time in a billion tosses. Despite these shortcomings, however, the frequency method of estimating probability has much to recommend it, e.g., if future and past incidents belong to the same reference class, as the number of trials increases the estimated probability becomes more and more likely to fall within a negligible distance of the true one; and this method appears to be favored by more theorists than any other. Probability estimates obtained by this method are sometimes called *a posteriori* or *empirical* probabilities.

5-4-3 The Subjective Estimation Method

A final method of estimating probabilities is simply to make an educated guess. The type of probability being guessed at or estimated could be anything from radically objective probability to a completely personal, vague, intuitive degree of belief; however, we shall assume that it is simply what we have called objective probability that is being guessed at, since that is what is commonly meant by probability. In arriving at the guess, one may be influenced by whatever factors he deems relevant and, indeed, by whatever factors affect his judgment: past experience, such as previously observed relative frequencies of occurrence of the outcome in question; logical deductions, mathematical calculations, and considerations of "symmetry," such as those involved in the a priori method; inductive generalizations and inferences from vaguely analogous situations; and a host of "psychological" factors classifiable as intuitions, hunches, etc.

The outstanding advantage of the subjective method is that it permits one to base his estimate upon *all* of the relevant information to which he has access and to *weight* the various sources of information according to his judgment of their reliability and degree of influence. Thus, he may use all of the information used by the other two methods plus whatever additional information he may have. There are many situations in which there are no "symmetries" and where relative frequency information (with reference to the particular class of incidents in which one is interested) simply does not exist and is impossible to obtain. For example, a new spaceship has just been constructed, employing radical new principles of construction and propulsion which surely must be relevant to its probability of success. What is the probability that it will crash on its first flight? There are no symmetries and no past relative frequencies for the appropriate reference class of incidents having *all* the characteristics known or suspected to be relevant. Yet there is an abundance of relevant information. The whole history of maiden voyages, the prestige and past record of the manufacturer and designers, the skill and training of the crew, etc., etc., are all relevant to the success of the voyage. There is no difficulty in finding relevant information, only in converting it into an estimated relative frequency within the appropriate reference class of similar incidents. This involves a complex and subjective integrative process, commonly called an educated guess.

There are several complications or difficulties associated with the subjective method. The objection has frequently been made that different people, when presented with the same objective information, arrive at different subjectively estimated probabilities. But the information upon which one bases his subjective estimate consists not only of the objective information with which he is presented but also of the "subjective" information (past experiences, etc.) that he carries in his head. So it does not follow that the same totality of information leads to different subjectively estimated probabilities, although it may well be true since people may be expected to differ in their ability to assess and integrate information so as to produce an appropriate estimate of relative frequency.

Another difficulty is the tendency for a person to assign to the various possible mutually exclusive outcomes of an event probabilities that do not add up to exactly 1.00. This has two consequences, the first of which is that it means that the subjective estimates cannot all be correct estimates of the actual proportions of the time that the event has the various outcomes; it introduces an unnecessary source of error. The second evil is that we cannot use such probabilities in calculations to obtain other more complicated probabilities; they do not meet the mathematical requirements. The difficulty, however, is easily corrected. If the inconsistency is pointed out to them, most people will willingly make the appropriate adjustments. Thus, if a subjective estimator believes that the probability that a monochromatic ball drawn from an urn will be red is .25, that it will be green is .60, and that it will be neither is .10, he has implied that its probability of being *any*thing is only .95 rather than 1.00. If this is pointed out to him, he may reconsider and conclude that his estimate that it will be neither is .15.

5-4-4 Critique

In the situation where a unit (or sample) is to be drawn randomly from a population having a known distribution, we know the exact numerical probability that the unit (or a given sample statistic) will have a specified value. And this includes the case of a random variable having a known distribution. In all other cases, numerical probabilities can only be estimated.

For each of the three methods of estimating numerical probabilities, there is a type of situation where it is maximally appropriate. When nearly perfect symmetries exist, the a priori method is quite effective. Such symmetries are very carefully built into many games of chance for the express purpose of creating equal likelihood and making the probabilities of molecular outcomes precisely calculable. So the a priori method is particularly appropriate to games of chance.

When the future may be expected to repeat the past, and there is a vast backlog of data on the outcomes of appropriately similar incidents, the frequency method works beautifully. Thus, the frequency method is widely used in weather forecasting, quality control of manufactured products, insurance underwriting, etc.

When there are neither data nor symmetries, the subjective method wins by default. Probability statements are often made about events that contain no exploitable symmetries and for which relative frequency data are not available: "The odds are 2 to 1 that within 20 years most of our food will come from sea farms." But even when there is a wealth of data, the subjective method may be the only appropriate one, especially if the probability estimate is to apply to a particular already-identified unit or event about which we have a great deal of relevant, but nearly idiosyncratic, information. For example, records kept by insurance companies may tell us that the probability that a 25-year-old frame house will burn down within the next 12 months is .0018. But we want to know the probability that John Jones' 25-year-old frame house will burn down within the year. And we know (a) that the house is 100 yards from a fire hydrant, (b) that John is a bachelor, (c) that John is an alcoholic, and (d) that John smokes in bed. All of this information is relevant and its influence upon the outcome is foreseeable. But insurance companies don't keep frequency records on 25-year old frame houses 100 yards from a fire hydrant owned by alcoholic bachelors who smoke in bed. And even if they did, John Jones' house would probably be the only one that fitted the description; so data would be far too sparse to yield a good estimate even if Jones had provided data by being previously insured. We cannot use the relative frequency estimate of .0018 because it ignores relevant information to which we are privy and therefore applies to the wrong reference class of similar incidents. We cannot use relative frequency data for the appropriate class of incidents because they don't exist, and even if they did, the total number of cases would be too small for the estimate to have sufficient reliability to trust. The only way we can take account of information too relevant to ignore is to use the subjective method.

We may regard the first two methods as special cases of the third, subjective, method, to which the subjective method reduces when *all* of our relevant information consists *exclusively* of knowledge about symmetries or past relative frequencies, respectively. If the symmetries are exact (and are accompanied by no influential asymmetries) and if the relative frequencies are based upon a very large number of occurrences within the appropriate reference class of similar incidents, then the intelligent application of the three methods should yield essentially the same probability values. However, the symmetries encountered in practice are seldom perfect, the relative frequencies are seldom based upon enough data, and their reference class is often not quite the right one. Furthermore, our information seldom consists *exclusively* of symmetries or frequencies. In such cases it is appropriate to use the subjective method in order to take account of partial symmetries or insufficiently or inappropriately based relative frequencies and to combine them with each other and with additional relevant information in arriving at an estimated probability. If this is done judiciously, one would expect the subjectively estimated probability to be more accurate than estimates based exclusively upon symmetries or frequencies since more information has gone into it.

The following true situation will illustrate the use of the three methods of estimation, their advantages and shortcomings, and the comforting fact that they often lead to approximately the same numerical values. Seeking a pair of fair dice, I entered a toy store, found a red die and a green die, and bought them for a total price of twenty cents. Consider first the a priori method of estimation. The overall shape of each die was cubical. Each die was translucent and, holding it up to the light, I could see no lead weights or other evidence of interior loading. However, the numerical value of each face of the die was denoted by "dots" which were actually hemispherical holes drilled into the face, their interior then being painted white. Since opposite faces contained unequal numbers of holes, complete symmetry was lacking. Thus, the conditions assumed by the a priori method apparently were met grossly but not precisely. In such a situation there are only two reasonable alternatives: refuse to use the method at all, or use it but regard the estimates only as first approximations to the real values. If we take the latter course, we can say that we estimate the probability that any given side will come up to be approximately 1/6.

Next consider the frequency method. Holding both dice in my hand, I shook them and then cast them upon a cleared table top. I performed this operation 3,600 times. The data are given in Table 5-1. The relative frequencies range from 566/3,600 for "2" to 646/3,600 for "6" for the red die and from 557/3,600 for "1" to 681/3,600 for "6" for the green die. Even if both dice had been exactly symmetric, it would have been very unlikely that every face of each die would come up exactly 600 times—we would have expected frequencies only "close" to 600. For the same reason, we cannot believe very strongly that the empirical relative frequency of 566/3,600 represents the exact probability of a "2" for the red die. But in that case, can't we regard all 12 frequencies as simply chance fluctuations about a true value of 600?

Table 5-1

Outcomes of 3,600 Tosses of a Pair of Dice

Number coming up on red die

		1	2	3	4	5	6	Total
Number	1	94	80	85	106	91	101	557
coming	2	100	106	99	93	84	91	573
up	3	97	87	104	87	98	116	589
on	4	86	102	94	98	106	121	607
green	5	100	78	95	109	114	97	593
die	6	112	113	102	108	126	120	681
	Total	589	566	579	601	619	646	3,600

Later, we will be able to answer that question mathematically, but even just on an intuitive basis the hypothesis of chance fluctuations about 600 seems tenable for the red die but dubious for the green one. This leaves us in an unsatisfactory state. The a priori and frequency estimates agree that each side has a probability of *approximately* 1/6; and both methods give us a *hint* that the probabilities, although grossly similar, are not exactly equal; but *neither* method, strictly applied, really gives us sufficient precision of estimate. (The frequency method *could* do so if we wanted to cast the dice, say, 36,000 times instead of merely 3,600, but even the latter was a tedious chore. Thus, in order to yield highly accurate estimates, the frequency method must be based upon extremely large amounts of data.)

Now consider the subjective method. Because I had purchased the dice in a toy store rather than in a novelty shop, because they were not advertized as loaded, and because I had paid only twenty cents for them, I strongly doubted that the dice were intentionally loaded. But because the dots were actually holes, and because opposite faces contained different numbers of dots, it seemed doubtful that the center of gravity of a die was at the center of the cube, i.e., at the intersection of its diagonals. Since 4 is the face opposite 3, 5 is the face opposite 2, and 6 is the face opposite 1, it seemed reasonable to conclude that in each of the three cases cited, the center of gravity should lie closer to the latter (smaller) number than to the former (larger) number. Furthermore, the 4 face contains only one more hole than the 3 face, but the 5 face contains three more than the 2 face, and the 6 face contains five more holes than the 1 face. Therefore, if the inequality in number of holes displaces the center of gravity a distance d toward the 3 face, it must be displaced $3d$ toward the 2 face and $5d$ toward the 1 face. From this it follows that the outcome order 1, 2, 3, 4, 5, 6 represents the outcomes in order of increasing likelihood. Only the red 1 and the green 5 have frequency-estimated probabilities that violate this order, and the violations can be fairly convincingly ascribed to chance fluctuation. Therefore, I doubt the accuracy of the frequency estimates for the red 1 and the green 5. (I assume that it is the green 5 rather than 4 that is in error because while both should have frequencies greater than 600 it is only the green 5 that does not. And I assume that it is the red 1 rather than 2 or 3 that is in error because the relative frequency of the red 1 differs considerably from that of the green 1.)

Another thing that seems to follow, from the assumed displacement (toward the lower-numbered side) of center of gravity of an otherwise symmetric die, is that to whatever extent the higher-numbered side tends to come up *more* frequently than 600 times out of 3,600 throws, to that same extent the lower-numbered side should come up *less* frequently. Therefore, in general, for opposite-face outcomes the *average* of their absolute deviations from 600 should provide a better estimate of absolute deviation than do either of the two individual values from which it is obtained. Following this line of thought I first constructed Table 5-2 (top) giving the deviations of outcome frequencies from 600, except that I omit the figures for the red 1 and green 5 which I distrust because they violate the "proper" order. Next I obtained a presumably better estimate of frequency-deviation-from-600 by simply averaging their absolute values for opposite faces, except that I did not use the distrusted data for the red 1 and green 5, preferring to use the single trusted value. This is shown in Table 5-2 (upper middle). The two dice yielded highly similar figures for the opposite-face pairs 3-4 and 2-5 and reasonably close for 1-6, so I was encouraged to assume that the differences were attributable to chance fluctuations and to average the figures for the two dice. This gave me the figures shown in Table 5-2 (lower middle).

Now recalling that if the center of gravity is displaced a distance d toward the 3 face, it should be displaced $3d$ toward the 2 face and $5d$ toward the 1 face, I observed that the figures 54, 27, and 10 in Table 5-2 (lower middle) had approximately the same relative values as the corresponding center of gravity displacements. This moved me to assume that the difference was due to chance fluctuations and that the correct estimates should be 50, 30, and 10 as shown. I then simply converted these absolute deviations to the appropriate frequencies by adding them to 600 for the higher numbered face and subtracting them from 600 for the lower. Lastly I converted these estimated frequencies to estimated relative frequencies, or probabilities, by dividing by 3,600, as shown in Table 5-2 (bottom).

The reader may object that I have made use of "symmetries" and that I have therefore merely employed the a priori method. However, this is not the case. I have made use not of symmetries but rather of systematic *a*symmetries. The key postulate of the a priori method is that outcomes must

Table 5-2

Frequency of Outcome Minus 600

Die		Outcome 1	2	3	4	5	6
	Red	—	− 34	− 21	1	19	46
	Green	− 43	− 27	− 11	7	—	81

Average Absolute Deviation of Frequency From 600
(or More Trustworthy of Two Values)

Die		Outcome 1 or 6	2 or 5	3 or 4
	Red	(46)	26.5	11
	Green	62	(27)	9

	Outcome 1 or 6	2 or 5	3 or 4
Average die	54	27	10
Subjective estimate of correct value	50	30	10

Subjectively Estimated Outcome Probabilities

Die		Outcome 1	2	3	4	5	6
	Red	$\dfrac{550}{3{,}600}$	$\dfrac{570}{3{,}600}$	$\dfrac{590}{3{,}600}$	$\dfrac{610}{3{,}600}$	$\dfrac{630}{3{,}600}$	$\dfrac{650}{3{,}600}$
	Green	$\dfrac{550}{3{,}600}$	$\dfrac{570}{3{,}600}$	$\dfrac{590}{3{,}600}$	$\dfrac{610}{3{,}600}$	$\dfrac{630}{3{,}600}$	$\dfrac{650}{3{,}600}$

be equally likely, but my approach has assumed the contrary. I *have* made use of inspection of structure, deduction, and empirical frequencies. [Without the empirical frequencies I could have arrived at the conclusion that outcome probabilities increase for outcomes in the order 1, 2, 3, 4, 5, 6 and that their values are $(600 - 5x)/3600$, $(600 - 3x)/3600$, $(600 - x)/3600$, $(600 + x)/3600$, $(600 + 3x)/3600$, and $(600 + 5x)/3600$, but I would have had to guess at x or obtain it by far more complicated calculations, without benefit of actual data.] I have also made use of a lot of assumptions. I do not *know* that there is a linear relationship between displacement of center of gravity and increase in relative frequency of outcome. Nor do I know that probabilities are the same for the red die as for the green one, etc. Actually, I do not claim that the probabilities that I obtained are *right*. What I claim is (a) that they make use of all information I believe to be important and represent the best subjective estimate of which I am capable—in the sense that I would consider a gamble based on these probabilities to be "fair," and (b) so far as I am aware, they are the result of a coldblooded, dispassionate analysis of the available evidence, uncontaminated by whim or psychological quirk. The latter characteristic is not necessarily absent in subjective probability estimates, but neither is it necessarily present. The fact that an estimate is subjective does not necessarily mean that it is unanalytic nor that it is simply a primitive feeling conjured out of the murky depths of one's subconscious mind.

5-5 Problems

1. By what method were the probabilities involved in each of the following statements most likely to have been obtained:
   (a) The probability of losing in a single game of Russian roulette is 1/6.
   (b) The probability that an apple will have a worm in it is .031.
   (c) The probability that the second hand on your watch is between 10 and 25 is 1/4.
   (d) The probability that the next plane you travel on will crash is .00002.
   (e) The probability that the real author of Shakespeare's works is Francis Bacon is .015.
   (f) If Karl Marx had been a peasant, rather than a city boy, the odds are 3 to 1 that there would be no collective farms today.

2. What is the reference class of similar incidents in each of the above cases?

3. The symmetry method of estimating probabilities presupposes that each different possibility is equally likely. Consider the following situations:
   (a) Is a farmer equally likely to paint his barn red, green, or purple?
   (b) Are the four tires on a car equally likely to blow out?
   (c) Are people who are equally exposed to a disease (doctors, nurses, orderlies, etc.) equally likely to contract it?
   (d) Is a shoelace equally likely to break at all points along its length?
   (e) Five letters of equal size and shape arrive in your mailbox. Is each letter equally likely to be the one you open first?

(f) Is the second hand of your watch equally likely to be in each of the four quadrants of the circle?

In each case explain why the assumption of equal likelihood is or is not justified.

6

general probability laws

The preceding chapter discussed the concept of probability and the various definitions of probability. In this chapter we shall take up the laws of probability, that is, the mathematical formulas by which probabilities are manipulated, i.e., derived from basic numerical data or from each other. No matter which definition of probability one chooses, the probability is always a proportion or can be conceptualized as one. Therefore, mathematically speaking, *the laws of probability are nothing more nor less than the laws governing proportions*. Adherents of the various definitions of probability do not differ as to how probabilities should be manipulated—only as to how they should be obtained and interpreted. Thus, while there are numerous schools of probability, there is only one, common, set of probability laws.

6-1   The Urn Model

Consider an urn containing N balls, all of which have characteristics $C_1, C_2, \ldots, C_j$, and N_k of which also have characteristic C_k. We may regard the balls as the units comprising a population, defined by the characteristics $C_1, C_2, \ldots, C_j$, from which a unit is to be randomly selected. In that event, the probability that the randomly selected unit will have characteristic C_k is N_k/N, the proportion of balls in the urn that do so (see Section 5-4-1-1). Alternatively, we may regard the balls as representing all of the potential occurrences of a random variable, defined by characteristics $C_1, C_2, \ldots, C_j$, which, of course, "selects" its own occurrences. Again, the probability that a particular occurrence of the random variable will have characteristic C_k is N_k/N, the proportion of balls in the urn having that characteristic (see Section 5-4-1-3). Finally, taking the most general case, which includes the two preceding cases, we may regard the balls as representing all of the incidents in the reference class of similar incidents, all of which have characteristics $C_1, C_2, \ldots, C_j$ in common with the incident in question. In that event, the probability that the incident in question will, when it occurs, have the additional characteristic C_k is N_k/N, the proportion of balls in the urn having that characteristic (see Section 5-3). Clearly, then, by deriving the laws pertaining to proportions of balls in the urn, we shall obtain the basic, i.e., general, laws of probability.

The characteristic C_k need not be unitary. It can, in fact, stand for the occurrence of all of a multiplicity of characteristics, such as C_r, C_s, C_t, so that $C_k = C_r$ *and* C_s *and* C_t. Furthermore, the incident whose probability we seek, and those comprising the reference class of similar incidents, also need not be unitary but may be composed of the joint occurrence of several subordinate incidents. Thus, "characteristics" C_r, C_s, and C_t may stand for the occurrence of subordinate incidents R, S, and T, whose joint occurrence creates characteristic C_k in a superordinate incident to which they contribute and which is a member of a reference class of superordinate incidents. And the subordinate incidents R, S, and T may or may not be closely, or even logically, related.

In deriving the general laws of proportions and probabilities from our urn model, we shall need a concrete example of the contents of the urn. For our example, we shall take an urn containing 200 balls each of which has a single color that is either red (R), white (W), or blue (B) and is stamped with a single numeral that is either I, II, III, IV, or V. (Thus, the balls constitute a bivariate population since there are two variables, color and numeral; a multivariate population would have been more general, but also more complicated.) The number of balls in each of the color-numeral categories is given in the cells of Table 6-1a where the three rows represent the three colors and the five columns represent the five numerals. The numbers in the margins of the table are simply row or column totals and represent the total number of balls of the corresponding row color or column numeral. Table 6-2a presents the more general case, of which Table 6-1a is a specific example: there are N balls, each of which belongs to a single row category R_i and a single column category C_j; the number of balls having both characteristic R_i and characteristic C_j is N_{ij}, the entry in the cell at the intersection of row R_i and column C_j; the number of balls having characteristic R_i is N_i. (where the dot subscript indicates summation over all values of the subscript letter that it replaces, i.e., $N_i. = \sum_j N_{ij}$) and the number having characteristic C_j

Table 6-1

Bivariate Population of Balls in an Urn (Specific Case)

(a) Numbers of Units

	I	II	III	IV	V	All numerals
Red	0	30	24	6	60	120
White	0	6	6	6	12	30
Blue	4	12	10	6	18	50
All colors	4	48	40	18	90	200

(b) Proportions of Units

	I	II	III	IV	V	All numerals
Red	.00	.15	.12	.03	.30	.60
White	.00	.03	.03	.03	.06	.15
Blue	.02	.06	.05	.03	.09	.25
All colors	.02	.24	.20	.09	.45	1.00

is $N_{.j}$. If we divide each of the entries in Table 6-2a by N, we get *proportions* of balls rather than numbers of balls, and these are presented in Table 6-2b where

$$p_{ij} = \frac{N_{ij}}{N}, \quad p_{i.} = \frac{N_{i.}}{N}, \quad p_{.j} = \frac{N_{.j}}{N},$$

and $1.00 = N/N$. Table 6-1b does the same thing for the specific numbers given in Table 6-1a; that is, Table 6-1b simply converts the numbers in Table 6-1a to proportions by dividing every entry by 200, the total number of balls in the urn.

Before deriving the general laws of proportions and probabilities, we need some convenient notation. Let P stand for proportion and let P(   ) stand for the proportions of units (or balls in the present case) having the characteristics listed within the parentheses. So $P(R)$ represents the proportion of balls in the urn that are red, and equals .60. When we write the symbol in this way, it is *understood* that we are talking about a proportion with reference to the entire set of units under consideration, e.g., $P(R)$ means the pro-

Table 6-2

Bivariate Population of Balls in an Urn (General Case)

(a) Numbers of Units

	C_1	C_2	......	C_j	......	C_s	$C.$
R_1	N_{11}	N_{12}	......	N_{1j}	......	N_{1s}	$N_{1\cdot}$
R_2	N_{21}	N_{22}	......	N_{2j}	......	N_{2s}	$N_{2\cdot}$
$\vdots$	$\vdots$	$\vdots$	$\vdots$	$\vdots$	$\vdots$	$\vdots$	$\vdots$
R_i	N_{i1}	N_{i2}	......	N_{ij}	......	N_{is}	$N_{i\cdot}$
$\vdots$	$\vdots$	$\vdots$	$\vdots$	$\vdots$	$\vdots$	$\vdots$	$\vdots$
R_r	N_{r1}	N_{r2}	......	N_{rj}	......	N_{rs}	$N_{r\cdot}$
$R.$	$N_{\cdot1}$	$N_{\cdot2}$	......	$N_{\cdot j}$	......	$N_{\cdot s}$	N

(b) Proportions of Units

	C_1	C_2	......	C_j	......	C_s	$C.$
R_1	p_{11}	p_{12}	......	p_{1j}	......	p_{1s}	$p_{1\cdot}$
R_2	p_{21}	p_{22}	......	p_{2j}	......	p_{2s}	$p_{2\cdot}$
$\vdots$	$\vdots$	$\vdots$	$\vdots$	$\vdots$	$\vdots$	$\vdots$	$\vdots$
R_i	p_{i1}	p_{i2}	......	p_{ij}	......	p_{is}	$p_{i\cdot}$
$\vdots$	$\vdots$	$\vdots$	$\vdots$	$\vdots$	$\vdots$	$\vdots$	$\vdots$
R_r	p_{r1}	p_{r2}	......	p_{rj}	......	p_{rs}	$p_{r\cdot}$
$R.$	$p_{\cdot1}$	$p_{\cdot2}$	......	$p_{\cdot j}$	......	$p_{\cdot s}$	1.00

portion of the entire 200 balls in the urn that are red. Sometimes, however, we wish to indicate a proportion within a smaller reference group. When we wish to indicate a qualifying condition, we can do so within the parentheses by inserting, immediately after the list of characteristics, a vertical line which stands for "given" or "given that" and by inserting the qualification (or the symbols standing for it) immediately to the right of the vertical line. So if we wish to indicate a reference group smaller than the entire set, we can do so by inserting the vertical line and then identifying the new reference group by listing the characteristics of the units composing it. Thus, if we want a symbol standing for the proportion of red balls within the group of balls stamped III, we write $P(R \,|\, \text{III})$.

Since probabilities are proportions and the laws of probability are the laws of proportions, we shall use the same symbols to refer to probabilities. Whether they refer to probabilities or to proportions will be clear from the *context* in which they are used, rather than from the symbols themselves. Thus, if we are talking about probabilities rather than proportions as such (e.g., if we are talking about the likelihood that some single future incident will have a certain outcome, rather than about the proportion of incidents, in a reference group of similar incidents, that have that outcome), we shall interpret the same symbols in the following different fashion. P stands for probability. $P(\quad)$ stands for the probability that the *incident* in question will have the characteristic specified within the parentheses; or, equivalently, that the *random variable* in question will, on a specified occasion, take the value listed within parentheses; or, that a *random selection* of a unit from the population in question will have the outcome specified within parentheses. The two latter situations are special cases of the former. That is, the occurrence of a random variable and its outcome, or the making of a random selection and its outcome, are incidents, belonging to a reference class of similar incidents; and the specified value of the random variable or the specified outcome of the random selection is the specified characteristic in which we are interested. Therefore, henceforth we shall use the general term "incident" to refer to the entity in the probability of whose characteristics we are interested. Again, qualifying conditions are introduced to the right of a vertical line and within the parentheses. Such conditions are generally a reduction in the size of the reference set of units, or (equivalently) the fact that certain relevant (e.g., subordinate or contributory) incidents are known or assumed already to have occurred. Probabilities so qualified are known as conditional probabilities since they are conditional upon the qualifying conditions.

6-2   The Range of Proportions or of Probabilities

Every proportion, and therefore every probability, is a fraction whose numerator represents a portion of (i.e., a part "totally enclosed within") whatever is represented by its denominator; if the denominator represents a set, the numerator represents a subset of it. So if the denominator is N, the numerator must lie in the region from zero to N. Consequently, every proportion, and therefore every probability, lies in the region from 0 to 1.

The first law of proportions, or of probabilities, is therefore that

$$0 \leq P(\text{anything}) \leq 1 \tag{6-1}$$

If X is a characteristic of no unit in the reference set (and is therefore "impossible") or if it is a characteristic of a finite number of units in an infinite reference set (and therefore has a relative frequency of zero), $P(X) = 0$. If X is a characteristic of every unit in the reference set (and is therefore "certain") or if it is a characteristic of all but a finite number of units in an infinite reference set (and therefore has a relative frequency of 1), $P(X) = 1$.

6-3 The Proportion of Units Having a Certain Characteristic Or the Probability That an Incident Will Have It

If N_A of the N units in the reference set have characteristic A, then by the definition of a proportion

$$P(A) = \frac{N_A}{N} \tag{6-2}$$

and this is also the probability that an incident beloging to such a reference set will have characteristic A.

Thus, 4 of the 200 balls in the urn are stamped with a I, so the proportion of balls stamped with a I is $P(\text{I}) = 4/200 = .02$ (see Tables 6-1a and 6-1b). And if we randomly selected a ball from the urn, the *probability* that it would be stamped with a I is $P(\text{I}) = 4/200 = .02$.

Illustrative Problems: If in 100,000 commercial airplane flights from Denver to Chicago three of the airplanes crashed, what proportion of the flights terminated in a crash?

ANSWER: $P(C) = 3/100,000 = .00003$.

What is the probability that a certain man taking one of those flights would be on a plane that crashed?

ANSWER: $P(C) = 3/100,000 = .00003$.

6-4 The Proportion of Units Having the Complementary Characteristic or the Probability That an Incident Will Have It

If N_A of the N units in the reference set have characteristic A, then $N - N_A$ of the units in the reference set do *not* have it and therefore have the complementary characteristic of "not A," symbolized $\bar{A}$. The proportion of units having the complementary characteristic $\bar{A}$ is $(N - N_A)/N$ which reduces to $1 - (N_A/N)$. But, by Formula 6-2, N_A/N is $P(A)$. So the proportion of units having the complementary characteristic "not A" or $\bar{A}$ is

$$P(\bar{A}) = 1 - P(A) \qquad (6\text{-}3)$$

and this is also the probability that an incident belonging to such a reference set will have characteristic $\bar{A}$. It is understood, of course, that $P(A)$ and $P(\bar{A})$ refer to, i.e., are based upon, the same reference set.

Thus, of the 200 balls in the urn, 4 are stamped with a I and 196 are not; so the proportion of balls having the characteristic of not being stamped with a I is $P(\bar{I}) = 1 - P(I) = 1 - .02 = .98$ which is easily confirmed since, by direct calculation, $P(\bar{I}) = N_{\bar{I}}/N = 196/200 = .98$. Likewise, if the balls in the urn are incidents in the reference class of all similar incidents, the probability that an incident will occur that has all of the specified common characteristics of the incidents in the reference class as well as the characteristic of "not I" is $P(\bar{I}) = .98$.

Illustrative Problems: If the proportion of coins in your pocket that are quarters is .2, what is the proportion of coins in your pocket that are not quarters?

ANSWER: $$P(\bar{Q}) = 1 - P(Q) = 1 - .2 = .8$$

If the probability that a commercial airplane will crash in a flight from Denver to Chicago is .00003, what is the probability that it will not crash?

ANSWER: $$P(\bar{C}) = 1 - P(C) = 1 - .00003 = .99997$$

6-5 The Proportion of Units Having Either or Both of Two Characteristics or the Probability That an Incident Will Do So

The proportion of balls in the urn that are either R's or V's or both is obtained from Table 6-1a by adding the numbers in the R row to the numbers, not already counted, in the V column and dividing by the total number of balls in the urn:

$$P(R \text{ or } V \text{ or both}) = \frac{0 + 30 + 24 + 6 + 60 + 12 + 18}{200}$$

$$= \frac{(0 + 30 + 24 + 6 + 60) - 60 + (60 + 12 + 18)}{200}$$

$$= \frac{120 - 60 + 90}{200} = \frac{N_R - N_{RV} + N_V}{N}$$

$$= \frac{N_R}{N} + \frac{N_V}{N} - \frac{N_{RV}}{N} = P(R) + P(V) - P(R \text{ \& } V)$$

Alternatively, we could have worked with Table 6-1b:

$$P(R \text{ or } V \text{ or both}) = 0 + .15 + .12 + .03 + .30 + .06 + .09$$
$$= (0 + .15 + .12 + .03 + .30) + (.30 + .06 + .09)$$
$$- .30$$
$$= .60 + .45 - .30 = P(R) + P(V) - P(R \& V)$$

Both $P(R)$ and $P(V)$ contain $P(R$ and $V)$, so it is necessary to subtract it from their sum in order to prevent it from being counted twice. The subtraction is necessary precisely because it is possible for a ball to be both R and V, i.e., to have both the characteristics of being red and being stamped with a V, so that the subset of balls that are R's may "overlap" with the subset that are V's.

The formula obtained for the specific situation depicted in Tables 6-1a and 6-1b is completely generalizable to the generic case depicted in Tables 6-2a and 6-2b:

$$P(R_i \text{ or } C_j \text{ or both}) = \frac{\sum_{j=1}^{s} N_{ij} + \sum_{i=1}^{r} N_{ij} - N_{ij}}{N} = \frac{N_{i.} + N_{.j} - N_{ij}}{N}$$
$$= \frac{N_{i.}}{N} + \frac{N_{.j}}{N} - \frac{N_{ij}}{N} = P(R_i) + P(C_j) - P(R_i \& C_j)$$

Alternatively,

$$P(R_i \text{ or } C_j \text{ or both}) = \sum_{j=1}^{s} p_{ij} + \sum_{i=1}^{r} p_{ij} - p_{ij} = p_{i.} + p_{.j} - p_{ij}$$
$$= P(R_i) + P(C_j) - P(R_i \& C_j)$$

Sometimes two characteristics are known to be **mutually exclusive,** so that the having of one characteristic makes having the other characteristic impossible, as, for example, being a veteran of World War II and being born in 1960. When this is known to be the case, since there can be no overlap, the subtracted term representing the overlap can be omitted. Thus, since each ball in the urn has one and only one color,

$$P(R \text{ or } W \text{ or both}) = P(R) + P(W) - P(R \& W) = .60 + .15 - 0 = .75$$
$$= P(R) + P(W)$$

Thus, *in general,* if A and B are two characteristics, the proportion of units in a given (understood) population that have either characteristic A or characteristic B or have both characteristics is given by the following formula. The formula also gives the probability that an incident will have one or both of the two characteristics, where all terms in the formula are understood to refer to the same reference set.

$$P(A \text{ or } B \text{ or both}) = P(A) + P(B) - P(A \& B) \qquad (6-4)$$

And in the special case where A and B are *mutually exclusive* characteristics, the above formula reduces to

$$P(A \text{ or } B \mid A \text{ and } B \text{ are mutually exclusive}) = P(A) + P(B) \quad (6\text{-}5)$$

Illustrative Problems: Of the books in a certain library, 80% are over ten years old, 90% have been borrowed more than once, and 75% are both over ten years old and at least twice-borrowed. What proportion of the books in the library are either over ten years old or at least twice-borrowed or both?

ANSWER: P(over ten years old or borrowed more than once or both) $= P$ (over ten years old) $+ P$(borrowed more than once) $- P$(both) $= .80 + .90 - .75 = .95$.

What is the probability that a book randomly selected from the above library will either be over ten years old or borrowed more than once or both?

ANSWER: P(OTYO or BMTO or both) $= P$(OTYO) $+ P$(BMTO) $- P$(OTYO & BMTO) $= .80 + .90 - .75 = .95$.

Ten percent of the footwear manufactured by a certain company are boots and 5% are sandals. What proportion of their footwear are either boots or sandals?

ANSWER: A shoe cannot be both a boot and a sandal, so the two types of footwear are mutually exclusive and $P(B \text{ or } S \mid B \text{ and } S \text{ are mutually exclusive}) = P(B) + P(S) = .10 + .05 = .15$.

If the probability that a man will die before his 40th birthday is .08 and the probability that he will die after his 80th birthday is .30, what is the probability that he will die either before 40 or after 80?

ANSWER: The two probabilities are mutually exclusive, so the answer is $.08 + .30$ or $.38$.

6-6   The Proportion of Units Having One or More of Several Characteristics or the Probability That an Incident Will Do So

The formulas obtained in the preceding section apply only to the very simple case where we are considering only *two* characteristics. Consider now the more general case where we want to develop analogous formulas for the proportion of units having one or more of *several* characteristics.

If all of the characteristics in question are *mutually exclusive,* i.e., if it is *impossible* for a unit to have *more* than one of them, then the number of units having one *or more* of them is the same as the number of units having *any one* of them, and that is simply the sum of the number of units having *each* of them. A corresponding statement can be made substituting "proportion" for "number." For example, the proportion of balls in the urn that are either I's, II's, III's, or IV's is

$$P(\text{I}) + P(\text{II}) + P(\text{III}) + P(\text{IV}) = .02 + .24 + .20 + .09 = .55.$$

Therefore, if we have K *mutually exclusive* characteristics, $C_1, C_2, \ldots, C_k$, the proportion of units that have any of these characteristics, or the probability that an incident will have any, is

$$P(C_1 \text{ or } C_2 \text{ or } \ldots \text{ or } C_{k-1} \text{ or } C_k \,|\, \text{all } C_i\text{'s}$$
$$\text{are mutually exclusive}) = \sum_{i=1}^{k} P(C_i) \qquad (6\text{-}6)$$

where, of course, all terms in the formula refer to the same reference set.

If the several characteristics are *not* known to be mutually exclusive, we must subtract out the "overlap" and this can become both tedious and complicated when there are more than a very few characteristics. For example, consider the case in which there are just three characteristics, A, B, and C:

$$P(A) \text{ includes } P(A \,\&\, B), P(A \,\&\, C) \text{ and } P(A \,\&\, B \,\&\, C)$$
$$P(B) \text{ includes } P(A \,\&\, B), P(B \,\&\, C) \text{ and } P(A \,\&\, B \,\&\, C)$$
$$P(C) \text{ includes } P(A \,\&\, C), P(B \,\&\, C) \text{ and } P(A \,\&\, B \,\&\, C)$$

So $P(A) + P(B) + P(C)$ includes each of the following twice, i.e., one more time than desired, $P(A \,\&\, B)$, $P(A \,\&\, C)$, and $P(B \,\&\, C)$; and it includes $P(A \,\&\, B \,\&\, C)$ three times which is twice more than desired. This suggests that we will have the proportion we wish if from $P(A) + P(B) + P(C)$ we subtract each of the "doubles" once and the "triple" twice. But there is a complication. Each of the "doubles" includes the "triple." So if we subtract three "doubles," we have automatically subtracted the "triple" three times, which is one more time than we intended. So to obtain the proper subtraction of overlap, we must subtract each "double" once and *add* a single "triple" obtaining

$$P(\text{units having any of the characteristics } A, B, C) =$$
$$P(\text{units having } A \text{ only or } B \text{ only or } C \text{ only or } A \,\&\, B \text{ only}$$
$$\text{or } A \,\&\, C \text{ only or } B \,\&\, C \text{ only or } A \,\&\, B \,\&\, C) = P(A) + P(B)$$
$$+ P(C) - P(A \,\&\, B) - P(A \,\&\, C) - P(B \,\&\, C) + P(A \,\&\, B \,\&\, C)$$

The proper procedure and the rationale for it is easier to see with the aid of a **Venn diagram** (see Figure 6-1). Each characteristic is represented by the area within an enclosure such as a circle. We may think of the enclosure as containing all the units having the given characteristic. The areas where the enclosures overlap represent the corresponding combinations of characteristics. It is clear from the crosshatching in the Venn diagram shown in Figure 6-1 that the sum of $P(A)$, $P(B)$, and $P(C)$ includes $P(A \,\&\, B)$, $P(A \,\&\, C)$, and

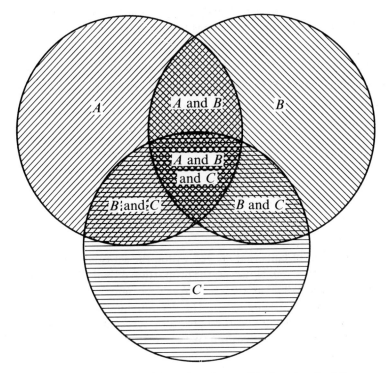

FIG. 6-1. Venn diagram illustrating the possible "overlaps" of three characteristics.

$P(B \& C)$ twice and includes $P(A \& B \& C)$ three times and that $P(A \& B)$, $P(A \& C)$, and $P(B \& C)$ each include it once so that if the "doubles" are each subtracted once, there remains no $P(A \& B \& C)$ at all, necessitating its being added back in.

Illustrative Problems: Of all U.S. battle deaths occurring in the seven wars prior to 1950, 28.41% occurred during the Civil War, 10.80% during World War I, and 59.00% during World War II. What proportion of all U.S. battle deaths occurred during the three wars named?

ANSWER: Although it is possible for the same man to be wounded in each of several wars separated in time, he can die in only one of them. Therefore, the three characteristics are mutually exclusive and P(died in Civil War or died in WWI or died in WWII) = P(died in CW) + P(died in WWI) + P(died in WWII) = .2841 + .1080 + .5900 = .9821.

Suppose that the probability that a randomly selected psychologist belongs to a psychological association (P) is .90, to a general science association (S) is .40, to a mathematical association (M) is .10, to both P and S is .35, to both P and M is .08, to both S and M is .05, and to P and S and M is .04. What is the probability that he belongs to one or more of the three types of society?

ANSWER: Substituting into the last formula given in this section we have,

$$P(P \text{ or } S \text{ or } M) = P(P) + P(S) + P(M) - P(P \& S) - P(P \& M)$$
$$- P(S \& M) + P(P \& S \& M)$$
$$= .90 + .40 + .10 - .35 - .08 - .05 + .04$$
$$= .96$$

6-7 Proportions Within a Subgroup, or Conditional Probabilities

The proportion of balls that are red within the set of all balls in the urn is $N_R/N = 120/200 = .60$. However, the proportion of balls that are red within the subgroup of balls that are V's is $N_{RV}/N_V = 60/90 = 2/3$. This follows from the definition of a proportion. The denominator of the fraction represents the reference group and the numerator represents that portion of it that has the characteristic in question. We could also have obtained the above answer this way:

$$P(R|V) = \frac{N_{RV}}{N_V} = \frac{\dfrac{N_{RV}}{N}}{\dfrac{N_V}{N}} = \frac{P(R \& V)}{P(V)} = \frac{.30}{.45} = \frac{2}{3}$$

The proportion of V-balls within the subgroup of balls that are red is obtained analogously, but it is a different proportion:

$$P(V|R) = \frac{N_{RV}}{N_R} = \frac{\dfrac{N_{RV}}{N}}{\dfrac{N_R}{N}} = \frac{P(R \& V)}{P(R)} = \frac{.30}{.60} = \frac{1}{2}$$

These specific formulas are particular cases of general formulas obtainable from Tables 6-2a and 6-2b:

$$P(R_i|C_j) = \frac{N_{ij}}{N_{\cdot j}} = \frac{\dfrac{N_{ij}}{N}}{\dfrac{N_{\cdot j}}{N}} = \frac{p_{ij}}{p_{\cdot j}} = \frac{P(R_i \& C_j)}{P(C_j)}$$

$$P(C_j|R_i) = \frac{N_{ij}}{N_{i\cdot}} = \frac{\dfrac{N_{ij}}{N}}{\dfrac{N_{i\cdot}}{N}} = \frac{p_{ij}}{p_{i\cdot}} = \frac{P(R_i \& C_j)}{P(R_i)}$$

Thus, *in general*, if A and B are any two characteristics, the proportion of units having characteristic A among only those units that have characteristic B is

$$P(A|B) = \frac{P(A \text{ \& } B)}{P(B)} \tag{6-7}$$

and this is also the probability that an incident will have characteristic A given that it is known that it has, or will have, characteristic B. Similarly, the proportion of units that are B's within the subset of units that are A's is

$$P(B|A) = \frac{P(A \text{ \& } B)}{P(A)} \tag{6-8}$$

and this formula also represents the probability that an incident will have characteristic B if it has characteristic A. Numerator and denominator on the right side of both equations, of course, refer to proportions or probabilities within the same original reference group. Such probabilities are called **conditional probabilities** because they involve a qualifying condition. The condition appears on the right of the vertical line, within the parentheses, and has the effect of reducing the size of the reference set.

It is easy to become confused by the semantics of qualifying statements. $P(A|B)$ stands for the proportion of units, within the subset of units having characteristic B, that also have characteristic A. It can also be referred to as "the proportion of A's among B's," which can be equivalently stated as "the proportion of B's that are A's." It is important to recognize that despite their differences in form these descriptions are synonymous in substance.

Illustrative Problems: Half of one percent of the students at a certain university are graduate students in the Engineering School. Twenty percent of the students in the university are in the Engineering School. What proportion of students in the Engineering School are graduate students?

ANSWER:

$$P(GS|ES) = \frac{P(GS \text{ \& } ES)}{P(ES)} = \frac{.005}{.200} = .025$$

Two dice are to be thrown. What is the probability that, if each die comes up an even number, the sum of the two numbers will be 8?

ANSWER: $P(\text{sum is } 8 \,|\, \text{both even}) = P(\text{sum is 8 and both even})/P(\text{both even})$. There are 36 equally likely atomic outcomes, in 9 of which both dice come up even numbers (2 & 2, 2 & 4, 2 & 6, 4 & 2, 4 & 4, 4 & 6, 6 & 2, 6 & 4, 6 & 6), so $P(\text{both even}) = 9/36$, and in three of which both dice come up even numbers whose sum is 8 (2 & 6, 4 & 4, 6 & 2), so $P(\text{sum is 8 \& both even}) = 3/36$. Therefore,

$$P(\text{sum is } 8 \,|\, \text{both even}) = \frac{3/36}{9/36} = 1/3$$

6-8 Bayes' Formula

Multiplying both sides of Formula 6-7 by $P(B)$, we obtain

$$P(A \mid B)P(B) = P(A \& B)$$

Multiplying both sides of Formula 6-8 by $P(A)$, we obtain

$$P(B \mid A)P(A) = P(A \& B)$$

Since two things equal to the same thing, $P(A \& B)$, are equal to each other, it follows that

$$P(A \mid B)P(B) = P(B \mid A)P(A)$$

Finally, dividing both sides of the last equation by $P(B)$, we obtain **Bayes' Formula**,

$$P(A \mid B) = \frac{P(B \mid A)P(A)}{P(B)} \tag{6-9}$$

which is a very useful general formula for the calculation of conditional probabilities or proportions.

The rationale for, and interrelationships between, Formulas 6-7, 6-8, and 6-9 can be seen more graphically in Table 6-3. Table 6-3a gives the proportion of units in the original reference set having, and the proportion not having, each of the characteristics A and B. Table 6-3b gives the proportion of units having, and the proportion not having, characteristic A in the restricted reference set of units having characteristic B. And Table 6-3c gives the proportion of units having, and the proportion not having, characteristic B in the restricted reference set of units having characteristic A. If one has the information in the first table, he can obtain the information in the second or third tables by dividing the entries in the first column by $P(B)$ or the entries in the first row by $P(A)$, respectively. Likewise, if he knows $P(A \mid B)$, then in effect, he has the information in the second table, and by multiplying each of its entries by $P(B)$ he obtains the information in the first column of the first table. Similarly, if he knows $P(B \mid A)$, he has all of the third table, and by multiplying each of its entries by $P(A)$ he obtains the information in the first row of the first table. In working problems it is sometimes easier to obtain solutions by constructing tables such as these, and working directly from them, than to obtain them by formal application of the formulas.

Thus, if the proportion of cars that are Fords $P(F)$ is .40, the proportion of cars that are convertibles $P(C)$ is .05, and the proportion of Fords that are convertibles $P(C \mid F)$ is .02, then the proportion of convertibles that are Fords is

$$P(F \mid C) = \frac{P(C \mid F)P(F)}{P(C)} = \frac{.02 \times .40}{.05} = \frac{.008}{.05} = .16$$

Table 6-3

Tabular Representation of Certain Probabilities

(a) Proportions (or probabilities) of units having
characteristics A or not-A, B or not-B

	B	$\overline{B}$	Total
A	$P(A \ \& \ B)$ which equals $P(A \mid B)P(B)$ or $P(B \mid A)P(A)$	$P(A \ \& \ \overline{B})$ which equals $P(A \mid \overline{B})P(\overline{B})$ or $P(\overline{B} \mid A)P(A)$	$P(A)$
$\overline{A}$	$P(\overline{A} \ \& \ B)$ which equals $P(\overline{A} \mid B)P(B)$ or $P(B \mid \overline{A})P(\overline{A})$	$P(\overline{A} \ \& \ \overline{B})$ which equals $P(\overline{A} \mid \overline{B})P(\overline{B})$ or $P(\overline{B} \mid \overline{A})P(\overline{A})$	$P(\overline{A})$
Total	$P(B)$	$P(\overline{B})$	1.00

(b) Entries in first column of table
6-3a divided by $P(B)$

	B
A	$P(A \mid B)$
$\overline{A}$	$P(\overline{A} \mid B)$
Total	1.00

(c) Entries in first row of table
6-3a divided by $P(A)$

	B	$\overline{B}$	Total
A	$P(B \mid A)$	$P(\overline{B} \mid A)$	1.00

Alternatively, we can obtain the same answer by constructing tables that incorporate the input information and deriving from them a table one of whose entries is the solution. This is done in Table 6-4. (Only the critical cell entries are filled in; except for the interior cells of the first table, all of the

Table 6-4

Example of "Tabular Solution" of Bayes' Formula

Step 1: Record input information in proper cells of proper tables.

	F	$\overline{F}$	Total			F
C			.05		C	.02
$\overline{C}$					$\overline{C}$	
Total	.40		1.00		Total	1.00

Step 2: Incorporate all information into the unconditional table (by multiplying entries in the conditional table by the factor, .40, that makes them equal the corresponding entries in the unconditional table).

	F	$\overline{F}$	Total
C	.008		.05
$\overline{C}$			
Total	.40		1.00

Step 3 Extract the required conditional table from the unconditional table (by multiplying entries in the unconditional table by the factor, 20, that makes the "total" entry of the conditional table 1.00)

	F	$\overline{F}$	Total
C	.16		1.00

empty cells can be filled in by subtraction of the "complementary" cell entry from the appropriate "total.")

The characteristics "not A" and "not B" need not be unitary. Suppose that what we have been calling characteristic A really is the ith value A_i of an A variable having several different possible values, $A_1, A_2, \ldots, A_i, \ldots, A_r$, so that the characteristic "not A" is simply the characteristic of having one of the values other than A_i. Then, by Formula 6-7,

$$P(A_i \mid B) = \frac{P(A_i \text{ \& } B)}{P(B)}$$

Now, in effect, we are dealing with a bivariate population of units each of which is either a B or a "not B" and *also* has one and only one of the possible A values $A_1, A_2, \ldots, A_i, \ldots, A_r$. Therefore, it follows (see Table 6-2b) that

$$P(B) = P(A_1 \& B) + P(A_2 \& B) + \cdots + P(A_i \& B) + \cdots + P(A_r \& B)$$
$$= \sum_i P(A_i \& B)$$

Making this substitution, we have

$$P(A_i \mid B) = \frac{P(A_i \& B)}{\sum_i P(A_i \& B)}$$

But Formula 6-8 tells us that $P(A_i \& B) = P(B \mid A_i)P(A_i)$, and making this final substitution we obtain the following modification of Bayes' Formula

$$P(A_i \mid B) = \frac{P(B \mid A_i)P(A_i)}{\sum_i P(B \mid A_i)P(A_i)} \qquad (6\text{-}10)$$

which is often the form of the formula that we need in working problems. Both forms of Bayes' Formula apply, of course, to probabilities as well as to proportions.

Illustrative Problems: If 40% of all degrees and 15% of Ph.D. doctoral degrees are awarded to women, and if 2% of all degrees are Ph.D.'s, what proportion of all degrees awarded to women are Ph.D.'s?

ANSWER:

$$P(D \mid W) = \frac{P(W \mid D)P(D)}{P(W)} = \frac{.15(.02)}{.40} = \frac{.003}{.400} = .0075$$

Bertrand's Box Problem: Each of three boxes has two drawers. One box (G) has a gold coin in each drawer, one (S) has a silver coin in each drawer, and one (B) has a gold coin in one drawer and a silver coin in the other. One of the six drawers is randomly selected and opened revealing a gold coin (g). What is the probability that the coin in the other drawer of the same box is also a gold one?

ANSWER: The question is equivalent to asking the probability that the *selected* drawer belonged to the G box. We may regard each possible selection as having two characteristics, a box characteristic and a coin characteristic. Of the three possible selections having the g characteristic, two also have the G characteristic, and since selection was random, each possibility was equally likely. Therefore, $P(G \mid g) = 2/3$. However, we may also obtain the answer by using Bayes' Formula

$$P(G \mid g) = \frac{P(g \mid G)P(G)}{P(g)} = \frac{(2/2)(2/6)}{3/6} = 2/3$$

or by using the modified Bayes' Formula

$$P(G|g) = \frac{P(g|G)P(G)}{P(g|G)P(G) + P(g|S)P(S) + P(g|B)P(B)}$$

$$= \frac{(2/2)(2/6)}{(2/2)(2/6) + (0/2)(2/6) + (1/2)(2/6)} = \frac{2}{2+0+1} = \frac{2}{3}$$

A ball is randomly drawn from an urn which had contained 2 red and 4 green balls and transferred to a second urn which had contained 4 red and 5 green balls. After the transfer, a ball is randomly drawn from the second urn and it turns out to be red. What is the probability that the ball drawn from the first urn, i.e., the transferred ball, was red?

ANSWER: Let capital letters stand for the possible colors of the ball drawn from the first urn and let lower-case letters stand for the possible colors of the ball drawn from the second urn. Then, at the time the first ball was drawn (before the information provided by the outcome of the second drawing became available) $P(R) = 2/6$ and $P(G) = 4/6$. After the transfer, the second urn contained 5 red and 5 green balls if the transferred ball was red, or 4 red and 6 green balls if the transferred ball was green. Therefore, $P(r|R) = 5/10$ and $P(r|G) = 4/10$. The problem asks for the conditional probability that the ball drawn from the first urn was red, *given* the relevant information that the ball drawn from the second urn was red. (The relevance would be completely obvious if the second urn had originally contained 0 red and 9 green balls.) The answer is given by use of the modified form of Bayes' Formula.

$$P(R|r) = \frac{P(r|R)P(R)}{P(r)} = \frac{P(r|R)P(R)}{P(r|R)P(R) + P(r|G)P(G)} = \frac{(\frac{5}{10})(\frac{2}{6})}{(\frac{5}{10})(\frac{2}{6}) + (\frac{4}{10})(\frac{4}{6})}$$

$$= \frac{\frac{10}{60}}{\frac{10}{60} + \frac{16}{60}} = \frac{10}{26}$$

One tenth of one percent of the items produced by a manufacturing process are defective. An inspection procedure detects 90% of the defective items subjected to it but falsely identifies 5% of the good items as defective. What is the probability that an item identified by the inspection process as defective actually is defective?

ANSWER: Let d stand for "item is identified as defective," D stand for "item is defective," and $\bar{D}$ stand for "item is nondefective." Then,

$$P(D|d) = \frac{P(d|D)P(D)}{P(d)} = \frac{P(d|D)P(D)}{P(d|D)P(D) + P(d|\bar{D})P(\bar{D})}$$

$$= \frac{(.90)(.001)}{(.90)(.001) + (.05)(.999)} = \frac{.0009}{.0009 + .04995}$$

$$= \frac{.00090}{.05085} = .0177$$

Suppose that one dog in 10,000 is rabid and that the probability that a dog will bite the postman is .07 if it is rabid but only .0002 if it is not rabid. While delivering the mail, a postman is bitten by a dog which then runs away and cannot be found. What is the probability that the dog is rabid?

ANSWER:

P(rabid dog | bite)

$$= \frac{P(\text{bite} \mid \text{rabid dog})P(\text{rabid dog})}{P(\text{bite})}$$

$$= \frac{P(\text{bite} \mid \text{rabid dog})P(\text{rabid dog})}{P(\text{bite} \mid \text{rabid dog})P(\text{rabid dog}) + P(\text{bite} \mid \text{nonrabid dog})P(\text{nonrabid dog})}$$

$$= \frac{(.07)(.0001)}{(.07)(.0001) + (.0002)(.9999)} = \frac{.000007}{.000007 + .00019998}$$

$$= \frac{.000007}{.00020698} = .0338$$

6-9 The Proportion of Units Having Both of Two Characteristics or the Probability That an Incident Will Do So

As we have already seen, we can obtain alternative formulas for $P(A \& B)$ by multiplying both sides of Formula 6-7 by $P(B)$ or of Formula 6-8 by $P(A)$. Doing so, we obtain the proportion of units having both characteristic A and characteristic B as well as the probability that an incident will have both of these two characteristics:

$$P(A \& B) = P(A \mid B)P(B) \tag{6-11}$$

$$= P(B \mid A)P(A) \tag{6-12}$$

Thus, if 7% of the people in the United States are foreign–born and 23% of the latter were born in English-speaking countries, the proportion of people in the U.S. who were born in English-speaking foreign countries is $P(E \& F) = P(E \mid F)P(F) = .23(.07) = .0161$. And this is also the probability that a randomly selected person in the U.S. would have been born abroad in an English-speaking country.

Two characteristics A and B are **independent** if the *proportion* of units having one characteristic is the same irrespective of whether the reference group is the entire population or is simply the subpopulation of units having the other characteristic. Alternatively, characteristics A and B are independent if the *probability* that an incident will have one characteristic, conditional upon its having the other, is the same as the unconditional probability that it will have the first characteristic. So A and B are independent if $P(A \mid B) =$

$P(A)$. Starting with this formula we can derive several other equivalent conditions under which A and B must therefore be independent. In the following derivation we shall substitute for $P(A \mid B)$ the value given for it in Formula 6-9:

$$P(A \mid B) = P(A)$$

$$\frac{P(B \mid A)P(A)}{P(B)} = P(A)$$

$$P(B \mid A) = P(B)$$

Alternatively, we could have substituted for $P(A \mid B)$ the value given for it in Formula 6-7 as is done in the next derivation.

$$P(A \mid B) = P(A)$$

$$\frac{P(A \ \& \ B)}{P(B)} = P(A)$$

$$P(A \ \& \ B) = P(A)P(B)$$

Furthermore, if the proportion of units having one of the two characteristics is zero, the proportion of units having both characteristics must be zero; so the left side of the above formula will be zero if either of the two terms on the right side is zero and the equation will be satisfied. Therefore, to summarize: A *and* B *are independent characteristics if any of the five following conditions is known to be met:*

$$P(A \mid B) = P(A)$$

$$P(B \mid A) = P(B)$$

$$P(A \ \& \ B) = P(A)P(B)$$

$$P(A) = 0$$

$$P(B) = 0$$

And, the proportion of units having both of two independent characteristics A and B, or the probability that an incident will have both of them, is

$$P(A \ \& \ B \mid A \text{ and } B \text{ are independent}) = P(A)P(B) \qquad (6\text{-}13)$$

Roughly speaking, two characteristics are independent if they are influenced neither by each other nor by a common external factor. Most dependence seems to be attributable to covariation with a common, third variable. For example, the characteristics of getting flat tires on the two rear wheels of your car are influenced by several common factors, including the driver, the terrain covered, etc., and are therefore dependent.

Illustrative Problems: Of the 26 letters in the alphabet, 10 contain curved lines, and of the latter, 2 are vowels. Using Formula 6-11, calculate the proportion of all letters that are curved-line vowels.

ANSWER: $P(V \ \& \ C) = P(V \mid C)P(C) = (2/10)(10/26) = 2/26$

Seventy percent of the people who contract a certain disease succumb to it; the rest recover and become immune to it. The probability, at birth, that one will ever contract the disease is .004. If one survives birth, what is the probability that one's death will be due to the disease?

ANSWER: $P(C \ \& \ S) = P(S \mid C)P(C) = .70(.004) = .0028$

George tosses a fair coin and Mike randomly draws a card from a bridge deck. What is the probability that George's coin comes up heads and Mike's card turns out to be a diamond?

ANSWER: Since the toss of a fair coin cannot influence the outcome of a truly random draw, the outcomes are independent and

$$P(H \ \& \ D) = P(H)P(D) = (1/2)(1/4) = 1/8$$

6-10   The Proportion of Units Having All of Several Characteristics or the Probability That an Incident Will Do So

Using the formula already obtained for the proportion of units having both of two characteristics, we can obtain the more general formula for the proportion of units having all of several characteristics. Suppose that we want the formula for the proportion of units having all of K characteristics $C_1, C_2, \ldots, C_k$. We start with the formula already obtained for the two-characteristic case,

$$P(A \ \& \ B) = P(A)P(B \mid A)$$

Letting $A = C_1$ and $B = C_2$, we obtain

$$P(C_1 \ \& \ C_2) = P(C_1)P(C_2 \mid C_1)$$

Now in the two-characteristic formula a characteristic can be as simple or as complex as we wish. Therefore, let A be the complex characteristic of having *both* characteristics $C_1 \ \& \ C_2$ and let $B = C_3$. Then,

$$P(A \ \& \ B) = P(A)P(B \mid A)$$
$$P(C_1 \ \& \ C_2 \ \& \ C_3) = P(C_1 \ \& \ C_2)P(C_3 \mid C_1 \ \& \ C_2)$$
$$= P(C_1)P(C_2 \mid C_1)P(C_3 \mid C_1 \ \& \ C_2)$$

which gives the required formula for the three-characteristic case. To obtain the formula for the four-characteristic case, let A be the complex characteristic of having the *combination* of characteristics C_1, C_2, and C_3, and let $B = C_4$. Then,

$$P(A \ \& \ B) = P(A)P(B \mid A)$$
$$P(C_1 \ \& \ C_2 \ \& \ C_3 \ \& \ C_4) = P(C_1 \ \& \ C_2 \ \& \ C_3)P(C_4 \mid C_1 \ \& \ C_2 \ \& \ C_3)$$
$$= P(C_1)P(C_2 \mid C_1)P(C_3 \mid C_1 \ \& \ C_2)P(C_4 \mid C_1 \ \& \ C_2 \ \& \ C_3)$$

The pattern is clear. If we always let A stand for the combination of characteristics C_1 & C_2 & ... & C_{k-1} and let B stand for C_k, then we have proved by induction that

$$P(C_1 \text{ \& } C_2 \text{ \& } \ldots \text{ \& } C_k)$$
$$= P(C_1 \text{ \& } C_2 \text{ \& } \ldots \text{ \& } C_{k-1})P(C_k | C_1 \text{ \& } C_2 \text{ \& } \ldots \text{ \& } C_{k-1})$$
$$= \prod_{i=1}^{k} P(C_i | C_1 \text{ \& } C_2 \text{ \& } \ldots \text{ \& } C_{i-1}) \tag{6-14}$$

The above formula is completely general and holds for any number K of characteristics. The following incomplete formula, however, may be easier to work with, using as much of it as is needed,

$$P(A \text{ \& } B \text{ \& } C \text{ \& } D \text{ \& } E \text{ \& } \ldots) =$$
$$P(A)P(B|A)P(C|A \text{ \& } B)P(D|A \text{ \& } B \text{ \& } C)P(E|A \text{ \& } B \text{ \& } C \text{ \& } D)\ldots$$

So if 85% of Americans are white, 40% of the whites have Anglo-Saxon ancestry, and 75% of the whites with Anglo-Saxon ancestry are Protestants, then the proportion of Americans who are white, Anglo-Saxon, Protestants is

$$P(W \text{ \& } AS \text{ \& } P) = P(W)P(AS|W)P(P|W \text{ \& } AS) = .85(.40)(.75) = .255$$

We have already defined independence for two characteristics. Several (two or more) characteristics are **mutually independent** if for *each* characteristic the proportion of units having that characteristic is the same within *every* possible (nonempty) reference group that can be formed from units having *any* combination of the other characteristics and is the same as its proportion in the general population. Thus, *the K characteristics* $C_1, C_2, \ldots,$ C_k *are mutually independent if for every value of* i, $P(C_i) = P(C_i | any combination of the other C's possessed by at least one unit$). So three characteristics A, B, and C (each of which is possessed by at least one unit in the unconditional reference set) are mutually independent if

$$P(A) = P(A|B) = P(A|C) = P(A|B \text{ \& } C)$$

and

$$P(B) = P(B|A) = P(B|C) = P(B|A \text{ \& } C)$$

and

$$P(C) = P(C|A) = P(C|B) = P(C|A \text{ \& } B)$$

The above formulas cannot, without qualification, define mutual independence in the case where one or more characteristics are possessed by no unit in the overall, unconditional reference set. For suppose that no units had characteristic B. Then $P(A|B) = P(A \text{ \& } B)/P(B) = 0/0$, which is undefined and cannot, therefore, be equated to $P(A)$, as required by the formulas. However, if we start with the formula equating $P(A)$ with $P(A|B)$, substitute $P(A \text{ \& } B)/P(B)$ for $P(A|B)$, and then multiply both sides of the resulting formula by $P(B)$, we obtain $P(A)P(B) = P(A \text{ \& } B)$ as an equivalent formula with which to define mutual independence. And this formula *does* hold true when there are no B units, for in that case both $P(B)$ and $P(A \text{ \& } B)$ are zero and the formula becomes $0 = 0$. Proceeding along these lines one can obtain an alternative set of formulas that define mutual independence in all cases without qualification. In the case of three characteristics A, B, and C, the characteristics are mutually independent if

$$P(A \text{ \& } B \text{ \& } C) = P(A)P(B)P(C)$$

and

$$P(A \text{ \& } B) = P(A)P(B)$$

and

$$P(A \text{ \& } C) = P(A)P(C)$$

and

$$P(B \text{ \& } C) = P(B)P(C)$$

In general, several characteristics are mutually independent if for every possible combination of them the proportion of units having that combination is the product of the proportions of units having each of the characteristics in the combination.

Actually, mutual independence is more often assumed than ascertained. If one can see no way in which any of several characteristics would influence each other, or vary systematically with fluctuations in some extraneous factor, it is common to assume that the characteristics are mutually independent. Thus, the last digit in a man's social security number is probably independent of whether or not he has had his tonsils removed. However, there is nothing in probability theory that justifies such an assumption. Rather, its validity depends upon the information, experience, imagination, and judiciousness of the person making the assumption. One should, in fact, be very careful about making such assumptions, since characteristics that seem convincingly independent at first thought often seem much less so upon further reflection.

We have already shown that in the general case of K characteristics C_1, $C_2, \ldots, C_k$,

$$P(C_1 \text{ \& } C_2 \text{ \& } \ldots \text{ \& } C_k) = \prod_{i=1}^{k} P(C_i | C_1 \text{ \& } C_2 \text{ \& } \ldots \text{ \& } C_{i-1})$$

It follows from our definitions that if C_1, C_2, $\ldots$, C_k are mutually independent, then for every value of i,

$$P(C_i | C_1 \text{ \& } C_2 \text{ \& } \ldots \text{ \& } C_{i-1}) = P(C_i)$$

so that

$$P(C_1 \text{ \& } C_2 \text{ \& } \ldots \text{ \& } C_k | \text{all } C_i\text{'s mutually independent}) = \prod_{i=1}^{k} P(C_i) \quad (6\text{-}15)$$

or if the characteristics are A, B, C, D, E, etc.,

$$P(A \text{ \& } B \text{ \& } C \text{ \& } D \text{ \& } E \text{ \& } \ldots | \text{mutual independence})$$
$$= P(A)P(B)P(C)P(D)P(E)\ldots.$$

So the proportion of units having all of several mutually independent characteristics is the product of the proportions of units having each of the individual characteristics. And an analogous statement applies to probabilities.

Illustrative Problems: In a bridge deck, half of the cards are red, 3/13 of the cards are face cards, 2/3 of the face cards depict males; whether a card is a face card or not and whether a face card depicts a male or not are both independent of the color of the card. What proportion of the cards in the deck are red face cards that depict a male?

ANSWER:

$$P(R \text{ \& } F \text{ \& } M) = P(R)P(F|R)P(M|F \text{ \& } R)$$

Because of independence

$$P(F|R) = P(F) \text{ and } P(M|F \text{ \& } R) = P(M|F)$$

So

$$P(R \text{ \& } F \text{ \& } M) = P(R)P(F)P(M|F) = (\tfrac{1}{2})(\tfrac{3}{13})(\tfrac{2}{3}) = \tfrac{1}{13}$$

Urns 1, 2, and 3 contain respectively 3 white and 4 red marbles, 1 white and 2 green marbles, and 1 red and 3 green marbles. A single marble is to be randomly drawn from Urn 1 and placed in Urn 2; then a single marble is to be randomly drawn from Urn 2 and placed in Urn 3; finally, a single marble is to be randomly drawn from Urn 3. What is the probability that the marble drawn from Urn 3 will be white?

ANSWER: Letting capital letters denote outcomes of draws and numerical subscripts denote the urn yielding the outcome, the two outcome sequences $W_1W_2W_3$ and $R_1W_2W_3$ are the only sequences possible if a white marble is to be drawn from Urn 3. These two sequences are mutually exclusive, since the first outcome of the first sequence prevents the second sequence from happening. The probability of obtaining a white marble from Urn 3 is therefore the sum of the probabilities of the two sequences. Now,

$$P(W_1W_2W_3) = P(W_1)P(W_2|W_1)P(W_3|W_1 \text{ \& } W_2)$$
$$= (3/7)(2/4)(1/5) = 3/70$$
$$P(R_1W_2W_3) = P(R_1)P(W_2|R_1)P(W_3|R_1 \text{ \& } W_2)$$
$$= (4/7)(1/4)(1/5) = 1/35$$

So the answer is

$$P(W_1W_2W_3) + P(R_1W_2W_3) = (3/70) + (2/70) = 1/14$$

6-11   Computational Strategems

A unit will either have one or more of the characteristics $C_1, C_2, \ldots, C_k$ or it will have none of them, and it must do one or the other. Therefore,

$$P(C_1 \text{ or } C_2 \text{ or } \ldots \text{ or } C_k, \text{ or any combination thereof})$$
$$+ P(\bar{C}_1 \text{ \& } \bar{C}_2 \text{ \& } \ldots \text{ \& } \bar{C}_k) = 1$$

and

$$P(C_1 \text{ or } C_2 \text{ or } \ldots \text{ or } C_k, \text{ or any combination thereof})$$
$$= 1 - P(\bar{C}_1 \text{ \& } \bar{C}_2 \text{ \& } \ldots \text{ \& } \bar{C}_k)$$

So, invoking Formula 6-14, we obtain

$$P(C_1 \text{ or } C_2 \text{ or } \ldots \text{ or } C_k, \text{ or any combination thereof})$$
$$= 1 - \prod_{i=1}^{k} P(\bar{C}_i | \bar{C}_1 \text{ \& } \bar{C}_2 \text{ \& } \ldots \text{ \& } \bar{C}_{i-1}) \tag{6-16}$$

and if the C_i's (and therefore the $\bar{C}_i$'s) are all mutually independent, it follows from Formula 6-15 that

$$P(C_1 \text{ or } C_2 \text{ or } \ldots \text{ or } C_k, \text{ or any combination}$$
$$\text{thereof} | \text{ all } C_i\text{'s mutually independent}) = 1 - \prod_{i=1}^{k} P(\bar{C}_i) \tag{6-17}$$

These two formulas may be much simpler to apply, in solving problems, than the formulas and methods given in Section 6-6. Roughly speaking, they can be conceptualized as

$$P(\text{any}) = 1 - P(\text{none})$$

Illustrative Problems: A ball is drawn randomly from an urn containing 10 white balls, of which 2 are large and heavy, 1 is large and light, 3 are small and heavy, and 4 are small and light, and 10 black balls of which 4 are large and heavy, 1 is large and light, 2 are small and heavy, and 3 are small and light. What is the probability that the randomly drawn ball will have one or more of the following characteristics: black, heavy, large?

ANSWER: The probability that the ball will *not* be black is 10/20; given that it is not, the probability that it will *not* be heavy, i.e., $P(\text{light}|\text{white})$, is $(1 + 4)/10$; given both circumstances, the probability that it will *not* be large, i.e., $P(\text{small}|\text{white \& light})$, is $4/(1 + 4)$. So

$P(\text{black or heavy or large, or any combination thereof})$

$$= 1 - P(\text{white})P(\text{light}|\text{white})P(\text{small}|\text{white \& light})$$
$$= 1 - (\tfrac{10}{20})(\tfrac{5}{10})(\tfrac{4}{5}) = 1 - \tfrac{4}{20} = \tfrac{16}{20} = .8$$

A game is played in which a fair coin is tossed, a fair die is thrown, and a card is randomly drawn from a bridge deck. If any of the following events occur, the player wins: coin lands heads; die comes up 1, 2, 3, or 4; card drawn is a club, heart, or diamond. What is the probability that the player will win?

ANSWER: The three outcomes are independent; so

$P(\text{coin lands heads, or die comes up 1, 2, 3, or 4, or card is a club, heart, or diamond; or any combination thereof})$

$$= 1 - P(\text{coin lands tails})P(\text{die comes up 5 or 6})P(\text{card is a spade})$$
$$= 1 - (\tfrac{1}{2})(\tfrac{1}{3})(\tfrac{1}{4}) = 1 - \tfrac{1}{24} = \tfrac{23}{24}$$

If, on the average, one passenger out of 200 is a hijacker, what is the probability that a plane carrying 100 passengers, who are all strangers to each other, will contain any hijackers?

ANSWER: There will either be one or more hijackers aboard the plane or there will not. Therefore, P(at least one hijacker) $= 1 - P$(no hijacker). Now in order for there to be no hijacker among the 200 passengers, each of them must *not* be a hijacker. The probability that any randomly selected one of them will not be a hijacker is $1 - (1/200) = 199/200$. And, assuming independence since the passengers are all strangers, the probability that all of them are not hijackers is $(199/200)^{100}$. Therefore,

$$P(\text{any hijackers}) = 1 - P(\text{no hijackers}) = 1 - (\tfrac{199}{200})^{100}$$

6-12   Problems

1.   A single ball is to be randomly drawn from an urn having the contents stated in Table 6-1a. What is the probability that the ball will have the characteristic(s): (a) W or II or both, (b) W or R or both, (c) both B and III, (d) both B and "not I" (i.e., $\bar{\text{I}}$), (e) both "not W" and "not either II or IV," (f) both II and V, (g) R given that it has characteristic III, (h) V given that it has characteristic $\bar{R}$ (i.e., "not red"), (i) $\bar{B}\,|\,$I.

2.   In the situation described above, for one particular numeral the probability that the ball will be stamped with that numeral is independent of the color of the ball. Which numeral is it?

3.   A card is to be randomly drawn from a bridge deck. What is the probability that it will be either a red card or a face card or both?

4.   On days when the stock exchange is open, the probability that the closing prices will be higher than on the previous occasion is .50, and the probability that there will be a meteor shower is .02. What is the probability that on a particular future market day one or the other or both of these events will happen?

5.   A red die and a green die are to be simultaneously thrown. (i) What is the probability that one or both of the following outcomes will occur: (a) a 1 or a 2 comes up on the red die, (b) a 3, 4, 5, or 6 comes up on the green die? (ii) What is the probability that one or both of the following outcomes will occur: (a) a 1 or a 2 comes up on the red die, (b) a 3, 4, 5, or 6 comes up on the red die? Explain why the two answers do or do not differ.

6.   The probability that Harry will play golf next Sunday is .3. The probability that he will play golf and get rained on while doing so next Sunday is .05. What is the probability that he will get rained on at the golf course *if* he plays golf next Sunday?

7. Urns A, B, C, D, and E contain, respectively, 5 white (W) and 2 red (R) marbles, 3 W's and 1 R, no W's and 2 R's, 4 W's and 6 R's, and 2 W's and 3 R's. In *succession* a single marble will be randomly drawn: from Urn A and placed in Urn B, then from B and placed in C, then from C and transferred to D, then from D and dropped into E. Finally, a single marble is to be randomly drawn from Urn E. What is the probability that in each of the five drawings the randomly drawn marble will be white?

8. Give an example of two *completely* independent events. Now reconsider your answer. Is there *any* way in which these events could be even faintly related—could one partially influence the other in any way however farfetched? In actual problems, where the various events have some common context, would you expect true independence to be frequent or rare?

9. Suppose that the probability of success in an attempted suicide: (a) by drowning is .4, (b) by shooting is .7, (c) by poison is .2. A man bent on destroying himself takes poison and then shoots himself as he jumps off a bridge into the bay. Assuming independence, what is the probability that these efforts will result in a successful suicide? Is the independence assumption a reasonable one?

10. If each time it is told or retold, the probability that an event in ancient history is told as it actually happened (in the first instance) or as it was actually said to have happened by the preceding teller is .9, how many times must it be told before the probability is $\leq .5$ that the event is told as it actually happened?

11. A manuscript contains 100 errors. On the nth proofreading, the proportion of errors in the manuscript at that time that are caught and corrected is $4/(5n)$. How many proofreadings are required to reduce the number of errors in the manuscript to less than 5?

12. In a certain group of women, 20% are blonds, 3/4 of the blonds are married, and 5/6 of the unmarrieds are not blond. What percentages of the entire group belong in each of the cells of the following table? Fill in all of the cells and margins with the proper percentages.

	Blond	Not blond	
Married			
Not married			

13. The proportion of men among adults is .45; the proportion of adults who are single is .25; the proportion of single adults who are men is .60. What proportion of men are single?

14. The following table gives the breakdown of a group of 300 women into four nonoverlapping marital status categories and three nonoverlapping eye-color categories. Each of the two sets of categories is exhaustive, and for this group of women marital status and eye color are independent. The table is only partly filled in. Fill the proper numbers into the empty cells and margins of the table.

Marital status

		Single	Married	Widowed	Divorced	All
	Blue					
Eye color	Brown					90
	Other					60
	All	30		90	120	

15. A player will toss a fair coin five times and roll a fair die five times. If in all this tossing and rolling a head ever comes up on the coin or a six ever comes up on the die, the play is stopped and he wins a prize. What is the probability that he will win the prize?

16. In Problem 7, what is the probability that the ball drawn from Urn C (subsequent to the drawings from the preceding urns and transfer to the urns immediately following them) will be red?

17. Two percent of women shoppers are shoplifters. Eighty percent of the women shoplifters carry oversized handbags, whereas only 10% of the nonshoplifting women do. A store detective sees a woman carrying an oversized handbag. What is the probability that she is a shoplifter?

18. Police estimate that two-tenths of one percent of the murders committed in a large city are attributable to the Mad Strangler. They are convinced that the Mad Strangler always uses a silk stocking whereas this modus operandi characterizes only 1% of the murders committed by others. A victim is found strangled with a silk stocking. If the police are correct in their beliefs, what is the probability that the crime was perpetrated by the Mad Strangler?

19. *Halley's problem:* On the average, of 1,000 persons alive at the age of one year, 445 will still be alive at age 40 and 377 will still be alive at age 47 (in Halley's time). What is the probability that a person celebrating his 40th birthday will still be alive at age 47?

20. *Lambert's problem:* n letters have been written and for each letter the corresponding envelope has been addressed. But the n letters are placed in the n envelopes at random. What is the probability that all of the letters are placed in the correct envelopes? What is the probability that all but one are placed in the correct envelopes and one is placed in the wrong envelope?

21. *Birthday problem:* What is the formula for the probability that, in a randomly selected group of n people, at least two people will have birthdays in the same month? How large must n be to make the probability at least 1/2?

22. *Galileo's problem:* In throwing three fair dice, a red die, a green die, and a white die, how many different *combinations* are there of the three numbers on the dice in which the three numbers add up to 9? In which the three numbers add up to 10? How many different *permutations* are there of the three numbers on the three different colored dice (e.g., 4 on red die, 3 on green, 2 on white) in which the three numbers add up to 9? In which they add up to 10? Which has the greater probability, a sum of 9 or a sum of 10? And what are their respective probabilities?

23. Whether or not an adult is hard of hearing is independent of whether or not he is nearsighted. Five percent of adults are hard of hearing and one percent of adults are both hard of hearing and nearsighted. What is the probability that an adult is neither hard of hearing nor nearsighted?

24. Four percent of Swampland mosquitoes carry malaria, and their sting insures the disease. A hunter is stung by 50 Swampland mosquitoes. What is the probability that they will give him malaria?

25. If the probability is .001 that a light bulb will fail during the first ten hours of use, what is the probability that a sign containing 1,000 such bulbs will have no bulb failures in the first ten hours? What is the probability that one particular, specified bulb will fail and the others will not? What is the probability that one *un*specified bulb will fail and the other 999 will not?

26. The cell entries in the tables below give $P(X = x_i$ and $Y = y_j)$. In which of the tables are X and Y independent?

$$
\begin{array}{c}
X \\
\begin{array}{cc} x_1 & x_2 \end{array}
\end{array}
$$

	x_1	x_2
y_1	.32	.23
y_2	.28	.17

(a)

	x_1	x_2
y_1	.32	.08
y_2	.48	.12

(b)

	x_1	x_2
y_1	.58	.22
y_2	.12	.08

(c)

	x_1	x_2
y_1	.21	.19
y_2	.27	.33

(d)

27. On the average, 10 of every 100,000 flights by Whiteknuckle Airlines have a bomb on board, and 400 of every 100,000 flights receive a bomb threat. Eighty percent of the time that a bomb is on board a bomb threat is received. In those cases when a bomb threat is received, what is the probability that a bomb is on board?

28. If we categorize the stock market (closing prices) only as "up" or "down," the probability is .6 that the market category on any given day will be the same as that on the preceding day and .4 that it will be different. If the market is up today, what is the probability that it will be up for the next two days and then down for the three following days?

29. An urn contains four black (B) and three white (W) marbles. If they are drawn randomly and without replacements, what is the probability that they will be drawn in the order $B\,W\,B\,W\,B\,W\,B$?

30. Doc's dog Prince has been his faithful companion ever since Doc delivered him a dozen years ago today. Given the information below, what is the probability that Prince will die at age 15?

Of 100,000 dogs alive at birth, x are still alive at age n years.

n	x	n	x	n	x
1	95,000	8	62,000	15	5,000
2	92,000	9	53,000	16	3,000
3	89,000	10	44,000	17	1,000
4	85,000	11	35,000	18	700
5	80,000	12	25,000	19	300
6	75,000	13	15,000	20	100
7	69,000	14	8,000	21	25

31. Using a six-shooting revolver, an intoxicated cavalry officer plays six games of Russian roulette with himself, spinning the chamber between games. What is the probability that he will survive?

7

specific probability laws

A *specific probability law* is a formula giving the probability that a named sample statistic will have a specified value (or, sometimes, that several statistics will have specified values) when the sample is drawn in a stated way from a population having certain predesignated characteristics. The statistic can have the specified value only if the sample is constituted in certain ways, so the probability law also gives the probability that the sample will be constituted in one of these ways.

7-1   Family of Laws to be Considered and Circumstances under Which They Apply

The specific probability laws that we shall consider all involve some form of the following paradigm: A *random* sample of n observations is to be drawn from a population containing N units each of which belongs to one and only one of K categories, the categories being, in effect, the possible values that an observation may have. Since a unit can belong to only one category, *the categories are mutually exclusive*, and since it *must* belong to one, *the categories are exhaustive*.

There are four major types of sampling situation with which we shall be concerned, depending upon the number of categories and the depletability of the population. The number of categories into which the population units are divided may either be two (in which case they form a *dichotomy*) or more than two. And for each of these cases, the population may either be *depletable* (as it is if N is finite and sampling is without replacements) or undepletable (as it is if there are an infinite number of units in each category or if sampling is with replacements.) If the population is depletable, the size of the population, and therefore the proportion of units belonging to each of its categories, varies as successive sample observations are drawn. However, if the population is undepletable, the proportions of units in each of its categories remain the same from one sample observation to the next. And in that case the characteristic of an observation belonging to a certain category is *independent* of the order of the draw. So the probability that an observation will have that characteristic is the same for each of the n sample observations. This simplifies the derivation of probability laws in these cases.

We may regard the individual observations as subordinate incidents and the entire sample as a superordinate incident. Likewise, we may regard the value of an individual observation as a subordinate characteristic and the composition of the entire sample (or that feature of it whose probability we seek) as a superordinate characteristic to which they contribute. Thus we may derive the specific probability laws from the general probability laws developed in the preceding section. To give a simple conceptual illustration, we may regard the category to which each of its observations belongs as being a separate characteristic of the sample. For the case where $n = 2$ and there are 4 population categories, Table 7-1 gives the probabilities of the various possible sample outcomes in a manner completely analogous to Table 6-2b, which applies to units that have two values. (For the case where $n = 3$, we could construct a table with 16 rows, representing the 16 different possible pairs of outcomes of the first two observations, and 4 columns, representing the 4 different possible outcomes of the third observation, etc.)

We said that a probability law is a formula giving the probability that, under certain prescribed circumstances, a sample statistic will take a particular stated value. If we solve the formula for each possible value that the statistic might take, the values and their probabilities constitute the **probability distribution** of the statistic under the specified circumstances. If the specified circumstances included the requirements that sampling be random and that sample size be a specified constant (the former is always the case and the latter usually is in this book), then the probability distribution of the statistic is the same as its sampling distribution (see Section 3-4). In that case, the probability that the statistic will have a certain value is simply the proportion of samples, in the set of all possible samples, that do so. We may use this fact to check the validity of a derived probability-law formula in those cases where the sampling distribution is easy to construct from the set of all possible samples.

The specific situations to which the probability laws apply are abstract *models* whose conditions are often closely approximated in practice. As abstractions the models may be interpreted in a variety of ways: the units

Table 7-1

Probabilities for Possible Outcomes when a Sample of Two Observations is
Drawn from a Population Whose Units Belong to One of Four Categories
C_1, C_2, C_3, C_4

Value of second observation drawn

		C_1	C_2	C_3	C_4
	C_1	p_{11}	p_{12}	p_{13}	p_{14}
	C_2	p_{21}	p_{22}	p_{23}	p_{24}
	C_3	p_{31}	p_{32}	p_{33}	p_{34}
	C_4	p_{41}	p_{42}	p_{43}	p_{44}

Value of first observation drawn

$p_{ij} = P$ (first observation belongs to category C_i and
second observation belongs to category C_j)

comprising a population may be a set of objects from which a sample is
literally removed or they may be the set of all outcomes of a random variable
or the set of all possible similar incidents. As analogues of practical situations, the better the fit the more accurate will be the probabilities obtained.
The more nearly random selection is (rather than simply haphazard) the
better. Also, *the model's specification that population categories are mutually
exclusive and exhaustive means in practice that categories must not overlap
and that all categories in which one is interested should be included in conceptually setting up one's population.*

7-2   The Hypergeometric Probability Law

The hypergeometric probability law applies to the following situation.
A *random* sample of n observations is to be drawn *without replacements* from
a *finite* population. The population contains N units, each of which belongs

to one or the other of two *mutually exclusive* and *exhaustive* categories, which therefore constitute a *dichotomy*. For convenience, we shall arbitrarily label the categories "success" and "failure," so the possible values of a unit are S and F where, of course, $F = \bar{S}$. Of the N population units, R are successes and $N - R$ are failures. Under these circumstances the probability that the sample will contain exactly r successes, no more and no less, is

$$P(r) = \frac{\binom{R}{r}\binom{N-R}{n-r}}{\binom{N}{n}} \tag{7-1}$$

Proof: Since we are not interested in the order in which certain observation-values are obtained, nor in which observations have which values, we may regard our sample as an unordered subset (i.e., a combination) of n units removed from N units. There are

$$\binom{N}{n}$$

different possible unordered subsets that could become our sample, and, since sampling is random, each of them is equally likely (actually, each of the $_NP_n$ *ordered* subsets is equally likely, but since the same number $n!$ of ordered subsets can be created from each *un*ordered subset, the latter must also be equally likely—they are molecular outcomes composed of the same number of atomic ones). The sample *will* be drawn, so the probability we seek is simply the proportion, among the

$$\binom{N}{n}$$

equally likely samples, of samples that contain exactly r successes and $n - r$ failures. Now we may regard the latter type of sample as an unordered subset of r successes obtained from R successes combined with another unordered subset of $n - r$ failures obtained from $N - R$ failures. There are

$$\binom{R}{r}$$

different ways of obtaining the former, and for each of these ways, there are

$$\binom{N-R}{n-r}$$

different ways of obtaining the latter. So there are

$$\binom{R}{r}\binom{N-R}{n-r}$$

different possible unordered samples of n observations containing r successes and $n - r$ failures. And these samples must be members of the unqualified set of

$$\binom{N}{n}$$

samples containing n observations. So the proportion of the

$$\binom{N}{n}$$

samples of n observations that happen to contain exactly r successes is

$$\frac{\binom{R}{r}\binom{N-R}{n-r}}{\binom{N}{n}}$$

and this is, therefore, the probability that a sample of n observations will contain exactly r successes. Notice that this probability is the ratio of the number of favorable ways of drawing the sample (i.e., the number of ways of drawing a sample of size n containing exactly r successes) to the number of ways of drawing the sample (i.e., the number of ways of drawing a sample of size n without further restrictions), where "ways" are equally likely molecular outcomes. It is therefore a valid application of a priori probability.

Example: If a random sample of three balls is to be drawn without replacements from an urn containing five balls of which three are white and two are black, the probability that the number r of white balls in the sample will be exactly 2 is

$$P(r=2) = \frac{\binom{R}{r}\binom{N-R}{n-r}}{\binom{N}{n}} = \frac{\binom{3}{2}\binom{2}{1}}{\binom{5}{3}} = \frac{3 \times 2}{10} = .6$$

This can be easily verified. Let W_1, W_2, W_3 be the white and B_4, B_5 be the black balls in the urn. Then the following list gives all of the different possible unordered samples of three balls that can be drawn without replacements from the urn:

$W_1W_2W_3$	$W_1W_3B_4$	$W_1B_4B_5$
$W_1W_2B_4$	$W_1W_3B_5$	$W_2B_4B_5$
$W_1W_2B_5$	$W_2W_3B_4$	$W_3B_4B_5$
	$W_2W_3B_5$	

Six of the ten possible unordered samples contain exactly two white balls. By listing all $3!$ permutations of each of the above unordered samples, we would obtain a list of all possible *ordered* samples of three observations (which, because sampling is random, are equally likely) and the list would contain $3!(10) = 60$ samples of which $3!(6) = 36$ contained exactly two white balls. And in either list the proportion of samples containing exactly two white balls is .6, the probability obtained by use of the formula.

Illustrative Problems: A chef opens a carton containing a dozen eggs, randomly chooses three of them and uses them to make an omelet. If two of the twelve eggs in the carton are rotten, and if the chef has a cold so that he couldn't detect a bad egg, what is the probability that the omelet has any bad eggs in it?

ANSWER: The omelet either contains no bad eggs or some bad eggs—which are complementary characteristics. So applying Formulas 6-3 and 6-16,

$$P(\text{any bad eggs}) = 1 - P(\text{no bad eggs}) = 1 - \frac{\binom{2}{0}\binom{10}{3}}{\binom{12}{3}}$$

$$= 1 - \frac{\dfrac{(1)10!}{3!7!}}{\dfrac{12!}{3!9!}} = 1 - \frac{10!9!}{12!7!}$$

$$= 1 - \frac{9 \times 8}{12 \times 11} = 1 - \frac{6}{11} = \frac{5}{11}$$

A scientist's assistant has performed some computations resulting in a data sheet containing 1,000 independent entries of which 20 are in error. The scientist "spot checks" by randomly selecting 5 of the entries and completely reperforming their computations. If the scientist's calculations are correct, what is the probability that the scientist will find no errors?

ANSWER: Let r be the number of errors he will find. Then

$$P(r = 0) = \frac{\binom{20}{0}\binom{980}{5}}{\binom{1,000}{5}} = \frac{\left[\dfrac{20!}{0!20!}\right]\left[\dfrac{980!}{5!975!}\right]}{\dfrac{1000!}{5!995!}} = \frac{995!980!}{1000!975!}$$

$$= \frac{(980)(979)(978)(977)(976)}{(1000)(999)(998)(997)(996)} = .904$$

7-3 The Multivariate Hypergeometric Probability Law

The multivariate hypergeometric probability law is a generalization of the hypergeometric probability law to the case where the sampled population contains more than two kinds of units. Specifically, it applies to the following situation. As before, a *random* sample of n observations is to be drawn *without replacements* from a *finite* population. The population contains N units each of which belongs to a single one of K different *mutually exclusive* and *exhaustive* categories. Of the N population units, R_1 belong to the first category, R_2 to the second, R_k to the Kth, and in general, R_i to the ith, so $\sum_{i=1}^{k} R_i = N$. Under these circumstances, the probability that of the n sample observations r_1 will belong to the first category, r_2 to the second, ..., r_i to the ith, ..., and r_k to the Kth (where, of course, $\sum_{i=1}^{k} r_i = n$) is

$$P(r_1, r_2, \ldots, r_k) = \frac{\binom{R_1}{r_1}\binom{R_2}{r_2}\cdots\binom{R_k}{r_k}}{\binom{N}{n}} = \frac{\prod_{i=1}^{k}\binom{R_i}{r_i}}{\binom{N}{n}} \qquad (7\text{-}2)$$

Proof: The proof is completely analogous to that for the simple hypergeometric case. There are

$$\binom{R_1}{r_1}$$

ways of obtaining the sample combination of r_1 units of the first type from the R_1 population units of the first type. For each of these ways, there are

$$\binom{R_2}{r_2}$$

ways of obtaining the combination of r_2 units of the second type from the population units of that type. For each of these ways, there are

$$\binom{R_3}{r_3}$$

ways of obtaining the combination of r_3 units for the sample from the R_3 units available in the population, etc. So there are

$$\binom{R_1}{r_1}\binom{R_2}{r_2}\cdots\binom{R_k}{r_k}$$

different unordered subsets of n units, of which r_1 are of type 1, r_2 of type 2, ..., and r_k of type K, that can be obtained from the population. And these samples belong to the group of

$$\binom{N}{n}$$

different possible equally likely unordered subsets of n units that can be drawn from the N units. So the proportion of the former within the latter is

$$\frac{\binom{R_1}{r_1}\binom{R_2}{r_2}\cdots\binom{R_k}{r_k}}{\binom{N}{n}}$$

which is, therefore, the probability that the sample will contain the specified numbers of units of each of the various types.

Example: If a random sample of three balls is to be drawn without replacements from an urn containing five balls of which three are white, one is black, and one is green, the probability that the sample will contain exactly two white, one black and zero green balls is

$$P(r_W = 2, r_B = 1, r_G = 0) = \frac{\binom{R_W}{r_W}\binom{R_B}{r_B}\binom{R_G}{r_G}}{\binom{N}{n}} = \frac{\binom{3}{2}\binom{1}{1}\binom{1}{0}}{\binom{5}{3}} = \frac{3}{10} = .3$$

To illustrate and verify this, let W_1, W_2, W_3, B, G be the three white, the black, and the green balls in the urn. Then all possible unordered samples of three balls that can be drawn without replacements are given in the following list:

$$
\begin{array}{lll}
W_1 W_2 W_3 & W_1 W_3 B & W_1 BG \\
W_1 W_2 B & W_1 W_3 G & W_2 BG \\
W_1 W_2 G & W_2 W_3 B & W_3 BG \\
 & W_2 W_3 G &
\end{array}
$$

Three of the ten possible unordered samples [or $3!(3) = 18$ of the $3!(10) = 60$ possible *ordered* samples] consist of two white and one black balls; so the proportion is $3/10 = .3$ [or $18/60 = .3$], and since each of the possible unordered [ordered] samples is equally likely, .3 is the probability sought. Notice, however, that if we had asked only for the probability that the number of white balls in the sample would be 2, *without specifying* the *particular* nonwhite color of the other ball, we would have obtained the same probability .6 obtained in the example for the hypergeometric probability law.

Illustrative Problems: In a pool of 12 potential subjects available for use in experiments, 3 of the subjects are of inferior intelligence, 5 are of average intelligence, and 4 are of superior intelligence. A random sample of 7 subjects is to be withdrawn from the pool to participate in an experiment. What is the probability that the sample will contain at least 2 subjects in each of the 3 categories, thereby being roughly representative of the pool?

ANSWER: There are only three ways of dividing the number 7 into three numbers each of which is ≥ 2. So we are interested in a sample constituted in one of the following three ways: $r_1 = 2, r_2 = 2, r_3 = 3$; or $r_1 = 2, r_2 = 3$, $r_3 = 2$; or $r_1 = 3, r_2 = 2, r_3 = 2$. Samples constituted in different ways are mutually exclusive, i.e., if a sample is constituted in one way, then it is *not* constituted in any of the others. Therefore, invoking Formulas 6-6 and 7-2, we obtain

$$
\begin{aligned}
P(r_1 \geq 2, r_2 \geq 2, r_3 \geq 2) = {}& P(r_1 = 2, r_2 = 2, r_3 = 3) \\
& + P(r_1 = 2, r_2 = 3, r_3 = 2) \\
& + P(r_1 = 3, r_2 = 2, r_3 = 2)
\end{aligned}
$$

$$= \frac{\binom{3}{2}\binom{5}{2}\binom{4}{3}}{\binom{12}{7}} + \frac{\binom{3}{2}\binom{5}{3}\binom{4}{2}}{\binom{12}{7}}$$

$$+ \frac{\binom{3}{3}\binom{5}{2}\binom{4}{2}}{\binom{12}{7}}$$

$$= \frac{[(3)(10)(4)] + [(3)(10)(6)] + [(1)(10)(6)]}{\binom{12}{7}}$$

$$= \frac{120 + 180 + 60}{792} = \frac{360}{792} = .4545$$

Three of the rats in a colony of 12 laboratory rats are presently being used in an experiment. Disease strikes the colony and kills 3, sickens 4, and leaves 5 rats unaffected. What is the probability that at least 2 of the rats being experimented upon are either killed or sickened? Work this two ways.

ANSWER:

$P(K + S \geq 2)$

$\quad = P(K = 3, S = 0, U = 0) + P(K = 2, S = 1, U = 0)$

$\quad + P(K = 2, S = 0, U = 1) + P(K = 1, S = 2, U = 0)$

$\quad + P(K = 1, S = 1, U = 1) + P(K = 0, S = 3, U = 0)$

$\quad + P(K = 0, S = 2, U = 1)$

$\quad = \dfrac{\binom{3}{3}\binom{4}{0}\binom{5}{0} + \binom{3}{2}\binom{4}{1}\binom{5}{0} + \binom{3}{2}\binom{4}{0}\binom{5}{1} + \binom{3}{1}\binom{4}{2}\binom{5}{0} + \binom{3}{1}\binom{4}{1}\binom{5}{1} + \binom{3}{0}\binom{4}{3}\binom{5}{0} + \binom{3}{0}\binom{4}{2}\binom{5}{1}}{\binom{12}{3}}$

$\quad = \dfrac{1 + 12 + 15 + 18 + 60 + 4 + 30}{\binom{12}{3}} = \dfrac{140}{220} = \dfrac{7}{11}$

Alternatively, the problem permits us to treat "killed" and "sickened" as a single category so that we have only two categories, "affected" and "unaffected," and the hypergeometric probability law applies. In the colony 7 rats were affected and 5 were unaffected. So

$$P(A \geq 2) = P(A = 3) + P(A = 2)$$

$$= \frac{\binom{7}{3}\binom{5}{0} + \binom{7}{2}\binom{5}{1}}{\binom{12}{3}} = \frac{(35)(1) + (21)(5)}{\binom{12}{3}}$$

$$= \frac{140}{220} = \frac{7}{11}$$

7-4   The Binomial Probability Law

The binomial probability law applies to the following situation. A *random sample* of n observations is to be drawn *either* from an *infinite* population or *with replacements* from a *finite* population, so the population is not depleted by sampling and the outcomes of the individual observations are *independent* of each other. Every unit in the population belongs to one or the other of two *mutually exclusive* and *exhaustive* categories, which therefore constitute a *dichotomy*. The proportion of units in the population belonging to one category, arbitrarily labeled "success" is p; so the proportion belonging to the other, arbitrarily called "failure," is $1 - p$. Under these conditions, the probability that the sample will contain exactly r successes, no more and no less, is

$$P(r) = \binom{n}{r} p^r (1 - p)^{n-r} \tag{7-3}$$

Proof: Since the sampled population is undepletable, the proportions p and $1 - p$ of successes and of failures in it remain constant as the n sample observations are drawn. Therefore, the probabilities that a given sample observation will be a success or will be a failure are *independent* of the point in sequence at which the observation was drawn and are p and $1 - p$, respectively. Let O_i stand for the value of the ith sample observation in the sequence in which the n observations were drawn. Then, from what has already been said, it follows that $P(O_i = S) = p$ and $P(O_i = F) = 1 - p$. Now consider the probability that r successes and $n - r$ failures will occur in the sample in a completely specified sequence. Because the various outcomes are independent (see Formula 6-15), the probability that all of the r successes will occur first, followed by the $n - r$ failures is:

$$P(O_1 = S \ \& \ O_2 = S \ \& \ldots \& \ O_r = S \ \& \ O_{r+1} = F \ \& \ \ldots \& \ O_n = F)$$
$$= P(O_1 = S)P(O_2 = S) \cdots P(O_r = S)P(O_{r+1} = F) \cdots P(O_n = F)$$
$$= pp \cdots p(1 - p) \cdots (1 - p)$$
$$= p^r(1 - p)^{n-r}$$

and for any other sequence containing r successes and $n - r$ failures we would obtain the same end result. Now there are as many different possible sequences of r successes and $n - r$ failures as there are distinguishable arrangements of r S's and $n - r$ F's within n fixed positions, i.e., as there are distinguishable permutations of n things r of which are of one indistinguishable kind and $n - r$ of which are of another. So there are

$$\frac{n!}{r!(n - r)!} = \binom{n}{r}$$

(see Formula 3-3 and Sections 3-1-3 and 3-2-3) different possible sequences in which the r successes and $n - r$ failures could occur. The probability we seek is the probability that some one of these sequences will occur. The S's and F's in the actually obtained sample can occur in only one distinguishable sequence, so the

$$\binom{n}{r}$$

distinguishable sequences are mutually exclusive, and the probability that some one of them will occur is the sum of their individual probabilities of occurrence (see Formula 6-6). But each of these individual probabilities has the same value $p^r(1 - p)^{n-r}$. So the sum of the

$$\binom{n}{r}$$

individual probabilities is

$$\binom{n}{r} p^r(1 - p)^{n-r}$$

which is the probability that some sequence of r successes and $n - r$ failures will occur and, therefore, that the sample will contain exactly r successes, no more and no less.

Example: If a random sample of two balls is to be drawn with replacements from an urn containing four balls of which three are white and one is black, the probability that the number r of white balls in the sample will be exactly 1 is

$$P(r = 1) = \binom{n}{r} p^r(1 - p)^{n-r} = \binom{2}{1}\left(\frac{3}{4}\right)^1\left(\frac{1}{4}\right)^1 = (2)\left(\frac{3}{4}\right)\left(\frac{1}{4}\right) = \frac{6}{16}$$

To verify this, let W_1, W_2, W_3 be the white balls and B be the black ball in the urn. Then the following list gives all of the different possible ordered samples of two balls that could be drawn with replacements from the urn:

W_1W_1	W_2W_1	W_3W_1	BW_1
W_1W_2	W_2W_2	W_3W_2	BW_2
W_1W_3	W_2W_3	W_3W_3	BW_3
W_1B	W_2B	W_3B	BB

There are 16 different possible ordered samples, which are equally likely because sampling is random, and 6 of them contain just 1 white ball; so the proportion of samples containing 1 white ball among all equally likely samples of the specified size, etc., is 6/16, the probability obtained by use of the formula.

An important special case, meeting the conditions under which the binomial probability law applies, is that of a random variable whose successive dichotomous outcomes are independent of each other. Suppose that the occurrences of a random variable have outcomes that can be dichotomized as either successes or nonsuccesses. Suppose further that the probability of an occurrence being a success is unaffected by the passage of time, the number of preceding occurrences, or the outcomes (success or nonsuccess) of actual preceding occurrences (such as the last one before it). Then we may regard all of the occurrences in a sequence of n future occurrences as belonging to a *single* potential (and therefore *infinite*) population of occurrences (so that we are dealing with a single p-value). Furthermore, any *predesignated n* future occurrences (such as the "next" n occurrences) of a random variable may be regarded as a *random sample* of n observations drawn from the infinite population of all potential occurrences of the random variable. Therefore, provided that we do not know their outcomes, we may regard any *non*randomly selected n future occurrences of a dichotomous random variable, a proportion p of whose "infinite" past occurrences were successes, as a *random* sample of n observations drawn from a single infinite dichotomous population a proportion p of whose units are successes. And the probability that r of these n future occurrences of the random variable will be successes is given by the binomial probability law.

Illustrative Problems: An expert marksman hits the bull's-eye in 75% of his shots at the target. In a competition he will fire five shots, separated by rest pauses, and will not be given knowledge of results until after the fifth shot. What is the probability that exactly four of them will be bull's-eyes?

ANSWER: The shots may be regarded as occurrences of a random variable and the outcomes may be dichotomized into hitting and not hitting the bull's-eye. Therefore, if outcomes are independent so that $p = .75$ for each of the five shots (as we would expect from the conditions stated in the problem—sequential physiological and psychological effects, such as fatigue and knowledge of preceding outcome, have been eliminated or minimized), the answer is

$$P(r = 4) = \binom{5}{4}\left(\frac{3}{4}\right)^4\left(\frac{1}{4}\right)^1 = 5\left(\frac{81}{256}\right)\left(\frac{1}{4}\right) = \frac{405}{1{,}024}$$

Ten percent of the items produced by a certain manufacturing process are defective. Three items will be randomly selected from those produced tomorrow and shipped to a retailer. What is the probability that the shipment will contain no defectives?

ANSWER: Presumably the process is a stable one in which the probability of a defective remains constant over time. Therefore, we may regard the items as units in an infinite, mostly future (and therefore potential) population in which the proportion of defectives is $p = .1$. The answer is therefore

$$P(r = 0) = \binom{3}{0}.1^0.9^3 = (1)(1)(.729) = .729$$

A warehouse contains 10,000 items, of which a randomly located 1,000, or 10%, are defective. Three items are to be selected from the warehouse and shipped to a retailer. What is the probability that the shipment will contain no defectives?

ANSWER: The *exact* probability is given by the hypergeometric probability law,

$$P(r = 0) = \frac{\binom{1,000}{0}\binom{9,000}{3}}{\binom{10,000}{3}} = \frac{\frac{9,000!}{3!\,8997!}}{\frac{10,000!}{3!\,9997!}} = \frac{9,000 \times 8,999 \times 8,998}{10,000 \times 9,999 \times 9,998}$$

$$= .72898$$

However, since the population is very large compared to the size of the sample, we can regard it as effectively infinite and use the binomial probability law to obtain a close *approximation* to the exact probability:

$$P(r = 0) = \binom{3}{0}.1^0.9^3 = (1)(1)(.729) = .729$$

7-5 The Multinomial Probability Law

The multinomial probability law is a generalization of the binomial probability law to the case where the sampled population contains more than two kinds of units. It applies to the following situation. As before, a *random* sample of n observations is to be drawn *either* from an *infinite* population or *with replacements* from a *finite* population; so the population is not depleted by sampling and the outcomes of the individual observations are *independent* of each other. Every unit in the population belongs to one of K *mutually exclusive* and *exhaustive* categories. The proportion of units in the population that belong to the first category is p_1, to the second is p_2, to the Kth is p_k, and, in general, to the ith is p_i (where $\sum_{i=1}^{k} p_i = 1$). Under these circumstances, the probability that of the n sample observations r_1 will belong to the first category, r_2 to the second, ..., r_i to the ith, ..., and r_k to the Kth (where $\sum_{i=1}^{k} r_i = n$) is

$$P(r_1, r_2, \ldots, r_k) = \frac{n!}{r_1!\,r_2!\cdots r_k!}p_1^{r_1}p_2^{r_2}\cdots p_k^{r_k} = n!\prod_{i=1}^{k}\frac{p_i^{r_i}}{r_i!} \qquad (7\text{-}4)$$

Proof: The proof is analogous to that for the binomial probability law. Because of independence, the probability of the particular sequence of observations in which all outcomes of a given type occur together in a block, the block of r_1 outcomes of type 1 occurring first, the block of r_2 outcomes of type 2 occurring second, and in general, the block of r_i outcomes of type i occurring ith is

$$p_1^{r_1}p_2^{r_2}\cdots p_i^{r_i}\cdots p_k^{r_k} = \prod_{i=1}^{k} p_i^{r_i}$$

and the same end result would be obtained for any other sequence containing the specified numbers r_i of outcomes of each of the K types. The number of different possible sequences is simply the number of distinguishable permutations of n things of which r_1 are of type 1, r_2 are of type 2, ..., r_i are of type i, ..., and r_k are of type K, where the r_i things of a given type are distinguishable from things of any other type but indistinguishable from each other, and that number (see Formula 3-3) is

$$\frac{n!}{r_1! r_2! \cdots r_i! \cdots r_k!} = \frac{n!}{\prod_{i=1}^{k} r_i!}$$

Since the obtained sample can have only one sequence, the various possible sequences are mutually exclusive. Therefore, the probability that the obtained sample sequence will be one of them is (see Formula 6-6) the sum of the probabilities for the $n!/\prod_{i=1}^{k} r_i!$ different possible sequences. And since they all have the same probability, the sum of their probabilities is

$$\frac{n!}{r_1! r_2! \cdots r_i! \cdots r_k!} p_1^{r_1} p_2^{r_2} \cdots p_i^{r_i} \cdots p_k^{r_k} = n! \prod_{i=1}^{k} \frac{p_i^{r_i}}{r_i!}$$

which is therefore the probability that the sample will contain the specified numbers r_i of observations of each type.

Example: If a random sample of two balls is to be drawn with replacements from an urn containing four balls of which two are white, one is black, and one is green, the probability that the sample will contain one white, one black, and no green balls is

$$P(r_1 = 1, r_2 = 1, r_3 = 0) = \frac{n!}{r_1! r_2! r_3!} p_1^{r_1} p_2^{r_2} p_3^{r_3} = \frac{2!}{1! 1! 0!} \left(\frac{2}{4}\right)^1 \left(\frac{1}{4}\right)^1 \left(\frac{1}{4}\right)^0$$

$$= \frac{4}{16}$$

To verify this, let W_1, W_2 be the white, B the black, and G the green balls in the urn. Then all of the different possible ordered samples of two balls that could be drawn, with replacements, from the urn are given in the following list:

$W_1 W_1$	$W_2 W_1$	$B W_1$	$G W_1$
$W_1 W_2$	$W_2 W_2$	$B W_2$	$G W_2$
$W_1 B$	$W_2 B$	BB	GB
$W_1 G$	$W_2 G$	BG	GG

Since sampling is random, each of the 16 possible samples is equally likely. And since exactly 4 of these equally likely samples contain 1 white and 1 black ball, the probability that the random sample will contain 1 white and 1 black ball is 4/16, the value obtained by the use of the formula.

An important special case, meeting the conditions under which the multi-nomial probability law applies, is that of a random variable whose successive outcomes are independent of each other and belong to one of K mutually exclusive and exhaustive categories. The reasons for this are similar to those given for the analogous special case to which the binomial probability law applies.

Illustrative Problems: An expert marksman hits the bull's-eye in 75% of his shots at the target, hits the target but not the bull's-eye in 24% of his shots, and misses the target entirely in 1% of his shots. If he fires five independent shots, what is the probability that exactly four will hit the bull's-eye and one will miss the target entirely?

ANSWER:

$$P(r_1 = 4, r_2 = 0, r_3 = 1) = \frac{5!}{4!0!1!}(.75)^4(.24)^0(.01)^1 = (5)(.3164)(1)(.01)$$

$$= .01582$$

This probability is different from the probability that exactly four shots will hit the bull's-eye (see first illustrative problem for the binomial probability law) because the latter has a single nonbull's-eye category whereas the former specifies exactly how many shots are to fall into each of two nonbull's-eye categories. Since the former probability applies to a more elaborately specified situation (which is therefore less likely to be encountered), the former probability is the smaller of the two.

Ten percent of the items produced by a certain manufacturing process are defective. Of the defectives, three-tenths are damaged beyond repair and are worthless, while seven-tenths are repairable. What is the probability that a random sample of three items will contain no unrepairable defectives, no repairable defectives, and three good items?

ANSWER:

$$P(r_1 = 0, r_2 = 0, r_3 = 3) = \frac{3!}{0!0!3!}(.03)^0(.07)^0(.90)^3 = (1)(1)(1)(.729)$$

$$= .729$$

This probability is the same as the probability that the sample will contain no defectives (see second illustrative problem for the binomial probability law) because the only way that zero defectives can be distributed to the categories "unrepairable defective" and "repairable defective" is for there to be zero units in each of the two categories. So the two cases, binomial and multinomial, are logically equivalent. Notice, however, that there is *more* than one way to distribute any *non*zero number of defectives to the two categories. So the multinomial probability that there will be x unrepairable defectives, y repairable defectives, and z good items can be expected to differ from the binomial probability that there will be $x + y$ defectives and z good items whenever $x + y \neq 0$.

A warehouse contains 10,000 items, of which 1,000 or 10% are defective, 7,000 or 70% are in good condition, and 2,000 or 20% are in excellent condition. What is the probability that a random sample of three items will contain no defectives, one good, and two excellent items?

ANSWER: The *exact* probability is given by the multivariate hypergeometric law,

$$P(r_1 = 0, r_2 = 1, r_3 = 2) = \frac{\binom{1,000}{0}\binom{7,000}{1}\binom{2,000}{2}}{\binom{10,000}{3}}$$

$$= \frac{(7,000)(2,000)(1,999)(3)(2)(1)}{(10,000)(9,999)(9,998)(2)(1)} = .083983$$

But, since the population is very much larger than the sample, we can *approximate* the answer using the multinomial probability law,

$$P(r_1 = 0, r_2 = 1, r_3 = 2) = \frac{3!}{0!\,1!\,2!}.1^0.7^1.2^2 = (3)(1)(.7)(.04) = .084$$

This probability differs from the probability that the sample will contain no defectives (see third illustrative example for the binomial probability law) since it specifies how the nondefectives are to be distributed to two non-defective categories.

7-6   Relationships between Preceding Specific Laws; and Some Special Features

All of the preceding probability laws may be thought of as special cases of the multivariate hypergeometric probability law, those other than the multivariate hypergeometric probability law itself being obtained when $K = 2$ or $N \longrightarrow \infty$ (and all $R_i \longrightarrow \infty$) or both. The number of categories K can equal 2 because there are only two kinds of units in the population (e.g., white balls and black balls) or because, although the population contains a variety of categories (e.g., white, red, green, and black balls), one is interested in only one kind (e.g., white) and therefore has recategorized the population into units that are of that kind and those that are not (e.g., white and "not white," the latter being any of the nonwhite colors red, green, or black— in this case, however, the various nonwhite colors need *not* be mutually exclusive with *each other*, but only with white).

By deriving the multivariate hypergeometric probability law in a different way, we can show that the multinomial distribution is a special case of it that occurs when N and each of the R_i's approach infinity (and therefore that the binomial is a special case of the hypergeometric that occurs when N, R, and $N - R$ all approach infinity). When the jth sample observation is drawn, the population has already been depleted by $j - 1$ units and therefore contains $N - (j - 1)$ units. Suppose that the already-removed $j - 1$ units include x units from the ith population category so that the ith population category has been depleted by x units. Then the probability that the jth sample observation will belong to the ith population category is $(R_i - x)/(N - j + 1)$. Thus, if the population consists of $R_1 = 1,000$ white balls, $R_2 = 2,000$ black balls, and $R_3 = 3,000$ green balls, the probability of obtaining the sequence $W\ B\ B\ W\ G\ B\ B$ is

$$\frac{1,000}{6,000} \times \frac{2,000}{5,999} \times \frac{1,999}{5,998} \times \frac{999}{5,997} \times \frac{3,000}{5,996} \times \frac{1,998}{5,995} \times \frac{1,997}{5,994}$$

$$\cong \frac{1,000}{6,000} \times \frac{2,000}{6,000} \times \frac{2,000}{6,000} \times \frac{1,000}{6,000} \times \frac{3,000}{6,000} \times \frac{2,000}{6,000} \times \frac{2,000}{6,000}$$

$$\cong \frac{R_1}{N} \times \frac{R_2}{N} \times \frac{R_2}{N} \times \frac{R_1}{N} \times \frac{R_3}{N} \times \frac{R_2}{N} \times \frac{R_2}{N} = \left(\frac{R_1}{N}\right)^{r_1}\left(\frac{R_2}{N}\right)^{r_2}\left(\frac{R_3}{N}\right)^{r_3}$$

$$\cong p_1^{r_1} p_2^{r_2} p_3^{r_3}$$

where p_i is the proportion R_i/N of balls of the ith type originally in the population. But this is the approximate probability of obtaining the r_1 observations of type 1, the r_2 of type 2, and the r_3 of type 3 in a specified sequential order. There are as many different orders as there are distinguishable permutations of $n = \sum_{i=1}^{3} r_i$ things of which r_i are of type i and are distinguishable from things of a different type but not from each other, the number of which is $n!/(r_1!r_2!r_3!)$, and since a single sample can be obtained in only one order, the different orders are mutually exclusive. Therefore, the probability of obtaining a sample containing specified numbers r_i's of observations in each of K categories but obtained in any, unspecified, order is the sum of the probabilities associated with each individual order

$$P(r_1, r_2, r_3) \cong \frac{n!}{r_1!r_2!r_3!} p_1^{r_1} p_2^{r_2} p_3^{r_3}$$

So, by substituting the approximation $p_i = R_i/N$ for the correct value $(R_i - x)/(N - j + 1)$, we have obtained the multinomial probability law as an approximation to the multivariate hypergeometric probability law. Since x and j are both depletions due to drawing the sample observations, they both must be $\leq n$. So if N and each R_i become "infinitely" large relative to n, $(R_i - x)/(N - j + 1)$ must differ negligibly from R_i/N, the value substituted for it.

If we have K population categories and we specify exactly how many observations in a sample of stated size n are to belong to each of $K - 1$ of the K categories, we have, in effect, specified the composition of the entire sample. For the number of sample observations belonging to the remaining category is already determined, being the difference between the sample size n and the sum of the specified numbers of observations for the $K - 1$ categories, e.g., $r_k = n - \sum_{i=1}^{k-1} r_i$. Therefore, although we speak of the hypergeometric and binomial laws as giving us the probability of r successes, what they really give us is the probability of r successes *and* $n - r$ nonsuccesses (failures), and this is explicitly brought out in their derivation. Likewise, the multivariate hypergeometric and multinomial laws really give us the probability that each of the K r_i's will have a specified value, although we need actually specify only $K - 1$ of the r_i's and the size n of the sample.

Now if we wish to obtain a probability where more than one of the K r_i's is left unspecified, we have two options. The generally less attractive option is to solve the probability formula once for each of the different possible sets of values the *un*specified r_i's could have, while giving the other r_i's their specified values, and then sum all of these probabilities to obtain the probability required. By far the more efficient method is to combine all of the mutually exclusive categories whose corresponding r_i's are unspecified into a *single* category, which may be appropriately labeled "other," and then solve the probability formula appropriate to the reduced-category situation. For example, suppose that a sample of $n = 5$ balls is to be drawn without replacements from a population containing $N = 10$ balls, of which one is red, four are yellow, three are green, and two are blue. Then the probability that the sample will contain no red balls and three yellow balls, the number of green and blue balls being left unspecified, is

$$P(r_1 = 0, r_2 = 3) = \sum_{x=0}^{2} P(r_1 = 0, r_2 = 3, r_3 = x, r_4 = 2 - x)$$

$$= \sum_{x=0}^{2} \frac{\binom{1}{0}\binom{4}{3}\binom{3}{x}\binom{2}{2-x}}{\binom{10}{5}}$$

$$= \frac{\binom{1}{0}\binom{4}{3}\binom{3}{0}\binom{2}{2} + \binom{1}{0}\binom{4}{3}\binom{3}{1}\binom{2}{1} + \binom{1}{0}\binom{4}{3}\binom{3}{2}\binom{2}{0}}{\binom{10}{5}}$$

$$= \frac{4 + 24 + 12}{252} = \frac{40}{252}$$

Or we can combine the green and blue categories into a single "other" category containing $3 + 2 = 5$ balls in the population. Then, letting r_3 now stand for the number of sample balls in the "other" category,

$$r_3 = n - \sum_{i=1}^{2} r_i = 5 - (0 + 3) = 2$$

and

$$P(r_1 = 0, r_2 = 3, r_3 = 2) = \frac{\binom{1}{0}\binom{4}{3}\binom{5}{2}}{\binom{10}{5}} = \frac{(1)(4)(10)}{252} = \frac{40}{252}$$

which is the same as the probability obtained by summing but which involves much less computation.

7-7 The Negative Binomial Probability Law

The negative binomial probability law applies under exactly the same conditions as the binomial probability law. However, instead of giving the probability of r successes in a sample of predetermined size n, as does the binomial, the negative binomial gives the probability that a sample of n observations will have to be taken in order to obtain a predetermined number r of successes. The probability that n observations will have to be drawn in order to obtain r successes (or, equivalently, the probability that the rth success will occur on the nth observation, or, equivalently, the probability that exactly $n - r$ failures will occur before the rth success) is

$$P(n) = \binom{n-1}{r-1} p^r (1-p)^{n-r} \qquad (7\text{-}5)$$

Proof: Consider the observations to be drawn one at a time sequentially. In order for n observations to *have* to be drawn in order to obtain r successes, the nth observation must be the rth success. And in order for this to happen (a) exactly $r - 1$ of the first $n - 1$ observations must be successes and (b) the nth observation must be a success. So, the probability we seek is

$P(n$th observation is rth success$) = P(r - 1$ of the first $n - 1$ observations are successes and nth observation is a success$) = P(r - 1$ of the first $n - 1$ observations are successes$)P(n$th observation is a success $| r - 1$ of the first $n - 1$ observations are successes$)$

Because of the independence of observation outcomes when sampling from an undepletable population, the last, conditional, probability can be replaced with its unconditional counterpart. So

$P(n$th observation is rth success$) = P(r - 1$ of the first $n - 1$ observations are successes$)P(n$th observation is a success$) = P(r - 1$ successes in a sample of $n - 1$ observations$)P(1$ success in a sample of 1 observation$)$

Both of the last two probabilities are given by binomial probability laws. So

$P(n$th observation is rth success$)$

$$= \left[\binom{n-1}{r-1} p^{r-1}(1-p)^{n-r} \right]\left[\binom{1}{1} p^1 (1-p)^0 \right] = \binom{n-1}{r-1} p^r (1-p)^{n-r}$$

Example: In sampling randomly and with replacements from an urn containing one white and two black balls, the probability that the ball drawn fifth in sequence will be the third white ball drawn is

$$P(\text{5th observation is 3rd success}) = \binom{n-1}{r-1} p^r (1-p)^{n-r}$$

$$= \binom{4}{2}\left(\frac{1}{3}\right)^3 \left(\frac{2}{3}\right)^2$$

$$= (6)\left(\frac{1}{27}\right)\left(\frac{4}{9}\right) = \frac{8}{81}$$

Illustrative Problems: Only one-third of men applying for the Police Academy pass the exams. Four successful candidates are wanted. What is the probability that the quota will be filled only when the seventh applicant is examined?

ANSWER:

$$P(\text{7th observation is 4th success}) = \binom{7-1}{4-1}\left(\frac{1}{3}\right)^4 \left(\frac{2}{3}\right)^3$$

$$= \binom{6}{3}\left(\frac{1}{81}\right)\left(\frac{8}{27}\right)$$

$$= \frac{160}{2,187} = .0731$$

A certain state revokes the license of a driver after the third accident in which he is at fault. If bad drivers are at fault in 90% of the accidents in which they are involved, what is the probability that a bad driver's license will not be revoked until after he has been involved in at least five accidents?

ANSWER: This probability can be restated as the probability that the license of a bad driver will not be revoked prior to his fifth accident, or equivalently (since the third accident in which he is at fault cannot occur before his third accident) that the third time he is at fault does not occur on either his third or fourth accident. So P(5 or more observations required to produce 3rd "success") = P(3rd "success" does *not* occur on either 3rd or 4th observation) = $1 - P$(3rd "success" *does* occur on either 3rd or 4th observation). Now the 3rd success cannot occur on both the 3rd and the 4th observation, so these two eventualities are mutually exclusive and the probability we seek is equal to

$1 - [P(\text{3rd success occurs on 3rd observation})$

$\qquad\qquad + P(\text{3rd success occurs on 4th observation})]$

$$= 1 - \left[\binom{3-1}{3-1}.9^3.1^0 + \binom{4-1}{3-1}.9^3.1^1\right]$$

$$= 1 - [(1)(.729)(1) + (3)(.729)(.1)]$$

$$= 1 - [.729 + .2187] = .0523.$$

7-8   The Geometric Probability Law

The geometric probability law is simply an interesting special case of the negative binomial probability law that occurs when $r = 1$. It applies under the same conditions as the negative binomial probability law and therefore under the same conditions listed for the binomial probability law. The probability that exactly n observations will have to be drawn in order to obtain the first success (or, equivalently, the probability that the first success will occur on the nth observation or trial, or, equivalently, the probability that exactly $n - 1$ failures will occur before the first success) is

$$P(n) = p(1 - p)^{n-1} \qquad (7\text{-}6)$$

Proof: Formula 7-5 reduces to the above formula when $r = 1$.

Example: In sampling randomly and with replacements from an urn containing one white and two black balls, the probability that the ball drawn fifth in sequence will be the first white ball drawn is

$$P(\text{5th observation is 1st success}) = p(1 - p)^{n-1} = \left(\frac{1}{3}\right)\left(\frac{2}{3}\right)^4 = \frac{16}{243}$$

Illustrative Problems: A poorly adjusted vending machine fails to deliver and robs the customer of his coin on random occasions that amount to 10% of the attempts to operate it. What is the probability that a new customer will be robbed for the first time on the third occasion on which he tries to operate the machine?

ANSWER: $P(\text{3rd observation is 1st "success"}) = .1(.9)^2 = .081.$

A certain phone number is "busy" a randomly distributed 75% of the time. In trying to place a call to that number, what is the probability that the number would have to be dialed at least three times before one got an answer.

ANSWER: Since the probability of getting an answer on a single try is not zero, if one had the patience to keep trying he would eventually get an answer. So the probability of eventually getting an answer is 1. That first answer will occur on either the first trial, the second trial, . . . , etc., up to the "infiniteth" trial, and since it can occur on only one of them, these possibilities are mutually exclusive. So

$$1 = \sum_{n=1}^{\infty} P(\text{1st answer on } n\text{th trial})$$

$$= \sum_{n=1}^{D-1} P(\text{1st answer on } n\text{th trial}) + \sum_{n=D}^{\infty} P(\text{1st answer on } n\text{th trial})$$

$$\sum_{n=D}^{\infty} P(\text{1st answer on } n\text{th trial}) = 1 - \sum_{n=1}^{D-1} P(\text{1st answer on } n\text{th trial})$$

Now the probability that at least three trials will be required is the probability that n equals 3 or more, so it is given by the summation from $n = D$ $= 3$ to $n = \infty$. The answer is therefore

$$\sum_{n=3}^{\infty} P(\text{1st answer on } n\text{th trial}) = 1 - \sum_{n=1}^{2} P(\text{1st answer on } n\text{th trial})$$
$$= 1 - [P(\text{1st answer on 1st trial})$$
$$+ P(\text{1st answer on 2nd trial})]$$
$$= 1 - [(1/4)^1(3/4)^0 + (1/4)^1(3/4)^1]$$
$$= 1 - \left[\frac{4+3}{16}\right] = \frac{9}{16}$$

7-9   The Poisson Approximation to the Binomial Probability Law

Binomial probabilities are laborious to calculate when n is large. However, if certain conditions are met, a formula requiring much less calculation provides a close approximation to the desired binomial probability.

Specifically, if the following conditions are met:

(a) n is large (≥ 50, but the larger the better)
(b) p is small ($\leq .10$, but the closer to zero the better)
(c) np is small (≤ 10, but the smaller the better)
(d) r is small (say ≤ 6 when $np \leq 3$, otherwise $\leq 2np$, but the smaller the better)

then,

$$P(r) = \binom{n}{r} p^r (1 - p)^{n-r} \cong \frac{(np)^r}{r! \, e^{np}} \tag{7-7}$$

and, if all r's $\leq C$ meet criterion (d),

$$P(r \leq C) = \sum_{r=0}^{C} \binom{n}{r} p^r (1 - p)^{n-r} \cong \sum_{r=0}^{C} \frac{(np)^r}{r! \, e^{np}} \tag{7-8}$$

where $e = 2.71828 \ldots$ is the base for natural logarithms. The expression to the right of the $\cong$ sign in the first formula is the Poisson approximation to the binomial probability law.

Proof: Starting with the binomial probability law

$$P(r) = \frac{n!}{r!(n-r)!} p^r (1 - p)^{n-r}$$

then dividing $n!$ by $(n - r)!$ and expressing $(1 - p)^{n-r}$ in the equivalent form $(1 - p)^n / (1 - p)^r$, we obtain

$$P(r) = \frac{n(n-1)(n-2) \cdots (n-r+1)}{r!} \frac{p^r (1 - p)^n}{(1 - p)^r}$$

then multiplying both numerator and denominator by n^r, shifting the position of the $r!$ and expressing $(1 - p)^n$ in the equivalent form $[(1 - p)^{-1/p}]^{-np}$, we obtain

$$P(r) = \frac{n(n - 1)(n - 2) \cdots (n - r + 1)}{n^r} \frac{n^r p^r [1 - p)^{-1/p}]^{-np}}{r!(1 - p)^r}$$

Now the numerator of the first fraction is the product of r terms of the form $n - i$ where i goes from 0 to $r - 1$, and the denominator is, in effect, the product of r individual n's; so we may express the first fraction as the product of r separate fractions each having a denominator of n and a numerator in the form $n - i$, and we may express $n^r p^r$ as $(np)^r$:

$$P(r) = \left(\frac{n}{n}\right)\left(\frac{n - 1}{n}\right)\left(\frac{n - 2}{n}\right) \cdots \left(\frac{n - r + 1}{n}\right) \frac{(np)^r [(1 - p)^{-1/p}]^{-np}}{r!(1 - p)^r}$$

All steps up to this point have been simple algebraic manipulations and the right side of the above formula is the exact value of $P(r)$. We are now ready to introduce approximations. If n is very large and r is very small, then the fraction $(n - r + 1)/n$ will have a value very close to 1, and so will each of the fractions to the left of it. Furthermore, the *product* of all these fractions will have a value very close to 1 if the total *number* of fractions involved in the product is small. The total number of fractions from n/n to $(n - r + 1)/n$ is r. So if n is very large relative to r and r is very small, the product of all of these fractions is approximately 1 and the series of fractions can be replaced by a 1. Now if p is very close to zero, $1 - p$ will be nearly 1, and if r is very small $(1 - p)^r$ will also be very close to 1 and we can therefore approximate $(1 - p)^r$ by replacing it with a 1. Finally, a theorem from calculus tells us that the limiting value of an expression in the form $(1 - p)^{-1/p}$ as p approaches zero is e, the base $2.71828 \ldots$ for natural logarithms. Therefore, if p is very small, the expression inside square brackets is approximately equal to e. The product of np such expressions, however, cannot be relied upon to approximate e^{np} unless np is small, since a small error of approximation in the original expression can produce a large error in the product of a large number of such expressions. So, if p is close to zero and np is very small, we can approximate $[(1 - p)^{-1/p}]^{-np}$ by replacing it with e^{-np}. We have shown, therefore, that if

(a) n is large enough relative to r so that

$$\frac{n - r + 1}{n} \cong 1$$

and r is small enough so that the product of the r terms

$$\left(\frac{n}{n}\right)\left(\frac{n - 1}{n}\right)\left(\frac{n - 2}{n}\right) \cdots \left(\frac{n - r + 1}{n}\right) \cong 1$$

(b) p is small enough so that

$$1 - p \cong 1$$

and r is small enough so that

$$(1 - p)^r \cong 1$$

(c) p is small enough so that

$$(1 - p)^{-1/p} \cong e$$

and np is small enough so that

$$[(1 - p)^{-1/p}]^{-np} \cong e^{-np}$$

then,

$$P(r) = \left(\frac{n}{n}\right)\left(\frac{n-1}{n}\right)\left(\frac{n-2}{n}\right) \cdots \left(\frac{n-r+1}{n}\right) \frac{(np)^r[(1-p)^{-1/p}]^{-np}}{r!(1-p)^r}$$

$$\cong (1)\frac{(np)^r e^{-np}}{r!(1)} = \frac{(np)^r}{r!e^{np}}$$

Furthermore, if the above approximation for $P(r)$ is good for every value of $r \leq C$, then the sum of the approximations for all r's $\leq C$ should be at least an adequate approximation to $P(r \leq C)$ provided that the sum is composed of only a few individual terms. So, if r is small as already required, then

$$P(r \leq C) \cong \sum_{r=0}^{C} \frac{(np)^r}{r!e^{np}}$$

Actually, the conditions, as stated in the above proof, may be a bit conservative, since the errors incurred in using approximations (a) and (b) in the numerator and denominator, respectively, must be in the same direction (overestimation of the true value) and therefore tend, at least partially, to cancel each other out.

There is another compensatory feature in the Poisson approximation. As we shall see in a later chapter, np is the average value of r in all possible samples of size n, and this means that

$$\left|\sum_{r<np} (r - np)P(r)\right| = \left|\sum_{r>np} (r - np)P(r)\right|$$

Now r can only take on integer values from 0 up to n. Therefore, when we specify that np must be small, both absolutely and relative to n, we are greatly restricting the number of values r can take in the left side of the above formula while permitting it to assume many more values on the right. However, since the two sides of the formula are equal, this means that the average value of $P(r)$ when $0 \leq r < np$ greatly exceeds its average value when $np < r \leq n$. In fact $P(r)$ takes its greatest values in the vicinity of np, dropping off rapidly as r increases above np (at least when np is small), so that the relatively large values of r for which the approximation is worst have very little probability of occurrence either individually or collectively. As a result, provided that conditions (a), (b), and (c) are met, the error in the approximation

$$P(r \leq C) \cong \sum_{r=0}^{C} \frac{(np)^r}{r!e^{np}}$$

is small, both absolutely and relative to the true value, for *all* values of $C \leq n$, and the error in the approximation

$$P(r \geq C) \cong \sum_{r=C}^{n} \frac{(np)^r}{r! \, e^{np}}$$

is small absolutely for all values of $C \leq n$, and is small relative to the true value for all values of r, and therefore of C, that meet condition (d).

Example: If a random sample of $n = 100$ observations is to be drawn from an infinite dichotomous population, a proportion $p = .01$ of whose units are successes, the probability that the number of successes r in the sample will be 2 is given by the approximate formula

$$P(r = 2) \cong \frac{(np)^r}{r! \, e^{np}} = \frac{(100 \times .01)^2}{2! \, 2.71828^{100(.01)}} = \frac{1}{2(2.71828)} = \frac{1}{5.43656} = .1839$$

The exact probability, obtained by using the binomial formula is .1849. The Poisson probability that r will be 2 or less is

$$P(r \leq 2) \cong \sum_{r=0}^{2} \frac{(np)^r}{r! \, e^{np}} = \sum_{r=0}^{2} \frac{1^r}{r! \, e^1} = \frac{1}{e}\left(\frac{1}{0!} + \frac{1}{1!} + \frac{1}{2!}\right)$$

$$= \frac{1}{e}(1 + 1 + .5) = \frac{2.5}{2.7183} = .9197$$

This approximation contrasts with an exact cumulative probability of .9206, obtained by using the cumulative binomial formula.

Illustrative Problems: Suppose that the Prussian army contains a million men. And suppose that the probability that in any given year any given Prussian soldier will be killed by a kick from a horse is 1/500,000. What is the probability that in the coming year three or more Prussian soldiers will be killed by a kick from a horse?

ANSWER: $n = 1,000,000 \qquad p = 1/500,000 \qquad np = 2$

$$P(r \geq 3) = 1 - P(r \leq 2) \cong 1 - \sum_{r=0}^{2} \frac{(np)^r}{r! \, e^{np}} = 1 - \sum_{r=0}^{2} \frac{2^r}{r! \, e^2}$$

$$= 1 - \frac{1}{e^2}\left(\frac{2^0}{0!} + \frac{2^1}{1!} + \frac{2^2}{2!}\right) = 1 - \frac{1}{2.718^2}(1 + 2 + 2)$$

$$= 1 - \frac{5}{7.388} = \frac{2.388}{7.388} = .3232$$

A beginning typist has a tendency to type "hte" whenever she encounters the word "the," and actually does so on 3% of such occasions. In a proficiency examination she is given a letter to type that contains 50 "the's." What is the probability that she will type all of them correctly?

ANSWER: The question is equivalent to asking the probability that there will be zero errors of the type mentioned. Let r be the number of errors that will be committed. Then $n = 50$, $p = .03$, $np = 1.5$ and

$$P(r = 0) \cong \frac{(np)^r}{r!e^{np}} = \frac{1.5^0}{0!2.718^{1.5}} = \frac{1}{(1)2.718\sqrt{2.718}} = \frac{1}{2.718(1.649)}$$

$$= \frac{1}{4.482} = .223$$

7-10   The Poisson Probability Law

Suppose that occurrences of a certain type happen at random within a continuum of time or space (i.e., of duration, length, area, or volume), being just as likely to happen (a given number of times) in any portion of the continuum as in any other portion of equal size. Suppose further that the average number of occurrences in portions of the continuum of size c is known to be m and that the point at which every occurrence happens can be unambiguously classified as lying either inside or outside of a specified portion of size c. Then the probability that there will be exactly x occurrences in a specified portion of size c is given by the Poisson probability law

$$P(x) = \frac{m^x}{x!e^m} \qquad (7\text{-}9)$$

Furthermore, if the value of m is not known directly, but it is known that the mean number of occurrences in portions of the continuum of size C is M, then

$$m = M\left(\frac{c}{C}\right)$$

An important implication of the randomness of "location" of the occurrences is that *occurrences must be independent of each other*—the happening of one must not, for example, enhance (or reduce) the likelihood that another will happen "close" to it.

Proof: All that we actually know, or need to know, is that in portions of the continuum of a specified size c (or C) the average number of occurrences is m (or M) and that every point within such a portion is as likely to be the locus for an occurrence as any other point. However, we may *imagine* the continuum containing the portion of size c in which we are interested to be extended (under identical circumstances) to a huge size K, large enough so that no matter how rare occurrences may be (provided that $m > 0$) the number n of occurrences within the extended portion of size K will be enormous. Let these occurrences be called successes if they happen in the specified portion of size c and failures otherwise. Then every one of the n occurrences will be either a success or a failure, and for each occurrence the probability that it will be a success is $p = c/K$. The conditions of the binomial probability model are therefore satisfied and the probability that the "sample" of size n will contain x successes is (as derived in the proof given in Section 7-9)

$$P(x) = \binom{n}{x} p^x (1 - p)^{n-x}$$

$$= \left(\frac{n}{n}\right)\left(\frac{n-1}{n}\right)\left(\frac{n-2}{n}\right) \cdots \left(\frac{n-x+1}{n}\right) \frac{(np)^x [(1-p)^{-1/p}]^{-np}}{x!(1-p)^x}$$

Now m, the average number of occurrences in portions of size c, must be the total number n of occurrences divided by the number K/c of portions of size c into which the extended continuum can be divided. So

$$m = \frac{n}{\dfrac{K}{c}} = n\left(\frac{c}{K}\right) = np$$

and we may substitute m for np in the above formula for $P(x)$. The portion of the continuum in which we are interested is "actual" and therefore both its size c and the average number of occurrences m that occur within portions of that size are *finite*. But the extended continuum of size K is imaginary. So we are free to imagine K to approach infinity, thereby causing n to approach infinity and $p = c/K$ to approach zero, while $m = np$ remains constant. Consequently, for finite values of x

$$\left(\frac{n}{n}\right)\left(\frac{n-1}{n}\right)\left(\frac{n-2}{n}\right) \cdots \left(\frac{n-x+1}{n}\right)$$

approaches 1 as n approaches infinity

$(1 - p)^x$ approaches 1 as p approaches 0

$(1 - p)^{-1/p}$ approaches e as p approaches 0

Making the appropriate substitutions, we obtain

$$P(x) = 1\frac{m^x e^{-m}}{x!\,1} = \frac{m^x}{x!\,e^m}$$

a result entirely analogous to the Poisson approximation to the binomial probability law (Formula 7-7), except that we now have an equality sign where before we had only an approximation sign. The reason is that the binomial involves *actual* n's and p's, which must therefore be "noninfinite" and "noninfinitesimal" so that the formula holds only as an approximation, whereas in the present case n and p are unknown and unneeded, being only conceptual entities which can therefore be imagined to have whatever values *completely* justify the substitutions of 1 for the fractions involving n and 1 or e for the terms involving p.

Example: If an event of negligible duration occurs randomly but on the average 3 times per hour, the probability that it will occur exactly twice during a specified 10-minute interval is obtained as follows:

$$m = \frac{t}{T} M = \frac{10 \text{ min}}{60 \text{ min}} \times 3 = \frac{1}{2}$$

$$P(x = 2) = \frac{m^x}{x!\,e^m} = \frac{(1/2)^2}{2!\,e^{1/2}} = \frac{1/4}{2\sqrt{e}} = \frac{1}{8\sqrt{2.718}} = \frac{1}{8(1.649)} = \frac{1}{13.192}$$

$$= .0758$$

Illustrative Problems: Suppose that in making a bolt of cloth the manufacturing process produces flaws randomly but at an average rate of 1 flaw per 4 ft. If a housewife buys a randomly selected 10 ft of cloth, what is the probability that the purchased cloth contains no flaws?

ANSWER:

$$m = \frac{10}{4}(1) = 2.5 \text{ flaws per 10 ft.}$$

$$P(x = 0) = \frac{m^x}{x!\,e^m} = \frac{2.5^0}{0!\,2.718^{2.5}} = \frac{1}{(1)(2.718)^2\sqrt{2.718}} = \frac{1}{7.3875(1.649)}$$

$$= .082$$

During the busiest period at a telephone exchange, telephone calls are initiated randomly at an average rate of 30 per minute. The system overloads if 3 or more calls are initiated within a 1-sec interval. What is the probability that the system will be overloaded by calls initiated during a particular 1-sec interval within the busiest period?

ANSWER:

$$m = \frac{t}{T} M = \frac{1 \text{ sec}}{1 \text{ min}} 30 = \frac{1 \text{ sec}}{60 \text{ sec}} 30 = .5$$

$$P(\text{overload}) = P(x \geq 3) = 1 - P(x \leq 2) = 1 - \sum_{x=0}^{2} \frac{m^x}{x!\,e^m}$$

$$= 1 - \sum_{x=0}^{2} \frac{.5^x}{x!\,e^{.5}} = 1 - \frac{1}{\sqrt{e}}\left(\frac{.5^0}{0!} + \frac{.5^1}{1!} + \frac{.5^2}{2!}\right)$$

$$= 1 - \frac{1}{\sqrt{2.718}}(1 + .5 + .125) = 1 - \frac{1.625}{1.649} = \frac{.024}{1.649} = .0146$$

7-11 Computational Strategems

Certain mathematical or logical relationships are helpful in obtaining probabilities, either by calculation or by looking them up in tables. The probability obtained may be either a ***point probability***, that is, the probability that the sample statistic will have a single specified value, or a ***cumulative***

probability, i.e., the probability that the sample statistic will have any one of the values in a specified interval of possible values.

7-11-1 Point Probabilities

Most binomial tables give probabilities only for cases where $p \leq .5$. The reason is as follows. If r represents the number of successes in the sample, n the number of observations in the sample, and p the proportion of successes in the population, then

$$P(r = n - x \,|\, n, p = 1 - \theta) = \binom{n}{n - x}(1 - \theta)^{n-x}[1 - (1 - \theta)]^{n-(n-x)}$$

$$= \frac{n!}{(n - x)!\,x!}(1 - \theta)^{n-x}\theta^x$$

$$= \binom{n}{x}\theta^x(1 - \theta)^{n-x}$$

$$= P(r = x \,|\, n, p = \theta) \tag{7-10}$$

So if one wishes to obtain the probability that in a random sample of size $n = 10$ drawn from an infinite population in which $p = .70$ there will be exactly 4 successes, he can find it by looking up in binomial tables the probability that in a random sample of size $n = 10$ drawn from an infinite population in which $p = 1 - .70 = .30$ there will be exactly $10 - 4 = 6$ successes. In effect, what one does if $p > .5$ is to interchange the labels "success" and "failure" so that former "failures" whose population proportion was $1 - p$, and therefore less than .5, are now "successes."

An analogous relationship holds for hypergeometric probabilities:

$$P(r = n - x \,|\, N, n, R = N - K) = \frac{\binom{N - K}{n - x}\binom{N - (N - K)}{n - (n - x)}}{\binom{N}{n}}$$

$$= \frac{\binom{N - K}{n - x}\binom{K}{x}}{\binom{N}{n}} = \frac{\binom{K}{x}\binom{N - K}{n - x}}{\binom{N}{n}}$$

$$= P(r = x \,|\, N, n, R = K) \tag{7-11}$$

As shown in Section 7-6 and in the illustrative problems, a multinomial (or multivariate hypergeometric) probability specifying the exact value of only a single one of the r_i's, letting the others take whatever values they will, is equivalent to a binomial (or hypergeometric) probability formula, e.g., in the case where there are three categories,

$$P(r_1 = x, r_2 = \,?, r_3 = \,?) = P(r_1 = x, r_2 + r_3 = n - x)$$

where 2's and 3's are now treated as a single category. Another important relationship is exemplified in the following equation:

$$P(r_1 = x, r_2 = 0, r_3 = 0, r_4 = n - x) = P(r_1 = x, r_{2\&3} = 0, r_4 = n - x)$$

That is, the probability that there will be no sample observations in each of several specified categories is equivalent to the probability that there will be no sample observations in a category combining all of the several categories.

The negative binomial probability that the rth success will occur on the nth trial can be obtained by looking up the binomial probability that there will be r successes in a sample of size n and multiplying it by r/n. This is because the negative binomial probability law can be written

$$P(n) = \binom{n-1}{r-1} p^r (1-p)^{n-r} = \frac{(n-1)!}{(r-1)!(n-r)!} p^r (1-p)^{n-r}$$

$$= \left(\frac{r}{n}\right)\left[\frac{n!}{r!(n-r)!} p^r (1-p)^{n-r}\right] \qquad (7\text{-}12)$$

$$= \left(\frac{r}{n}\right)[\text{binomial } P(r \mid n, p)]$$

Likewise, the geometric probability that the first success will occur on the nth trial can be obtained by looking up the binomial probability of one success in n trials and multiplying it by $1/n$, since the geometric probability law can be written

$$P(n) = p(1-p)^{n-1} = \frac{1}{n}\left[\binom{n}{1} p^1 (1-p)^{n-1}\right]$$

$$= \left(\frac{1}{n}\right)[\text{binomial } P(r = 1 \mid n, p)] \qquad (7\text{-}13)$$

7-11-2 Recursion Formulas

A formula that gives the probability of a possible value of a sample statistic in terms of the probability of the next smaller possible value is called a *recursion formula*. With it one can start with the calculated probability of the smallest possible value and calculate the probability of each successive value, often with much less computation than if it were calculated directly from the probability law. Such formulas often provide efficient means of calculating cumulative probabilities (when the cumulation starts with the smallest possible value), since each of the successive probabilities must be calculated no matter what method is used.

Suppose that we want a recursion formula for the Poisson probability law, giving us the probability of $x + 1$ successes in terms of the probability of x successes. The probability of $x + 1$ successes is

$$P(x + 1) = \frac{m^{x+1}}{(x + 1)! e^m}$$

and the probability of x successes is $P(x) = m^x/(x!e^m)$. Therefore, taking the ratio of the former to the latter we have

$$\frac{P(x+1)}{P(x)} = \frac{\dfrac{m^{x+1}}{(x+1)!\,e^m}}{\dfrac{m^x}{x!\,e^m}} = \frac{x!\,e^m m^{x+1}}{(x+1)!\,e^m m^x} = \frac{m}{x+1}$$

$$P(x+1) = \frac{m}{x+1}P(x)$$

which is a very simple and useful recursion formula. To illustrate its application, suppose that $m = 2$. Then the probability of the smallest possible number of successes, zero successes, is given by the Poisson probability law:

$$P(0 \text{ successes}) = \frac{m^x}{x!\,e^m} = \frac{2^0}{0!\,e^2} = \frac{1}{2.7183^2} = \frac{1}{7.389} = .135336$$

Now, invoking the recursion formula,

$$P(1 \text{ success}) = P(0 + 1 \text{ success}) = \frac{m}{0+1}P(0) = \frac{2}{1}(.135336) = .27067$$

$$P(2 \text{ successes}) = P(1 + 1 \text{ successes}) = \frac{m}{1+1}P(1) = \frac{2}{2}(.27067) = .27067$$

$$P(3 \text{ successes}) = P(2 + 1 \text{ successes}) = \frac{m}{2+1}P(2) = \frac{2}{3}(.27067) = .18045$$

Of course, if the recursion formula is to be used successively to obtain the probability of a large number of successes, accuracy in the final result will require that a large number of decimal places be used throughout.

Recursion formulas for probability laws requiring dichotomous populations are given in Table 7-2. All of them were derived in a way completely analogous to the derivation already given—by forming a ratio between probabilities for successive values, then solving for the probability for the larger value.

7-11-3 Cumulative Probabilities

In any particular instance a sample statistic can have only one value. So the various *possible* values that it might have are *mutually exclusive*. And the probability that it will take some one of several designated possible values is therefore (see Formula 6-6) the sum of the individual probabilities of the designated values. It follows therefore that since the number r of successes in a sample can only take integer values in the range from 0 to n,

$$P(r \le C) = P(r = 0) + P(r = 1) + \cdots + P(r = C)$$

$$= \sum_{r=0}^{C} P(r) \tag{7-14}$$

$$P(r \ge D) = P(r = D) + P(r = D + 1) + \cdots + P(r = n)$$

$$= \sum_{r=D}^{n} P(r) \tag{7-15}$$

and if $C < D$,

$$P(C \le r \le D) = P(r = C) + P(r = C + 1) + \cdots + P(r = D)$$
$$= \textstyle\sum_{r=C}^{D} P(r) = \sum_{r=0}^{D} P(r) - \sum_{r=0}^{C-1} P(r) \qquad (7\text{-}16)$$
$$= P(r \le D) - P(r < C)$$

Now, since r must take some one of its possible values, $\sum_{r=0}^{n} P(r) = 1$. We can break the summation into two parts

$$\textstyle\sum_{r=0}^{C} P(r) + \sum_{r=C+1}^{n} P(r) = 1$$

from which it follows that

$$\textstyle\sum_{r=0}^{C} P(r) = 1 - \sum_{r=C+1}^{n} P(r) \qquad (7\text{-}17)$$

or, contrarywise, letting $C + 1 = D$ so that $C = D - 1$,

$$\textstyle\sum_{r=D}^{n} P(r) = 1 - \sum_{r=0}^{D-1} P(r) \qquad (7\text{-}18)$$

Such formulas are computationally convenient. For example, it is much easier to obtain $P(r \le 8)$, when $n = 9$, by simply calculating $P(r = 9)$ and subtracting it from 1, than by calculating $P(r)$ for each of the nine values from 0 to 8 and adding them.

When there is no upper limit upon the possible values that the sample statistic might take (as is the case with the sample statistics involved in the negative binomial, geometric, and Poisson probability laws), there is no alternative to obtaining upper-tail cumulative probabilities by subtraction of a lower-tail cumulative probability from 1. Thus, for n the trial upon which the rth success occurs,

$$P(n \le D - 1) = P(n = 1) + P(n = 2) + \cdots + P(n = D - 1)$$
$$= \textstyle\sum_{n=1}^{D-1} P(n)$$
$$P(n \ge D) = P(n = D) + P(n = D + 1) + \cdots = \textstyle\sum_{n=D}^{\infty} P(n)$$

and, since the last summation contains an infinite number of terms it cannot be calculated directly; instead, one obtains it by calculating the complementary probability and subtracting it from 1:

$$P(n \ge D) = 1 - P(n \le D - 1) = 1 - \textstyle\sum_{n=0}^{D-1} P(n) \qquad (7\text{-}19)$$

A similar approach is of course required for the Poisson statistic x, which also has no upper limit upon the values it can take.

The summations involved in both lower-tail and upper-tail cumulative probabilities are given in Table 7-2 for the probability laws requiring dichotomous populations.

Semantic considerations are important in the statement, and in the solution, of problems. In particular, one should keep in mind the following

Table 7-2

Formulas Concerning Probability Laws in the Two-Category Case

Law	Formula for probability	Formula for cumulative probability		Recursion formula
		Lower-tail	Upper-tail	
Hypergeometric	$P(r) = \dfrac{\binom{R}{r}\binom{N-R}{n-r}}{\binom{N}{n}}$	$P(r \leq C) = \sum_{r=0}^{C}\dfrac{\binom{R}{r}\binom{N-R}{n-r}}{\binom{N}{n}}$	$P(r \geq D) = \sum_{r=D}^{n}\dfrac{\binom{R}{r}\binom{N-R}{n-r}}{\binom{N}{n}}$	$P(r+1) = \dfrac{(R-r)(n-r)}{(r+1)(N-R-n+r+1)}P(r)$
Binomial	$P(r) = \binom{n}{r}\rho^r(1-\rho)^{n-r}$	$P(r \leq C) = \sum_{r=0}^{C}\binom{n}{r}\rho^r(1-\rho)^{n-r}$	$P(r \geq D) = \sum_{r=D}^{n}\binom{n}{r}\rho^r(1-\rho)^{n-r}$	$P(r+1) = \dfrac{(n-r)\rho}{(r+1)(1-\rho)}P(r)$
Poisson approximation to binomial	$P(r) \cong \dfrac{(n\rho)^r}{r!e^{n\rho}}$	$P(r \leq C) \cong \sum_{r=0}^{C}\dfrac{(n\rho)^r}{r!e^{n\rho}}$	$P(r \geq D) \cong \sum_{r=D}^{n}\dfrac{(n\rho)^r}{r!e^{n\rho}}$	$P(r+1) \cong \dfrac{n\rho}{r+1}P(r)$
Poisson	$P(x) = \dfrac{m^x}{x!e^m}$	$P(x \leq C) = \sum_{x=0}^{C}\dfrac{m^x}{x!e^m}$	$P(x \geq D) = \sum_{x=D}^{\infty}\dfrac{m^x}{x!e^m}$ $= 1 - \sum_{x=0}^{D-1}\dfrac{m^x}{x!e^m}$	$P(x+1) = \dfrac{m}{x+1}P(x)$
Negative binomial	$P(n) = \binom{n-1}{r-1}\rho^r(1-\rho)^{n-r}$	$P(n \leq C) = \sum_{n=1}^{C}\binom{n-1}{r-1}\rho^r(1-\rho)^{n-r}$	$P(n \geq D) = \sum_{n=D}^{\infty}\binom{n-1}{r-1}\rho^r(1-\rho)^{n-r}$ $= 1 - \sum_{n=1}^{D-1}\binom{n-1}{r-1}\rho^r(1-\rho)^{n-r}$	$P(n+1) = \dfrac{n(1-\rho)}{n-r+1}P(n)$
Geometric	$P(n) = \rho(1-\rho)^{n-1}$	$P(n \leq C) = \sum_{n=1}^{C}\rho(1-\rho)^{n-1}$	$P(n \geq D) = \sum_{n=D}^{\infty}\rho(1-\rho)^{n-1}$ $= 1 - \sum_{n=1}^{D-1}\rho(1-\rho)^{n-1}$	$P(n+1) = (1-\rho)P(n)$

153

equivalences between mathematical and verbal symbols:

$$P(r \leq C) = P(r < C + 1) = P(r \text{ will be less than or equal to } C)$$
$$= P(\text{there will be } C \text{ or fewer successes})$$
$$= P(r \text{ will not exceed } C)$$
$$= P(\text{there will be at most } C \text{ successes})$$
$$= 1 - P(r \geq C + 1)$$
$$= 1 - P(r > C)$$

$$P(r < C) = P(r \leq C - 1) = P(r \text{ will be less than } C)$$
$$= P(\text{there will be fewer than } C \text{ successes})$$
$$= P(r \text{ will not be as great as } C)$$
$$= P(\text{there will not be as many as } C \text{ successes})$$
$$= 1 - P(r \geq C)$$
$$= 1 - P(r > C - 1)$$

$$P(r \geq D) = P(r > D - 1) = P(r \text{ will be greater than or equal to } D)$$
$$= P(\text{there will be } D \text{ or more successes})$$
$$= P(r \text{ will be no smaller than } D)$$
$$= P(\text{there will be at least } D \text{ successes})$$
$$= 1 - P(r \leq D - 1)$$
$$= 1 - P(r < D)$$

$$P(r > D) = P(r \geq D + 1) = P(r \text{ will be greater than } D)$$
$$= P(\text{there will be more than } D \text{ successes})$$
$$= P(r \text{ will not be as small as } D)$$
$$= P(\text{there will not be as few as } D \text{ successes})$$
$$= 1 - P(r \leq D)$$
$$= 1 - P(r < D + 1)$$

$$P(\text{any successes}) = 1 - P(\text{no successes})$$

7-11-4 Tables of Probabilities

Tables in Appendix B give solutions of the binomial and Poisson probability laws in both point and cumulative form. Cumulative probabilities are also given for use of the hypergeometric probability law in a specialized context. Table 7-3 summarizes the conditions upon which the validity of the various specific probability laws depend.

Table 7-3

Summary of Conditions Required by Specific Probability Laws

Law	Conditions under which law applies					
	Random sampling	Independent observations*	K mutually exclusive and exhaustive categories $K = 2$	$K > 2$	Sample observations Deplete source	Do not deplete source
Hypergeometric	X		X		X	
Multivariate hypergeometric	X			X	X	
Binomial	X	X	X			X
Multinomial	X	X		X		X
Negative binomial	X	X	X			X
Geometric	X	X	X			X
Poisson approximation to binomial	X	X	X			X
Poisson	X	X				X

*Independence is guaranteed by random sampling when sample observations do not deplete their source.

7-12   Problems

1. A colorblind bachelor pulls two socks out of a drawer containing nine socks identical except for color. If three of the original nine socks are red and six are green, what is the probability that his two socks will match?

2. What would be the answer to the above problem if four of the original nine socks were red and five were green?

3. A random sample of ten subjects is to be drawn from the subject pool given below. What is the probability that the sample will contain at least one man and one woman who have both the characteristics of right-handedness and normal vision?

	Left-handed Normal Vision	Poor Vision	Right-handed Normal Vision	Poor Vision
Men	0	2	1	3
Women	1	3	2	4

4. In the above problem, what is the probability that at least 90% of the sample will be either right-handed or have normal vision or both?

5. Ten percent of the items received by a retail merchant from a certain supplier have been defective in the past. If the retailer orders a half-dozen more items from that supplier, what is the probability

   (a) that the shipment will contain no defective items?
   (b) that it will contain one or more defectives?
   (c) that it will contain two or more defectives?
   (d) that all six items will be defective?

6. The Economics Department offers ten graduate courses, of which three are very difficult. A newly arrived graduate student randomly selects four of the ten courses and enrolls in them. What is the probability that he has enrolled in (a) exactly two of the very difficult courses, (b) at least one very difficult course?

7. A pond contains 12 fish, of which 7 are less than 1 ft long, 3 are between 12 in. and 18 in. and 2 are over 18 in. long. If a fisherman catches a randomly selected 7 fish, what is the probability that he catches at least 2 of each variety?

8. An urn contains ten balls, of which five are stamped with the number 0, three with the number 1, and two with the number 10. A random sample of three balls is to be drawn without replacements and the sum Z of the numbers stamped on the three balls is to be calculated.

   (a) Tabulate the probability distribution of Z.
   (b) Tabulate the probability distribution of the number of balls in the sample that are stamped with a 1.

9. Over the years, 25% of the patients admitted to a certain mental hospital have recovered within six months. Today eight patients were admitted. What is the probability that at least half of them will have recovered within six months?

10. Over the years, half of the patients admitted to a certain rest home have been neurotic, a third have been psychotic, and a sixth have been "other." Five patients will be admitted today.

    (a) What is the probability that four will be neurotic, one psychotic, and none "other"?
    (b) What is the probability that at least three will be neurotic and at least two psychotic?
    (c) What is the probability that at least three will be neurotic?
    (d) What is the probability that at least two will be neurotic and at least two will be psychotic?

11. A doctor has just read of a new type of shock treatment which allegedly

produces instantaneous recovery in 10% of schizophrenics upon whom it is tried. He decides to check the validity of this information by trying the treatment on some randomly selected schizophrenic patients.

(a) What is the smallest number of patients he must plan to try it on in order that there be a probability of at least .5 that at least one patient will instanteneously recover if the information he read is correct?

(b) If his information is correct, what is the probability that the first instantaneous recovery will occur on the seventh schizophrenic treated?

(c) Suppose the doctor decides he must have at least two instantaneous recoveries to be convinced. If his information is correct, what is the probability that he will have to try the treatment on exactly ten schizophrenics in order to obtain two instantaneous recoveries?

12. Tabulate the probability distribution of the number of successes r in a sample of size $n = 5$ drawn randomly from an infinite dichotomous population in which the proportion of successes is $p = 1/3$.

13. Army intelligence has determined that 10% of captured enemy soldiers can be tricked into giving information, 20% can be frightened into doing so, and 70% will not give information in any situation. Five enemy soldiers are captured. What is the probability

(a) that one can be tricked and two frightened into giving information?

(b) that at least one can be tricked and at least one can be frightened into giving information?

(c) that at least one can be induced to give information?

(d) that none of them will give any information?

14. An infinite population consists of units belonging to one of three mutually exclusive and exhaustive categories. One-half of all units belong to the first category, one-third of all units belong to the second category, and one-sixth of all units belong to the third category. A random sample of three units is to be drawn from the population. Let X_1, X_2, and X_3 be the number of sample units belonging to the first, second, and third categories, respectively. And let $Z = X_1 + 10X_2 + 100X_3$.

(a) Tabulate the probability distribution of Z.

(b) Tabulate the probability distribution of X_1.

15. In the past, half of the items received by a retail merchant from a certain supplier have been in good working order, a third have been defective but easily repairable, and the rest have been so badly defective that it is either impossible or uneconomical to repair them. If the retailer orders a half-dozen more items from that supplier, what is the probability

(a) that three items will be good (*G*), two defective but easily repairable (*R*), and one defective but unrepairable (*U*)?

(b) that there will be at least one *R* and one *U*?

(c) that there will be at least one *R*?

(d) that there will be at least one item that is either an *R* or a *U*?

16. A subject pool contains 20 subjects, of whom 11 will both understand and follow instructions (*B*'s), 5 will follow the portion of the instructions that they understand, but will not understand all of the instructions (*F*'s), and 4 will neither completely understand nor completely follow what they do understand (*N*'s). A random sample of 8 subjects is drawn without replacements from the pool to serve in an experiment. What is the probability that the sample will contain:

(a) no *F*'s and no *N*'s?

(b) no *F*'s?

(c) at least one undesirable subject (*F* or *N*)?

(d) at least 3 *F*'s and at least 3 *N*'s?

17. A random sample of four subjects is to be drawn without replacements from the subject pool in the above problem. What is the probability that the sample will contain

(a) 4 *F*'s?

(b) 2 or more *F*'s?

(c) at least 1 *F*?

18. The probability that a cigarette lighter will work on any given try is .2. What is the probability that

(a) a flame will be first produced on the fourth attempt?

(b) four or more attempts will be necessary to produce a light?

19. On the average, of every 10,000 bombs dropped by the Myopic Air Force, 10 hit their target. In a saturation bombing raid over an enemy munitions factory, 500 bombs are dropped. What is the probability that at least one of them will hit the target?

20. A random sample of 10,000 observations is to be drawn from an infinite dichotomous population in which the proportion of successes is .9999. If *Z* is the number of failures in the sample, tabulate the probability distribution of *Z* for values of $Z \leq 6$.

21. A psychiatrist is convinced that the development of a healthy ego requires early successes in coping with crises. If the probabiltiy of success in coping with an early crisis is the constant value 1/3, what is the probability that

(a) a child's first success will occur on the fourth crisis?

(b) a child's first success will occur on or after the fourth crisis?

22. A company recruiter is interviewing applicants for five identical jobs.

The probability that any one applicant will be acceptable is 2/3. What is the probability that

(a) exactly eight applicants will have to be interviewed in order to fill all five positions?

(b) eight or fewer interviews will have to be conducted in order to fill all five positions?

23. On the average, one in every million rocks in the Southwestern desert is an arrowhead. A rockhound will observe ten thousand rocks during an arrow hunt. What is the probability that any of these rocks will be an arrowhead?

24. Units are to be randomly drawn, in sequence, from an infinite dichotomous population, in which the proportion of successes is $p = 2/3$, until the first success is obtained. Let n be the size of the sample immediately after the first success is obtained. Tabulate the probability distribution of n for all values of n whose relative frequency in the probability distribution exceeds .01.

25. At a certain beach the active services of a lifeguard are required to rescue a swimmer only once per 30 days, on the average. During one 10-day period, the lifeguard is hung over on 3 days. What is the probability that his active services will be needed at any time during those 3 days?

26. Stressful events occur in one's life at the rate of about a dozen per year. If five or more such events occur within a one-month period, the person is a candidate for a nervous breakdown. What is the probability of five or more such events occurring within a particular one-month period?

27. Drivers scoring at or above 90 on their driver's test are labeled "good" (G) and those scoring below 90 are labeled "bad" (B). The accompanying table gives statistics on the involvement of good and bad drivers in two-car accidents. The courts always find one driver at fault and the other blameless in such accidents, and they always revoke a driver's license immediately after he has been found at fault in three two-car accidents (other kinds of accidents are ignored). Calculate (a) for good drivers, and (b) for bad drivers, the probability that the driver's license will be revoked immediately after his fifth accident.

Driver at Fault	Driver Blameless	Number of Accidents
G	G	5,000
G	B	15,000
B	G	25,000
B	B	55,000
		100,000

28. The probability of an international political crisis occurring on any one day is the same as that of its occurring on any other day. Such crises occur at the rate of three every ten years. The President plans to take a vacation during which he will be gone for $36\frac{1}{2}$ days. What is the probability that one or more political crises will arise while he is gone?

29. Along the length of an elephant's trunk, the average number of wrinkles is 30 per ft. What is the probability that a randomly chosen inch (along the length of the trunk) will contain one or more wrinkles?

30. In a vigilance experiment lasting 5 hours, an instantaneous visual signal appears at random at an average rate of 20 per hour. The subject blinks his eyes at random at an average rate of once per minute and for an average duration of .6 seconds per blink. What is the probability that two or more of the signals will occur while the subject is blinking?

8

games

This is the first of several consecutive chapters about Decision Theory, which deals with appropriate methods of choosing from among alternative actions. This chapter concerns games—situations in which the choice among actions is complicated by the machinations of a counterstrategizing opponent. The next chapter concerns the much simpler case in which the only complicating feature is one's ignorance about which of several possible "states" a passive and nonstrategizing "Nature" may be in. This is like a game except that one's opponent is Nature who "chooses" her actions without regard to those of the human player or their effect upon him. A final category is a special case of the "game against Nature" in which the actions are to announce which state Nature is in, that is, to estimate the true situation or to choose between hypotheses about it.

Before taking up games, however, we need to consider ways of assessing the values, to a player, of the possible situations that he may be in in particular cases and the value of an action in general.

8-1 Value

The value of a thing can be measured in two different ways, objectively and subjectively, often with strikingly different results. The value, on the open art market, of a portrait of my spade-bearded great-grandfather would be approximately equal to the secondhand price of the frame and canvas. To my mother, however, it is a cherished and priceless possession and perhaps even a great work of art.

8-1-1 Objective Value

As implied above, a thing's objective value may be regarded as its economic value or price on the open market—as measured by some standard medium of exchange such as dollars, wampum, or packages of cigarettes during World War II. Thus the objective value of a pound of coffee to a housewife is the price she must pay the grocer to obtain it.

8-1-2 Expected Value

When one gambles, an issue that is bound to arise is "What are fair stakes?" For example, how much should one pay for the privilege of playing a game in which a fair die is tossed once and one is paid the same number of dollars as dots coming up on the die? The answer supplied by mathematicians during the seventeenth century was that it was fair for one to stake on a game of chance (with equally likely atomic outcomes) the *average* amount of money that he would win under all possible outcomes (i.e., if every possible outcome occurred just once). Thus, in the case considered, the fair stake would be $(1 + 2 + 3 + 4 + 5 + 6)/6 = 3.5$ dollars, since this is the amount he could "expect" to win on the average, thereby breaking even in the long run.

From Chapter 2 we know that if a population consists of N units whose values are $u_1, u_2, \ldots, u_j, \ldots, u_N$, and if there are k *different* values $x_1, x_2, \ldots, x_i, \ldots, x_k$ whose frequencies of occurrence are $f_1, f_2, \ldots, f_i, \ldots, f_k$, respectively, then the average value of a unit is the mean of the population:

$$\frac{\sum_{j=1}^{N} u_j}{N} = \frac{\sum_{i=1}^{k} f_i x_i}{N} = \sum_{i=1}^{k} \frac{f_i}{N} x_i = \sum_{i=1}^{k} p_i x_i$$

where p_i is the proportion of units having the value x_i. Now if we let X represent the value of a unit, let p_i represent the proportion of times, or the probability, that X takes the value x_i, then the **expected value** of X, which we denote $E(X)$, is, by definition, simply the average value

$$E(X) = p_1 x_1 + p_2 x_2 + \cdots + p_k x_k = \sum_{i=1}^{k} p_i x_i \qquad (8\text{-}1)$$

Thus, if the price X of a pound of coffee fluctuated randomly being \$1.00 half of the time, \$1.50 a third of the time, and \$1.80 a sixth of the time, the "expected" price of a pound of coffee, for the housewife, would be

$$E(X) = \left(\frac{1}{2}\right)(1.00) + \left(\frac{1}{3}\right)(1.50) + \left(\frac{1}{6}\right)(1.80) = .50 + .50 + .30 = \$1.30$$

8-1-3   Hazards in Using Expected Value as a Criterion for Action

It is important to distinguish between two different usages of expected value. One is simply to describe a population by providing an index of location, namely the mean. The other is to serve as an objective basis for deciding whether or not to take certain gambles. In the former case, expected value is synonymous with average value and is quite innocuous. In the latter case, expected value implies a sort of expectation of average payoff in a long-run gambling (or other risk-taking) situation. In this context, as a criterion for choice behavior, it is not always reliable and can be treacherous.

This is illustrated by the celebrated *St. Petersburg Paradox*. A fair coin is to be tossed repeatedly until a head comes up. If it comes up heads on the first toss, your opponent pays you two dollars. If a head first comes up on the second toss you get four dollars. And, in general, if the first head comes up on the nth toss, you receive 2^n dollars. The probability that the first head will come up on the first toss is $1/2$, on the second toss is $1/2^2$, and on the nth toss is $1/2^n$ (see Formula 7-6). So the expected value of the payoff to you is

$$E(X) = \sum_i p_i x_i = \sum_{n=1}^{\infty} \left(\frac{1}{2^n}\right) 2^n = \sum_{n=1}^{\infty} \left(\frac{2}{2}\right)^n = \sum_{n=1}^{\infty} 1^n$$

$$= 1 + 1 + 1 + \cdots = \infty$$

Therefore, if your fair stake is the expected value of your winnings, you should pay an infinite number of dollars for the privilege of playing the game once. But no one would put up an "infinite" stake to play a game in which recouping one's stake depended upon the occurrence of events having "infinitesimal" probabilities. (In the St. Petersburg game there is less than one chance in a million of winning more than two million dollars). To be "fair," such a game would have to be played eternally by immortal players possessed of superinfinite capital. Thus, the average payoff can serve as a rational basis for "expectation" if the game is to be played often enough for all possible outcomes (or at least payoffs) to occur with relative frequencies approximately equal to their probabilities. But it is of questionable value as a guide to action when the game is to be played less often than that, especially so when the game is a one-shot affair to be played only once.

A more sophisticated example is to be found in a common theological argument. If one leads an irreligious life, at best he enjoys a life of finite duration and finite pleasure, whose value we shall call L, and he abandons all hope of heaven. However, if he chooses to lead a religious life, he stakes L on a game whose payoffs are x_1, a heaven of both infinite duration and infinite happiness, whose value is therefore $L(\infty)(\infty)$, and x_2, the nonexistence of heaven, whose value is therefore 0. Now even if $p_1 = 1/\infty$ and

$p_2 = 1 - (1/\infty)$,

$$E(X) = p_1 x_1 + p_2 x_2 = \left(\frac{1}{\infty}\right)[(L)(\infty)(\infty)] + \left(1 - \frac{1}{\infty}\right)0 = L(\infty)$$

and since the expected value $L(\infty)$ infinitely exceeds the stake L, the intelligent choice is to lead the religious life. An argument similar to this was published by Blaise Pascal, a religious ascetic and early probabilist, in an effort to convert his former gambling friends, who might have countered, "Yes, but you only play the game once."

One does not have to invoke the concept of infinity to see that expectation is not always a reliable guide to action. Consider a lottery in which there are 10,001 tickets, of which 10,000 are winners and only one is a loser. It costs $100 to buy a ticket; if you win you get $100.01 and if you lose you get nothing. The expected value of the game is

$$\frac{10,000}{10,001}(\$100.01) + \frac{1}{10,001}(\$0) = \$100$$

the amount of your stake; so by the criterion of expectation it is a fair game. But you probably wouldn't take the gamble even if the "profit" on a winning ticket were doubled to 2 cents, because the expected gain is too small. An additional two cents in your pocket to accompany your $100 wouldn't appreciably increase your happiness. But the loss of your $100 probably would appreciably diminish it.

Now consider another lottery in which there are a trillion and one tickets of which a trillion are losers and one is a winner. Again, it costs $100 to buy a ticket; if you win you get a hundred trillion and 100 dollars, and if you lose you get nothing. The expected value of the game is again the amount of your stake but you probably wouldn't accept this gamble even if the winning payoff were increased to a billion trillion dollars. This time the amount of gain is appreciable enough, but the probability of obtaining it is too small. Furthermore, the amount of gain far exceeds your needs, so willingness to accept increments of risk is not proportional to the corresponding increments of "fair" gain.

Finally, consider the plight of a businessman who has $1,000 cash and no credit and who must raise an additional $10,000 immediately or lose his $100,000 business. His only hope is a gambler who offers him the following gamble: the businessman makes a single throw with a pair of fair dice; if the dice come up 2 or 12, the gambler pays the businessman $10,000, but if any other number comes up, the businessman pays the gambler $1,000. The probability that the businessman will win is 1/18 and the expected value is

$$10,000\left(\frac{1}{18}\right) + (-1,000)\left(\frac{17}{18}\right) = \frac{-7,000}{18} = -\$388.89$$

so the game is "unfair" to the businessman. Yet it is definitely to his advantage to take the gamble. The paradox is quickly resolved when we realize that the "fairness" of gambling odds is based upon *objective* values of the stakes and not upon their subjective desirability or usefulness to the players.

To summarize, expected value as a criterion for action in choice behavior

(a) is based upon long-run considerations and is not necessarily a valid criterion in the short run or one-shot case, especially when either the possible consequences or their probabilities have extreme values.

(b) assumes that the subjective values of the possible outcomes are proportional to their objective values, which is not necessarily the case, especially when the values involved are large.

8-1-4 Utility or Subjective Value

The disproportionality between desirability and objective value, especially at extremes of the latter, suggests the substitution of subjective values for objective ones. Subjective values are called utilities. *Utilities* are numerical quantities assigned by a particular person to objects or situations (which can be ranked in order of preference—tied ranks being admissible) to represent their current degree of desirability to him. They vary from person to person and from context to context. If the utility to me of one pound of salt is designated X, then the utility of two pounds might be 1.5 X, the utility of ten pounds might be 2 X, and the utility of a forty-foot-high mound of salt in my backyard might be $-1,000$ X. Likewise, if to a mother the utility of her first child is 1, the utility of her fourteenth might be -50. Clearly, utilities need not be proportional to physical quantities and need not increase with increasing amounts of the physical quantity.

Ideally, we would like a scale of utilities to have the following characteristics: (a) complete indifference is represented by zero, (b) desirable things are represented by positive numbers and undesirable things by negative ones, (c) if U_s is the numerical utility value of some standard item S and U_o is the numerical utility value of some other item O, then if both S and O are desirable and O is X times as desirable as S, $U_o = XU_s$ (or, in general, if O is X times as desirable-or-undesirable as S is desirable-or-undesirable, $|U_o| = X|U_s|$). Such a scale would, of course, depend upon the desirability of the standard item and the numerical utility value assigned to it.

Unfortunately, utilities are subject to large errors in self-assessment and therefore may be as disproportionate to actual desirability as objective values are, or even more so, unless certain rigorous procedures are followed. These procedures improve matters greatly but do not guarantee perfect proportionality.

8-1-5 Measuring Utility

Consider three objects or things O_1, O_2, and O_3 where both O_2 and O_3 are more desirable than O_1, and O_3 is more desirable than O_2. Then if U_1, U_2, and U_3 are their respective utilities, $U_1 < U_2 < U_3$. Now if we know the numerical values of two of these three U_i's, we can obtain the third in the following manner. The person whose utilities are in question is presented with a choice between (a) receiving O_2 for certain and (b) taking a gamble in which he puts up no stake but wins either O_3 or O_1, the probability of winning the most desirable object O_3 being p and that of winning the least desirable object O_1 being $1 - p$. We vary p until he is indifferent between the two choices, i.e., finds choices (a) and (b) equally attractive. At this point we have forced the utility of O_2 to equal the expected utility of the gamble so that

$$U_2 = (1 - p)U_1 + pU_3 \qquad (8\text{-}2)$$

And since we know the values of p and two of the U_i's, we can solve the equation for the unknown U_i.

Of course, we need to start with two objects having known utilities. However, they are often easily obtained. In most cases it is not unreasonable to suppose that physical nothingness has a neutral desirability, being neither desirable nor undesirable. So we can often let one of the comparison objects be "nothing" and assign it a utility of zero. And in other cases we may know from research what physical stimulus object corresponds to psychological indifference. The second utility value can be obtained by deciding upon an object to serve as a "standard," arbitrarily assigning it a numerically convenient and appropriately "standard" utility value, such as 1, and using it as one of the three objects whose utilities appear in Formula 8-2. If the two known utilities in Formula 8-2 have the values zero and one, the unknown utility may be U_1 (< 0 since $U_1 < U_2 < U_3$), U_2 (between 0 and 1), or U_3 (> 1), in which cases Formula 8-2 simplifies respectively to $U_1 = -p/(1 - p)$, $U_2 = p$, and $U_3 = 1/p$. It is often convenient, when dealing with a set of objects having nonnegative utilities, to let the most desirable of these objects be the standard with a utility of 1. In that case, if one of the objects is the "empty object" having a utility of zero, the above method yields a utility scale extending from zero to one on which any object has a calculated utility value of p.

For example, if for me zero dollars has zero utility and a million dollars has a utility of 1, I can determine the utility to me of $100,000. I simply ask myself what value p must have to make me indifferent between receiving $100,000 for certain and a gamble in which I receive either $1,000,000, the probability for which is p, or $0, the probability for which is $1 - p$. After some reflection I conclude that the two choices would be equally attractive only if $p = .96$. Therefore, $U_1 = 0$, $U_3 = 1$, $p = .96$ and, from Formula

8-2, the utility to me of \$100,000 is

$$U_2 = (1 - p)U_1 + pU_3 = (.04)0 + (.96)1 = .96$$

That is, $U_2 = p$ as shown in the preceding paragraph.

If one wishes to determine an entire spectrum of utilities, i.e., a utility function, rather than a single value, he can take advantage of the fact that most people find it easier to conceptualize gambles that have equal probabilities for winning and losing. Thus, if Q_0 and Q_1 are quantities having utilities of 0 and 1 respectively, and Q is some intermediate quantity, he asks himself what value Q would have to have to make him indifferent between receiving Q for certain and taking a gamble in which he had equal chances of receiving Q_0 and Q_1. The resulting value of Q has a utility of .5 and will be designated $Q_{.5}$. He can then ask what value Q must have to make him indifferent between Q for certain and a gamble in which he has equal chances of receiving Q_0 and $Q_{.5}$. This value of Q has a utility of .25 and is designated $Q_{.25}$. Using $Q_{.5}$ and Q_1 he can determine $Q_{.75}$. Using $Q_{.25}$ and $Q_{.75}$ he can *again* determine $Q_{.5}$ and thereby check its reliability and his own consistency, *making* himself more consistent by modifying his entire sequence of responses if necessary. Then using Q_0 and $Q_{.25}$, he can determine $Q_{.125}$; using $Q_{.25}$ and $Q_{.5}$, he can determine $Q_{.375}$, etc., continuing this "fractionation method" until he has enough points to plot his utilities as a function of quantities, i.e., subscripts as a function of the corresponding subscripted Q values. Once the graph is obtained, utilities can be read from it for any quantity between Q_0 and Q_1, *so long as the attitudes* (and the factors causing them, such as financial condition) *of the person involved do not change.*

In order to infuse accuracy into self-assessed utility values, a great deal of cross-checking of the type mentioned in the preceding paragraph is necessary. Thus, while accurate utilities are more appropriate indices of desirability than are objective or economic values, it is often tedious to evaluate them correctly and sometimes difficult to evaluate them at all. Consequently, in the many problems where utilities are almost directly proportional to objective quantities, it is often more convenient and just as accurate to use the latter. One should always be alert to the exception, however, and be suspicious of a problem that treats quintuplets as 2.5 times as desirable as twins.

8-1-6 Expected Utility

We already know that if a random variable X has k different possible values or outcomes $x_1, x_2, \ldots, x_i, \ldots, x_k$ that it takes proportions of the time $p_1, p_2, \ldots, p_i, \ldots, p_k$, respectively, then the expected value of X is defined by the formula $E(X) = \sum_{i=1}^{k} x_i p_i$. Now a utility is a special type of value—the subjective value attached to an objective situation, object, or outcome—and (as implied earlier) it too can have an expectation in complete analogy with expected objective value. So, if the outcomes or values $x_1, x_2, \ldots, x_i, \ldots, x_k$ have utilities $U_1, U_2, \ldots, U_i, \ldots, U_k$, respectively, the **expected utility** of X is

$$E(\text{utility of } X) = \sum_{i=1}^{k} U_i p_i \qquad (8\text{-}3)$$

Thus, if X is humidity and takes the values "dry," "moderate," and "damp," 60%, 30%, and 10% of the time, respectively, and if these three conditions have utilities for me of 1, 0, and -27, then for me the expected utility of the humidity is

$$E(\text{utility of the humidity}) = 1(.60) + 0(.30) - 27(.10)$$
$$= .6 + 0 - 2.7 = -2.1$$

8-2   Two-Person Zero-Sum Games

8-2-1   General Theory

Much of decision making can be conceptualized as a game between two players, one of which is the decision maker, the other being either another decision maker or the forces of Nature. Decision-making is at its trickiest when "the other player" is a thinking adversary intent upon thwarting one's objectives. The theory of decision making under such circumstances has been most satisfactorily worked out for the ruthlessly competitive situation in which each player tries to maximize his well-being at the other's expense under the following highly idealized conditions illustrated in Table 8-1: Each player has available to him a limited number of possible actions and each player knows not only what his own possible actions are but also what those of his opponent are; furthermore, each player is rational and will not intentionally take action contrary to his self-interest. For any given pairing of action a_i, chosen by Player A, with action b_j chosen by Player B, there is a consequence or outcome O_{ij}, and both players know what these outcomes are. The O_{ij} are amounts (e.g., dollars) given up (or lost) by Player B to Player A, so the O_{ij} are **payoffs** received by A from B. When an O_{ij} is positive, it represents an amount lost by B to A, i.e., a gain to A; when it is negative, it represents an amount lost by A to B, i.e., a loss to A. What A gains B loses and vice versa. Therefore, the *net* gain to the two players combined (but not necessarily individually) is zero. For this reason the situation is called a **Two-Person Zero-Sum Game**. The actual play of the game consists of each player simultaneously selecting a single one of the actions available to him. And the outcome is the payment from B to A of the O_{ij} corresponding to (i.e., identified by) A's action a_i and B's action b_j. Practically speaking, the simultaneity with which A and B choose their actions means simply that

Table 8·1

Payoff Matrix

Player B
(minimizing player)

	b_1	b_2	b_3	...	b_j	...	b_s
a_1	O_{11}	O_{12}	O_{13}	...	O_{1j}	...	O_{1s}
a_2	O_{21}	O_{22}	O_{23}	...	O_{2j}	...	O_{2s}
a_3	O_{31}	O_{32}	O_{33}	...	O_{3j}	...	O_{3s}
$\vdots$	$\vdots$	$\vdots$	$\vdots$		$\vdots$		$\vdots$
a_i	O_{i1}	O_{i2}	O_{i3}	...	O_{ij}	...	O_{is}
$\vdots$	$\vdots$	$\vdots$	$\vdots$		$\vdots$		$\vdots$
a_r	O_{r1}	O_{r2}	O_{r3}	...	O_{rj}	...	O_{rs}

Player A
(maximizing player)

neither knows the other's choice until after his own choice is made and that both choices once made are irrevocable.

Since each player knows all the options and all the consequences, he may be in an excellent position to prevent his opponent from obtaining the outcome he desires the most. (For every action by one player there is generally an optimal counter-action by the other.) And since one player's gain is the other's loss, a player can hardly expect his opponent to be sympathetic to his objectives. Under these circumstances a conservative strategy has much to recommend it. The most cautious and pessimistic strategy of all would be to choose that action that would have the least disastrous consequences even if one's opponent knew in advance that one had chosen it and could therefore choose the most appropriate counter-action. Such a strategy is illustrated in Table 8-2. If Player A took action a_1 and B knew it, B would take action b_2 since that would require the smallest payment 4 to A. If A chose a_2 and B knew it, B would choose b_1 since the outcome 5 of a_2 and b_1 is the smallest of the four payments 5, 7, 11, and 21 in the a_2 row. And if B knew that A had chosen a_3, B would choose b_3 since the payment under b_3 is the smallest payment associated with action a_3. Therefore, if B knew his actions, A could expect to receive 4 under action a_1, 5 under action a_2, and 1 under action a_3. While A cannot control B's actions, he can control his own and he wants to

Table 8-2

Illustration of Maximin and Minimax Strategies

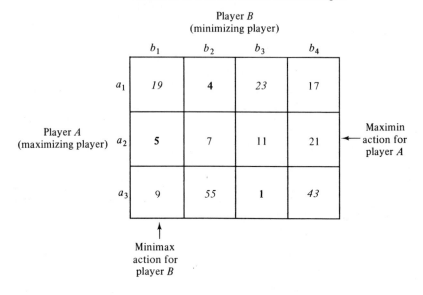

Player *B*
(minimizing player)

	b_1	b_2	b_3	b_4
a_1	*19*	**4**	*23*	17
a_2	**5**	7	11	21
a_3	9	*55*	**1**	*43*

Player *A*
(maximizing player)

Maximin action for player *A* ← (a_2 row)

↑ Minimax action for player *B* (b_1 column)

Minimum entry in each row is **boldface**. *A*'s maximin strategy is to choose the row (action) whose minimum entry is largest, thereby maximizing the minimum payoff he can receive from *B*. Maximum entry in each column is *italicized*. *B*'s minimax strategy is to choose the column (action) whose maximum entry is smallest, thereby minimizing the maximum payoff he might have to give to *A*.

receive as large a payment as possible; so he chooses action a_2 associated with 5, the largest of the three values.

This strategy amounts to determining the minimum payoff or gain associated with each action a_i and then selecting the action associated with the maximum of these minimum gains. Since it "maximizes the minimum gain," it is called a *maximin* strategy.

Since the outcomes are payments from B to A and since B wishes to pay as little as possible, the analogous pessimistic strategy for Player B is to determine the maximum payoff associated with each action b_j and then select the action associated with the minimum of these maxima. Since B is "minimizing his maximum loss," his strategy is called a *minimax* strategy. The maximum losses to B in Table 8-2 are: 19 for b_1, 55 for b_2, 23 for b_3, and 43 for b_4 of which the smallest is 19; so b_1 is the action called for if B follows a minimax strategy.

8-2-2 Two-Action Games

In the process of determining his maximin or minimax strategy, a player is really only considering whether one payoff is larger or smaller than another.

That is, he need pay attention only to the size *ranks* of the payoff values, not to the values themselves. Thus, in a table such as Table 8-2, if one replaces the payoffs by their size ranks, the maximin and minimax strategies will call for the same actions as before. Now let us confine our attention to games in which Player A will follow a maximin and Player B a minimax strategy, in which each player has only two possible actions, and in which the four payoffs are four different amounts (no ties). Then the payoffs can be replaced by the ranks, 1, 2, 3, and 4, and *all possible situations* are represented by the $4! = 24$ permutations of these four ranks within the four cells of the square payoff matrix. However, half of these permutations differ from some permutation in the remaining half only in that the entries under b_1 are switched with those under b_2 (i.e., whole columns of entries are switched). Since it doesn't really matter which action we call b_1 and which b_2, for our purposes such permutations are redundant and can be eliminated. Of the remaining 12 permutations, half of them differ from some permutation in the remaining half only in that entries in row a_1 are switched with those in row a_2, and since the labeling of rows is arbitrary, these permutations are also redundant and can be eliminated.

This leaves us with six matrices that represent all situations of interest under the conditions mentioned. They are given as Table 8-3. For convenience, the matrices are arranged so that a_1 is always Player A's maximin action and b_1 is always Player B's minimax action. For games in general (multi-action as well as two-action), an action is said to be **dominated** when some other single action exists that is sometimes better and always at least as good. (Thus, for Player A action a_d is dominated by action a_x if, for every action that B might take, i.e., in every column, the payoff to A under a_d is $\leq$ that under a_x and for at least one action that B might take it is less. And for Player B action b_d is dominated by action b_x if, no matter what action A takes, i.e., in every row, the payoff from B to A under b_d is $\geq$ that under b_x and for at least one action that A might take it is greater.) In Table 8-3 when one action is dominated by another, the latter, superior, action is indicated by boldface type.

If a player is determined to follow a maximin or minimax strategy, the arrow in Table 8-3 tells him which action to choose. However, let us now inquire into the *desirability* of such a strategy. In games I and II of Table 8-3 both players have a dominating action that is *optimal irrespective* of what the other player may do or how many times the game is played. Therefore, it behooves both players to pick, and stay with, their respective dominating action. Doubly-dominated two-action games are therefore completely stable in that there is no motivation for either player to depart from his dominating action. And since that action corresponds to the maximin action by Player A and the minimax action by Player B, the maximin-minimax criteria are completely satisfactory in such cases.

In games III and IV one player has a dominating action and the other does

Table 8-3

The Six Possible Games

when each of two players has only two actions, maximin and minimax
strategies are followed, and there are no ties among payoffs
(Dominating actions and their payoffs are in boldface;
maximin and minimax actions are indicated by arrow.)

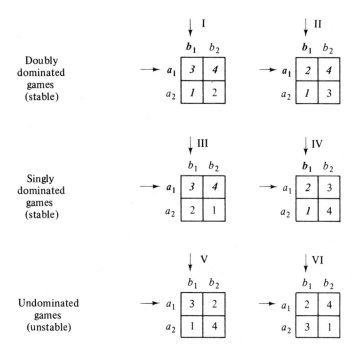

not. The player who does can be expected to use it, since it is universally optimal irrespective of the other player's behavior either in a single game or in repeated games. And if the player with the dominating action chooses it, the other player can only lose by departing from his maximin or minimax action. Again, therefore, the game is stable and the maximin-minimax criteria are completely satisfactory.

In games V and VI there are no dominating actions and therefore no actions that are optimal *irrespective* of what the other player does. This means that each player's motivations and strategies are relevant to the other player's choice of actions and introduces a vastly complicating psychological factor. These games are highly unstable in that if both players "knew" (or cunningly foresaw) what actions would be chosen, one of them could always improve his position by changing his choice. For example, in game V if Player B knew that Player A would choose the maximin action a_1, then B should abandon the minimax strategy and choose action b_2, thereby reducing his payment from 3 to 2. But if A knew that B would do so, then A should switch to a_2, thereby increasing his gain from 2 to 4. However, if B became aware of this plan, he should then switch from b_2 back to the minimax action b_1. And if A knew of this, he should switch back to a_1, which brings the two players back to where they were when both were following the maximin-minimax criterion. At this point, of course, the whole procedure can recycle.

8-2-3 Mixed Strategies

In the case of unstable games, A can still maximize his minimum gain by playing a maximin strategy and B can still minimize his maximum loss by a minimax strategy. However, when they do so the benefits often seem to be inequitably distributed so that one player gets by far the best of it. Consider, for example, the game shown in Table 8-4. The maximin action for Player A is a_1, the minimax action for Player B is b_2, and if the players choose these respective actions, A will receive a payoff of 1, which is only slightly greater than the worst of the four payoffs but is considerably less than either of the two best payoffs. This situation is pleasant for B but not for A. If A is determined not to risk receiving a payoff of 0, there is nothing better that he can do.

However, if he is willing to abandon the maximin strategy, he can greatly improve his expected gain by following a ***mixed strategy***. This consists of letting "chance" determine his choice of actions: a_1 is given probability p of being chosen and a_2 therefore has probability $1 - p$, the value of p being such that A's expected gain is unaffected by B's choice of action. Thus, if A will randomly draw a marble from a jar full of marbles, a proportion p of which are white, and choose a_1 only if the drawn marble is white, then the expected value of the payoff is

$$10p + 0(1 - p) \qquad \text{if B chooses } b_1$$

and

$$1p + 6(1 - p) \qquad \text{if B chooses } b_2$$

Table 8-4

Player B
(minimizer)

	b_1	b_2
a_1	10	1
a_2	0	6

Player A
(maximizer)

Since Player A doesn't know which action B will choose (B can see that A may gain by using certain nonmaximin strategies and can be expected to consider counter-strategies), he can "protect himself" against this uncertainty by choosing p so that it doesn't matter *what* B does, i.e., by letting p have the value that makes A's expected gain the same under b_1 as it is under b_2. This value of p is obtained by equating the expected gain under b_1 with the expected gain under b_2 and solving for p:

$$10p + 0(1 - p) = 1p + 6(1 - p)$$
$$15p = 6$$
$$p = \frac{2}{5} = .4$$

So if A lets chance determine his choices in such way that a_1 has a probability of .4 of being chosen, the expected value of the payoff he will receive is

$$E(\text{payoff} | b_1) = 10(.4) + 0(.6) = 4 + 0 = 4$$
$$E(\text{payoff} | b_2) = 1(.4) + 6(.6) = .4 + 3.6 = 4$$

and therefore, since the conditional expected payoffs are equal, the unconditional $E(\text{payoff}) = 4$. When the smallest possible payoff is 0, an expected payoff of 4 probably looks better to A than a guaranteed payoff of 1 even if the game is to be played only once. And if the game is to be played repeatedly, A would have to be cursed with abominable luck to fare worse under the mixed strategy than under the maximin strategy.

Player B, of course, cannot count upon Player A to choose between a maximin and a mixed strategy. Player A might, instead, follow a psychological "poker strategy" of trying to outguess Player B. For example, if A thought B would follow a minimax strategy, A would be motivated to choose a_2, and if A guessed correctly he would collect 6 from B. To protect himself against such contingencies, B might resort to an analogous mixed strategy in which there is a probability π of choosing b_1 and $1 - \pi$ of choosing b_2, where π is whatever value makes the expected payoff the same under all possible actions by A, and that value is obtained as follows:

$$E(\text{payoff} | a_1) = E(\text{payoff} | a_2)$$
$$10\pi + 1(1 - \pi) = 0\pi + 6(1 - \pi)$$
$$15\pi = 5$$
$$\pi = \frac{1}{3}$$

The expected payoff, for unstable two-person two-action games, will always be the same for B's mixed strategy as for A's (provided only that the "mixture" is such that the expected payoff is the same under all of the opponent's actions):

$$E(\text{payoff}\,|\,a_1) = 10\left(\frac{1}{3}\right) + 1\left(\frac{2}{3}\right) = \frac{12}{3} = 4$$

$$E(\text{payoff}\,|\,a_2) = 0\left(\frac{1}{3}\right) + 6\left(\frac{2}{3}\right) = \frac{12}{3} = 4$$

So the unconditional E(payoff) = 4 for B's mixed strategy, just as it was for A's. Thus, in the case of unstable games a mixed strategy has attractive features for either player and should often prove desirable to the more vulnerable of the two players, especially if the game is to be played repeatedly.

8-2-4 Multi-action Games

Not all of what we have said about two-action games is generalizable to multi-action games. The procedure is somewhat analogous, but there are complications. The selection of maximin or minimax strategies *is* directly generalizable. And if, for either player, a single action dominates all others, then the game is stable, and a maximin strategy for Player A and a minimax strategy for Player B will be optimal. But there is another situation that will also produce a sort of equilibrium, even if no actions are dominated. For example, if the outcome corresponding to A's maximin action and B's minimax action is such that all other payoffs in the same row are at least as large and all other payoffs in the same column are at least as small (in which case the outcome is called a *saddlepoint*), then if both players believe the other player will follow the appropriate maximin-minimax strategy, neither player is motivated to do other than what his opponent anticipates, and the game is in equilibrium (see Table 8-5). Another complication is encountered in the selection of mixed strategies. Even after all dominated actions have been eliminated, it may not be possible to select action probabilities resulting in equal expected payoffs under all remaining actions by one's opponent. And sometimes an action, though dominated by no other single action, is "dominated" by a mixture of other actions.

Thus, multi-action zero-sum games are more complicated than two-action zero-sum games (and *non*zero-sum games are far more complex than zero-sum games). However, this need not concern us, since our primary objective has been simply to show the basic elements of decision making when one's opponent is a strategizing and counterstrategizing adversary.

8-2-5 Summary and Critique

The theory of nonzero-sum games has not yet been very satisfactorily developed. That of zero-sum games has been, but the assumptions of zero-sum

Table 8-5

Illustration of Saddle Point and Dominated Actions

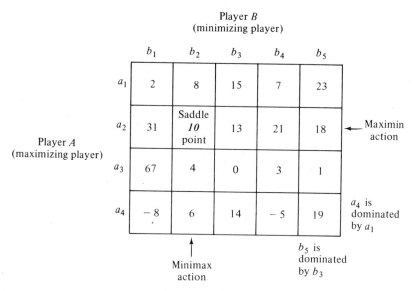

Player *B*
(minimizing player)

		b_1	b_2	b_3	b_4	b_5	
	a_1	2	8	15	7	23	
Player *A* (maximizing player)	a_2	31	Saddle *10* point	13	21	18	← Maximin action
	a_3	67	4	0	3	1	
	a_4	− 8	6	14	− 5	19	a_4 is dominated by a_1

↑
Minimax
action

b_5 is dominated by b_3

Payoff corresponding to *A*'s maximin action and *B*'s minimax action is lowest value in its row and highest value in its column, which makes it a saddle point. This provides a certain equilibrium in that neither player is motivated to depart from the maximin-minimax strategies unless he believes the other will do so.

Every payoff under a_4 is worse, from *A*'s point of view, than the corresponding payoff (in the same column) under a_1, so a_4 is dominated by a_1. Every payout under b_5 is worse, from *B*'s point of view, than the corresponding payout (in the same row) under b_3, so b_5 is dominated by b_3.

games are seldom met or even closely approximated by practical real-world decision-making situations. Competitors are rarely completely informed of all possible actions accessible to their opponent and all possible consequences of pairs of actions. They are seldom completely ignorant of each other's choices (or their likelihoods) and their actions are seldom both irrevocable and unmodifiable. *A*'s *objective* (e.g., dollar) gain is unlikely to be entirely at *B*'s expense. And a zero-sum *utility* matrix (in which B's utility loss becomes A's utility gain) strains credulity. Yet it is utilities that reflect choice tendencies and that are therefore the appropriate entries in the payoff matrix. For all these reasons it would be well to regard the zero-sum game as theoretically instructive but as only a very gross and approximate model, at best, for the solution of practical problems.

8-3 Problems

1. A variable X takes the value $x_1 = -5$ a proportion $p_1 = .7$ of the time, the value $x_2 = 4$ a proportion $p_2 = .2$ of the time, and the value $x_3 = 21$ a proportion $p_3 = .1$ of the time. What is the expected value of X?

2. A pair of fair dice are to be tossed. One die is red, the other green. The value of the outcome is the sum of (a) the number coming up on the red die plus (b) the number coming up on the green die only if it is an odd number, otherwise zero. What is the expected value of the outcome?

3. Of the items produced by a certain factory, 50% are flawless, 20% have one flaw, 15% have two, 10% have three, and 5% have four. What is the expected number of flaws in an item?

4. If ultimately the percentages of marriages producing 0, 1, 2, 3, 4, 5, 6, 7, 8, 9, and 10 children are, respectively, 10, 20, 30, 15, 8, 6, 4, 3, 2, 1, 1, what is the expected number of children resulting from a marriage?

5. A man is going to throw a fair die, draw a card at random from a bridge deck, and toss a fair coin. If the die comes up 2 and the drawn card is an ace and the coin lands heads, he wins $1,000; otherwise he loses X. What must X be in order for his expected gain to be zero?

6. Mike has no means of transportation. To him, no vehicle at all has a utility of zero, a bicycle has a utility of 1, and a motorcycle is more desirable than a bicycle. He is indifferent between (a) receiving a bicycle for certain and (b) randomly drawing a marble from an urn containing two white and eight black marbles and receiving nothing if he draws out a white marble or a motorcycle if he draws out a black one. What is the utility of the motorcycle for Mike?

7. What would be the answer to the above problem if a white marble meant receiving a motorcycle and a back marble meant receiving nothing?

8. Using the fractionation method, construct your own utility function for money, from zero to one million dollars. What is the shape of the function and why is it that shape?

9. In the two-person zero-sum game shown in Table 8-6
   (a) What action should Player A select if he wishes to follow a maximin strategy?

Table 8-6

Payoff Matrix for Problem 9

Player B

		b_1	b_2	b_3	b_4	b_5	b_6
	a_1	35	33	30	2	50	60
	a_2	110	140	50	100	105	115
Player A	a_3	1,500	1,000	90	3	1,110	1,700
	a_4	30	43	25	1	100	75
	a_5	100	90	40	2	45	90

(b) What action should Player B select in order to follow a minimax strategy?

(c) Which of Player A's actions are dominated and, for each of them, by what actions are they dominated?

(d) Which of Player B's actions are dominated and, for each of them, by what actions are they dominated?

(e) Construct the payoff matrix remaining when all dominated actions are eliminated.

(f) In this reduced payoff matrix, if Player A followed a mixed strategy, what would be the action probabilities and the expected payoff?

(g) In the reduced payoff matrix, if Player B used a mixed strategy, what would be the action probabilities and the expected payoff?

(h) Which of the two players has the greater inducement to follow a mixed strategy and why?

10. In the two-person zero-sum game shown in Table 8-7a what is the payoff to Player A if he follows a maximin strategy and Player B follows a minimax strategy? Which player would benefit in the long run by abandoning the maximin or minimax strategy in favor of a mixed strategy? What are the action probabilities and the expected payoff under that mixed strategy?

11. Answer the questions in Problem 10 but with regard to the game shown in Table 8-7b.

Table 8-7

Payoff Matrices for Problems 10 Through 13

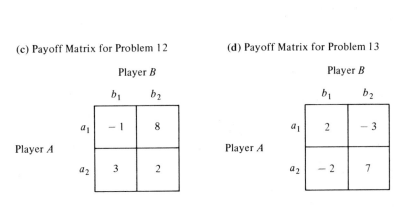

(a) Payoff Matrix for Problem 10

Player B

	b_1	b_2
a_1	1,000	100
a_2	0	600

Player A

(b) Payoff Matrix for Problem 11

Player B

	b_1	b_2
a_1	2	8
a_2	10	0

Player A

(c) Payoff Matrix for Problem 12

Player B

	b_1	b_2
a_1	− 1	8
a_2	3	2

Player A

(d) Payoff Matrix for Problem 13

Player B

	b_1	b_2
a_1	2	− 3
a_2	− 2	7

Player A

12. Answer the questions in Problem 10 but with regard to the game shown in Table 8-7c.

13. Answer the questions in Problem 10 but with regard to the game shown in Table 8-7d.

14. Two supermarkets compete for business in a town having no other grocery stores. Each supermarket decides to spend $50,000 on improvements. The options considered by supermarket A are: (a_1) increase the size of the parking lot, (a_2) play continous music inside the store, (a_3) give trading stamps. Those considered by supermarket B are: (b_1) buy more colorful uniforms for the salesclerks, (b_2) have lotteries, (b_3) give free balloons to children, (b_4) renovate the store. From past records kept by the central office of each supermarket chain, it is known what percentage of the market is won by A for each of the $a_i b_j$ combinations.

Table 8-8

Payoff Matrix for Problem 14

Supermarket B

		b_1	b_2	b_3	b_4
	a_1	30	− 10	25	− 5
Supermarket A	a_2	15	− 15	10	− 25
	a_3	18	− 15	10	20

This payoff matrix is given in Table 8-8. What should each supermarket do, and why?

15. A will repeatedly toss a fair coin until it comes up heads or for 100 tosses, whichever comes first. If the coin first comes up heads on the nth toss (where $n \leq 100$), B will pay A 2^n dollars. But if in 100 tosses it never comes up heads, B will pay A nothing. What is the expected value of the amount A will win from B? What is the probability that a single game will last longer than 10 tosses? What is the probability that a head will not come up in 100 tosses?

9

decisions

9-1 Decision Making Under "States of Nature"

In many, if not most, decision-making situations, the contingencies with which one must cope are not the hostile and counterstrategizing actions of a thinking antagonist but, rather, various "states of nature" whose occurrence or nonoccurrence is essentially independent of the action one chooses or might choose. For a farmer the relevant states of nature might be adequate rain, flood, drought, etc.; for a businessman they might be recession, inflation, interest rates, etc.; for a scientist they might be "theory works," "theory doesn't work," etc. Thus, states of nature are simply potential situations that could affect the outcome of the decision maker's actions but that are not *controlled* by any person who might use that control to influence the outcome for or against the decision maker.

9-1-1 The Game Against Nature

Such situations can be regarded as games in which nature is Player B and the relevant states of nature are Player B's "actions." "Nature" need not be inanimate, but, if animate, it must be uninfluenced by the possible strategies of the decision maker, Player A. The game, of course, need not be, and probably *is* not, a zero-sum game. The theory of proper decision making is far better worked out for this type of "game" than for games against a thinking antagonist. At this point, we shall abandon game terminology and rationale and substitute the terminology and rationale appropriate to decision making under states of nonstrategizing nature. The two have much in common.

In making such decisions we wish to take the action most appropriate to (i.e., having the most favorable outcome under) the situation that we are in. The vital ingredients of a decision-making predicament are: (a) a limited number of possible actions $a_1, a_2, \ldots, a_i, \ldots, a_r$ that are relevant and available to the decision maker and from which he must choose a single one, (b) a limited number of possible situations $\theta_1, \theta_2, \ldots, \theta_j, \ldots, \theta_s$, one of which is the *true* situation (the state of nature prevailing) on this particular occasion, (c) a set of outcomes O_{ij} of taking action a_i when the true situation is θ_j, together with their *utilities* $U(O_{ij})$, or more briefly U_{ij}, which represent the consequences, and therefore reflect the appropriateness, of the actions under the various situations. These ingredients are shown in their proper relationship in Table 9-1, which presents a sort of general format for decision problems when the true situation is a state of nature.

Table 9-1

Payoff Matrix for Decision Making Under States of Nature

Possible true situations or "States of Nature"

		θ_1	θ_2	θ_3	$\cdots$	θ_j	$\cdots$	θ_s
	a_1	$U(O_{11})$	$U(O_{12})$	$U(O_{13})$	$\cdots$	$U(O_{1j})$	$\cdots$	$U(O_{1s})$
	a_2	$U(O_{21})$	$U(O_{22})$	$U(O_{23})$	$\cdots$	$U(O_{2j})$	$\cdots$	$U(O_{2s})$
Actions available to the decision maker	$\vdots$	$\vdots$	$\vdots$	$\vdots$	$\vdots$	$\vdots$	$\vdots$	$\vdots$
	a_i	$U(O_{i1})$	$U(O_{i2})$	$U(O_{i3})$	$\cdots$	$U(O_{ij})$	$\cdots$	$U(O_{is})$
	$\vdots$	$\vdots$	$\vdots$	$\vdots$	$\vdots$	$\vdots$	$\vdots$	$\vdots$
	a_r	$U(O_{r1})$	$U(O_{r2})$	$U(O_{r3})$	$\cdots$	$U(O_{rj})$	$\cdots$	$U(O_{rs})$

9-1-2 Degrees of Knowledge About the State of Nature

The decision maker may be in any one of four different situations, corresponding to his degree of knowledge about the state of nature. These situations will be termed "certainty," "risk," "uncertainty," and "ignorance."

In *decision making under certainty*, the decision maker *knows* which of its possible states nature is in at present. He therefore simply takes the action most appropriate to the known value of θ.

In *decision making under risk*, the decision maker knows the *objective* probability for each possible state of nature but does not know which of these states nature is in at present. Thus, although he does not know the present value of θ, he knows the objective probability distribution of θ, i.e., he knows the objective $P(\theta = \theta_j)$ for every value of j.

In *decision making under uncertainty*, the decision maker knows the possible states of nature and is willing and able to supply and use *subjective* probabilities for each of them; however, he does not know the present state of nature nor the objective probability for each possible state. Thus, although he does not know either the present value of θ or the objective $P(\theta = \theta_j)$'s, he "knows" his subjective $P(\theta = \theta_j)$ for every value of j and is willing to use these subjective probabilities in making decisions.

In *decision making under ignorance*, the decision maker knows the possible states of nature, does not know the objective probabilities for the states, and is either unable or unwilling to supply and use subjective probabilities. In effect, therefore, he is completely ignorant of the $P(\theta = \theta_j)$'s, irrespective of whether they be objective or subjective. (Customarily the last two situations, "uncertainty" and "ignorance," are regarded as a single category "decision making under uncertainty"; however, we shall distinguish between them in this book.)

9-1-3 Criteria and Methods for Choosing Actions

There are a variety of *circumstances* in which the decision maker may find himself, depending upon combinations of factors such as his degree of knowledge about the state of nature, the frequency with which similar decisions will have to be made (i.e., the number of times he will play this game against nature), his willingness to gamble on the basis of long-run considerations, the relative importance of the *present* decision, his willingness or ability to obtain sample information, etc. There are also a (much smaller) variety of established *methods* for arriving at a decision, each based upon a different criterion for choosing among possible actions. Each method is highly appropriate to *some* combination of circumstantial factors. But there are *many* combinations for which *no* method is entirely satisfactory. In such cases the decision maker may either select the most nearly appropriate of the inappropriate methods or reject them all (relying perhaps upon pure intuition in making his decision).

All of the methods to follow presuppose that the decision maker "knows" Table 9-1—that he knows what the possible *states* of nature are, knows what the possible *actions* are, and knows the quantitative (e.g., numerical) *utility* value corresponding to each possible pairing of an action with a state of nature. Also, in what follows, these utilities for the outcome of the *i*th action under the *j*th state of nature, previously symbolized $U(O_{ij})$, will be abbreviated to U_{ij}, and $P(\theta = \theta_j)$ will be abbreviated to $P(\theta_j)$.

9-2   The Maximum Utility Criterion

The simplest of all decision-making situations is that in which the decision maker *knows* what the present state of nature is. This is decision making under certainty and the appropriate method is to use the **maximum utility criterion**: Knowing that $\theta = \theta_x$, the decision maker simply takes the action that produces the outcome with the largest utility when $\theta = \theta_x$. Thus, he need only inspect the utilities in the column headed θ_x, find the largest one, and take the action corresponding to the row in which it is found. This method is optimal for decision making under certainty.

Example: An employee of a certain company must decide between taking his vacation at: (a_1) a fancy resort, (a_2) his cabin in the mountains, (a_3) a rented cottage by the seashore, (a_4) at home. Each year his company follows one of three policies: (θ_1) it pays all of every employee's vacation expenses, (θ_2) it pays half, (θ_3) it pays none. The employee's utilities for the various possible actions under the various possible states are given in Table 9-2.

Table 9-2

Vacation Example of Decision Making Under Certainty
Using Maximum Utility Criterion

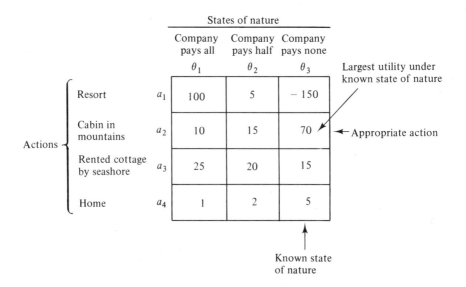

			States of nature			
			Company pays all	Company pays half	Company pays none	
			θ_1	θ_2	θ_3	Largest utility under known state of nature
	Resort	a_1	100	5	-150	
Actions	Cabin in mountains	a_2	10	15	70	←Appropriate action
	Rented cottage by seashore	a_3	25	20	15	
	Home	a_4	1	2	5	

Known state
of nature

The directors of the company meet and announce that this year the company will pay none of an employee's vacation expenses. The employee therefore is certain that $\theta = \theta_3$. Looking at the utility values in the column headed θ_3, he finds the largest utility 70 to correspond to action a_2. So he selects action a_2 and takes his vacation at his cabin in the mountains.

9-3   The Maximum Expected Utility Criterion

9-3-1   Method

If in addition to the actions a_i's, states θ_j's, and utilities U_{ij}'s, the decision maker (a) also knows the probabilities for each of the states of nature, the $P(\theta_j)$'s, and (b) is willing to base his decision on long-run considerations, he can use the **maximum expected utility criterion**, as follows. For each action a_i its *expected utility*

$$E[U(a_i)] = \sum_j U_{ij} P(\theta_j) \tag{9-1}$$

is calculated and the action having the largest expected utility is chosen.

9-3-2   Examples

Example 1:   A good example of this type of decision making situation is that of a merchant who must stock perishable items to supply a demand that fluctuates randomly but whose probability distribution is known from past experience. If bananas cost the merchant 20 cents a bunch and sell for 60 cents, he makes a profit of 40 cents on each bunch sold but suffers a loss of 20 cents on each bunch that rots on his shelf. The number of bunches he should stock obviously depends upon the state of demand. From past records of his banana purchases and subsequent sales he knows that, before they start to rot, he has been able to sell: none of them 5% of the time, one bunch 10% of the time, two bunches 45%, three bunches 30%, four bunches 10%, and more than four bunches never. He therefore has enough information to use an expected utility criterion, and if he can afford to suffer the occasional losses involved in particular cases, he should probably use this criterion since it maximizes his long-run expected profits. Table 9-3 summarizes his information. The cell entries are his profits (assumed to equal his utilities U_{ij}) if he chooses stock action a_i when the demand state of nature is θ_j. The probability that θ_j is the state of demand is given below each column. Thus, the expected utility for action a_4 (stock 3 bunches) is

$$E[U(a_4)] = \sum_j U_{4j} P(\theta_j)$$
$$= (-60)(.05) + (0)(.10) + (60)(.45) + (120)(.30) + (120)(.10)$$
$$= -3 + 0 + 27 + 36 + 12 = 72$$

Table 9-3

Supply-and-Demand Example of Decision Making Under Risk Using Maximum Expected Utility Criterion

(Each item costs 20¢, sells for 60¢, and is perishable; so utilities, i.e., cell entries, are total price of items sold [60 times number of items that are both stocked and demanded] minus total cost of items stocked [20 times number stocked].)

Number of items demanded

	θ_1 0	θ_2 1	θ_3 2	θ_4 3	θ_5 4	Expected utility $\sum_j$ (cell entry in column j)$P(\theta_j)$
a_1 : Stock 0	0	0	0	0	0	0
a_2 : Stock 1	− 20	40	40	40	40	37
a_3 : Stock 2	− 40	20	80	80	80	68
a_4 : Stock 3	− 60	0	60	120	120	72 ⟵ Appropriate action
a_5 : Stock 4	− 80	− 20	40	100	160	58

Supplying actions

Demand probability, $P(\theta_j)$　.05　.10　.45　.30　.10

which is larger than the expected utility for any other action; therefore, if he employs an expected utility criterion, the merchant should take action a_4 and stock three bunches of bananas.

Example 2: A second and more elaborate example is afforded by the following industrial problem, in which it is supposed that utilities can be represented adequately by profits. During a production run, a certain machine turns out 1,000 items, of which the good items can be sold for $10 each and the defectives must be scrapped. The machine has three states, depending upon its condition just prior to the beginning of a run. If a certain critical part does not need either adjustment or replacement, the machine is in "good" condition and will produce, on the average, only 10 defective items during the 1,000 item run. If the crucial part needs to be adjusted but does not need to be replaced, the machine is in "fair" condition and will produce, on the average, 50 defectives per 1,000 items. Finally, if the crucial part needs to be replaced, in which case it must also be out of adjustment, the machine is in "poor" condition and will produce an average of 250 defectives per 1,000 items. So corresponding to states θ_1 (good), θ_2 (fair), and θ_3 (poor) the probabilities that a just-produced item will be a defective are $p_1 = .01$, $p_2 = .05$, and $p_3 = .25$, respectively.

The critical part is not readily accessible, and only by overhauling the machine can one tell whether or not the part needs adjustment or replacement. Furthermore, the cost of overhauling is so great, relative to the value of the part, that whenever the machine is overhauled the part is replaced and adjusted, thereby placing the machine in "good" condition. Therefore, management is willing to consider only two actions prior to a new 1,000 item run: a_1 "overhaul machine," which costs $500, and a_2 "don't overhaul machine," which costs nothing. The total cost to the company of making a 1,000 item run is $9,000 plus the cost of whichever action is taken: $500 under a_1 and nothing under a_2. Therefore, the average profit under action a_i and *initial* state θ_j (i.e., the state of the machine at the time the action was chosen) is $10\ G_j - (9,000 + C_i)$ where C_i is the cost of taking action a_i and G_j is the average number of good items per thousand turned out by the machine in whatever state it is in subsequent to action a_i ("good condition" if action is to overhaul, the same condition as the initial condition if action is not to overhaul). Using this formula, we obtain the average profits, which appear as cell entries in the payoff matrix shown in Table 9-4.

From past records it is known that the condition of the unoverhauled machine at the beginning of a run is "good" 50% of the time, "fair" 30%, and "poor" 20%. So $P(\theta_1) = .5$, $P(\theta_2) = .3$, and $P(\theta_3) = .2$. With this information we can calculate the expected utility $E[U(a_i)]$ for each action a_i by multiplying each utility in a_i's row by the probability for the θ_j in whose

Table 9-4

Machine Example of Decision Making Under Risk Using
Maximum Expected Utility Criterion

(Cell entries for a_1 are $10(1,000)(.99) - (9,000 + 500)$; cell entries for a_2 are $10(1,000)(1 - p_j) - 9,000$)

		State of machine before choosing action				
		θ_1 Good $p_1 = .01$	θ_2 Fair $p_2 = .05$	θ_3 Poor $p_3 = .25$	Expected utility $\sum_j$ (cell entry in column j)$P(\theta_j)$	
Action	a_1 Overhaul machine	400	400	400	400	Appropriate action
	a_2 Don't overhaul machine	900	500	$-1,500$	300	
State probability, $P(\theta_j)$		.5	.3	.2		

column it lies and summing the products:

$$E[U(a_1)] = \sum_j U_{1j}P(\theta_j) = 400(.5) + 400(.3) + 400(.2) = 400$$

$$E[U(a_2)] = \sum_j U_{2j}P(\theta_j) = 900(.5) + 500(.3) - 1500(.2) = 300$$

The larger of these expected utilities, 400, corresponds to the action a_1. So if the decision maker employs an expected utility criterion, he will take action a_1 and overhaul the machine. In the long run this choice, under these circumstances, should maximize the company's profits. The expected utility criterion is particularly appropriate for problems such as this one, in which the cell entries in the payoff matrix are *average* utility values, since calculating expected utilities involves averaging anyway. (Average utilities would be of questionable value if one wished to pursue a maximin strategy.)

9-3-3   Use of Sample Information to Revise the $P(\theta_j)$s

In the machine example, if the decision maker could take a small sample of the machine's output, the sample results would give him helpful additional information about the state of the machine. It is true that he already knows the $P(\theta_j)$'s, but these are *unconditional* probabilities about the state of the machine *in general*—unconditional upon any sample results. If the decision maker has data on the *present* state of the machine in the form of sample results, the proper state probabilities to use are the probabilities that the machine is in state θ_x conditional upon those sample results. Thus, if D represents the obtained data, the proper probabilities are obtained by use of Bayes' formula

$$P(\theta_x \mid D) = \frac{P(D \mid \theta_x)P(\theta_x)}{P(D)} = \frac{P(D \mid \theta_x)P(\theta_x)}{\sum_j P(D \mid \theta_j)P(\theta_j)} \tag{9-2}$$

Suppose that before choosing an action the decision maker ran the machine just long enough to obtain a sample of two items and that both of them turned out to be good so that there were 0 defectives in the sample. Assuming randomness of defective output by the machine in a given state and letting the proportion of defectives turned out in state θ_j be p_j,

$$P(D \mid \theta_j) = P(0 \text{ defectives in a binomial sample of } 2 \mid p_j)$$

$$= \binom{2}{0} p_j^0 (1 - p_j)^2 = (1 - p_j)^2$$

Substituting the known values of p_j into this formula, we get

$$P(D \mid \theta_1) = (1 - .01)^2 = .99^2 = .9801$$

$$P(D \mid \theta_2) = (1 - .05)^2 = .95^2 = .9025$$

$$P(D \mid \theta_3) = (1 - .25)^2 = .75^2 = .5625$$

and the unconditional probability of the data is

$$P(D) = \sum_j P(D|\theta_j)P(\theta_j) = P(D|\theta_1)P(\theta_1) + P(D|\theta_2)P(\theta_2) + P(D|\theta_3)P(\theta_3)$$
$$= .9801(.5) + .9025(.3) + .5625(.2)$$
$$= .49005 + .27075 + .11250 = .8733$$

Substituting the appropriate figures above into Bayes' formula, we obtain state probabilities conditional upon the sample data—that is, *new* state probabilities *revised* on the basis of relevant additional sample information about the *present* state of the machine, which we might call "updated" state probabilities

$$P(\theta_1|D) = \frac{P(D|\theta_1)P(\theta_1)}{P(D)} = \frac{.49005}{.8733} = .56$$

$$P(\theta_2|D) = \frac{P(D|\theta_2)P(\theta_2)}{P(D)} = \frac{.27075}{.8733} = .31$$

$$P(\theta_3|D) = \frac{P(D|\theta_3)P(\theta_3)}{P(D)} = \frac{.11250}{.8733} = .13$$

We now simply substitute these revised probabilities $P(\theta_j|D)$'s for the original $P(\theta_j)$'s in calculating a *revised* set of *expected utilities* for a production run of 1,000 items (see Table 9-5). Thus, our new formula is

$$E[U(a_i|D)] = \sum_j U_{ij}P(\theta_j|D) \qquad (9\text{-}3)$$

and our new solutions are:

$$E[U(a_1|D)] = \sum_j U_{1j}P(\theta_j|D) = 400(.56) + 400(.31) + 400(.13) = 400$$

$$E[U(a_2|D)] = \sum_j U_{2j}P(\theta_j|D) = 900(.56) + 500(.31) - 1,500(.13) = 464$$

Table 9-5

Machine Example of Decision Making Under Risk Using Maximum
Expected Utility Criterion and State Probabilities
Revised on the Basis of Sample Data

		State of machine before choosing an action			
		θ_1 Good $\rho_1 = .01$	θ_2 Fair $\rho_2 = .05$	θ_3 Poor $\rho_3 = .25$	Revised expected utility $\sum_j$ (cell entry in column j)$P(\theta_j|D)$
Action	a_1 Overhaul machine	400	400	400	400
	a_2 Don't overhaul machine	900	500	−1,500	464 ← Appropriate action
	$P(\theta_j|D)$	.56	.31	.13	

The larger expected utility now occurs for the action a_2 "don't overhaul machine" whereas using the original or *prior* state probabilities it had occurred for a_1. Therefore, if he followed an expected utility strategy, with the information provided by the sample the decision maker would not overhaul the machine. These revised expected utilities, however, are only indices for action in this particular production run. They should not be regarded as long-run expected profits in the general case; actually, they are long-run expected profits only for those production runs in which a preliminary sample of two items contained zero defectives. Therefore, if the decision maker wished to reap the benefits of revised probabilities (in decision making under risk), he should sample before each production run.

The logic of using revised probabilities may be more easily seen as follows. Suppose that we took a sample of two items before each of an infinite number of production runs, and let D_r be the outcome that r of the two sample items are defectives. Then $P(D_0 \& \theta_1)$, $P(D_0 \& \theta_2)$, $P(D_0 \& \theta_3)$ are the proportions of the time that the sample outcome is zero defectives and, respectively, $\theta = \theta_1, \theta = \theta_2, \theta = \theta_3$. The three θ values are mutually exclusive and exhaustive, so the proportion of the time that the sample outcome is zero and θ is *anything* is $P(D_0) = P(D_0 \& \theta_1) + P(D_0 \& \theta_2) + P(D_0 \& \theta_3)$. Therefore, within the subset of cases in which the sample outcome is D_0, the proportion of the time that θ is θ_1, and therefore the conditional probability that $\theta = \theta_1$ given D_0, is

$$P(\theta_1 | D_0) = \frac{P(D_0 \& \theta_1)}{P(D_0 \& \theta_1) + P(D_0 \& \theta_2) + P(D_0 \& \theta_3)} = \frac{P(D_0 \& \theta_1)}{\sum_j (D_0 \& \theta_j)}$$

But from Formula 6-11,

$$P(D_0 \& \theta_j) = P(D_0 | \theta_j)P(\theta_j)$$

so

$$P(\theta_1 | D_0) = \frac{P(D_0 | \theta_1)P(\theta_1)}{\sum_j P(D_0 | \theta_j)P(\theta_j)}$$

which is the Bayesian formula for the probability of θ_1 revised on the basis of data D_0.

9-3-4   The Value of the Sample

Of course, the decision maker cannot know in advance of sampling how many defectives will turn up in the sample. Therefore, in order to evaluate the effect of taking a sample of two items, he needs to know the *revised* expected utility under the *optimal* action called for by *each* of the possible sample outcomes. We have already calculated it for the case where, in a sample of two items, the obtained data D is 0 defectives. Using methods already described, we can also calculate it for the cases where the obtained data D

is 1 defective and 2 defectives, obtaining revised expected utilities of 400 and 400, respectively.

At this point, our calculations have told us that if the decision maker wishes to follow an expected utility strategy, then if the number of defectives in a sample of two items is:

0 he should choose action a_2 for which $E[U(a_2 | D$ is 0 defectives$)] = 464$

1 he should choose action a_1 for which $E[U(a_1 | D$ is 1 defective$)] = 400$

2 he should choose action a_1 for which $E[U(a_1 | D$ is 2 defectives$)] = 400$

However, since the decision maker has no control over the number of defectives that will appear in the sample, we would like to obtain an overall (unconditional) expected utility for the future sample whatever its outcome—that is, for the *strategy* of following action a_2 if there are no defectives, a_1 if there is 1 defective, and a_1 if there are 2 defectives in the intended sample of two items. The expected utility of this strategy would be the sum of (a) the revised expected utility of taking action a_2 when there are 0 defectives times the probability (i.e., the proportion of the time) that there *will be* 0 defectives in the sample, (b) the revised expected utility of taking action a_1 when there is 1 defective times the probability that there will be 1, (c) the revised expected utility of taking action a_1 when there are 2 defectives times the probability that there will be 2 defectives in the sample of two items. Now, these probabilities (or proportions of the time) that the sample will contain D defectives must be the *overall* probabilities of D *unqualified* by the state of nature

$$P(D = x) = \sum_j P(D = x | \theta_j)P(\theta_j),$$

where the $P(\theta_j)$'s are the *original* probabilities of the states. We already obtained $P(D = 0) = .8733$ in the previous section. Analogously, we obtain $P(D = 1) = .1134$ and $P(D = 2) = .0133$. Therefore, the overall expected utility of taking a sample of two items and following the optimal strategy of choosing action a_2 if there are 0 defectives, a_1 if there is 1 defective, and a_1 if there are 2 defectives in the sample is

$$
\begin{aligned}
E[U(\text{sample and strategy})] = {} & E[U(a_2 | D = 0)]P(D = 0) \\
& + E[U(a_1 | D = 1)]P(D = 1) \\
& + E[U(a_1 | D = 2)]P(D = 2) \\
= {} & 464(.8733) + 400(.1134) + 400(.0133) \\
= {} & 405.21 + 45.36 + 5.32 = 455.89
\end{aligned}
$$

The value 455.89 is the expected utility (under the optimum set of actions) if it *costs nothing* to take the sample of two items. The expected utility (for the optimum action) without sampling was 400. Therefore, if sampling costs nothing, the decision maker will increase his expected utility by $455.89 - 400$

$= 55.89$ by deciding to take the sample. This, then, is the *expected value of the sample information.*

9-3-5 Taking the Cost of the Sample into Account

Generally, however, it does cost something to take a sample. Let C_2 stand for the cost (expressed in utility units) of taking the sample of two items. Then the expected utility of taking the sample and following the optimal strategy is $455.89 - C_2$ and it is profitable to take the sample if C_2 is less than 55.89, the expected value of the sample information calculated under the assumption of zero cost.

However, it may be more profitable to take a sample of a different size. The optimum sample size may be determined as follows. Let n be the size of a sample, let C_n be its cost, and let ΔU_n be the increase in expected utility caused by deciding to take the sample and follow the optimal strategy (associating an action with each possible sample outcome) when the sample costs nothing (55.89 in the machine problem). Then the net gain in expected utility to be expected from the sample is $\Delta U_n - C_n$. The optimal sample size is the n value for which $\Delta U_n - C_n$ becomes maximum.

Considering the amount of labor expended just in calculating ΔU_2, it is clear that determining the optimal sample size may be an arduous process since it may involve calculating ΔU_n and C_n for a variety of n values. That the set of sample sizes that need be considered is limited, however, is shown by the following considerations. As the size n of the sample increases, one obtains better and better information about the true state of nature (e.g., of the machine). But the *most* that a sample could *possibly* do for the decision maker is to place him in the situation of decision making under certainty— where he would *know* the true state and take the action having the largest utility under that state. The *expected* utility of such perfect information in general would simply be

$$\sum_j (\text{largest utility under } \theta_j)P(\theta_j)$$

where the $P(\theta_j)$'s are the original probabilities of the states unqualified by sample outcomes. In our machine example this expected utility would be

$$900(.5) + 500(.3) + 400(.2) = 450 + 150 + 80 = 680$$

The expected utility without sampling was 400 for the appropriate action, so the *expected value of perfect information* about the true state of nature is $680 - 400 = 280$ if it costs nothing to obtain it. Clearly then, a sample for which the cost exceeds 280 should not be considered at all. So in the majority of cases where the cost C_n of taking a sample of size n increases with increasing n, one need only consider increasing values of n until C_n exceeds the expected value of perfect information; and in other cases one need only consider sample sizes for which $C_n <$ EVPI.

The cost of sampling is a very important consideration in business and industrial decisions where costs can easily be subtracted from utilities because both are measured in dollars. In these areas "cost accounting" for sampling is often *explicitly* worked out. This is seldom the case in scientific areas, however, where the "cost" of sampling may be partly dollars, partly delay, partly diverted effort, etc., whereas the basic utility of an action may be due to knowledge expanded, efficiency promoted, lives saved or enriched, etc. The ingredients of gains and of costs are so dissimilar and so varied as to discourage scientific decision makers from attempting to assess them in common utility units. Although it could be done (in theory, anyway), it is not an easy task, and the effort might contribute greatly to the "cost." Therefore, it is the general practice in scientific decision making to be vaguely influenced by cost considerations but not in the explicitly mathematical fashion described above. In short, the cost of sampling is often a complex but relatively minor consideration and is therefore virtually ignored in scientific decision making.

9-3-6 Use of Subjective Probabilities

The situation in which it is known that $\theta_1, \theta_2, \ldots, \theta_j, \ldots, \theta_s$ are various possible values of the state of nature θ, but where the decision maker does not know the *objective* probability distribution of θ over these values, was called decision making under *uncertainty* if the decision maker is willing and able to supply subjective probabilities. The "uncertainty" may be of various types: (a) the decision maker may be certain that θ is a random variable and that $\theta_1, \theta_2, \ldots, \theta_j, \ldots, \theta_s$ are its values, but uncertain as to the exact numerical values for the $P(\theta = \theta_j)$'s, (b) the decision maker may be certain that θ is a constant and that one of the values $\theta_1, \theta_2, \ldots, \theta_j, \ldots, \theta_s$ is the correct constant value of θ, but uncertain as to which one it is, (c) the decision maker may be certain that θ takes no values other than $\theta_1, \theta_2, \ldots, \theta_j, \ldots, \theta_s$ but uncertain as to whether θ is a constant or a variable—if θ is a constant, he does not know which one; if θ is a variable, he does not know the objective $P(\theta = \theta_j)$'s.

In all three cases the decision maker may have, or be able to produce, subjective probabilities that $\theta = \theta_j$ for every value of j. In case (a), the subjective probabilities may be regarded simply as subjective estimates of the objective probabilities. In case (b), the subjective probabilities are better regarded as degrees of belief, and perhaps in case (c) also. In all these situations, problems may be worked by the same *method* employed in decision making under risk with the maximum expected utility criterion, subjective $P(\theta_j)$'s being used instead of objective ones (and, when sample information is obtained, being revised in the light of that information, using Bayes' formula, just as the objective probabilities were in Section 9-3-3.) The *interpretation*, however, may need modification. For example, in case (b) "expected utilities" are merely guides to action appropriate to the decision maker's state of mind and no longer have the meaning of average payoffs in the long run. For unless the decision maker modifies his subjective probabilities in the light of experience or sample information (as he should), his (objective) long-run average payoff is simply the utility under the constant, true value of θ for the action taken. Thus, when subjective probabilities are used, the expected utilities may become equally subjective and, indeed, fictitious and misleading if the subjective probabilities depart greatly from the objective ones. Only when the subjective $P(\theta = \theta_j)$'s are nearly perfect estimates of the objective ones will the "expected utilities" correspond to average long-run payoffs.

If initial subjective probabilities (that are neither 0 nor 1.00) are revised on the basis of a sufficiently large sample of data, however, the revised subjective probabilities tend to converge upon the correct revised objective probabilities. And this is the case *even if they are badly in error.* If the sample is sufficiently large, the obtained data will tend to have appreciable probability under the true state of nature and negligible probability under all others. (The smaller the "difference" between the true and "adjacent" states, the less difference there will generally be between the corresponding $P(\text{data} | \theta_j)$'s at a given sample size and the larger the sample size will have to be.) Furthermore, the larger the sample size, the more faithfully the sample tends to reflect the true situation. Thus, the objective sample information tends to "overwhelm" the erroneous initial probabilities and to produce more nearly correct revised probabilities.

To illustrate, suppose that a manufacturing process has only two possible states, θ_1 in which the proportion p of defectives produced is .2 and θ_2 in which it is .9. The objective probabilities are $P(\theta_1) = .99$ and $P(\theta_2) = .01$ and the present state of nature is θ_1. The decision maker's subjective probabilities for θ, however, are just the opposite, $P(\theta_1) = .01$ and $P(\theta_2) = .99$, and therefore very badly in error. He takes a random sample of ten items of which two turn out to be defective (which is reasonable since the proportion of defectives under the present state of nature is .2). The probabilities of this result under the two states are

$$P(D | \theta_1) = P(r = 2 | n = 10, p = .2) = \binom{10}{2}.2^2.8^8 = .3019899$$

$$P(D | \theta_2) = P(r = 2 | n = 10, p = .9) = \binom{10}{2}.9^2.1^8 = .0000004$$

And the revised probabilities for the states are obtained by using Bayes' formula with subjective $P(\theta_j)$'s

$$P(\theta_1 | D) = \frac{P(D | \theta_1)P(\theta_1)}{P(D | \theta_1)P(\theta_1) + P(D | \theta_2)P(\theta_2)} = \frac{.3019899(.01)}{.3019899(.01) + .0000004(.99)}$$

$$= \frac{.003019899}{.003019899 + .000000396} = \frac{.003019899}{.003020295} = .99987$$

And, of course, $P(\theta_2 | D) = 1 - .99987 = .00013$

These revised probabilities are far from the erroneous initial subjective probabilities and compare favorably with the revised objective probabilities obtained by using the objective $P(\theta_j)$'s in Bayes formula

$$P(\theta_1 | D) = \frac{.3019899(.99)}{.3019899(.99) + .0000004(.01)} = \frac{.298970001}{.298970005} \cong 1$$

$$P(\theta_2 | D) \cong 0$$

Thus, a revision of probabilities on the basis of a sufficiently large sample not only serves to correct the effect of erroneous initial subjective probabilities but also may bring the decision maker very close to decision making under certainty.

9-3-7 Critique

This method is a *feasible* method for decision making under either risk or uncertainty since both situations provide the required $P(\theta_j)$'s. Since the criterion is based upon expectation, it is *appropriate* only if the decision maker is willing to let his present decisions be determined by the average payoff in the infinitely long run. It is *particularly appropriate* when a long run of decisions will have to be made under similar circumstances, the decision maker can afford to experience some large initial losses, and the probabilities are objective (or, if subjective, are accurate estimates of the objective probabilities). For, in that case it tends to maximize the long-run payoff. It does not, of course, protect one against disaster in the short run or single case (nor absolutely guarantee that one will escape it in the finite run). It may be a reasonable guide to action in the short run or single case if the decision maker is willing and able to take the risk. However, the method can be disastrous in the case of momentous one-shot decisions (and even in the case of long runs of decisions if highly inaccurate subjective probabilities are used without modification by sample information).

9-4 The Maximum Utility Under Most Probable State Criterion

9-4-1 Method

If in addition to the information in the payoff matrix the decision maker (a) knows the probabilities for each of the states of nature, the $P(\theta_j)$'s, and (b) is willing to gamble that nature is presently in its most probable state, he can use the *maximum utility under most probable state criterion*, as follows. He simply chooses the action having the largest utility under the θ value for which $P(\theta_j)$ is maximum. (That is, he inspects the various $P(\theta_j)$'s and finds the largest one, corresponding to a θ-value which we shall call θ_x. He then inspects the utility values U_{ix} in the θ_x column of the payoff matrix, finds the largest one, and takes the action in whose row it lies.)

9-4-2 Example

A hunter is bitten by a fox, which then disappears. There are two possible states of nature: either the fox was rabid or it was not. The hunter can choose between two actions: either take rabies shots or do not. His utilities are given in Table 9-6. After discussing the prevalence of rabid foxes in that area with forest rangers and public health officials, the hunter subjectively estimates the

Table 9-6

Rabid Fox Example of Decision Making Under Uncertainty Using Maximum Utility Under Most Probable State of Nature Criterion

(a) No sample information available

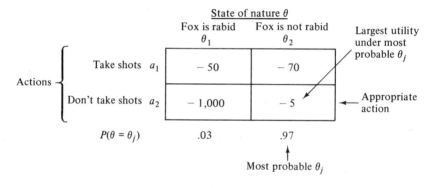

(b) $P(\theta_j)$'s revised on basis of sample information

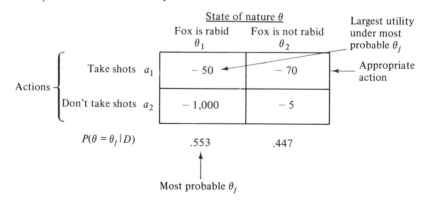

probability that the fox was rabid to be .03. If this is correct, then by far the most probable state of nature is that the fox was nonrabid. Considering the favorable odds, the hunter is willing to gamble that the overwhelmingly probable state "nonrabid" is the actual situation. He therefore takes the action having the largest utility under that state, which is not to take rabies shots (see Table 9-6a).

9-4-3 Use of Sample Information

Although the hunter's probabilities for θ are subjective, it is entirely appropriate (indeed, *especially* appropriate) to revise them in the light of sample information. If the decision maker has sample information relevant to the present state of nature, he should revise the state probabilities on the basis of that information, using Bayes' formula (Formula 9-2). He should then take the action having the greatest utility under the state with the largest revised probability.

Suppose, for example, that the doctor who treated the hunter's wound applied a test to it that was known to give positive results in 40% of cases where the biting animal was rabid and in only 1% of cases where the animal was nonrabid. And suppose that the test results were positive. Then, letting R stand for rabid, $\bar{R}$ for not rabid, and D for the data (i.e., the positive test result), Bayes formula tells us that

$$P(R \mid D) = \frac{P(D \mid R)P(R)}{P(D)} = \frac{P(D \mid R)P(R)}{P(D \mid R)P(R) + P(D \mid \bar{R})P(\bar{R})}$$

where $P(R)$ and $P(\bar{R})$ are the hunter's original subjective probabilities that the fox was rabid or nonrabid, respectively, and $P(R \mid D)$ is his revised probability that it was rabid. Applying this formula, we have

$$P(R \mid D) = \frac{.40(.03)}{.40(.03) + .01(.97)} = \frac{.0120}{.0120 + .0097} = \frac{.0120}{.0217} = .553$$

And, of course,

$$P(\bar{R} \mid D) = \frac{P(D \mid \bar{R})P(\bar{R})}{P(D \mid R)P(R) + P(D \mid \bar{R})P(\bar{R})} = \frac{.01(.97)}{.40(.03) + .01(.97)} = .447$$

The larger of the two revised probabilities is .553, the revised probability for "rabid." So the hunter should take the action with the largest utility under the "rabid" state, which is to take the shots.

9-4-4 Critique

This method is *feasible* whenever the decision maker knows the states, actions, utilities, and $P(\theta_j)$'s, so it is feasible under risk or uncertainty. It is *appropriate* when, in addition, he is willing to gamble that the present state of nature is the most probable of the states. It is *particularly appropriate* when, in addition, one state of nature is overwhelmingly probable (i.e., has a probability very close to 1.00), a single decision is to be made under these or similar circumstances, and utilities do not vary to an extreme degree. Obviously, the maximum utility method is a special case of the present method which occurs when one state has a probability of 1.00. The present method is widely used under conditions of near-certainty. None of us would ever cross a street if we were not willing to ignore the minute probability that we will be killed in the attempt and gamble on the overwhelming likelihood that we will arrive safely on the other side.

9-5   The Maximum Likelihood Criterion

9-5-1   Method

If the decision maker is ignorant of the state probabilities $P(\theta_j)$s but is willing and able to take a sample of data and knows or can calculate the probability $P(D|\theta_j)$ of the obtained data D for each state of nature θ_j, he can use the **maximum likelihood criterion**: A sample is obtained and its data outcome is noted. The probability that the actually obtained data outcome had before the sample was obtained is calculated for each state of nature. The state of nature for which this probability is maximum is selected, the utilities under it are examined, and the largest of these utilities is determined. The action corresponding to this largest utility is the action chosen. More formally, the first step is to obtain some data D. The second is to calculate $P(D|\theta_j)$ for every θ_j. Finding *max* $P(D|\theta_j)$ is the third step. Let θ_x stand for the value of θ for which $P(D|\theta_j)$ is maximum. Then the utilities U_{ix} in the θ_x column of the payoff matrix are examined and *max* U_{ix} is determined. The action a_i corresponding to *max* U_{ix} is then chosen.

9-5-2   Rationale

The most probable present state of nature is the θ_j involved in the maximum $P(\theta_j|D)$ rather than the θ_j involved in the maximum $P(D|\theta_j)$. The preceding method (based on sample information) correctly takes it to be the former, whereas the present method takes it to be the latter. The present method therefore is justified only under circumstances which render the two methods equivalent, i.e., only when both methods "identify" the same state of nature. The maximum utility under most probable state (MUUMPS) method calculated $P(\theta_j|D)$ using Bayes' formula

$$P(\theta_j|D) = \frac{P(D|\theta_j)P(\theta_j)}{P(D)}$$

Now the unconditional $P(D)$ has the same value irrespective of θ_j and is therefore a constant. So, if each of the $P(\theta_j)$'s had the same constant value, $P(\theta_j)/P(D)$ would be a positive constant (positive because all probabilities are positive), which we shall call C, and

$$P(\theta_j | D) = CP(D | \theta_j)$$

In that event, $P(D | \theta_j)$ is directly proportional to $P(\theta_j | D)$, and the two probabilities will therefore reach their maxima for the same θ_j. Clearly then, if all the $P(\theta_j)$'s are equal, the present method is equivalent to the previous (MUUMPS) one and is therefore valid.

Actually, the $P(\theta_j)$'s are unknown and the present method, by using $P(D | \theta_j)$'s in place of $P(\theta_j | D)$'s, is *treating* the $P(\theta_j)$'s *as if* they all had the same value. (This is analogous to using equal subjective $P(\theta_j)$'s in place of the unknown objective ones.) So, in effect, the $P(\theta_j)$'s are replaced by constants (the reciprocal of the number of states). Now if the decision maker revises these erroneous initial probabilities on the basis of a sufficiently large sample of data (i.e., one large enough to cause one $P(D | \theta_j)$ to be much larger than all others), the realistic sample information will overwhelm the erroneous initial probabilities (see Section 9-3-6) with the result that the $P(\theta_j | D)$'s will be nearly as accurate as those obtained by the previous (MUUMPS) method using objective $P(\theta_j)$'s in Bayes' formula. Thus, the present method is roughly equivalent to the preceding one if the initial state probabilities are about equal or if sample information is sufficiently "discriminating."

9-5-3 Example

A wholesaler receives a shipment of 10,000 items. Normally, such shipments are labeled as having one of three qualities A, B, or C, depending upon the manufacturing process used to produce them. The proportion of defective items turned out by processes A, B, and C are .01, .05, and .20, respectively. However, due to a clerical error, the quality of this particular shipment is unspecified and unknown. The wholesaler is willing to consider four possible actions: (a_1) market shipment as A quality, (a_2) market as B quality, (a_3) market as C quality, (a_4) inspect all 10,000 items, remove the defectives, and market the remainder as "superquality." His utilities for these actions under the possible states are given in Table 9-7. The maximum like-lihood method would be equivalent to the maximum utility for most probable state method (using the same sample information) if $P(A) = P(B) = P(C)$, would be more conservative if $P(A) > P(B) > P(C)$, and would be more reckless if $P(A) < P(B) < P(C)$. The wholesaler is willing to gamble that the latter situation is not the case and therefore to use the maximum likelihood method. He draws a random sample of 100 items from the shipment, inspects them, and finds that 7 of them are defective. So his actual data outcome D

Table 9-7

Wholesaler Example of Decision Making Under Ignorance
Using the Maximum Likelihood Criterion

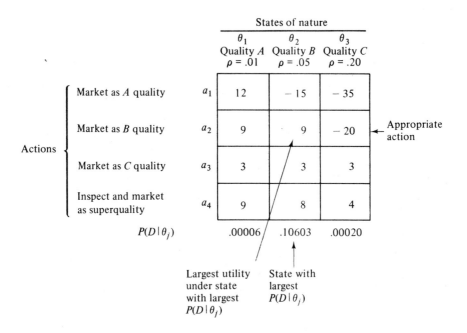

		θ_1 Quality A $p = .01$	θ_2 Quality B $p = .05$	θ_3 Quality C $p = .20$	
Market as A quality	a_1	12	-15	-35	
Market as B quality	a_2	9	9	-20	← Appropriate action
Market as C quality	a_3	3	3	3	
Inspect and market as superquality	a_4	9	8	4	
$P(D\mid\theta_j)$		.00006	.10603	.00020	

States of nature

Actions

Largest utility under state with largest $P(D\mid\theta_j)$

State with largest $P(D\mid\theta_j)$

is 7 "successes" in a binomial sample of 100 observations. He now obtains $P(D\mid\theta_j) = P(r = 7\mid n = 100, p_j)$ from binomial tables, for each value of θ_j,

$$P(D\mid\theta_1) = P(r = 7\mid n = 100, p = .01) = \binom{100}{7}.01^7.99^{93} = .00006$$

$$P(D\mid\theta_2) = P(r = 7\mid n = 100, p = .05) = \binom{100}{7}.05^7.95^{93} = .10603$$

$$P(D\mid\theta_3) = P(r = 7\mid n = 100, p = .20) = \binom{100}{7}.20^7.80^{93} = .00020$$

By far the largest of these three $P(D\mid\theta_j)$'s is the .10603 corresponding to $P(D\mid\theta_2)$. So he finds the largest utility 9 in the θ_2 column of the payoff matrix, and takes the action corresponding to it, a_2 "market as B quality."

9-5-4 Critique

This method requires data and a knowledge of data probabilities, but it does not require a knowledge of probabilities of the states of nature. It is therefore a *feasible* method for decision making under ignorance, risk, or uncertainty, when samples and their probabilities are available. It is an *appropriate* method when the decision maker is willing to infer that the most likely present state of nature is the one most conducive to the obtained sample result and is willing to use this as his sole guide to action. It is *particularly appropriate* in the case of single decisions under ignorance when the states of nature are sufficiently different and the sample size is sufficiently large that there is virtually no likelihood of "misclassifying" θ.

9-6 The Maximin Expected Utility Criterion

9-6-1 Method

Suppose that the decision maker does not know (or does not wish to use) $P(\theta_j)$'s, but that he is willing to take a sample of data and be guided by its results, and, finally, that for each θ_j he knows (i.e., can calculate) the probability for each possible data outcome in the intended sample. A strategy is a plan that tells for each possible data outcome what action will be taken. Since for each θ_j the decision maker knows the probabilities of all possible data outcomes, he also knows the probability of taking each action associated with an outcome by a strategy. Using these probabilities, and the utilities in the payoff matrix, he can calculate for each θ_j the expected utility for any strategy. He can do this for every possible strategy and then select that strategy whose smallest expected utility is maximum. He can then take the sample, observe its data outcome, and take the action called for by the obtained data outcome in the selected strategy.

Thus, if there are r different possible actions, s different possible states of nature, and t different possible data outcomes or sample results, and the decision maker wishes to use the **maximin expected utility criterion**, he will: (a) formulate each of the r^t possible strategies associating one of the r possible actions with each of the t possible sample results, (b) for each possible strategy under each possible state of nature, calculate the expected utility of that strategy under that state of nature

$$E[U(\text{strategy}\,|\,\theta_j)] = \sum_{k=1}^{t} U(a_k^*\,|\,\theta_j)P(D_k\,|\,\theta_j) \qquad (9\text{-}4)$$

where a_k^* is the action associated with data outcome D_k by the strategy under consideration, (c) determine for each strategy the minimum of the s different expected utilities under the s different possible states of nature, (d) select the strategy whose minimum expected utility is largest, (e) obtain the sample, observe the result, and take the action associated with that result by the selected strategy.

9-6-2 Rationale

The formal rationale for Formula 9-4 is as follows. A given strategy associates with each different possible data outcome D_k an action a_k^* and commits the decision maker to take action a_1^* when D_1 occurs, to take action a_2^* when D_2 occurs, etc. This commitment forces the probability that an outcome D_k and its associated action a_k^* *both* will happen to equal the probability that the outcome D_k will happen, and it attaches to the paired outcome and action the utility of the action. We therefore have the situation that

$$P(D_k \ \& \ a_k^* | \theta_j) = P(D_k | \theta_j) \quad \text{and} \quad U(D_k \ \& \ a_k^* | \theta_j) = U(a_k^* | \theta_j)$$

So the expected utility of a strategy associating a_k^*'s with D_k's is, when $\theta = \theta_j$,

$$E[U(\text{strategy} | \theta_j)] = \sum_k U(D_k \ \& \ a_k^* | \theta_j) P(D_k \ \& \ a_k^* | \theta_j)$$

$$= \sum_k U(a_k^* | \theta_j) P(D_k | \theta_j)$$

9-6-3 Example

For the captain of a fishing boat there are two states of nature. Either a school of fish is nearby, θ_1, or it is not, θ_2. He has two possible actions. He can stop the boat and (after considerable time-consuming preparation) start to fish, a_1, or he can move on to another location, a_2. His utilities for each possible action under each possible state of nature are given in Table 9-8a. Being motivated by long-term profits, he is willing to use expected utilities. But since he is fishing in unfamiliar waters characterized by rapidly changing ecological conditions, he has no objective $P(\theta_j)$'s and does not trust his subjective ones. However, he has a scientific device that he does trust probabilistically to detect schools of fish by echo of sound impulses (i.e., by supplying data). The device gives only three different possible indications, positive $(+)$, neutral (0), and negative $(-)$, which may be regarded as data outcomes D_1, D_2, and D_3, respectively. The probability for each possible indication under each possible state of nature, i.e., the $P(D_k | \theta_j)$'s, are given in Table 9-8b. There are $2^3 = 8$ different ways of assigning one of the two possible actions to each of the three possible data outcomes and each such assignment is a possible strategy. These eight strategies are listed in Table 9-8c.

For each strategy he calculates an expected utility under each possible state of nature. The computations are shown for strategy S_2 in Table 9-9 and more compactly (but perhaps less insightfully) for all strategies in Table 9-10. Section D of Table 9-10 lists for each possible state of nature the expected utilities under each strategy. The smaller of the two expected utilities for strategy S_5 is 2.2 and this is larger than the smaller of the two expected utilities under any other strategy. So strategy S_5 is the maximin strategy and therefore the one that the captain will use. The captain now uses his sound-echo device to obtain the information with which to choose an action. If the device

Table 9-8

Fishing-boat Example of Decision Making Under Ignorance
Using Maxim Expected Utility Criterion

(Preliminary information)

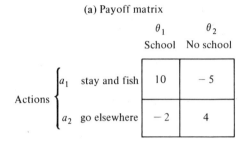

(a) Payoff matrix

Actions		θ_1 School	θ_2 No school
a_1	stay and fish	10	-5
a_2	go elsewhere	-2	4

(b) Probabilities of data outcomes
under states of nature

	θ_1	θ_2
$P(+\mid\theta_j)$	.6	.2
$P(0\mid\theta_j)$	.3	.3
$P(-\mid\theta_j)$	.1	.5

(c) Strategies

Action to be taken
if data outcome is

Strategy	D_1 $+$	D_2 0	D_3 $-$
S_1	a_1	a_1	a_1
S_2	a_1	a_1	a_2
S_3	a_1	a_2	a_1
S_4	a_2	a_1	a_1
S_5	a_1	a_2	a_2
S_6	a_2	a_1	a_2
S_7	a_2	a_2	a_1
S_8	a_2	a_2	a_2

gives him a $+$ indication, he takes action a_1, as strategy S_5 commits him
to do, and stays to fish; if he gets either a 0 or a $-$ indication, he takes action
a_2, as determined by strategy S_5, and goes elsewhere.

9-6-4   Critique

This method requires only the information in the payoff matrix and
the sample and a knowledge of the sample probabilities $P(D_k\mid\theta_j)$'s (which

Table 9-9

Fishing-Boat Example of Decision Making Under Ignorance Using Maximin Expected Utility Criterion

(Computation of expected utilities for strategy S_2)

Data outcome D_k	Associated action a_k^*	θ_1 Probability of outcome $P(D_k\|\theta_1)$	θ_1 Utility of action $U(a_k^*\|\theta_1)$	θ_1 $P(D_k\|\theta_1)U(a_k^*\|\theta_1)$	θ_2 Probability of outcome $P(D_k\|\theta_2)$	θ_2 Utility of action $U(a_k^*\|\theta_2)$	θ_2 $P(D_k\|\theta_2)U(a_k^*\|\theta_2)$
$D_1(+)$	a_1	.6	10	6	.2	-5	-1
$D_2(0)$	a_1	.3	10	3	.3	-5	-1.5
$D_3(-)$	a_2	.1	-2	$-.2$	.5	4	2
				$E[U(S_2\|\theta_1)] = 8.8$			$E[U(S_2\|\theta_2)] = -.5$

Table 9-10

Fishing-boat Example of Decision Making Under Ignorance Using Maximin Expected Utility Criterion

(Computation of expected utilities for all strategies)

	A			B — $U(a_k^*\mid\theta_j)$ Utilities associated with action called for by strategy when state of nature is						C — $U(a_k^*\mid\theta_j)P(D_k\mid\theta_j)$ when state of nature is						D — $\sum_k U(a_k^*\mid\theta_j)P(D_k\mid\theta_j)$ Expected utility for strategy when state of nature is	
	Action to be taken if data outcome is			θ_1 and data outcome is			θ_2 and data outcome is			θ_1 and data outcome is			θ_2 and data outcome is				
Strategy	+	0	−	+	0	−	+	0	−	+	0	−	+	0	−	θ_1	θ_2
S_1	a_1	a_1	a_1	10	10	10	−5	−5	−5	6	3	1	−1	−1.5	−2.5	10	−5
S_2	a_1	a_1	a_2	10	10	−2	−5	−5	4	6	3	−.2	−1	−1.5	2	8.8	−.5
S_3	a_1	a_2	a_1	10	−2	10	−5	4	−5	6	−.6	1	−1	1.2	−2.5	6.4	−2.3
S_4	a_2	a_1	a_1	−2	10	10	4	−5	−5	−1.2	3	1	.8	−1.5	−2.5	2.8	−3.2
S_5	a_1	a_2	a_2	10	−2	−2	−5	4	4	6	−.6	−.2	−1	1.2	2	5.2	2.2
S_6	a_2	a_1	a_2	−2	10	−2	4	−5	4	−1.2	3	−.2	.8	−1.5	2	1.6	1.3
S_7	a_2	a_2	a_1	−2	−2	10	4	4	−5	−1.2	−.6	1	.8	1.2	−2.5	−.8	−.5
S_8	a_2	a_2	a_2	−2	−2	−2	4	4	4	−1.2	−.6	−.2	.8	1.2	2	−2	4
$P(D_k\mid\theta_j)$:				.6	.3	.1	.2	.3	.5								

← Maximin strategy

205

must be sensitive to the state of nature—varying with the θ_j's). It is therefore a *feasible* method for decision making under ignorance, uncertainty, or risk, whenever the above requirements are met. It is an *appropriate* method when the decision maker is willing to let his present decisions be guided by a combination of maximin and long-run expected utility criteria, the former applying across states and the latter within states. It is *particularly appropriate* when, in decision making under ignorance, a long run of decisions will have to be made under similar circumstances, the decision maker can afford to experience some large initial losses, sample information and objective sample probabilities are available, and ignorance of the $P(\theta_j)$'s prevents the use of the maximum expected utility criterion. It may be a reasonable guide to action in the short run or single case if the decision maker is willing and able to take the risk. However, it can be extremely hazardous in one-shot decisions where the worst outcomes are intolerable.

9-7   The Maximin Criterion

9-7-1   Method

We have already encountered the **maximin criterion** in the chapter on games. Essentially, the decision maker determines the worst that could happen to him under each action and takes the action whose worst outcome is least disastrous (has the optimum worst consequence). More formally, for each action (or row in the payoff matrix) a_i the decision maker determines the minimum utility (or cell entry) U_{ij} and selects the action (or row) whose minimum U_{ij} is largest. (Dominated *actions* can be eliminated but dominated *states* cannot since inanimate nature cannot be counted upon to be a "rational opponent.")

9-7-2   Example

The owner-operator of a ferryboat considers two possible states of nature and two possible actions. Either the boat will sink during the coming year, costing him $50,000, or it will not. And he can either insure it for $50,000 at a cost of $100 or he can not insure it at all. His utilities are given in Table 9-11. The owner has very little capital and his livelihood depends upon the boat. Since he cannot afford to risk large losses, he decides to use the maximin criterion. The minimum utility if he insures is -100 whereas the minimum if he does not is $-50,000$. The larger of these two minima occurs for the action "insure boat," so he takes that action.

Table 9-11

Ferryboat Example of Decision Making Under Ignorance
Using the Maximin Criterion

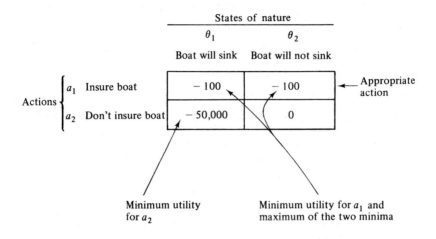

	States of nature	
	θ_1	θ_2
	Boat will sink	Boat will not sink
a_1 Insure boat	− 100	− 100
a_2 Don't insure boat	− 50,000	0

Actions {

Appropriate action ←

Minimum utility for a_2

Minimum utility for a_1 and maximum of the two minima

9-7-3 Critique

The method requires only the information embodied in the payoff matrix: possible states, possible actions, and their utilities. It is therefore a *feasible* method for decision making under ignorance as well as risk and uncertainty. It is an *appropriate* method when the decision maker is primarily concerned with protecting himself against the worst, and the worst outcome under the minimax action is tolerable whereas that under some other action is not. If the "game" is to be played repeatedly, or if a sample is taken, the decision maker will obtain some empirical information about the $P(\theta_j)$'s. Furthermore, "average" outcomes may be much more tolerable than "worst" outcomes. Therefore, the method is *particularly appropriate* to decision making under ignorance when no sample information is available, a single decision is to be made, the worst possible outcome is intolerable whereas the worst outcome under the maximin action is not, and the decision maker is most concerned with protecting himself against the worst disaster that could befall him. Since nature does not counterstrategize, this method tends to be expensively overcautious, even in single decisions, if the decision maker can afford to experience the worst consequence in the payoff table. And its use in a long run of similar situations with relatively mild consequences may have ridiculous results. For example, following a maximin criterion, the storekeeper in the supply-and-demand example (see Table 9-3) would never stock any perishable items, and the captain in the fishing-boat example (see Table 9-8) would always be "going elsewhere" to fish.

9-8   Problems

1. In the Vacation Example (Section 9-2), what would be the appropriate action, and why, if the company had announced that this year it would pay half of each employee's vacation expenses?

2. Suppose that in the Vacation Example (Section 9-2), the company had decided upon a policy but was unwilling to announce it until after all vacations had been taken. And suppose that on the basis of past experience it was known that $P(\theta_1) = .6$; $P(\theta_2) = .1$; and $P(\theta_3) = .3$.

   (a) What would be the four expected utilities and the appropriate action if the maximum expected utility criterion were used?

   (b) What action would be chosen, and why, if the maximum utility under most probable state criterion were used?

   (c) What action would be chosen, and why, if the maximin criterion were used?

   (d) Are any actions dominated and, if so, which ones?

   (e) May the decision maker ignore dominated actions in all three of the above cases? Why or why not?

3. In the Supply and Demand Example (Section 9-3-2, Example 1), what action would be called for by the maximum utility under most probable state method. Under what circumstances would this be a reasonable method? An unreasonable method?

4. What would be the five expected utilities and the appropriate action in the Supply and Demand Example (Section 9-3-2, Example 1) if each item cost 10¢ rather than 20¢?

5. What would be the five expected utilities and the appropriate action in the Supply and Demand Example (Section 9-3-2, Example 1) if, instead of those given, the demand probabilities were: $P(\theta_1) = .20$, $P(\theta_2) = .35$, $P(\theta_3) = .25$, $P(\theta_4) = .15$, $P(\theta_5) = .05$?

6. What would be the six utilities in the Machine Example (Section 9-3-2, Example 2) if instead of $10, each good item could be sold for $12? Using these utilities, what would be the expected utilities and the appropriate action if the state probabilities, instead of those given, were $P(\theta_1) = .5$, $P(\theta_2) = .4$, $P(\theta_3) = .1$?

7. In the Machine Example with data (Section 9-3-3), what would the revised state probabilities be if a random sample of three items had been taken of which one was defective? Using these revised probabilities, what are the expected utilities for the two actions and which action is called for?

8. In Problem 7, before it was actually drawn, what was the value of the sample and what was the expected value of the sample information? If the cost of taking a sample of n observations is n^4 dollars, what is the optimal sample size?

9. Of 100 urns, one is of Type A, the rest of Type B. Type A urns contain 700,000 black beads and 300,000 white ones. Type B urns contain 100,000 black beads and 900,000 white ones. A single urn is drawn randomly from the set of 100 urns. Then a random sample of 5 beads is drawn from that urn. If 4 of the 5 beads are black, what is the probability that the drawn urn is of Type A?

10. In the Rabid Fox Example (Section 9-4-2), what would be the expected utilities and the appropriate action if the hunter used the maximum expected utility criterion? What action would the hunter take if he used the maximin criterion?

11. In the Rabid Fox Example (Section 9-4-3), what would the revised state probabilities and the appropriate action be if the probability of a positive result were .99 when the biting animal was rabid, rather than .40, and .01 when the animal was nonrabid, as before?

12. In the Wholesaler Example (Section 9-5-3), suppose that, instead of drawing a random sample of 100 items of which 7 were defective, he drew a random sample of 5 items of which 3 were defective. What would be the values of the $P(D|\theta_j)$'s and what action would be called for? What would be the maximin action? Are any actions dominated and, if so, what are they?

13. In the Fishing-Boat Example (Section 9-6-3), suppose that, reading from top row to bottom row, (a) the entries within a column of Table 9-8a were 9 and 0 for the left column and -6 and 5 for the right, (b) the entries within a column of Table 9-8b were .7, .2, .1 for the left column and .1, .4, .5 for the right. Using these substitute utilities and probabilities, obtain the expected utility for each strategy under each state. Which strategy has the largest minimum expected utility? What are its expected utilities under θ_1 and θ_2?

14. In the Ferryboat Example (Section 9-7-2), the owner is using a maximin criterion (and utilities are assumed proportional to money). What decision criterion is the insurance company using? Within what range of values must $P(\theta_1)$ lie in order for the insurance company to be motivated to insure the boat for $50,000 for a premium of $100? If $P(\theta_1)$ lies in that range, what action would the owner take if he used a maximum expected utility criterion?

15. Two million subscribers are expected to enter a magazine-sponsored lottery in which there is one first prize of $100,000, five second prizes

of $10,000 each, a hundred third prizes of $100 each, and a thousand fourth prizes of $5 each. One subscriber figures that the nuisance of making out the necessary papers and mailing them in costs him $1 worth of his time. Using monetary values as utilities, make out the pay-off matrix for this subscriber. What are the expected utilities for the actions? Will the subscriber enter or not enter the lottery if he uses (a) a maximum expected utility criterion, (b) a maximum utility under most probable state criterion, (c) a maximin criterion?

16. Professor Plume has written an article for the *Journal of Applied Metaphysics*. As a courtesy he will receive 50 free reprints of the article from the *Journal*. And if he orders them *before* the article is published, he can obtain an additional 100 reprints from the *Journal* for $50 or an additional 200 for $75 or an additional 300 for $90. After the article is published, the type is destroyed and the *Journal* can no longer print additional copies of the article. However, Plume can have additional copies printed by a local job printer in lots of 100 at a flat rate of $70 per hundred. In the past the number N of requests Plume has received for reprints of an article of this type has had the following distribution: $N \leq 50$ ten percent of the time; $50 < N \leq 150$ forty percent; $150 < N \leq 250$ thirty percent; $250 < N \leq 350$ twenty percent. Plume is determined to supply reprints to all requestors, so if the demand exceeds the number of copies initially ordered from the *Journal*, the excess demand will be met with copies supplied in lots of 100 from the job printer. Expressing costs as negative numbers and using them as utilities, make out the payoff matrix for the four demand states and the four actions (where the actions are the number of reprints initially ordered from the *Journal*). What are the four expected utilities and the action called for if Plume uses the maximum expected utility criterion? What action would be called for if he used the maximum utility under most probable state criterion? What is the maximin action?

17. A doctor must decide whether or not to operate upon a patient who is in an unknown one of three possible conditions A, B, or C. The number of additional years the patient may be expected to live, (a) if the operation is performed, is 10 if he is in condition A, 2 if in condition B, and 12 if in condition C, (b) if the operation is not performed, is 0 for condition A, 5 for B, and 30 for C. Of all patients who are in one of the three conditions, 60% are in condition A, 10% are in B, and 30% are in C. This particular patient shows a syndrome of symptoms that occurs in 10% of patients in condition A, 25% of those in condition B, and 30% of those in condition C. Using further life expectancies as utilities, make out the payoff matrix for the three states and two actions. What are the expected utilities and what action will the doctor

take if he uses the maximum expected utility criterion? What action will he take if he uses (a) the maximum utility under most probable state criterion, (b) the maximum likelihood criterion, (c) the maximin criterion?

18. A gourmet shop has imported a dozen cans of a rare delicacy for which it paid a total of $50. In the past the number of cans containing spoiled food among the dozen imported has been: zero in 80% of such lots, 1 in 4%, 2 in 1%, and all 12 in 15%. The proprietor considers three actions: (a_1) advertise the cans as "guaranteed top quality," charge $15 per can, and promise to pay to the customer a penalty of $50 for each can purchased that turns out to contain spoiled food, (a_2) advertise the cans as "high quality" with "double your money back if not fresh" and charge $12 per can, (a_3) advertise the cans as "good quality" with "money refunded in occasional cases of spoilage" and charge $10 per can. Assuming that all cans will be sold, and treating profits as utilities, make out the payoff matrix for the four possible states and three possible actions. What are the expected utilities and the action called for if the proprietor uses the maximum expected utility criterion? What action is called for if (a) he uses a maximum utility under most probable state criterion (b) he uses a maximin criterion?

19. In Problem 18, suppose that the proprietor took a random sample of two cans from the dozen, opened them, and found that the contents of one can were spoiled. Opened cans cannot, of course, be sold. Make out a revised payoff matrix for this new situation. What are the revised state probabilities? What are the expected utilities and the action called for if the proprietor uses the maximum expected utility criterion?

20. In Problem 19, taking both the cost of the sample and the effect of the sample into account, what is the expected increment in utility resulting from the decision to take the sample? What is the optimal sample size?

21. The Epicurean Bakery bakes a fresh batch of a fancy bread early each morning and throws away all unsold loaves each night. It costs them 50 ¢ per loaf plus a fixed cost of $2 to produce the bread, and each loaf sells for $1.50. In the past they have sold 0 loaves 2% of the time, 1 loaf 3%, 2 loaves 5%, 3 loaves 10%, 4 loaves 30%, 5 loaves 20%, 6 loaves 15%, 7 loaves 10%, and 8 loaves 5%. How many loaves should they bake if they wish to maximize expected profits? What is their expected profit if they do so?

22. A businessman is threatened with a lawsuit by a crank who offers to settle out of court for $10,000. It will cost him $5,000 just to go to court, and if he loses it will cost him an additional $50,000. The business-

man estimates the probability of losing to be .10. Treating costs as negative utilities, what action will he take if he uses (a) the maximum utility under most probable state criterion, (b) the maximin criterion, (c) the maximum expected utility criterion?

decisions about the state of nature

A special type of decision-making situation is that in which the actions are to announce which of its possible states nature is in on a particular occasion. The specified occasion may lie in the past, present, or future. However, we shall use the present tense when speaking of the general case. Thus, a_1 is to announce that θ is in state θ_1, a_2 is to announce that θ is in state θ_2, etc. This chapter is concerned with such decisions.

10-1 Expedient Announcements of the State of Nature

If the decision maker wishes to let his announcements be influenced by their possible consequences, i.e., if he wishes to make the most expedient, rather than simply the most truthful, announcement, then utilities should be taken into account. In that event, announcing the state of nature is just one of the many types of decisions that can be handled by the methods outlined in Chapter 9. Any one of those methods *could* be used if the necessary information were available. However, since it is the state of nature on a particular occasion, rather than the general state, that is to be announced, it is desirable to select a method that makes use of sample information germane to that occasion. A southwestern weatherman could, of course, always announce sunny weather, since this is the most likely condition in general. However, it would obviously be foolish to do so, especially in the middle of a thunderstorm.

Example: A large and dangerous hurricane has developed in the Caribbean Sea. The mayor of Gulf Harbor considers three possible states of nature: either the center, or the edge, or none, of the hurricane will pass over Gulf Harbor. If he announces that the eye of the storm will pass over the city, many of the most vulnerable inhabitants will drive inland, thereby reducing the number of fatalities that would occur near the waterfront if the announcement were correct. But, because of greatly increased traffic under poor weather conditions, the number of fatal traffic accidents would sharply increase. If he announces that the edge of the storm will strike the city, fewer of the vulnerable will flee and there will be fewer additional traffic fatalities, but there will be more deaths near the waterfront if the eye, rather than the edge, of the storm strikes. If he announces that none of the hurricane will pass over the city, the inhabitants will neither flee nor prepare for the onslaught of the storm. There will be no additional traffic fatalities, but if the storm actually strikes, there will be a maximum number of deaths near the waterfront. Table 10-1 summarizes the situation. Each utility's absolute value is the expected number of fatalities due directly to the storm or to the increase in traffic caused by the announcement. Meteorological data D have been obtained. In the past when similar data have been obtained, the eye of the storm has passed over Gulf Harbor in 10% of the cases, the edge in 30%, and none in 60%. So, $P(\theta_1 \mid D) = .1$, $P(\theta_2 \mid D) = .3$, and $P(\theta_3 \mid D) = .6$. The mayor decides to take consequences into consideration and to use the maximum expected utility criterion, which tends to minimize fatalities in the long run. He therefore makes the expedient announcement (see Table 10-1) that the edge of the storm will pass over the city, even though this is not the most probable outcome.

Table 10-1

Hurricane Example of Expedient Announcement of the State of Nature

		Portion of hurricane that will strike			
		Center θ_1	Edge θ_2	None θ_3	Expected utility
Action	Announce center a_1	-8	-5	-3	$-\ .8 - 1.5 - 1.8 = -4.1$
	Announce edge a_2	-15	-4	-1	$-1.5 - 1.2 - .6 = -3.3$
	Announce none a_3	-37	-6	0	$-3.7 - 1.8 + \ \ 0 = -5.5$
$P(\theta_j \mid D)$		.1	.3	.6	

10-2   Estimating the State of Nature

10-2-1   Estimation

If the decision maker wishes his announcement of the state of nature to be "truthful," rather than simply expedient, he must claim θ to have the value that is most credible on the basis of the available information, rather than most comfortable on the basis of the utilities. An *estimate* is an announcement of the state of nature based solely upon the available evidence (i.e., ignoring consequences and therefore utilities). And since it is the value of θ on a particular occasion that is being assessed, it is highly desirable that the evidence include sample information that is relevant to that particular occasion. If the announcement states a single value that θ is alleged to have, i.e., a single point at which θ is said to be, that value is called a *point estimate*. (This contrasts with another type of estimate called an *interval estimate*, which states that θ has a value that lies somewhere within a specified interval of values. Interval estimates will be taken up in a later chapter; at present we are concerned only with point estimates.)

To obtain a point estimate of θ, then, the decision maker must determine the most probable state θ_x on the basis of the available information (preferably including sample information) and then estimate that $\theta = \theta_x$. Thus, he estimates θ to have the value yielding the largest $P(\theta_j \mid D)$, or if there are no data the largest $P(\theta_j)$, or if there are data but no $P(\theta_j)$'s the largest $P(D \mid \theta_j)$. The first two estimates will be called *conditionally most probable state* and *most probable state* estimates, respectively; the third is called a *maximum likelihood* estimate. (Such estimates are equivalent to using the maximum utility under most probable state method from Chapter 9 in the first two cases, or the maximum likelihood method from the same chapter in the last case, in connection with a payoff matrix in which the largest utility under each state corresponds to the action of announcing θ to be in that state.)

We have already said that sample information is highly desirable in order to sensitize the estimate to the particular occasion to which it refers. Furthermore, if the decision maker wishes his estimate to be almost certainly correct, he should take a sample discriminating enough to make one $P(\theta_j \mid D)$ almost 1.00 or to make the largest $P(D \mid \theta_j)$ huge relative to its nearest rival. Such discrimination can generally be achieved by taking a sufficiently large number n of sample observations.

10-2-2   The Most Probable State Estimate

If the decision maker knows only the unconditional state probabilities $P(\theta_j)$'s and is unable to obtain any sample information about the state of nature on the occasion in question, his best estimate on the basis of this limited information is that θ is in the state having the maximum $P(\theta_j)$. This estimate may be highly unconvincing. The maximum $P(\theta_j)$ may not be reassuringly close to 1.00. And if there are more than two possible states, the maximum state probability $P(\theta = \theta_x)$ may be far smaller than the probability $P(\theta \neq \theta_x)$ that nature is *not* in the estimated state. Analogous statements are true of the other two methods. But in those methods the decision maker can generally *make* the maximum probability convincingly large (in fact, very close to 1.00) simply by taking a sufficiently large sample of data. If he uses the present method, the decision maker may find himself in the uncomfortable position of estimating that $\theta = \theta_3$ because $P(\theta_1) = .33$, $P(\theta_2) = .33$, and $P(\theta_3) = .34$ and of having no recourse.

Example: Shirley's mother sends her to the grocery to buy a pound of her favorite coffee. On the way Shirley forgets what brand she was supposed to buy. When she arrives she obtains from the grocer the information that 40% of his coffee-purchasing customers buy Pride of Brazil, 25% buy Café Olé, 25% buy Despierto, 10% buy Espresso, and none buy more than one brand. The most probable state of coffee demand in general, and therefore presumably on the occasion of Shirley's purchase, is Pride of Brazil. So Shirley estimates Pride of Brazil to be the brand demanded by her mother.

10-2-3 The Conditionally Most Probable State Estimate

If on the basis of some data D, relevant to the occasion in question, the decision maker knows the numerical value of $P(\theta_j \mid D)$ for each state, he finds the maximum of the $P(\theta_j \mid D)$'s and estimates that θ is in the state associated with that maximum. There are three distinguishable situations which we now consider.

10-2-3-1 $P(\theta_j \mid D)$'s KNOWN DIRECTLY. The decision maker may know the $P(\theta_j \mid D)$'s directly, rather than obtaining them by derivation. In that case, if the maximum of the $P(\theta_j \mid D)$'s is $P(\theta_x \mid D)$, he simply estimates that θ is in state θ_x.

Example: A weatherman considers three possible states of tomorrow's weather, sunny, cloudy, and rain (defined so as to be mutually exclusive and exhaustive), and wishes to estimate in which state it will be. His instruments give him a set of readings, data D. In the past such data have been followed the next day by: (θ_1) sunny weather 25% of the time, (θ_2) cloudy weather 15%, and (θ_3) rain 60%. So $P(\theta_1 \mid D) = .25$, $P(\theta_2 \mid D) = .15$, $P(\theta_3 \mid D) = .60$. The largest of these conditional probabilities is $P(\theta_3 \mid D)$, so the weatherman forecasts that tomorrow's weather will be θ_3, i.e., that it will rain.

10-2-3-2 $P(\theta_j|D)$'S Obtained by Simple Application of Bayes' Formula. The decision maker may know the $P(\theta_j)$'s and know the $P(D|\theta_j)$'s based upon some relevant data. In that case he can calculate the required $P(\theta_j|D)$'s by using, Bayes' formula

$$P(\theta_j|D) = \frac{P(D|\theta_j)P(\theta_j)}{\sum_j P(D|\theta_j)P(\theta_j)}$$

and then estimate θ to be in the state for which $P(\theta_j|D)$ is a maximum.

Example: A candy company sells chocolate-covered mixed nuts of three different qualities, depending upon the proportion of nuts that are peanuts. Quality A is an expensive deluxe mixture in which only 10% of the nuts are peanuts. Quality B is an intermediate mixture in which 25% are peanuts. And quality C is a cheap mixture 50% of which are peanuts. Five percent of the batches they produce are A quality, 20% are B quality, and 75% are C quality. Due to a clerical error, the quality of their most recently produced batch is unknown, except that it must be either A, B, or C. The candy makers wish to obtain a highly reliable estimate of its quality so that they can label it truthfully. They know, of course, that *in general* $P(A) = .05$, $P(B) = .20$, and $P(C) = .75$. But since they want their estimate to be highly sensitive to the *present* state of nature, they decide to take a sample and use the sample results to obtain revised state probabilities more in tune with the current situation. So they take, and inspect, a random sample of 40 chocolate-covered nuts, of which 9 turn out to be peanuts. They then determine the probability of obtaining such a result, under each of the possible states of nature, using the binomial probability law (since a batch contains a huge number of nuts), and obtain:

$$P(D|A) = P(r = 9 | n = 40, p = .10) = \binom{40}{9}.10^9.90^{31} = .01043$$

$$P(D|B) = P(r = 9 | n = 40, p = .25) = \binom{40}{9}.25^9.75^{31} = .13971$$

$$P(D|C) = P(r = 9 | n = 40, p = .50) = \binom{40}{9}.50^9.50^{31} = .00025$$

The unqualified probability of the data D is

$$P(D) = P(D|A)P(A) + P(D|B)P(B) + P(D|C)P(C)$$
$$= .01043(.05) + .13971(.20) + .00025(.75)$$
$$= .0005215 + .0279420 + .0001875$$
$$= .028651$$

So, applying Bayes' formula they obtain the following state probabilities, revised on the basis of the sample information:

$$P(A|D) = \frac{P(D|A)P(A)}{P(D)} = \frac{.01043(.05)}{.028651} = \frac{.0005215}{.028651} = .0182$$

$$P(B|D) = \frac{P(D|B)P(B)}{P(D)} = \frac{.13971(.20)}{.028651} = \frac{.0279420}{.028651} = .9753$$

$$P(C|D) = \frac{P(D|C)P(C)}{P(D)} = \frac{.00025(.75)}{.028651} = \frac{.0001875}{.028651} = .0065$$

The largest of these .9753 is associated with state B; so they estimate that the batch is of quality B.

10-2-3-3 $P(\theta_j|D)$'s OBTAINED BY GENERALIZED BAYESIAN METHOD. Sometimes the relevant data D consist of several separate sets of data d_1, $d_2, \ldots, d_k$, which may be separate because they are qualitatively different or because they were obtained on different occasions. If the decision maker knows the $P(D|\theta_j)$'s and $P(\theta_j)$'s, he may obtain the required $P(\theta_j|D)$'s by applying Bayes' formula in the form in which it appears in the preceding section. Otherwise, it may be far more convenient to have a form of Bayes' formula whose terms involve d_i's rather than D. If, under each different possible state of nature θ_j, *the* d_i's *are all mutually independent* (so that the probability that d_y will take any one of its possible values, when $\theta = \theta_z$, is unaffected by the values taken by any other d_i, or combination of d_i's, when $\theta = \theta_z$), the $P(\theta_j|D)$'s may be obtained by using either of the following formulas:

$$P(\theta_j|d_1 \& d_2 \& \cdots \& d_k) = \frac{P(d_1|\theta_j)P(d_2|\theta_j) \cdots P(d_k|\theta_j)P(\theta_j)}{\sum_j P(d_1|\theta_j)P(d_2|\theta_j) \cdots P(d_k|\theta_j)P(\theta_j)} \quad (10\text{-}1)$$

$$= \frac{P(d_k|\theta_j)P(\theta_j|d_1 \& d_2 \& \cdots \& d_{k-1})}{\sum_j P(d_k|\theta_j)P(\theta_j|d_1 \& d_2 \& \cdots \& d_{k-1})} \quad (10\text{-}2)$$

If the decision maker has all K sets of data available when he *first* decides to calculate conditional state probabilities, Formula 10-1 will be appropriate if he knows all of the required $P(d_i|\theta_j)$'s and $P(\theta_j)$'s. However, if the d_i's become available sequentially, the decision maker may not even know how many d_i's will ultimately appear. Yet he may wish always to know the conditional state probabilities based upon the data that *are* available. His motive may be that he wishes always to have a current, up-to-date, estimate of the actual state of nature. In that case, after either obtaining $P(\theta_j|d_1)$ directly or calculating it by using Formula 10-1, he needs, for the sake of efficiency, a formula that simply *revises* his previously calculated conditional state probabilities on the basis of the set of data acquired since then. Formula 10-2 meets that need, obtaining $P(\theta_j|d_1 \& d_2)$'s by revising the already obtained $P(\theta_j|d_1)$'s on the basis of the subsequent set of data d_2, and then obtaining $P(\theta_j | d_1 \ \& \ d_2 \ \& \ d_3)$'s by revising the $P(\theta_j | d_1 \ \& \ d_2)$'s on the basis of the subsequent set of data d_3, etc.

Thus, if the decision maker has obtained K sets of data $d_1, d_2, \ldots, d_k$ that are mutually independent under each θ_j, and if he either knows all of the $P(\theta_j)$'s and $P(d_i | \theta_j)$'s (as required by Formula 10-1) or knows all of the $P(\theta_j | d_1 \ \& \ d_2 \ \& \ \cdots \ \& \ d_c)$'s and all of the $P(d_i | \theta_j)$'s for $i > c$ (as required for successive applications of Formula 10-2), he simply calculates $P(\theta_j | d_1 \ \& \ d_2 \ \& \ \cdots \ \& \ d_k)$ for each θ_j and estimates the actual state of nature to be the θ_j that maximizes the latter probability.

Proof of Formulas and Extension to the Case of Nonindependence: We shall now derive two general formulas for $P(\theta_j | d_1 \ \& \ d_2 \ \& \ \cdots \ \& \ d_k)$ which do *not* require that the d_i's be mutually independent. Formulas 10-1 and 10-2 are simply special cases to which these general formulas are reduced when the d_i's are mutually independent of each other under every θ_j. Since the relevant data D consist of K subsets of data $d_1, d_2, \ldots, d_k$, we may substitute the latter for D in Bayes' formula for $P(\theta_j | D)$, obtaining

$$P(\theta_j | d_1 \ \& \ d_2 \ \& \ \cdots \ \& \ d_k)$$

$$= \frac{P(d_1 \ \& \ d_2 \ \& \ \cdots \ \& \ d_k | \theta_j)P(\theta_j)}{\sum_j P(d_1 \ \& \ d_2 \ \& \ \cdots \ \& \ d_k | \theta_j)P(\theta_j)}$$

$$= \frac{P(d_1 | \theta_j)P(d_2 | \theta_j \ \& \ d_1) \cdots P(d_k | \theta_j \ \& \ d_1 \ \& \ d_2 \ \& \ \cdots \ \& \ d_{k-1})P(\theta_j)}{\sum_j P(d_1 | \theta_j)P(d_2 | \theta_j \ \& \ d_1) \cdots P(d_k | \theta_j \ \& \ d_1 \ \& \ d_2 \ \& \ \cdots \ \& \ d_{k-1})P(\theta_j)}$$

$$(10\text{-}3)$$

Now, dividing both numerator and denominator of the fraction on the right by

$$\sum_j P(d_1 | \theta_j)P(d_2 | \theta_j \ \& \ d_1) \cdots P(d_{k-1} | \theta_j \ \& \ d_1 \ \& \ d_2 \ \& \ \cdots \ \& \ d_{k-2})P(\theta_j)$$

which, because it involves a summation over all values of j, is a constant having the same value for all particular values of j, we obtain

$$P(\theta_j | d_1 \ \& \ d_2 \ \& \ \cdots \ \& \ d_k) =$$

$$\frac{P(d_k | \theta_j \ \& \ d_1 \ \& \ d_2 \ \& \ \cdots \ \& \ d_{k-1}) \left[\frac{P(d_1 | \theta_j)P(d_2 | \theta_j \ \& \ d_1) \cdots P(d_{k-1} | \theta_j \ \& \ d_1 \ \& \ d_2 \ \& \ \cdots \ \& \ d_{k-2})P(\theta_j)}{\sum_j P(d_1 | \theta_j)P(d_2 | \theta_j \ \& \ d_1) \cdots P(d_{k-1} | \theta_j \ \& \ d_1 \ \& \ d_2 \ \& \ \cdots \ \& \ d_{k-2})P(\theta_j)} \right]}{\sum_j P(d_k | \theta_j \ \& \ d_1 \ \& \ d_2 \ \& \ \cdots \ \& \ d_{k-1}) \left[\frac{P(d_1 | \theta_j)P(d_2 | \theta_j \ \& \ d_1) \cdots P(d_{k-1} | \theta_j \ \& \ d_1 \ \& \ d_2 \ \& \ \cdots \ \& \ d_{k-2})P(\theta_j)}{\sum_j P(d_1 | \theta_j)P(d_2 | \theta_j \ \& \ d_1) \cdots P(d_{k-1} | \theta_j \ \& \ d_1 \ \& \ d_2 \ \& \ \cdots \ \& \ d_{k-2})P(\theta_j)} \right]}$$

But Formula 10-3 tells us that the expression inside square brackets is $P(\theta_j | d_1 \ \& \ d_2 \ \& \ \cdots \ \& \ d_{k-1})$; so substituting the latter for the former, we obtain

$$P(\theta_j | d_1 \ \& \ d_2 \ \& \ \cdots \ \& \ d_k)$$

$$= \frac{P(d_k | \theta_j \ \& \ d_1 \ \& \ d_2 \ \& \ \cdots \ \& \ d_{k-1})P(\theta_j | d_1 \ \& \ d_2 \ \& \ \cdots \ \& \ d_{k-1})}{\sum_j P(d_k | \theta_j \ \& \ d_1 \ \& \ d_2 \ \& \ \cdots \ \& \ d_{k-1})P(\theta_j | d_1 \ \& \ d_2 \ \& \ \cdots \ \& \ d_{k-1})}$$

$$(10\text{-}4)$$

And when the d_i's are mutually independent under each θ_j, Formulas 10-3 and 10-4 reduce to Formulas 10-1 and 10-2, respectively.

Example: Suppose that in addition to the information used in the preceding example $[P(A) = .05, P(B) = .20, P(C) = .75, P(d_1 | A) = .01043,$ $P(d_1 | B) = .13971, P(d_1 | C) = .00025]$, it is discovered that the batch in question was made on a Friday, and it is known that in general half of the A batches, three-fourths of the B batches, and one-fiftieth of the C batches, are made on Fridays. The candy makers then have an additional relevant and independent datum d_2 and they know that $P(d_2 | A) = .5, P(d_2 | B) = .75$, and $P(d_2 | C) = .02$. If, when they learn of this they have not yet calculated any state probabilities, it will be convenient to use Formula 10-1 to calculate

$P(A | d_1 \ \& \ d_2)$

$$= \frac{P(d_1 | A)P(d_2 | A)P(A)}{P(d_1 | A)P(d_2 | A)P(A) + P(d_1 | B)P(d_2 | B)P(B) + P(d_1 | C)P(d_2 | C)P(C)}$$

$$= \frac{(.01043)(.5)(.05)}{(.01043)(.5)(.05) + (.13971)(.75)(.2) + (.00025)(.02)(.75)}$$

$$= \frac{.00026075}{.00026075 + .02095650 + .00000375}$$

$$= \frac{.00026075}{.02122100} = .0123$$

$$P(B | d_1 \ \& \ d_2) = \frac{.02095650}{.02122100} = .9875$$

$$P(C | d_1 \ \& \ d_2) = \frac{.00000375}{.02122100} = .0002$$

On the other hand, if they have already calculated state probabilities conditional upon the first set of data, which we found to be $P(A | d_1) = .0182,$ $P(B | d_1) = .9753, P(C | d_1) = .0065$, it will be convenient to use Formula 10-2 to obtain

$$P(A | d_1 \ \& \ d_2) = \frac{P(d_2 | A)P(A | d_1)}{P(d_2 | A)P(A | d_1) + P(d_2 | B)P(B | d_1) + P(d_2 | C)P(C | d_1)}$$

$$= \frac{(.5)(.0182)}{(.5)(.0182) + (.75)(.9753) + (.02)(.0065)}$$

$$= \frac{.009100}{.009100 + .731475 + .000130}$$

$$= \frac{.009100}{.740705} = .0123$$

$$P(B | d_1 \ \& \ d_2) = \frac{.731475}{.740705} = .9875$$

$$P(C | d_1 \ \& \ d_2) = \frac{.000130}{.740705} = .0002$$

which, of course, are the same values we obtained by using Formula 10-1. The largest of the state probabilities conditional upon both sets of data is .9875 which corresponds to state *B*. So the company labels the batch "*B* quality."

10-2-4   The Maximum Likelihood Estimate

If the decision maker has some data *D* and knows the $P(D|\theta_j)$'s but has no information about state probabilities (i.e., does not know, and cannot obtain, either $P(\theta_j)$'s, $P(\theta_j|D)$'s, or $P(\theta_j|\delta)$'s, where δ is an entirely different set of data), then he has little choice but to make a maximum likelihood estimate. This is accomplished by estimating θ to have the value for which $P(D|\theta_j)$ is largest. In doing so, the decision maker is implicitly assuming that the $P(\theta_j)$'s all have the same value (see Section 9-5-2), which, of course, is unlikely to be the case. However, if *D* is based upon a sufficiently large or (at least) discriminating sample, and the states differ appreciably from each other, the largest $P(D|\theta_j)$ will vastly exceed its closest rival, which will be negligible by comparison. In that event the maximum likelihood estimate will be almost certain to be correct even if the implicit assumption is badly in error (see Sections 9-5-2 and 9-3-6).

Example: A housewife opens a carton of a dozen eggs, breaks three, and discovers that one of them is rotten. She now wishes to estimate how many rotten eggs were originally in the carton. From her sample she knows that at least one of the original dozen eggs was rotten and at least two were not. So the possible values for the number of rotten eggs are 1, 2, 3, 4, 5, 6, 7, 8, 9, 10. Letting *R* stand for the number of rotten eggs in the carton of 12 and *r* the number in the sample of 3, she can calculate the probability of the sample data for each possible value of *R* using the hypergeometric probability law

$$P(D|\theta) = P(r = 1 | N = 12, n = 3, R) = \frac{\binom{R}{r}\binom{N-R}{n-r}}{\binom{N}{n}} = \frac{\binom{R}{1}\binom{12-R}{2}}{\binom{12}{3}}$$

$$= \frac{(12-R)!\,R}{(10-R)!\,440}$$

The solutions of the last expression for *R*'s from 1 to 10 are, in sequence, 110/440, 180/440, 216/440, 224/440, 210/440, 180/440, 140/440, 96/440, 54/440, 20/440. The largest of these occurs for $R = 4$, so the housewife estimates that there were 4 rotten eggs among the original dozen.

This estimate is not especially convincing since the probability of the sample result is not much different under the estimated *R*-value of 4 than under the "adjacent" values 3 and 5 or "close" values, such as 2 or 6. This is likely to be the case when all possible values of a variable, such as R, within an interval, such as 1 to 10, are regarded as possible states of nature. It need not be the case, however, when *only certain ones* of these values are possible states of nature and they are reasonably separated from each other. For example, suppose that we have a letter containing 1,000 words of which 6 are incorrectly typed. The letter is known to have been typed by one of three typists A, B, and C. The average number of mistyped words per thousand is known to be 1 for Typist A, 5 for B, and 15 for C. The Poisson probability law gives us the probability of the obtained data under each typist.

$$P(D|A) = P(x = 6 | m = 1) = \frac{1^6}{6! \, e^1} = .0005$$

$$P(D|B) = P(x = 6 | m = 5) = \frac{5^6}{6! \, e^5} = .1462$$

$$P(D|C) = P(x = 6 | m = 15) = \frac{15^6}{6! \, e^{15}} = .0048$$

The m-values for the three typists are fairly widely separated and consequently the data probabilities differ greatly. The largest of these probabilities, by far, occurs for Typist B. So the maximum likelihood estimate that the letter was typed by B is reasonably convincing. (It would be more convincing, if there were good reason to believe that each typist types about one-third of all letters, however, since the data probabilities under the other typists are not negligible.)

10-3 Choosing Between Conflicting Hypotheses

10-3-1 Rationale

A common type of decision problem is that in which the decision maker must choose one of several competing hypotheses as the correct one. If the following conditions are met, this problem reduces to a special, but somewhat elaborated, case of announcing the state of nature. (a) One, and only one, of the hypotheses must be true. (b) Each of the hypotheses $H_1, H_2, \ldots, H_j, \ldots,$ H_s must logically imply an associated state of nature $\theta_1, \theta_2, \ldots, \theta_j, \ldots, \theta_s,$ respectively, with which it is inexorably linked. H_j may simply state that $\theta = \theta_j$, i.e., that θ_j is the state of nature. Or, the implied state of nature θ_j may be an inevitable logical consequence of H_j—so that it must be the state of nature if H_j is true. (c) Each of the implied states of nature must differ from all the others.

If the above conditions are met, we have a number of different hypotheses, one of which is correct, each of which is associated with a different value of the state of nature, one of which is the actual value. The θ-values are deduced from, or stated by, their associated hypotheses. Since we cannot validly deduce a false consequence from a true hypothesis, the actual state of nature must be associated with the true hypothesis. Therefore, if we can identify the actual state of nature, we can identify the true hypothesis as being the one associated with it. In effect, then, the entire problem resolves itself into one of announcing the state of nature.

If the hypotheses state θ-values directly, one way to insure that one of the hypotheses is correct is to make the hypotheses exhaustive, i.e., to have an hypothesis for each different θ-value that is possible under the circumstances. Unfortunately, this ploy does not work if the hypotheses are not simply allegations about θ-values; it is quite possible for the true θ-value to be obtained by a valid deduction from a false hypothesis. And we should especially beware of making the false assumption that, if none of the hypotheses is true, the most nearly correct of the deduced θ-values will be associated with the most nearly correct hypothesis.

In many, if not most, cases, the single true hypothesis is, always was, and always will be, true, whereas the other hypotheses are, always were, and always will be, false. These circumstances present a special case in announcing the associated state of nature. In this special case θ, instead of being a random variable that takes each of the values $\theta_1, \theta_2, \ldots, \theta_j, \ldots, \theta_s$ a certain proportion of the time, is a constant that has an unknown one of those values *all* of the time—so that the probability for that state is 1.00. Thus, ignorance of the correct state is equivalent to ignorance of which state has the probability of 1.00, and this tends to rule out all decision methods that make use of objective (or even subjective) state probabilities. By default, therefore, attempts to identify the correct hypothesis generally rely upon $P(D|\theta_j)$'s as indices of relative likelihood.

Suppose, therefore, that some sample statistic S is "sensitive" to the state of nature in that it has a known and calculable probability distribution that depends upon θ, being different for each different θ-value. The decision maker may then take a sample of data and let the results, and their $P(D|\theta_j)$'s, determine which of the states of nature, and therefore which of the hypotheses, to choose as correct.

Sample statistics are sensitive only to characteristics of the sampled population or of the sampling procedure. Therefore, the requirement that S be sensitive to the state of nature virtually forces θ to be either a population or a sampling characteristic. Often, but not always, the θ-values are the possible values of a population parameter, such as the proportion p of units that are "successes" in an infinite dichotomous population [in which case if S is the number r of successes in a random sample of size n, $P(D|\theta_j)$ is given by the binomial probability law

$$P(r|n, p_j) = \binom{n}{r} p_j^r (1 - p_j)^{n-r}]$$

Decision problems of the type discussed above are frequently encountered in science where the scientist must choose between competing explanations for an observed phenomenon. "The truth" is not regarded as sometimes residing in one explanation and sometimes in another (with rare pragmatic exceptions such as the simultaneously "accepted" wave and corpuscular theories of light). Rather, the truth is regarded as a constant whose value is unknown, so that no more than one of the possible explanations is the (exactly) correct one. The scientist is prevented from using subjective probabilities and subjective values (utilities) by the consensus in the scientific community that the status of theories (i.e., acceptance of explanations) should be based upon objective rather than subjective factors. (It is, of course, virtually impossible to eliminate subjectivity and personal bias from the scientific process; but subjective probabilities and utilities are too blatantly peculiar to an individual scientist's "personality" to be presently acceptable to the majority of the scientific community in so important a decision situation as selection of the correct theory.) Instead, discrimination between theories is supposed to be based upon data. Furthermore, scientific objectivity seems to demand the distribution of equal degrees of credence to all explanations prior to the collection of data and prior to the general acceptance of a theory. These are precisely the conditions under which the maximum likelihood method of estimation is appropriate. On the other hand, it is customary in science not to abandon an old established theory until the evidence against it is quite impressive. Were this not the case science would be in a chronic state of chaotic upheaval since new alternative explanations abound and since it is far easier to perceive flaws in the familiar than in the novel. For the sake of its own stability, then, science tends to favor an established theory against an upstart challenger. One way of doing this is to impose a handicap upon the challenger, "announcing" that the new theory is true only if the ratio

$$\frac{P(\text{data} \mid \text{old established theory is true})}{P(\text{data} \mid \text{new challenging theory is true})}$$

falls below some specified value far smaller than 1. Such "unfair" procedures are called *tests* since they impose the burden of convincing proof upon the challenger. This method of choosing between hypotheses is called the Likelihood Ratio Test. We shall discuss these methods further in the following sections.

10-3-2 Choosing Accurately: The Maximum Likelihood Estimate

10-3-2-1 METHOD.   Consider first the case in which the decision maker wishes to identify correctly, without regard for consequences, the single true hypothesis he knows to exist in a set of otherwise false and conflicting hypotheses. Suppose that each hypothesis alleges nature to be in a different state, hypothesis H_j claiming that $\theta = \theta_j$. Alternatively, suppose that from each of the hypotheses $H_1, H_2, \ldots, H_j, \ldots, H_s$ one can deduce a different value $\theta_1, \theta_2, \ldots, \theta_j, \ldots, \theta_s$, respectively, that θ would be compelled to take if the corresponding hypothesis were true. In either case the true hypothesis must be associated with the true state of nature. Finally, suppose that the state of nature is some characteristic of a population from which a sample can be drawn and that the probabilities of the various possible sample outcomes depend upon the population characteristic. Then the decision maker may draw a random sample of data D from the population, calculate $P(D|\theta_j)$ for each of the θ_j's, determine the largest of them $P(D|\theta_x)$ occurring for the state of nature we shall call θ_x, and then estimate that the corresponding hypothesis H_x is the true one. This is a truthful maximum likelihood estimate.

θ may be any one of a variety of population characteristics. However, in the case of maximum likelihood estimates, θ is almost always a population parameter such as the proportion of successes, the median, mean, or variance, etc. The parameter we shall use in our examples is the proportion p of successes in an undepletable dichotomous population.

It should be clear that what the maximum $P(D|\theta_j)$ really permits us to estimate is the value of θ and that the estimation of true hypothesis is simply a further extension of inference. Therefore, especially when H_j simply states that $\theta = \theta_j$, it is just as legitimate to say that we are making a maximum likelihood estimate of θ as of true hypothesis.

10-3-2-2 EXAMPLE. A biologist is interested in the proportion p of animals in a certain rare species that have both of two characteristics A and B. From genetic theory, he knows that if A is a dominant characteristic its probability is 3/4, whereas if it is a recessive characteristic its probability is 1/4; and likewise for B. Furthermore, he knows that each characteristic must be either dominant or recessive and that A and B are independent. Under these conditions, the proportion p must have one of three values: $(3/4)^2$ or 9/16 if both characteristics are dominant, $2(3/4)(1/4)$ or 6/16 if one characteristic is dominant and the other is recessive, and $(1/4)^2$ or 1/16 if both characteristics are recessive. The biologist therefore formulates three hypotheses, H_1 which says that $p = 9/16$, H_2 that $p = 6/16$, and H_3 that $p = 1/16$. (Had his interest been focused upon the reason for the proportion rather than upon the proportion itself, his hypotheses could have been H_1: both characteristics are dominant, H_2: one characteristic is dominant, H_3: neither characteristic is dominant. This substitution would be acceptable because one of these hypotheses is true and each hypothesis is necessarily associated with a different one of the replaced proportions.)

The biologist knows that one of these conflicting hypotheses is correct. In order to decide which one it is, he intends to take a random sample of 8 animals and base his decison upon the outcome. The probability that in a random sample of eight animals, r of them will have both characteristics is

$$P(r \mid p, n = 8) = \binom{8}{r} p^r (1 - p)^{8-r}$$

He takes the sample and finds that four of the eight animals have both characteristics. He therefore calculates

$$P(r = 4 \mid n = 8, p = 9/16) = \binom{8}{4} (9/16)^4 (7/16)^4 = .25674$$

$$P(r = 4 \mid n = 8, p = 6/16) = \binom{8}{4} (6/16)^4 (10/16)^4 = .21122$$

$$P(r = 4 \mid n = 8, p = 1/16) = \binom{8}{4} (1/16)^4 (15/16)^4 = .00082$$

The largest of these probabilities .25674 occurs for $p = 9/16$, so the biologist estimates that the proportion of animals having both characteristics is 9/16 and therefore that H_1 is true. He might feel a bit nervous about his estimate, however, since the likelihood .21122 of the obtained data under $p = 6/16$ is almost as large as its likelihood .25674 under $p = 9/16$. Ways of dealing with this type of difficulty will be discussed in a later section.

10-3-2-3 EQUIVALENT DECISION RULE. Table 10-2 gives $P(r \mid n = 8, p)$ for all possible data outcomes r and all three possible values of p. For each possible outcome r the maximum of the three probabilities is printed in boldface. It is clear from Table 10-2 that if the biologist decides to make a maximum likelihood estimate on the basis of a sample of eight, he is really following the equivalent *decision rule*: if $r \leq 1$ estimate $p = 1/16$; if $2 \leq r \leq 3$ estimate $p = 6/16$; if $r \geq 4$ estimate $p = 9/16$. Furthermore, by

Table 10-2

Probabilities of All Possible Data Outcomes Under All
Possible Hypotheses in Genetics Example of
Maximum Likelihood Estimation of True Hypothesis

	r	H_1 ($p = 9/16$)	H_2 ($p = 6/16$)	H_3 ($p = 1/16$)
			Possible hypotheses	
	0	.00134	.02328	*.59672*
	1	.01381	.11176	*.31825*
Possible	2	.06212	*.23470*	.07426
data	3	.15976	*.28163*	.00990
outcomes	4	*.25674*	.21122	.00082
	5	*.26408*	.10139	.00005
	6	*.16976*	.03041	.00000
	7	*.06237*	.00522	.00000
	8	*.01002*	.00039	.00000

simply adding the boldface probabilities in each column, we obtain the probabilities of correctly estimating p if $p = 9/16$, if $p = 6/16$, and if $p = 1/16$, as shown in Table 10-3a.

10-3-2-4 EFFECT OF SAMPLE SIZE. Clearly, there is a substantial probability of making an incorrect estimate on the basis of a sample of only 8 observations. However, if we increase the sample size to 100 observations, we find that the maximum likelihood of r occurs for $p = 1/16$ for all $r \leq 18$, for $p = 6/16$ when $19 \leq r \leq 46$, and for $p = 9/16$ for all $r \geq 47$. Furthermore,

$$P(r \leq 18 \,|\, n = 100, \, p = 1/16) = .99999$$
$$P(19 \leq r \leq 46 \,|\, n = 100, \, p = 6/16) = .96727$$
$$P(r \geq 47 \,|\, n = 100, \, p = 9/16) = .97489$$

and, since we shall estimate that $p = 1/16$ if $r \leq 18$, that $p = 6/16$ if $19 \leq r \leq 46$, and that $p = 9/16$ if $r \geq 47$, we have greatly improved our probabilities of making correct estimations, as shown in Table 10-3b. Thus, no matter which of the three hypothesized values of p is the correct one, the probability that a maximum likelihood estimate based on a sample of 100 will correctly identify the true value of p is at least .96. By taking a sufficiently large sample we can make this minimum probability of correct estimation as close to 1.00 as we wish, although we cannot make it exactly 1.00. And we can always do this provided that the hypothesized values of p are separated by gaps, rather than forming a continuum.

Table 10-3
Probabilities of Correct and Incorrect Estimation if H_j: $p = p_j$ Is

True in Genetics Example of Maximum Likelihood Estimation

(a) When sample size is $n = 8$

	Hypothesis that is true		
	H_1	H_2	H_3
	$p = 9/16$	$p = 6/16$	$p = 1/16$
P (correct estimation)	.76297	.51633	.91497
P (incorrect estimation)	.23703	.48367	.08503

(b) When sample size is $n = 100$

	Hypothesis that is true		
	H_1	H_2	H_3
	$p = 9/16$	$p = 6/16$	$p = 1/16$
P (correct estimation)	.97489	.96727	.99999
P (incorrect estimation)	.02511	.03273	.00001

10-3-2-5   CASE OF AN INFINITE NUMBER OF POSSIBILITIES.   If the values of p under consideration do form a continuum, then there are an infinite number of possible values of p and we cannot calculate the probability of the obtained data under each of them. However, we can *still* make maximum likelihood estimates. A fairly common situation is that in which we want to estimate the proportion p of units in an infinite dichotomous population that are "successes" where the possible values of p extend continuously from 0 to 1. The probability of obtaining r successes in a random sample of n observations is, of course,

$$P(r\,|\,n,\,p) = \binom{n}{r} p^r (1 - p)^{n-r}$$

Now it is a fact that as p increases from 0 to 1 while r and n remain constant, $P(r\,|\,n,\,p)$ increases to a maximum, which it reaches when $p = r/n$, and then decreases. This means that the value of p that maximizes the likelihood of the obtained data is r/n, which is therefore the maximum likelihood estimate of p. Thus, if one person in a random sample of four people is found to be senile, the maximum likelihood estimate of the proportion of senile persons in the general population is $p = r/n = 1/4 = .25$. Furthermore, although p may have any value in the interval from 0 to 1, the *estimate* must be one of the values 0, .25, .50, .75, 1.00. The wisdom of taking large samples when making such maximum likelihood estimates should be apparent.

10-3-3   Choosing Cautiously: The Likelihood Ratio Test

10-3-3-1   RATIONALE.   If there were only two hypotheses, the maximum likelihood method would choose the one under which the obtained data had the larger probability, no matter by how small an amount it exceeded the probability of the data under the other hypothesis. However, the decision maker may wish to pursue a conservative policy of favoring one particular hypothesis unless the evidence against it is overwhelming. He may be motivated to do so by unequal *consequences* of choosing incorrectly, even though he may have equal degrees of credence for the competing hypotheses. As already mentioned, in science it is customary to favor the old established theory over its upstart challenger since overthrowing the former is more disruptive than overthrowing the latter; in business and industry it is prudent to favor the hypothesis calling for the least costly action or least hazardous policy. If H_1 is the favored hypothesis, the decision maker may follow a *likelihood ratio* criterion of choosing H_1 if the ratio of $P(\text{data}\,|\,\theta_1)$ to $P(\text{data}\,|\,\theta_2)$ exceeds some predesignated value K. That is, the decision rule is

$$\text{Act as if } H_1 \text{ is true if } L = \frac{P(\text{data}\,|\,\theta_1)}{P(\text{data}\,|\,\theta_2)} > K$$

$$\text{Act as if } H_2 \text{ is true if } L = \frac{P(\text{data}\,|\,\theta_1)}{P(\text{data}\,|\,\theta_2)} \leq K$$

And H_1 can be favored and protected by making K a small fraction, i.e., a positive value much smaller than 1.00.

The conditions under which the test is valid may be briefly recapitulated: H_1 and H_2, a single one of which is correct, state or imply that $\theta = \theta_1$ and $\theta = \theta_2$, respectively. θ is a characteristic of a population from which a random sample of data D can be drawn, and the various possible values of D have different probabilities if $\theta = \theta_1$ than they do if $\theta = \theta_2$. The test is performed by drawing the sample, calculating $P(D|\theta_1)$ and $P(D|\theta_2)$ for the obtained value of D, and accepting H_1 or H_2 if $L = P(D|\theta_1)/P(D|\theta_2)$ is $> K$ or $\leq K$, respectively, where (before drawing the sample) K was made small enough to favor and protect H_1 to the desired degree.

10-3-3-2 NOTE. The likelihood ratio test, by favoring one hypothesis over another, takes expediency into account, which is analogous to re-introducing utilities. Furthermore, it treats the $P(D|\theta_j)$'s as if they were $P(\theta_j|D)$'s, as does the maximum likelihood method of which it is a modification. In fact, it can be shown that the likelihood ratio test is equivalent to choosing between θ-values using the maximum expected utility method (Section 9-3) with $P(D|\theta_j)$'s replacing $P(\theta_j|D)$'s and with utilities that meet a certain criterion. Consider a payoff matrix in which θ_1 and θ_2 are column headings, a_1 and a_2 are to announce that $\theta = \theta_1$ and that $\theta = \theta_2$, respectively, and U_{ij} is the utility of the ith action under the jth state. Then if the decision maker wished to use the maximum expected utility method, treating the $P(D|\theta_j)$'s as $P(\theta_j|D)$'s, he would calculate

$$E[U(a_1)] = U_{11}P(D|\theta_1) + U_{12}P(D|\theta_2)$$
$$E[U(a_2)] = U_{21}P(D|\theta_1) + U_{22}P(D|\theta_2)$$

and take the action having the larger expected utility. The utility values for which he would be indifferent between the two actions must satisfy the equation

$$U_{11}P(D|\theta_1) + U_{12}P(D|\theta_2) = U_{21}P(D|\theta_1) + U_{22}P(D|\theta_2)$$

which can be simplified as follows:

$$U_{11}P(D|\theta_1) - U_{21}P(D|\theta_1) = U_{22}P(D|\theta_2) - U_{12}P(D|\theta_2)$$
$$(U_{11} - U_{21})P(D|\theta_1) = (U_{22} - U_{12})P(D|\theta_2)$$
$$\frac{P(D|\theta_1)}{P(D|\theta_2)} = \frac{U_{22} - U_{12}}{U_{11} - U_{21}}$$

But the expression on the left is the likelihood ratio L. So the expression on the right, involving utilities, must equal K, the indifference point between the two actions. The advantages of the likelihood ratio test over the equivalent maximum expected utility method are that (a) it generally requires less computation, (b) it may be much easier to decide upon a single value for K than upon four properly related utility values, and (c) it is simpler and more standardized in the sense that there are an infinite number of sets of four utility values corresponding to each single value of K. The standardization feature makes it particularly attractive to scientists, both because it seems more efficient and because it seems more impersonal.

10-3-3-3 EXAMPLE.   Changing the preceding example a bit, suppose that according to presently accepted theory a certain biological characteristic in a rare species of animal is a dominant one with probability of .75. A young scientist doubts this believing that the characteristic is recessive having probability .25 of occurring. In order to determine which theory should be accepted, he decides to obtain a random sample of seven members of the species and let the sample data decide the issue. However, since he knows that the old theory will not be overthrown or even cast into serious doubt unless the evidence against it is overwhelming, he decides to use the likelihood ratio criterion favoring the old theory by accepting it as valid if the ratio of $P(\text{data}\,|\,\text{old theory})$ to $P(\text{data}\,|\,\text{new theory})$, i.e., the ratio of $P(D\,|\,p = .75)$ to $P(D\,|\,p = .25)$, exceeds .01. He obtains the sample and finds that two of the seven animals have the characteristic. He therefore calculates the likelihood ratio

$$L = \frac{P(\text{data}\,|\,\text{old theory})}{P(\text{data}\,|\,\text{new theory})} = \frac{P(r = 2\,|\,n = 7,\, p = .75)}{P(r = 2\,|\,n = 7,\, p = .25)} = \frac{\binom{7}{2}.75^2.25^5}{\binom{7}{2}.25^2.75^5}$$

$$= \frac{.25^3}{.75^3} = \left(\frac{1}{3}\right)^3 = .037$$

which exceeds .01; so the scientist must acquiesce to the old theory, even though the obtained data were much more likely under the new theory.

10-3-3-4 EQUIVALENT DECISION RULE.   The possible data outcomes, their probabilities of occurrence under H_1 and H_2, and the likelihood ratio corresponding to each possible outcome, are given in Table 10-4. The likelihood ratio exceeds .01 for all possible outcomes except $r = 0$ and $r = 1$. Clearly, therefore, the likelihood ratio criterion with $K = .01$ is equivalent to the decision rule: choose H_1 (accept that $p = .75$) if $r > 1$, choose H_2 (claim that $p = .25$) if $r \leq 1$. Furthermore, we can obtain the probabilities that r will be > 1 or ≤ 1 under each possible value of θ by simply summing the appropriate probabilities in the appropriate column of the table. And these cumulated probabilities will also be the probabilities of making the choice called for by the decision rule. Therefore, we can express the relevant information in Table 10-4 more concisely as shown in Table 10-5a.

Clearly, it is an error to choose the wrong hypothesis, so the actions listed in the lower left and upper right cells of the table are errors. As intended, the

Table 10-4

Probabilities of All Possible Data Outcomes Under Both Hypotheses in Example of Likelihood Ratio Test of Challenging Hypothesis

Possible data outcome	$P(D \mid H_1$ is true)	$P(D \mid H_2$ is true)	Likelihood ratio
r	$P(r \mid n = 7, \rho = .75)$	$P(r \mid n = 7, \rho = .25)$	$\dfrac{P(r \mid n = 7, \rho = .75)}{P(r \mid n = 7, \rho = .25)}$
0	.0000610	.1334839	.000457
1	.0012818	.3114624	.004115
2	.0115356	.3114624	.037037
3	.0576782	.1730347	.333333
4	.1730347	.0576782	3.000000
5	.3114624	.0115356	27.000000
6	.3114624	.0012818	243.000000
7	.1334839	.0000610	2187.000000

Table 10-5

Probabilities of Data Outcomes Calling For Correct and Incorrect Choices, Under True Hypotheses, in Example of Likelihood Ratio Test of Challenging Hypothesis

(a) When sample size is $n = 7$

Data outcome	Probability of data outcome (and, in parentheses, action called for) under H_1 $\rho = .75$	Probability of data outcome (and, in parentheses, action called for) under H_2 $\rho = .25$
$r > 1$	.9986572 (accept H_1 as true)	.5550537 (accept H_1 as true)
$r \leqslant 1$	.0013428 (accept H_2 as true)	.4449463 (accept H_2 as true)

(b) When sample size is $n = 35$

Data outcome	Probability of data outcome (and, in parentheses, action called for) under H_1 $\rho = .75$	Probability of data outcome (and, in parentheses, action called for) under H_2 $\rho = .25$
$r > 15$	.99995 (accept H_1 as true)	.00618 (accept H_1 as true)
$r \leqslant 15$	.00005 (accept H_2 as true)	.99382 (accept H_2 as true)

old theory H_1 is protected in the sense that if it is true, the probability of "overthrowing it" (actually seriously questioning it) by claiming that the new theory H_2 is true is very small (.0013428). However, H_2 is relatively vulnerable in that if it is true there is a large probability (.5550537) of wrongly accepting its rival.

10-3-3-5 EFFECT OF SAMPLE SIZE. This can be ameliorated by taking more data. For example, if we leave the likelihood ratio criterion at $K = .01$ but increase the sample size from $n = 7$ to $n = 35$, we find that the equivalent decision rule is to choose H_1 if $r > 15$ and to choose H_2 if $r \leq 15$, and we obtain high probabilities for correct choices and low probabilities for wrong ones. See either Table 10-5b or Table 10-6 which expresses the same information in different form. By taking a sufficiently large sample we can make the minimum probability of taking the correct action as close to 1.00 as we wish, although we can never make it exactly 1.00.

Table 10-6

Information Contained in Table 10-5(b) Expressed in a Different Format

Action	Probability of action if true hypothesis is	
	H_1	H_2
	$\rho = .75$	$\rho = .25$
Accept H_1	$P(r > 15 \mid n = 35, \rho = .75)$	$P(r > 15 \mid n = 35, \rho = .25)$
	$= .99995$	$= .00618$
Accept H_2	$P(r \leq 15 \mid n = 35, \rho = .75)$	$P(r \leq 15 \mid n = 35, \rho = .25)$
	$= .00005$	$= .99382$

10-3-3-6 SUMMARY AND GENERALIZATION. Let there be two hypotheses, one of which is true, H_1 stating or implying that $\theta = \theta_1$ and H_2 stating or implying that $\theta = \theta_2$, the former of which is to be given preferential treatment by the decision maker. Let S be a sample statistic whose probability distribution is different under the two possible θ-values and therefore "sensitive" to the validity of the hypotheses. And let S' be the value of S in the actually obtained sample. Then the likelihood ratio criterion is to:

$$\text{Accept } H_1 \text{ as true if } L = \frac{P(S' \mid \theta_1)}{P(S' \mid \theta_2)} > K$$

$$\text{Accept } H_2 \text{ as true if } L = \frac{P(S' \mid \theta_1)}{P(S' \mid \theta_2)} \leq K$$

where K is a value selected by the decision maker to give H_1 the desired degree of preferential treatment. Corresponding to K is a critical value C of the sample statistic S that divides the possible values of S into two nonoverlapping

Table 10-7

Probabilities of Making Correct and Incorrect Choices When Using Likelihood Ratio Test

Action	Probability of action if H_1 is true (implying that $\theta = \theta_1$)	H_2 is true (implying that $\theta = \theta_2$)
Choose H_1	$1 - \alpha = P(S \text{ in acceptance region} \mid \theta = \theta_1)$ $= P(\text{correct choice} \mid H_1 \text{ is true})$	$\beta = P(S \text{ in acceptance region} \mid \theta = \theta_2)$ $= P(\text{wrong choice} \mid H_2 \text{ is true})$
Choose H_2	$\alpha = P(S \text{ in rejection region} \mid \theta = \theta_1)$ $= P(\text{wrong choice} \mid H_1 \text{ is true})$	$1 - \beta = P(S \text{ in rejection region} \mid \theta = \theta_2)$ $= P(\text{correct choice} \mid H_2 \text{ is true})$

regions. When $L > K$, calling for the acceptance of H_1, S falls on one side of C—that is, S falls in a region we shall call the *acceptance* region since it calls for "acceptance" of the favored hypothesis. When $L \leq K$, calling for the choice of H_2, S either falls on the other side of C or equals C—that is, S falls in a region we shall call the *rejection* region, since it calls for "rejecting" the favored hypothesis. (Whether values of $S \leq C$ or $S \geq C$ is the rejection region depends upon a number of factors, such as the relative values of θ_1 and θ_2, the distribution of the test statistic, etc.). Now, "rejecting" H_1 when it is true is an error whose conditional probability we shall call α; likewise, "accepting" H_1 when it is false is an error whose conditional probability we shall call β. Table 10-7 summarizes the situation. The favored hypothesis H_1 is given preferential treatment by taking K to be a small fraction of 1—which causes C to have a value that makes α a small fraction of .5. Alternatively, preferential treatment of H_1 can be insured simply by fixing α at a sufficiently small value and using the corresponding values of K or C. The test can be made sufficiently discriminating by taking a sample large enough so that, for the chosen value of K, α will be tolerably small and β will be very close to zero. The Neyman-Pearson hypothesis-testing procedure, discussed in the next chapter, is a generalization and elaboration of the form of the likelihood ratio test in which the hypothesis chosen depends upon whether or not the test statistic S falls in the rejection region.

10-4   Problems

1. In the hurricane example (Section 10-1), suppose that an additional and independent set of data δ became available and that on the basis of past records it is known that

$$P(\delta|\theta_1) = .12, \ P(\delta|\theta_2) = .14, \ P(\delta|\theta_3) = .01$$

   What are the proper revised state probabilities, the expected utilities, and the expedient action on the basis of a maximum expected utility criterion? What state of nature is most probable on the basis of all available data?

2. At the beginning of each play, a One-Armed Bandit is in one of three states: θ_1, it will pay nothing; θ_2, it will pay back double the investment; θ_3, it will pay back 50 times the investment. Consider the following possible items of information about the machine. (A) On 80% of all plays it is in state θ_1, on 19% it is in state θ_2, and in 1% it is in the state θ_3. (B) At the end of the last play the machine made a faint clicking noise. This clicking noise on the preceding play occurs 10% of the time that the next state is θ_1, 25% of the time that it is θ_2, and 20% of the time

that it is θ_3. (C) At the end of the last play three lemons showed in the window of the machine. This event is independent of the clicking noise, and when it happens the next state of the machine becomes θ_1 50% of the time, θ_2 20%, and θ_3 30%. Estimate the state of the machine if the only information about it is (a) A, (b) B, (c) C. Estimate the state of the machine and give the state probabilities based upon all relevant information if the available information consists of (d) A and B, (e) A and C, (f) B and C, (g) A and B and C.

3. The output of a factory is produced by three simultaneously operating machines. Each machine produces 1,000 items per hour, of which on the average 30 are defective when the machine is in adjustment and 120 are defective when it is out of adjustment. The outputs of all three machines are dumped onto a single conveyor belt and intermixed so that it is impossible to determine which machine produced which item. A random sample of 10 items is drawn from the conveyor belt, of which 1 turns out to be defective. On the basis of this information, estimate the number of machines that are out of adjustment. What is the probability of this sample result if the estimate is correct?

4. In the above problem suppose that when out of adjustment, instead of each machine producing 120 defectives per 1,000, machines A, B, and C produce 120, 150, and 180 defectives per 1,000, respectively. Estimate which of the eight possible combinations of machines is the combination of machines that are out of adjustment.

5. Ninety percent of the time a machine is in adjustment, 8% of the time it is in need of minor adjustment, and 2% of the time it needs major overhauling. When it is in adjustment, 1% of the items that it produces are defective; when it needs minor adjustment, 10% are defective; and when it needs overhauling, 30% are defective. A random sample of six items is taken from the output of the machine and two of them are defective. Estimate the state of the machine and give the probability for each state based upon all relevant information.

6. What would be the answer to Problem 5 if, subsequent to the sample of six items, a second random sample had been taken in which four items were drawn and one of them was defective?

7. What would be the answer to Problem 6 if, instead of taking the sample of six items of which two were defective followed by the sample of four items of which one was defective, a single sample of ten items had been taken of which three were defective?

8. A patient has a set of symptoms that is found in 90% of people suffering from disease A, 70% of those having disease B, 40% of those with disease C, and 5% of those having none of these diseases. At any given

moment, .03% of the population has disease A, .06% has disease B, .01% has disease C, and 99.9% has none of these diseases. (The percentage having more than one of the diseases is negligible.) Estimate which of the four conditions, A, B, C, or none, the patient is in and give the probability for each, based on the relevant information.

9. On the average, one person in 10,000 at presidential campaign speeches is a potential assassin. At such gatherings, the probability that a potential assassin will be young is .9, will be male is .9, will have not attended college is .8, will be unemployed or in a temporary menial job is .9, will be carrying a weapon is 1.00; whereas the corresponding probabilities for those who are not potential assasins are .3, .5, .4, .1, and .01, respectively. At such a gathering, police pick up a suspicious character who has all of the characteristics listed above. If these characteristics are independent, what is the probability that he is a potential assassin?

10. A crime has been committed in a community of 5,000 people. It is definitely established that the criminal is a member of the community and that he is (i) right-handed, (ii) a middle-aged male, (iii) the owner of a blue station wagon, (iv) has a license plate the last two numbers of which are 97. In the community the proportion of people having these respective characteristics is (i) .9, (ii) .1, (iii) .005, (iv) .01. A suspect, having all of these characteristics, is picked up by the police. What is the probability that he is the criminal (a) based merely upon his membership in the community, (b) based upon all of the evidence if the characteristics are independent?

11. One airline passenger in 100,000 is a hijacker. However, among hijackers (i) 90% are male; (ii) 90% are young; (iii) 70% carry excess baggage; (iv) 99% carry a handbag with them; (v) 90% travel alone; (vi) 99% are obviously very nervous; (vii) 90% are irritable; (viii) 95% carry both a map and a schedule with them; and (ix) 98% survey the interior of the plane and the other passengers very carefully before taking their seat. Whereas the percentages of ordinary passengers having these respective characteristics are (i) 50%; (ii) 10%; (iii) 5%; (iv) 40%; (v) 20%; (vi) .1%; (vii) 1%; (viii) .1%; (ix) 2%. Passenger X has all of the listed characteristics except the third. If all the listed characteristics are independent, what is the probability that passenger X is a hijacker?

12. A fragment of a letter is found under circumstances indicating that either Shakespeare, Marlowe, or Fletcher must be the author. The fragment contains ten words, three of which are long words, i.e., contain more than seven letters, and one of which is the word contumely. Of the known letters written by one of the three authors, 30% were written by Shakespeare, 10% by Marlowe, and 60% by Fletcher. In those written

by Shakespeare 40% of the words are long, by Marlowe 30%, and by Fletcher 20%. And the word contumely appears, on the average, five times per 100,000 long words in the works of Shakespeare, three times per 100,000 in Marlowe, and four times per 100,000 in Fletcher. Estimate which man wrote the fragment, and for each man give the probability that he authored the fragment, based upon all relevant information.

13. In throwing the newspaper at a subscriber's porch, the paperboy misses in some proportion p of his throws. The subscriber hypothesizes that $p = .9$, whereas the route manager hypothesizes that $p = .2$, and one of them is correct. To settle the dispute, the subscriber keeps a record of hits and misses; in a total of nine throws the paper has missed the porch on three occasions.

   (a) What is the maximum likelihood estimate of p?
   (b) What is the maximum likelihood choice of hypothesis, and what is the probability of the obtained data under each hypothesis?
   (c) Suppose that they had agreed to accept the subscriber's hypothesis if the likelihood ratio P(data|subscriber correct)/P(data|route manager correct) exceeds .1 and to accept the route manager's hypothesis otherwise. Which hypothesis would be accepted, and what is the value of the likelihood ratio, L?

14. In the above problem, before collecting data the subscriber had subjective probability P that his hypothesis was correct and $1 - P$ that the route manager's was correct. After collecting data he uses Bayes' formula to revise his initial subjective probabilities, obtaining revised subjective probabilities P' and $(1 - P')$. What is the value of the revised probability P' if the initial probability P was (a) 0, (b) .01, (c) .2, (d) .5, (e) .8, (f) .99, (g) 1.00?

15. The new king entertains two hypotheses: that he is popular, and that he is unpopular. Historically, $33\frac{1}{3}\%$ of the meals served to unpopular kings have been poisoned, whereas only 1% of those served to popular kings have been. To protect himself, the king employs the services of clandestine food-tasters, one of whom perishes during the course of the first ten meals. (a) What is the probability of this result under each hypothesis? (b) Which hypothesis would be chosen if a maximum likelihood criterion were used? (c) Which hypothesis would be chosen if a likelihood ratio criterion with $K = .7$ were used? (d) What is the likelihood ratio?

16. A congress of medieval theologians is debating the virtues of three different hypotheses about the number N of teeth in a fully toothed horse's mouth. The hypotheses are contradictory statements by Aristotle, Plotinus, and Augustine, and one of them is correct. Each says that N

is a constant, but Aristotle says that $N = x$, Plotinus that $N = y$, and Augustine that $N = z$. A normal, fully toothed horse is found, its teeth are carefully counted, and the number of teeth is z. What is the probability of obtaining such a result under the Aristotelian, Plotinian, and Augustinian hypotheses, respectively? What is the maximum likelihood estimate of the correct hypothesis?

17. In the above problem, suppose that before counting teeth the theologians agreed upon subjective probabilities of .99, .009, and .001 for the Aristotelian, Plotinian, and Augustinian hypotheses, respectively, being correct. If they revise these subjective probabilities in the light of the obtained data, using Bayes' formula, what do the subjective probabilities become after counting teeth?

18. What would the answer to the above problem be if the initial subjective probabilities were 1.00, 0, and 0?

19. An anthropologist has just arrived at the village of a primitive tribe. It is a known characteristic of this tribe that orders are given only by three people, the chief who gives 60% of the orders, the witch doctor who gives 30%, and the grand vizier who gives 10%. It is considered a terrible diplomatic faux pas for a visitor not to bow first to the chief, then to the witch doctor, and last to the grand vizier, which, of course, requires that he correctly identify them. As he approached the village, the anthropologist noticed that six orders were given: three by native A, two by native B, and one by native C. The anthropologist therefore entertains six hypotheses corresponding to the six different possible assignments of the labels "chief," "doctor," and "vizier" to natives A, B, and C. For each hypothesis, calculate the probability of obtaining the above data in a random sample of six orders if the hypothesis is true. If he uses a maximum likelihood estimate, to whom should the anthropologist bow first, second, and third?

20. During an air raid, the enemy has dropped ten bombs in an area containing two possible targets, one of which is an obvious target, the other of which is a very important concealed target of whose existence the enemy was thought to be unaware. Allied intelligence has divided the bombed area into three parts A, B, and C, and they consider two hypotheses: H_1, the enemy was aiming at the concealed target; H_2, the enemy was aiming at the obvious target. If H_1 is true, then in general, 60% of the bombs should land in area A, 10% in area B, and 30% in area C. If H_2 is true, 10% should land in area A, 70% in area B, and 20% in area C. Actually, of the ten bombs dropped three landed in area A, three in area B, and four in area C.

(a) What is the probability of such a result if H_1 is true? If H_2 is true?

(b) Which hypothesis would be chosen by the maximum likelihood criterion?

(c) What is the likelihood ratio?

(d) Which hypothesis would be chosen if one used the likelihood ratio test with $K = .3$?

11

hypothesis testing—
the neyman-pearson procedure

As we pointed out in the preceding chapter, a common type of decision problem requires that a choice be made from a set of conflicting hypotheses, H_1 (implying that $\theta = \theta_1$), H_2 (implying θ_2), ..., H_j (implying θ_j), ..., H_s (implying θ_s), an unknown one of which is presumed to be correct. The maximum likelihood estimate was shown to be an appropriate method for choosing "truthfully," while ignoring utilities. However, we said that the various wrong choices may have unequal consequences. It may be more hazardous to believe some wrong estimates than others; the choices may logically call for subsequent actions having unequal utilities. Or, in order to give them greater credence, it may be desirable to make certain choices extremely unlikely unless they are correct. The likelihood ratio test was shown to be appropriate to this situation in certain cases, but not when the implied values of θ form a continuum or are so close together that the probabilities of the obtained data outcome under two adjacent θ-values are negligibly different.

Half a century ago Jerzy Neyman and E. S. Pearson devised a procedure that is appropriate to the above situation in the general case where the number of possible θ-values may be anything from 2 to ∞. Instead of formulating s different hypotheses $H_1, H_2, \ldots, H_j, \ldots, H_s$ implying s different states of nature $\theta_1, \theta_2, \ldots, \theta_j, \ldots, \theta_s$, the decision maker formulates just two hypotheses, one implying that the true value of θ lies in a certain subset of the above θ-values, the other implying that it lies in the complementary subset. Sample information is then used to determine which of the two subsets to accept as containing the true value of θ and, therefore, which of the two

240

hypotheses to accept as true. The unequal consequences or desirabilities of mistaken choices can be taken account of by making sure that the hypothesis requiring the greater degree of credence when chosen, or having the direr consequences when wrongly chosen, is highly unlikely to be chosen if not correct. In the case where there are just two possible values of θ, the Neyman-Pearson method is equivalent to use of the likelihood ratio test, although the actual procedure is slightly different. There are four important cases which we shall now illustrate.

11-1   Case in Which There Are Only Two Possible Values of θ

11-1-1   Illustration

A convicted confidence man in a newly constructed prison overhears the warden tell a guard that for prisons of that type the probability that an attempted escape will succeed is .9. Or does he? A door slammed during the statement and the convict is not sure whether the warden said "will succeed" or "will not succeed." He therefore has two contradictory hypotheses, one stating that the probability p of success in an escape attempt is .9, the other stating that $p = .1$. He decides to test which hypothesis is correct by conning four randomly selected prisoners into making separate and independent attempts at escape and observing the result. Under these conditions, the number of successful escapes r in $n = 4$ future attempts should be a random variable following the binomial probability law

$$P(r) = \binom{4}{r} p^r (1 - p)^{4-r}$$

Using this formula, he calculates the probability distribution of r for both $p = .1$ and $p = .9$ and plots them as shown in Figure 11-1. The graph shows that small values of r are much more probable when $p = .1$ than when $p = .9$ and that large values are much more likely if $p = .9$ than if $p = .1$. Therefore, the statistic r is sensitive to which hypothesis is true in somewhat the same way that litmus paper is sensitive to whether a liquid is acid or alkaline. Since escape attempts are dangerous, the convict resolves to accept the hypothesis that $p = .1$ and forego the escape attempt unless the evidence in favor of the alternative hypothesis is overwhelming—that is, unless, when the sample is actually taken, the obtained value of r has a value that was *both* highly *un*likely if $p = .1$ and *far more* likely if $p = .9$. Therefore, he decides to reject the hypothesis that $p = .1$ only if the obtained value of r falls at the upper end of the sequence of possible r-values in a region whose probability, if $p = .1$, of containing the sample value of r is $\leq .01$. The set of r-values that meet these criteria are 4, and 3, and the probability that r

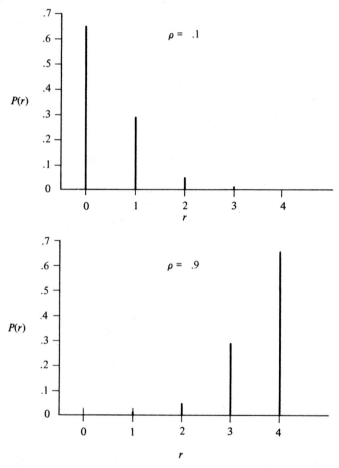

FIG. 11-1. Probability distribution of the number r of successful attempts to escape in 4 independent trials when the probability p of success is .1 and also when it is .9.

will have one of the values in this rejection region if $p = .1$ is

$$P(r \geq 3 \mid p = .1, n = 4) = \sum_{r=3}^{4} \binom{4}{r} .1^r .9^{4-r} = .00370$$

On the other hand, the probability that r will fall in this region when actually $p = .9$ is

$$P(r \geq 3 \mid p = .9, n = 4) = \sum_{r=3}^{4} \binom{4}{r} .9^r .1^{4-r} = .94770$$

Table 11-1 summarizes the situation. Accepting the hypothesis that $p = .1$

Table 11-1

Confidence Man's Decision Situation

Action	Criterion for action	Probability of taking action [and, in Brackets, appropriateness of action to ρ-value]	
		if $\rho = .1$	if $\rho = .9$
Accept hypothesis that $\rho = .1$ and Forego escape	Sample r-value falls in acceptance region: $r \leqslant 2$	$P(r \leqslant 2 \mid n = 4, \rho = .1) = .9963$ [Appropriate]	$P(r \leqslant 2 \mid n = 4, \rho = .9) = .0523$ [Inappropriate]
Reject hypothesis that $\rho = .1$ Accept hypothesis that $\rho = .9$ and Attempt escape	Sample r-value falls in rejection region: $r \geqslant 3$	$P(r \geqslant 3 \mid n = 4, \rho = .1) = .0037$ [Highly inappropriate]	$P(r \geqslant 3 \mid n = 4, \rho = .9) = .9477$ [Highly approriate]

243

means foregoing the escape attempt, and this has milder rewards if the hypothesis is correct and milder penalties if it is wrong than does accepting the hypothesis that $p = .9$ and trying to escape to freedom at the risk of being shot. Clearly, the probabilities of taking the appropriate actions are both high, those of taking the inappropriate actions are both fairly low, and that of committing the folly of attempting to escape when the probability of success is only .1 is at the tolerable level of .0037.

The convict puts his plan into action, conning four inmates into making independent attempts at escape, of which three succeed. So, since the obtained value of r has fallen into the rejection region, he rejects the hypothesis that $p = .1$, accepts the hypothesis that $p = .9$, and takes the appropriate action of trying to escape. (Note that the criterion of accepting the hypothesis that $p = .1$ unless $r \geq 3$ is equivalent to using a likelihood ratio criterion of accepting that $p = .1$ if

$$L = \frac{P(\text{obtained } r \mid n = 4, \, p = .1)}{P(\text{obtained } r \mid n = 4, \, p = .9)} > .0124)$$

Consider a second example (one that we have encountered before). D'Alembert was presented with the following problem: A fair coin is to be tossed twice and if a head appears on either toss Player A wins; what is the probability that A will win? D'Alembert believed that there were three possible outcomes: (a) head on first toss, so Player A wins; (b) tail on first toss, head on second, so Player A wins; (c) tails on both tosses, so Player B wins. He therefore concluded that the probability that A wins is 2/3 whereas modern probability theory says that it is 3/4. A student of probability agrees with D'Alembert's reasoning and decides to test which hypothesis is correct by twice-tossing a fair coin 1,000 times and being guided by the result. Since the mathematical world accepts the hypothesis that $p = 3/4$, he knows that contrary evidence will not be taken seriously unless it is extremely unlikely if $p = 3/4$ and far more likely if $p = 2/3$. Therefore, he decides to "reject" (actually to challenge) the popular hypothesis that $p = 3/4$ only if r, the number of times that "A wins" in 1,000 games, falls in a region whose probability of containing it was $\leq .0001$ if $p = 3/4$ but much greater if $p = 2/3$. From binomial tables he finds that a rejection region consisting of all r-values ≤ 697 meets these criteria, since

$$P(r \leq 697 \mid n = 1,000, \, p = 3/4) = .00009$$

and

$$P(r \leq 697 \mid n = 1,000, \, p = 2/3) = .98130$$

Table 11-2 summarizes his situation before the sample is actually drawn. The student now plays 1,000 times the game of tossing the fair coin twice and finds that in $r = 722$ of them heads comes up at least once. Therefore, since this falls in the acceptance region, he accepts the hypothesis that $p = 3/4$. He is curious, however, about the probability of obtaining an r-value as far

Table 11-2

Student's Decision Situation

Action	Criterion for action	Probability of taking action [and, in Brackets, appropriateness of action to ρ-value]	
		if $\rho = 3/4$	if $\rho = 2/3$
Accept hypothesis that $\rho = 3/4$ and Don't dispute present theory	Sample r-value falls in acceptance region: $r > 697$	$P(r > 697 \mid n = 1,000, \rho = 3/4) = .99991$ [Appropriate]	$P(r > 697 \mid n = 1,000, \rho = 2/3 =$ [Inappropriate]
Reject hypothesis that $\rho = 3/4$ and Accept hypothesis that $\rho = 2/3$ and Challenge present theory	Sample r-value falls in rejection region: $r \leqslant 697$	$P(r \leqslant 697 \mid n = 1,000, \rho = 3/4) = .00009$ [Highly inappropriate]	$P(r \leqslant 697 \mid n = 1,000, \rho = 2/3) = .98130$ [Highly approriate]

as that in the direction of the rejection region if actually $p = 3/4$ and calculates

$$P(r \leq 722 \,|\, n = 1,000, \, p = 3/4) = .02324$$

This would have been the probability of rejecting the hypothesis that $p = 3/4$ had he originally taken values of $r \leq 722$ as his rejection region.

11-1-2 Generalization

We shall now present the generalized Neyman-Pearson procedure in the case in which there are just two possible θ-values and shall provide some standard terminology. There are two conflicting hypotheses, one of which is true. One called the *null hypothesis* and denoted H_0 states or implies that $\theta = \theta_0$. The other called the *alternative hypothesis* and denoted H_a states or implies that $\theta = \theta_a$. The null hypothesis is given certain advantages and the test consists of determining whether or not sample evidence will nullify it (hence the name) by calling for its rejection. Precisely because the test is loaded in its favor, accepting the null hypothesis on the basis of the sample data may only weakly suggest that H_0 is true. However, for the same reason, rejecting it on the basis of the sample data and therefore accepting the alternative hypothesis tends much more strongly to suggest that H_a is true.

It is known that a certain statistic S based on a sample of n observations has a different probability distribution if H_0 is true than it does if H_a is true, and can therefore serve as a *test statistic*. Its range of possible values is divided by the decision maker into two regions: a *rejection region* into which S has only a fixed small probability α of falling if H_0 is true, and is more likely to fall if H_a is true, and a complementary *acceptance region* into which S has a large fixed probability $1 - \alpha$ of falling if H_0 is true and is less likely to fall if H_a is true. The probabilities that S will fall in the rejection or acceptance regions if the alternative hypothesis H_a is true are designated $1 - \beta$ and β, respectively. If when the sample is actually taken the *obtained value S'* of S falls in the acceptance region, H_0 is accepted as true; if it falls in the rejection region, H_0 is rejected as false, so H_a is accepted as true. Rejecting H_0 when actually it is true (and therefore wrongly accepting H_a) is an error called a *Type I Error* and having probability α. Similarly, accepting H_0 when actually it is false (and therefore H_a is true) is also an error, called a *Type II Error* and having probability β. Since the test is loaded in favor of accepting H_0, doing so only weakly implies that H_0 is true, but accepting H_a on the basis of the obtained sample has much stronger implications. The probability of accepting H_a if actually it is true is $1 - \beta$ and is called the *power* of the test. The power of the test and the probability of a Type II Error are complementary, and both vary with (a) the size n of the sample (power increases as n increases), (b) the size of the rejection region, and therefore of α (power decreases as α decreases), and (c) the difference between θ_0 and θ_a (power increases as the difference increases). Table 11-3 summarizes much of the

Table 11-3

Generalized Decision Situation

Action	Criterion for action	Probability of taking action [and, in Brackets, appropriateness of action to actual state of nature]	
		if H_o is true (so $\theta = \theta_0$)	if H_a is true (so $\theta = \theta_a$)
Accept H_o that $\theta = \theta_0$ and Act accordingly	Sample S-value falls in acceptance region	$P(S$ falls in acceptance region $\mid \theta = \theta_0)$ $= 1 - \alpha$ [Action called for is approriate to true H_o]	$P(S$ falls in acceptance region $\mid \theta = \theta_a)$ $= \beta$ $= P$(Type II error) [Action called for is inappropriate to true H_a]
Reject H_o Accept H_a that $\theta = \theta_a$ and Act accordingly	Sample S-value falls in rejection region	$P(S$ falls in rejection region $\mid \theta = \theta_0)$ $= \alpha$ $= P$(Type I error) [Action called for is highly inappropriate to true H_o]	$P(S$ falls in rejection region $\mid \theta = \theta_a)$ $= 1 - \beta$ $= $ Power of test [Action called for is highly appropriate to true H_a]

above. The probability α of a Type I Error is sometimes called the *significance level* at which the test is conducted. When the sample is actually drawn, if the actually obtained value S' of the test statistic S falls in the rejection region, the result is said to be *statistically significant* or the test outcome is said to have attained the α *level of significance*. The *critical value* C.V. of the test statistic S is the S-value in the rejection region that is closest to (i.e., borders on) the acceptance region. The *probability level* of an actually conducted test is the value that α would have had if the rejection region had extended just far enough to include the obtained value S' of the test statistic as its critical value. It is a cumulative probability of S-values starting with the most extreme (tailmost) S-value in the rejection region and ending with S', wherever S' may be (whether inside or outside the rejection region).

11-2   Case in Which There May Be Any Number of Possible Values of θ—Left-Tail Test

A manufacturer of light bulbs claims that at least 70% of his light bulbs will survive 1,000 hours of continuous use. A consumers' organization suspects that the claim is exaggerated and decides to test it by subjecting a random sample of 12 bulbs to the 1,000 hours of continuous use, using the number of survivors r as the test statistic. The manufacturer's claim is that the population proportion p of light bulbs that would survive is $\geq .70$ and the consumer-organization's suspicion is that it is $< .70$. Because it would reduce their prestige and invite lawsuits, the consumers do not wish to contradict the manufacturer if his claim is true. So to insure that there will be small likelihood of such embarrassment, they let the null hypothesis be that $p \geq .70$ which leaves $p < .70$ as the alternative hypothesis.

The test statistic r is binomially distributed with probability

$$P(r|n = 12, p) = \binom{12}{r}p^r(1 - p)^{12-r}$$

and the probability distribution of r when $n = 12$ depends upon p. Its probability distribution for values of p from .9 to .1 in steps of .2 is shown in Figure 11-2. If p is in the region $p \geq .70$ as claimed by the manufacturer, r always has small probability, $\leq .00949$, of taking a value in the region from 0 to 4; and the maximum of these probabilities .00949 occurs when $p = .70$, the value on the "boundary between H_0 and H_a." Whereas, for any value of $p < .70$, the probability that r will fall in the region from 0 to 4 is greater than it is for any value of $p \geq .70$. Clearly, therefore, the r values 0, 1, 2, 3, 4 form an appropriate rejection region for testing $H_0: p \geq .70$ against $H_a: p < .70$, the maximum probability α of a Type I Error being .00949.

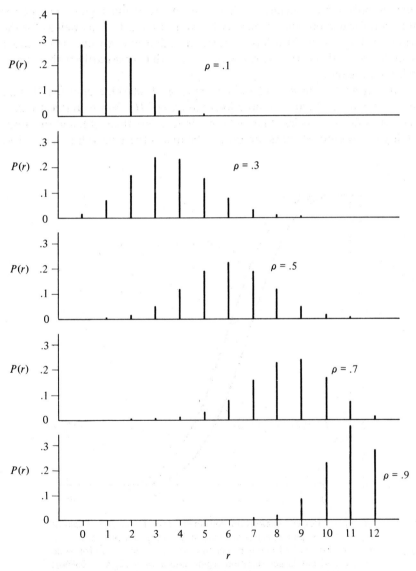

FIG. 11-2. Probability distribution of the binomially distributed test statistic *r* when $n = 12$ and p takes various values.

However, while we can keep this maximum probability α at an acceptably low level, we cannot do so for β the probability of a Type II error, as can be seen in Table 11-4.

Since power varies with the true value of p, what is needed is a graph that plots power on the vertical axis against true value of p on the horizontal axis. Such a graph is called a *power function* and, of course, the power function applies only for a specified sample size and rejection region. The power function for our present problem is shown in Figure 11-3, as well as for the same problem with (a) a larger rejection region and α-value, (b) a larger sample size with nearly the same value of α (but necessarily with a larger rejection region).

It is clear that for $n = 12$ and maximum α of .00949, even if the H_0 that $p \geq .70$ is false, the test is more likely to accept H_0 than to reject it unless the true value of p is less than .40 (and, in fact, less than .38), and the probability of rejecting H_0 does not exceed .90 unless the true value of p is less

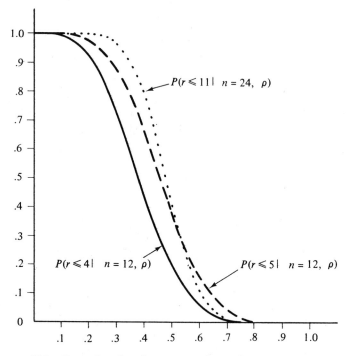

FIG. 11-3. Power functions for consumers' test of $H_0: p \geq .7$ against $H_a: p < .7$. Solid line is power function for actual test (where $n = 12$, $\alpha = .00949$); dashed line is power function test would have for same sample size but larger rejection region and α ($n = 12$, $\alpha = .03860$); dotted line is power function test would have if sample size were doubled while keeping α about the same ($n = 24$, $\alpha = .01150$).

Table 11-4

Action Probabilities Under Selected ρ-Values in Consumer's Decision Situation

	ρ-values within region specified by $H_a: \rho < .7$						ρ-values within region specified by $H_0: \rho \geq .7$		
	$\rho = .1$	$\rho = .2$	$\rho = .3$	$\rho = .4$	$\rho = .5$	$\rho = .6$	$\rho = .7$	$\rho = .8$	$\rho = .9$
$P(\text{accepting } H_0)$ $= P(r > 4 \mid n = 12, \rho)$	.00433 $P(\text{Type II error})$ if $\rho = .1$	.07256 $P(\text{Type II error})$ if $\rho = .2$	.27634 $P(\text{Type II error})$ if $\rho = .3$	.56182 $P(\text{Type II error})$ if $\rho = .4$	.80615 $P(\text{Type II error})$ if $\rho = .5$	.94269 $P(\text{Type II error})$ if $\rho = .6$	.99051	.99942	.99999 +
$P(\text{rejecting } H_0)$ $= P(r \leqslant 4 \mid n = 12, \rho)$	.99567 Power if $\rho = .1$	.92744 Power if $\rho = .2$	.72366 Power if $\rho = .3$	.43818 Power if $\rho = .4$	.19385 Power if $\rho = .5$	.05731 Power if $\rho = .6$	.00949 $P(\text{Type I error})$ if $\rho = .7$	.00058 $P(\text{Type I error})$ if $\rho = .8$	.00001 − $P(\text{Type I error})$ if $\rho = .9$

than .22. Therefore, the test is not impressively powerful unless the true value of p is far smaller than the minimum value claimed by the manufacturer. The consumers' organization ought to consider such facts in deciding upon the size of the sample, although the actual performance of the test does not require it.

The random sample is now drawn and of the 12 light bulbs, 5 survive the 1,000 hours of continuous use. Since $r = 5$ is not in the rejection region, $H_0: p \geq .70$ must be accepted. The probability level of the obtained value of the test statistic (actually the *maximum* probability level) is

$$P(r \leq 5 \mid n = 12, \ p = .7) = .0386$$

Since the rejection region lies at the lower or left end, or tail, of the sampling distribution of the test statistic r, the test is called a *lower-tail test* or *left-tail test*.

11-3   Case in Which There May Be Any Number of Possible Values of θ—Right-Tail Test

A psychologist believes that about 90% of adults can be hypnotized by him, but he will be satisfied if he can claim that over 60% can be. He knows that accepting a null hypothesis is a relatively weak confirmation, whereas accepting an alternative hypothesis tends to be much stronger evidence of its validity. Therefore, he lets H_0 be that $p \leq .60$ and lets H_a be that $p > .60$. Consulting binomial tables he observes that

$$P(r \geq 14 \mid n = 17, \ p = .9) = .91736$$

and that

$$P(r \geq 14 \mid n = 17, \ p = .6) = .04642$$

so if he uses a sample of size $n = 17$ and a rejection region consisting of r values ≥ 14, his maximum probability α of rejecting H_0 if true will be .04642 whereas his probability $1 - \beta$ of rejecting it if actually $p = .9$ is .91736. And his probability β of accepting $H_0: p \leq .60$ if actually $p = .9$ will be .08264. These probabilities are acceptable to him, so he randomly selects 17 adults and attempts to hypnotize them. Of these $n = 17$ attempts, $r = 15$ succeed. Since the obtained value 15 of the test statistic r lies in the rejection region, $H_0: p \leq .60$ is rejected in favor of $H_a: p > .60$. The (maximum) probability level of the test is $P(r \geq 15 \mid n = 17, \ p = .60) = .01232$. Figure 11-4 shows the power function for this test, i.e., plots $P(r \geq 14 \mid n = 17, \ p)$ on the vertical axis against p on the horizontal axis, as well as for (a) a test using the same H_0, H_a, and n but a smaller rejection region and therefore a smaller maximum α, (b) a test using the same H_0 and H_a and approximately the same α but a smaller n. Table 11-5 shows how the probabilities of accepting or rejecting

Table 11-5

Action Probabilities Under Selected ρ-Values in Hypnotist's Decision Situation

	ρ-values within region specified by $H_0: \rho \leq .6$						ρ-values within region specified by $H_a: \rho > .6$		
	$\rho = .1$	$\rho = .2$	$\rho = .3$	$\rho = .4$	$\rho = .5$	$\rho = .6$	$\rho = .7$	$\rho = .8$	$\rho = .9$
P(accepting H_0) $= P(r < 14 \mid n = 17, \rho)$	$1 - .1^{11}$	$1 - .1^7$	.99999	.99955	.99364	.95358	.79809 P(Type II error) if $\rho = .7$	.45112 P(Type II error) if $\rho = .8$	.08264 P(Type II error) if $\rho = .9$
P(rejecting H_0) $= P(r \geq 14 \mid n = 17, \rho)$	$.1^{11}$ P(Type I error) if $\rho = .1$	$.1^7$ P(Type I error) if $\rho = .2$	.00001 P(Type I error) if $\rho = .3$	.00045 P(Type I error) if $\rho = .4$	.00636 P(Type I error) if $\rho = .5$	.04642 P(Type I error) if $\rho = .6$	.20191 Power if $\rho = .7$	.54888 Power if $\rho = .8$	.91736 Power if $\rho = .9$

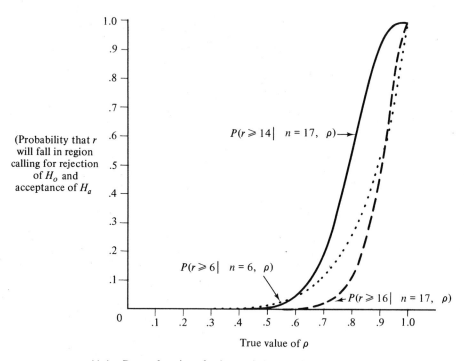

FIG. 11-4. Power functions for hypnotist's test of H_0: $p \leq .6$ against H_a: $p > .6$. Solid line is power function for actual test (where $n = 17$, $\alpha = .04642$); dashed line is power function test would have had for same sample size but smaller rejection region and α-value ($n = 17$, $\alpha = .00209$); dotted line is power function test would have had for smaller sample size but about the same α-value ($n = 6$, $\alpha = .04666$).

H_0 vary with the actual value of p. Since the rejection region lies at the upper or right tail of the probability distribution of the test statistic r, the test is called an *upper-tail test* or a *right-tail test*.

11-4 Case in Which There May Be Any Number of Possible Values of θ—The Conventional Two-Tail Test

11-4-1 Illustration

A doctor wishes to test whether or not the probability p of a boy-baby is exactly .5. He suspects that it is not, but he does not wish to hypothesize on which side of .5 the true value of p lies. Therefore, he lets the null hypothesis H_0 be that $p = .5$ and the alternative hypothesis H_a be that $p \neq .5$. Actually, the alternative hypothesis could be expressed as "p is either less than .5 or else it is greater than .5." This suggests that *two* rejection regions would be appropriate: one consisting of very small values of r, into which r will be more likely to fall if $p < .5$ than if $p = .5$; the other consisting of very large values of r, into which r will be more likely to fall if $p > .5$ than if $p = .5$. The doctor decides to take a random sample of 1,000 births and reject H_0 if the number r of boy-babies is either ≤ 468 or ≥ 532. His selection of rejection regions was determined by the facts that he didn't want his total α to exceed .05 and that $P(r \leq 468) = .02315$ and $P(r \geq 532) = .02315$ so that $P(r$ is either ≤ 468 or $\geq 532) = .04630$. Larger, symmetric, rejection regions would have caused α to exceed .05. The situation is summarized in Table 11-6. In the obtained sample, $r = 519$ which lies in the acceptance region, so $H_0: p = .5$ is accepted. The probability level of the test is

$$P(r \geq 519 | n = 1,000, p = .5) + P(r \leq 1,000 - 519 | n = 1,000, p = .5)$$

$$= .12098 + .12098 = .24196$$

Since both ends of the probability distribution of r are used as rejection region, the test is called a *two-tail test*.

Figure 11-5 gives the power function for the test, plotting

$$P(r \leq 468 | n = 1,000, p) + P(r \geq 532 | n = 1,000, p),$$

the probability that r will fall into one of the rejection regions, on the vertical axis against p on the horizontal axis. The figure also gives the power function for the test applied at a smaller α (corresponding to rejection regions of $r \leq 438$ and $r \geq 562$), and for the test applied at about the same α but at $n = 25$.

11-4-2 Critique

Despite the facts that a very large sample was taken and H_0 was accepted, the null hypothesis is wrong. It is a fact that the probability of a boy-baby is slightly greater than that of a girl-baby, p being perhaps somewhere between .51 and .52. However, even if $p = .52$, the power of the test to reject $H_0: p = .5$ is only

$$P(r \leq 468 | n = 1,000, p = .52) + P(r \geq 532 | n = 1,000, p = .52)$$

$$= .00056 + .23340 = .23396$$

We have already said that accepting a null hypothesis, on the basis of test results, is not very convincing evidence that it is true. However, it is far *less* convincing when the test is two-tailed than when it is one-tailed, if (a) the possible values of p form a continuum and (b) the *conventional* two-tail test is used in which H_0 implies only a single possible value for p. In the remarks that follow we shall be speaking of two-tail tests in the case where (a) and (b) are both true.

Table 11-6

Action Probabilities Under Selected ρ-Values in Doctor's Decision Situation

	ρ-values within lower portion of region specified by $H_a: \rho \neq .5$				Single ρ-value specified by $H_0: \rho = .5$	ρ-values within upper portion of region specified by $H_a: \rho \neq .5$			
	$\rho = .40$	$\rho = .45$	$\rho = .48$	$\rho = .49$	$\rho = .50$	$\rho = .51$	$\rho = .52$	$\rho = .55$	$\rho = .60$
$P(\text{accepting } H_0)$ $= P(468 < r < 532 \mid n = 1{,}000, \rho)$	.00001	.11988	.76604	.90881	.95370	.90881	.76604	.11988	.00001
	$P(\text{Type II error})$ if $\rho = .40$	$P(\text{Type II error})$ if $\rho = .45$	$P(\text{Type II error})$ if $\rho = .48$	$P(\text{Type II error})$ if $\rho = .49$		$P(\text{Type II error})$ if $\rho = .51$	$P(\text{Type II error})$ if $\rho = .52$	$P(\text{Type II error})$ if $\rho = .55$	$P(\text{Type II error})$ if $\rho = .60$
$P(\text{rejecting } H_0)$ $= P(r \leqslant 468 \mid n = 1{,}000, \rho)$ $+ P(r \geqslant 532 \mid n = 1{,}000, \rho)$	.99999	.88012	.23396	.09119	.04630	.09119	.23396	.88012	.99999
	Power if $\rho = .40$	Power if $\rho = .45$	Power if $\rho = .48$	Power if $\rho = .49$	$P(\text{Type I error})$	Power if $\rho = .51$	Power if $\rho = .52$	Power if $\rho = .55$	Power if $\rho = .60$

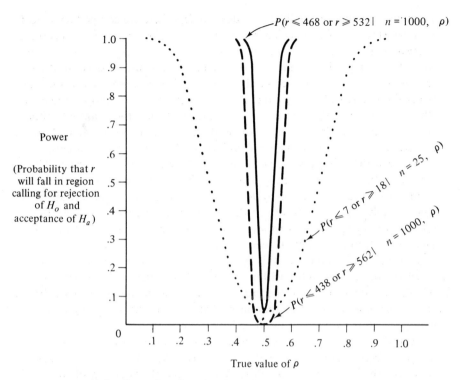

FIG. 11-5. Power functions for doctor's test of H_0: $p = .5$ against H_a: $p \neq .5$. Solid line is power function for actual test (where $n = 1000$, $\alpha = .04630$); dashed line is power function test would have had for same sample size but smaller rejection region and α-value ($n = 1000$, $\alpha = .00010$); dotted line is power function test would have had for much smaller sample size but about the same α-value ($n = 25$, $\alpha = .04328$).

Both one-tail and two-tail tests concentrate upon only a single value of p. But in two-tail tests H_0 says that p has this single point-value and no other. Whereas in one-tail tests H_0 says that p lies *somewhere* in an *entire region* of p-values and only concentrates upon the one that maximizes the probability of a Type I Error, i.e., the p-value bordering on the region of p-values specified by H_a. In the latter case, the *true* p-value may lie *far inside* the region specified by H_0 so that there is a great gap between the true p-value and the nearest p-value in the region specified by H_a. If this is the case and sample size is reasonably large, the test will be nearly certain to call for the correct action of accepting H_0. With a two-tail test the probability of accepting H_0 if true is always exactly $1 - \alpha$, never more. Furthermore, if the set of possible p-values form a continuum, then there are an infinite number of them and it strains credulity that H_0 has correctly identified the one true value among an infinite number of candidates. Indeed, unless there are compelling logical or a priori reasons for believing that H_0 is exactly correct, it is almost certain to be wrong—at least to a small extent—so that if a large enough sample were drawn it would be almost certainly rejected. Thus, most two-tail tests test an H_0 that is almost certainly wrong and would almost certainly be rejected if n were sufficiently large. But this means that one "knows" the truth (that H_0 is false) without testing and that the test, therefore, can only either confirm what one already knows or produce misleading results that contradict what one already knows, or would know if he analyzed the situation.

Despite these objections, such two-tail tests are quite common. Their defense is that what they really test is whether or not p differs *appreciably* from the one value specified by H_0. This is a bit vague, especially since, if "appreciably" is not quantitatively defined, there is a very real danger of taking a sample large enough to reject H_0 when p has a value that does *not* differ to a degree that anyone cares about.

The obvious solution to this dilemma is to formulate an H_0 implying that the true value of p lies within an *interval* of values in the continuum of possible p-values. Thus, in the example given, the doctor might have been more realistic and "doubted" that the probability p of a boy-baby was *approximately* .5. More specifically, he could have doubted the H_0 that $.48 \leq p \leq .52$ or even that $.45 \leq p \leq .55$, depending upon what he means by "approximately." Suppose that he had decided to test H_0: $.48 \leq p \leq .52$ against H_a: p is either $< .48$ or $> .52$, in such a way that the largest probability of rejecting a true H_0 is $\leq .05$, while using a sample size of $n = 1{,}000$, as before. He selects the r-values 0 to 453 and 547 to 1,000 as rejection region because the maximum probability that r will fall in this rejection region if H_0 is true (which occurs for either of the p-values .48 or .52) is

$$P(r \leq 453 \,|\, n = 1{,}000,\, p = .48) + P(r \geq 547 \,|\, n = 1{,}000,\, p = .48)$$
$$= .04664 + .00001 = .04665$$

whereas the probability that r will fall in a larger symmetric rejection region exceeds .05. His obtained value of r, 519, does not fall in the rejection region, so he accepts H_0 as before. This time H_0 is true. The important point, however, is that this time H_0 is *credible*. Accepting it does not commit us to the belief that of the infinite number of possible p-values between 0 and 1, p has exactly the value .5 followed by an infinite number of zeros—a value that we could never verify by actual measurement.

It should be pointed out, however, that some "point H_0's" are more credible than others. The H_0 that states that there is *no difference* in the birth rates of boy- and girl-babies, thereby implying that $p = .5$, seems more credible than an H_0 implying that there is a difference causing p to be exactly .517489251. The H_0 that states that the color of an experimenter's necktie has *no influence* upon his subjects' reaction times seems more credible than if it had stated that changing from a blue to a red necktie increased subjects' average reaction time by exactly .00729413 seconds. And, in general, "passive" H_0's implying "no influence," "no difference," or the negation of some positive effect, seem more credible than those claiming to have identified the exact value of a quantity associated with some active influence, real difference, or genuine effect. The reason is obvious. If, indeed, some variables exert no influence upon others, the increment caused by the former upon the values of the latter is specifiable as exactly zero. Likewise, if there is no difference between p and $1 - p$, it follows that p is exactly .5., etc. It is the experimentally-based opinion of the writer, however, that in those situations in which experiments or statistical tests are performed (i.e., when there is enough suspicion of a real influence to motivate a serious research worker to test for it), there is practically always *some* influence.

11-5 Possible Varieties of Cases

Types of test differ both as to the *number* of possible θ-values implied by each hypothesis and as to the relative *locations* of the sets of θ-values implied by each hypothesis. A *simple hypothesis* states or implies a single value that θ must have. A *composite hypothesis* states or implies that the value of θ is one of a set of several specified values. We have illustrated the four most common types of test in the case where θ is the proportion p of successes in an infinite dichotomous population and the sample statistic is r the number of successes in a random sample of n observations. Logically, at least seven different types of test could have been distinguished; they are listed below:

Test	Nature of H_0 and H_a	H_0	H_a	Location of Rejection Region
A	Simple-Simple	$p = p_0$	$p = p_a$	Left tail if $p_a < p_0$, Right tail if $p_a > p_0$
B	Composite-Composite	$p \geq p_0$	$p < p_0$	Left tail
C	Composite-Composite	$p \leq p_0$	$p > p_0$	Right tail
D	Simple-Composite	$p = p_0$	$p \neq p_0$	Both tails
E	Composite-Composite	$p' \leq p \leq p''$	$p < p'$ or $p > p''$	Both tails
F	Composite-Composite	$p \leq p'$ or $p \geq p''$	$p' < p < p''$	Interior
G	Composite-Simple	$p \neq p_a$	$p = p_a$	Interior

The following comments presuppose that all p-values in the continuum from 0 to 1 are possible. The difficulty with test G is that (a) a sample size of $n > 250$ is required in order to obtain an appropriate rejection region for which $\alpha \leq .05$ (if $p = .5$ and $n = 300$, this consists of the single r-value 150) and (b) the maximum power of the test to reject a false H_0 in favor of a true H_a is identical to the maximum probability of a Type I Error, α. Thus for small samples, α is much too large, and for large samples with reasonably small α's, power is much too small. These features make test G virtually worthless. Similar, but less drastic, objections hold for test F, especially when the interval between ρ' and ρ'' is a narrow one. This should be clear from the fact that when the width of the interval narrows to zero, i.e., when $\rho' = \rho''$, test F reduces to test G. For all tests involving at least one composite hypothesis, the minimum power if H_a is true is α, and for tests B, C, D, and E (where one of the p-values in the set implied by H_a is either 0 or 1) the maximum power if H_a is true is 1. However, as already mentioned, for test G it is α. And all we can say for test F is that the maximum power is between α and 1. If n is not very large and the interval from ρ' to ρ'' is not very wide, it tends to be closer to α than to 1, although for very large n-values it may be practically 1.00. For all tests the minimum probability of accepting a true H_0 is $1 - \alpha$. For tests A and D this is also the maximum probability. For test E the maximum probability lies between $1 - \alpha$ and 1. And for tests B, C, F, and G the maximum probability is 1.

Thus test G is worthless and test F tends to be weak unless n is very large. However, test E has characteristics that make it superior to the conventional two-tail test D.

11-6   Summary of Neyman-Pearson Hypothesis Testing Procedures

The set of all possible values of θ is divided into two subsets, a null subset and an alternative subset. The decision maker wishes to test which subset contains the true value of θ. So he formulates two hypotheses: a *null hypothesis* H_0, which states that θ lies in the null subset, and an *alternative hypothesis* H_a, which states that θ lies in the complementary, alternative subset. But he wants the test to indicate that θ lies in the alternative subset only if the evidence for that conclusion is reasonably convincing; otherwise, he is willing to assume, perhaps on weak evidence, that θ lies in the null subset. His reason for this may be that he really believes that θ lies in the alternative subset and wants to be able to announce it with great conviction. Or the consequences of falsely claiming (or believing, and acting accordingly) that θ lies in the alternative subset may be far harsher than those of falsely claiming that θ lies in the null subset, and he may simply wish to play safe.

In order to perform the test, it is necessary that there be a *test statistic S* whose probability distribution at any n-value will be "sensitive" to the true value of θ, i.e., whose probability distribution varies according to which one of the possible θ values is the true one. The set of all possible values of S is divided into two categories, a contiguous subset of possible S-values called the *acceptance region*, into which S is highly likely to fall if H_0 is true and less likely to fall if H_a is true, and the complementary subset called the *rejection region* (or regions), into which S is highly unlikely to fall if H_0 is true and more likely to fall if H_a is true. If (at any given n-value) small values of S have larger probabilities under *all* θ-values covered by H_a (i.e., in the alternative subset) than they have under *any* covered by H_0, the rejection region should be the lower end of S's probability distribution and the test is *lower-tailed* or *left-tailed*. If large values of S do, the rejection region should be the upper end of S's probability distribution, and the test is *upper-tailed* or *right-tailed*. And if under some θ-values covered by H_a small values of S, and under other θ-values covered by H_a large values of S, have larger probabilities than they have under the θ-value covered by H_0, both ends of S's probability distribution should be used as rejection regions and the test is *two-tailed*. If the obtained sample's S-value falls in the acceptance region, H_0 is accepted as true; if it falls in the rejection region, H_0 is rejected as false. The S value(s) in the rejection region(s) that lies (lie) closest to the acceptance region is (are) called the critical value(s), abbreviated C.V.

Now let θ_0 be that possible θ-value in the null subset under which the test statistic S has the largest probability of falling in the rejection region, and let that probability under θ_0 be designated α. Then,

$$\text{Maximum } P(\text{rejecting } H_0 \,|\, H_0 \text{ is true}) = P(S \text{ in rejection region} \,|\, \theta = \theta_0)$$

$$= \alpha$$

$$\text{Minimum } P(\text{accepting } H_0 \,|\, H_0 \text{ is true}) = P(S \text{ in acceptance region} \,|\, \theta = \theta_0)$$

$$= 1 - \alpha$$

Now let θ_a be some particular one of the possible values of θ in the alternative subset, and let β stand for the probability that S will fall in the acceptance region when actually H_0 is false and $\theta = \theta_a$. Then,

$$P(\text{rejecting } H_0 \,|\, H_0 \text{ is false and } \theta = \theta_a) = P(S \text{ in rejection region} \,|\, \theta = \theta_a)$$

$$= 1 - \beta$$

$$P(\text{accepting } H_0 \,|\, H_0 \text{ is false and } \theta = \theta_a) = P(S \text{ in acceptance region} \,|\, \theta = \theta_a)$$

$$= \beta$$

Rejecting H_0 when it is true is an error, called a *Type I Error*, whose maximum probability (i.e., whose probability if $\theta = \theta_0$) is α. Accepting H_0 when it is false is also an error, called a *Type II Error*, and its probability when

$\theta = \theta_a$ is β. Rejecting H_0 in favor of H_a when the latter is true is highly desirable, and its probability $1 - \beta$ is called the **power** of the test, and its value depends upon which particular one of the θ-values in the alternative subset is the true value of θ. These probabilities and error types are summarized in Table 11-7.

The maximum probability α of rejecting H_0 if it is actually true is called the **significance level** at which the test is conducted. And if, when the sample is drawn, the obtained value of S falls in the rejection region, it is said that the test results attained the α level of significance. "Conventional" values of α are .05, .01, and .001 or, if it is impossible to obtain the desired one of these values exactly, the closest *smaller* value obtainable. The decision maker, of course, need not be a slave to convention, and he may select the α-value that affords the degree of protection against a Type I Error that he feels is appropriate to his needs. The significance level tells us how likely it is, if $\theta = \theta_0$, that the test statistic will fall in the rejection region, and therefore that H_0 will be rejected, before the sample is actually drawn. The **probability level** of the test tells us the probability, if $\theta = \theta_0$, of obtaining a value of the test statistic "as extreme" as the value actually obtained in the drawn sample. By "as extreme" we mean "as far toward or into the rejection region or regions." More precisely, the probability level is the value that α would have had if the rejection region or regions had been just large enough to include the actually obtained value of the test statistic, which would therefore be a critical value.

The test of hypotheses is performed as follows. First, the decision maker sets up his hypotheses, letting the one that he wishes to accept only on strong evidence (and therefore the one that actually motivated him to perform the test) be the *alternative* hypothesis. The alternative hypothesis will determine the location of the rejection region, i.e., will determine whether the test will be left-tailed, right-tailed, or two-tailed. Second, he decides upon an acceptable α level. Third, ideally, he decides upon the power or (if there are more than two possible values of θ) the power function that he requires and determines (by trial-and-error calculation or by consulting tables) the sample size n that will give it to him when α has the value decided upon. It may not be possible to make power large for all possible θ_a's, (and will not be if α is small and there are an infinite number of possible θ-values, i.e., if the possible θ-values form a continuum). And even if it is possible, too large an n may be called for; so some compromise may be required. The decision maker may have to accept small power for those θ_a's that differ negligibly from θ_0 and simply use a large enough sample to reduce β to a tolerable level for θ_a's that are appreciably different from θ_0. This third step, although desirable, is often omitted, the proper value of n being obtained by guesswork. The alternative hypothesis, α, and n, together determine the exact set of S-values that comprise the rejection region. Lastly, the sample of n observations is drawn, the

Table 11-7

Probabilities of Taking Appropriate and Inappropriate Actions if $\theta = \theta_0$, Conforming to H_0, and if $\theta = \theta_a$, Conforming to H_a

Action	Probability of taking action under	
	θ_0 ("Boundary" θ-value in the null subset of θ-values specified by H_0)	θ_a (A particular θ-value in the alternative subset of θ-values specified by H_a)
Accept H_0	$P(S$ in acceptance region $\mid \theta = \theta_0)$ $= 1 - \alpha$	$P(S$ in acceptance region $\mid \theta = \theta_a)$ $= \beta$ Probability of a Type II error if $\theta = \theta_a$
Reject H_0 (Accept H_a)	$P(S$ in rejection region $\mid \theta = \theta_0)$ $= \alpha$ Maximum probability of a Type I error	$P(S$ in rejection region $\mid \theta = \theta_a)$ $= 1 - \beta$ Power of test if $\theta = \theta_a$

obtained value of S is calculated, and H_0 is either accepted or rejected in favor of H_a depending upon whether the obtained value of S falls in the acceptance region or the rejection region.

11-7   Problems

1. Suppose that, instead of deciding to take a sample of size $n = 4$ and to use r-values of 3 and 4 as the rejection region, the confidence man (in Section 11-1-1) had decided to take a sample of size $n = 8$ and to use r-values of 5, 6, 7, and 8 as the rejection region (thus doubling both the size of the sample and the size of the rejection region). In that case, in testing $H_0: p = .1$ against $H_a: p = .9$, what are (a) the probability α of committing a Type I Error, (b) the probability β of making a Type II Error, (c) the power of the test, (d) the significance level at which the test is conducted, (e) the probability level of the test if the obtained value of r is 7?

2. Suppose that the student (in Section 11-1-1) had decided to test the $H_0: p = 3/4$ against the $H_a: p = 2/3$ by rejecting H_0 only if in a sample of size $n = 10$ the obtained value of r is ≤ 4. In that event, what are the numerical values of (a) α, (b) β, (c) the power of the test, (d) the probability level of the test if the value of r in the obtained sample is 3?

3. In the above problem suppose that, instead of choosing r-values ≤ 4 as the rejection region, the student had decided to use whatever lower-tail rejection region gave him about the same α-value as before, i.e., the α-value closest to .00009. In that case (a) what values of r constitute the rejection region, (b) what is the probability of a Type II Error, (c) what is the power of the test, (d) what is the probability level of the test if the value of r obtained in the sample is 3?

4. An experimenter decides to test the H_0 that the proportion p of successes in an infinite dichotomous population is .1 against the H_a that it is .7, by taking a random sample of $n = 3$ observations and rejecting H_0 in favor of H_a only if the number of successes in the sample is 2 or 3. What are (a) the significance level of the test, (b) the power of the test, (c) the probability level if when the sample is drawn it contains just two successes?

5. In a test of hypotheses about the proportion of successes in an infinite dichotomous population, H_0 states that $p = 2/3$, H_a states that $p = 1/4$, the size of the sample is $n = 4$, and α is to be the largest possible value $\leq .20$ for the appropriately located rejection region. What are (a) the

significance level at which the test is conducted, (b) the probability of a Type II Error, (c) the power of the test, (d) the probability level if the obtained value of r is 2?

6. In testing the H_0 that the proportion p of successes in an infinite dichotomous population is .9 against the H_a that it is .2, the experimenter decides to take a sample of size $n = 3$ and calculates that the power of the test is .992. What is the significance level of the test?

7. In a test of hypotheses about the proportion p of successes in an infinite dichotomous population, H_a is that $p = 1/3$, the size of the sample is $n = 5$, and the r-values comprising the rejection region are 0 and 1. What is the power of the test if (a) $\alpha = .05$, (b) $\alpha = .01$?

8. In a test of hypotheses about the proportion p of successes in an infinite dichotomous population, H_0 is that $p = 1/3$, H_a is that $p = 3/4$, $n = 4$, and α is the largest possible value $\leq .25$ for the appropriate rejection region. What is the value of β?

9. A victim has been killed by a single shot through the heart, known to have been one of three shots all fired from a distance d by an unknown one of two suspects A and B. Both suspects belong to a rifle club and past records show that the proportion of hits in firing at a heart-sized target from a distance d is .90 for A and .25 for B. Subsequent to the shooting, suspect B was killed in an automobile accident. Suspect A has been arrested but denies guilt and has a plausible alibi. The district attorney knows that one of two hypotheses is correct: either A was the killer or B was the killer. He therefore decides to test these hypotheses at the largest available α-value $\leq .05$, using the already obtained data, and to do so with a test that treats A as innocent unless the evidence against him is quite strong. In this test (a) which hypothesis is H_0 and which is H_a, (b) what values constitute the rejection region, (c) which hypothesis will the D. A. accept, (d) what is the probability level of the test, (e) what is the power of the test?

10. A father whose child is missing receives a note demanding ransom. Police detectives are certain that it was written by one of two people, a known kidnapper or a known crank who does no kidnapping himself but attempts to profit from those done by others. From past notes, the kidnapper and the crank are known to misspell an average of one word out of 2 and 5, respectively. The ransom note contains 20 words of which 2 are misspelled. The father decides to test which man wrote the note and to pay the ransom unless the evidence strongly points to the crank as the author. He decides to use the largest appropriate α-value $\leq .05$. What are (a) the test statistic, (b) the probability law giving its probability distribution, (c) H_0 and H_a, (d) the θ_0 and θ_a implied

by H_0 and H_a, (e) the rejection region, (f) α, (g) the probability level, (h) β, (i) the power of the test, (j) the accepted hypothesis?

11. Fishing is forbidden in Pond A but allowed in neighboring Pond B. Each pond has just been stocked with fish. After being stocked, Pond A contained 10 fish of which 3 were trout, and Pond B contained 8 fish of which 5 were trout. A game warden stops a fisherman who has just caught 3 fish 2 of which are trout. The game warden accuses him of having caught them in Pond A but the fisherman claims to have caught them in Pond B. Favoring the fisherman, and using the largest appropriate α-value $\leq .05$, test which of the two claims should be accepted. What are (a) the test statistic, (b) the probability law giving its probability distribution, (c) H_0 and H_a, (d) the θ_0 and θ_a implied by H_0 and H_a, (e) the rejection region, (f) α, (g) the probability level, (h) β, (i) the power of the test, (j) the hypothesis that should be accepted?

12. Suppose that the consumers' organization in Section 11-2 had decided to test the H_0: $p \geq .70$ against the H_a: $p < .70$ by taking a random sample of $n = 6$ bulbs and using r-values of 0 and 1 as the rejection region. What are (a) the value of α, (b) the probability of a Type I Error if actually $p = .9$, (c) the probability of a Type II Error if actually $p = .4$, (d) the power of the test if actually $p = .3$, (e) the probability level of the obtained result if, in the sample, $r = 2$?

13. Suppose that the hypnotist in Section 11-3 had decided to test H_0: $p \leq .60$ against H_a: $p > .60$ by using a sample of size $n = 8$ and an upper-tail rejection region corresponding to the largest possible α-value $\leq .05$. In that case, what would be (a) the value of α, (b) the probability of a Type I Error if actually $p = .40$, (c) the value of β if actually $p = .75$, (d) the power of the test if actually $p = .75$, (e) the probability level of the obtained result if, in the sample, $r = 1$?

14. Suppose that the doctor in Section 11-4-1 had decided to test H_0: $p = .5$ against H_a: $p \neq .5$ by taking a sample of size $n = 10$ and using the largest symmetric rejection region whose α-value is $\leq .05$. What would be (a) the value of α, (b) the value of β if actually $p = .6$, (c) the power of the test if actually $p = .4$, (d) the probability level of the obtained result if, in the sample, $r = 3$?

15. In a test of hypotheses about the proportion p of successes in an infinite dichotomous population, H_0 is that $p \leq .1$, H_a is that $p > .1$, the largest appropriate α-value $\leq .06$ is used, and sample size is $n = 9$. What are (a) the r-values that constitute the rejection region, (b) the exact value of α, (c) the probability of a Type I Error if actually $p = .01$, (d) the probability of a Type II Error if actually $p = .25$, (e) the power

of the test if actually $p = .25$, (f) the probability level of the obtained data if the obtained value of r is 8?

16. What would be the answers to Problem 15 if sample size were $n = 12$?

17. What would be the answers to Problem 15 if the largest appropriate α-value $\leq .01$ had been used?

18. On the same sheet of graph paper plot the power functions of the following three tests based on a sample of size $n = 6$ (where p is the proportion of successes in an infinite dichotomous population): (a) H_0: $p \geq .6$, H_a: $p < .6$, rejection region: $r \leq 1$, (b) H_0: $p \leq .6$, H_a: $p > .6$, rejection region: $r = 6$, (c) H_0: $p = .6$, H_a: $p \neq .6$, rejection region r-values: 0, 1, 6.

19. In a test of hypotheses about the proportion p of successes in an infinite dichotomous population, H_0 is that $p \geq .40$, H_a is that $p < .40$, the largest appropriate α-value $\leq .05$ is used and sample size is $n = 10$. What are (a) the r-values that comprise the rejection region, (b) the exact value of α, (c) the probability of a Type I Error if actually $p = .8$, (d) the probability of a Type II Error if actually $p = .1$, (e) the power of the test if actually $p = .40$, (f) the probability level if the obtained value of r is 2?

20. What would be the answers to Problem 19 if sample size were 8 instead of 10?

21. What would be the answers to Problem 19 if the largest possible α-value $\leq .01$ had been used?

22. In a test of hypotheses about the proportion p of successes in an infinite dichotomous population, H_0 is that $p = .40$, H_a is that $p \neq .40$, and sample size is $n = 10$. Also, the rejection region corresponding to the largest $\alpha \leq .05$ is constructed without consideration for symmetry. What are (a) the r-values comprising the rejection region, (b) the exact value of α, (c) the power of the test if actually $p = .5$?

23. What would be the answers to Problem 22 if sample size were 8 instead of 10?

24. What would be the answers to Problem 22 if we changed $\alpha \leq .05$ to $\alpha \leq .01$?

25. The inventor of a radical new manufacturing process claims that the proportion p of defective items turned out by the process is less than .02. This figure is considerably smaller than the proportion of defectives turned out by the conventional process, so if the claim is true it would be economical to adopt the new process. But it would be folly to adopt

it if the claim is false since doing so would involve expensive retraining of personnel and replacement of old machines with a new type of machine—expenses which must be recouped by savings resulting from the new technique. During a demonstration run at an industrialists' convention, the new process produced 250 items of which 3 were defective. Using these data, the management of a factory decides to test which of two hypotheses to accept, "claim is true" or "claim is false," adopting the costly new process if they accept the first hypothesis. If they perform the appropriate test, using the largest available α-value $\leq .05$, (a) what p-values are implied by H_0, (b) what p-values are implied by H_a, (c) what r-values constitute the rejection region, (d) what is the actual value of α, (e) what is the probability level of the obtained results, (f) which hypothesis should be accepted?

26. A psychologist believes that blood pressure increases in over 70% of adults while they are telling a lie. And he wishes to test this hypothesis in such a way that a favorable verdict by the test will be highly convincing. He takes a random sample of 12 subjects, measures their blood pressure while they are telling a lie, and uses the number r of subjects whose blood pressure increases as the test statistic. In his test, he uses the largest available α-value $\leq .05$. What are (a) the p values implied by H_0, (b) the p values implied by H_a, (c) the rejection region r-values, (d) the exact value of α, (e) the probability level and the accepted hypothesis if the obtained value of r is 10?

27. A recurrent disease has two forms. If a particular epidemic has been caused by the mild form of the disease, 1% of those contracting it will die of it. Whereas, if it has been caused by the virulent form, 30% of the infected will perish. And it is impossible to determine which form is involved until after lengthy tests which are not completed until after the epidemic is over. An epidemic has just started, and of 11 people infected long enough to have died from it, none have done so. Medical officials decide to test which disease is involved; and, in order to be safely conservative, they wish the test to favor the hypothesis that the disease is in its virulent form since if they prepare for the virulent form, they will automatically be prepared for the mild form. They also wish to use the largest available α-value $\leq .10$. What are (a) the rejection region r-values, (b) the exact value of α, (c) the probability level, (d) the accepted hypothesis, (e) the power of the test?

28. A certain disease is fatal if untreated. For many years the only known treatment was a drastic one that cured in 40% of cases and was fatal in the remainder. A biochemist has developed a new treatment that he claims to be better than the old one. To support his claim he administers

the new treatment to a randomly selected 10 patients suffering from the disease, of which 7 are cured and the rest die. Using this sample information he performs the appropriate test at the largest available α-value $\leq .05$. What are: (a) the p-values implied by H_0, (b) the p-values implied by H_a, (c) the rejection region r-values, (d) the exact value of α, (e) the probability level, (f) the accepted hypothesis?

29. As a buried cable deteriorates, it develops minute flaws or ruptures on its surface at randomly located places. When the average number m of flaws per yard (for the entire cable) exceeds 1, it becomes economical to dig up the old cable and replace it with a new one. However, the digging itself is expensive, so decisions must be based upon inspection of very small portions of the cable. The standard procedure is to dig up a yard of cable and count the number of flaws x. If x equals or exceeds a critical value C, the entire cable is dug up and replaced; otherwise, the old cable is left in the ground. What should C be in order to insure that the error of replacing a good cable has a probability no greater than .10? If C has this value, what is the maximum probability of not replacing a bad cable in which $m \geq 2$?

30. The bodies comprising the solar system either rotate about an internal axis or revolve about a larger body or both. In Laplace's time the direction (whether clockwise or counterclockwise) of 43 of these circular motions was known, and all 43 of them were in the same direction. Laplace considered two hypotheses concerning the formation of the solar system: H_0 that each circular motion was independently determined with P(circular motion in same direction as that of the sun's rotation) $= \rho = 1/2$ and H_a that the motions, including the sun's, were determined by similar or identical causes so that $\rho > 1/2$. If he used the sample information that 42 of 42 nonsolar motions were in the same direction as that of the sun to test H_0 against H_a, what would be the probability level of the test?

31. A manufacturer of rat poison believes that a certain dosage of a particular drug will prove fatal in more than 75% of cases. He decides to test the H_0: $\rho \leq .75$ against the H_a: $\rho > .75$, using the largest available $\alpha \leq .05$, by administering the drug to a random sample of 11 rats. He takes as his test statistic r the number of rats that *survive*. (a) What values of r constitute the rejection region? (b) What is the exact value of α? (c) If 2 rats survive, should he accept H_0 or reject it, and what is the probability level of the result? (d) What is the power of the test if actually $\rho = .90$?

32. A scientist has just prepared a graph. For each of ten values of a manipulated variable X, he has obtained a value of a measured variable Y

under a "treatment" condition and also under a "control" condition. Each of the ten points on the treatment curve lies above the corresponding point on the control curve. (a) What was the probability of this result if for each pair of points $P(Y_t > Y_c) = P(Y_t < Y_c) = .5$, i.e., if treatment and control really have identical effects? Test at the largest available $\alpha \leq .05$ the H_0 that $P(Y_t > Y_c) = .5$ against the H_a that $P(Y_t > Y_c) \neq .5$. What are (b) the rejection region r-values, (c) the exact value of α, (d) the accepted hypothesis, (e) the probability level of the obtained result, (f) the power of the test if actually $P(Y_t > Y_c) = .8$?

12

generally applicable tests and estimates

In this chapter we shall present some standard tests and estimates that are used to answer some common statistical questions, such as within what range of values a population's median may be expected to lie, whether or not there is a sequential trend in a variable's values, whether or not two variables are independent, and whether several different treatments in an experiment have had the same or different effects. In all of the tests the sampling distribution of the test statistic will be discrete and easily calculable, at least when sample sizes are small. The tests will not be burdened by elaborate qualifications concerning the conditions under which they are applicable. No assumptions will be required about the distributions of the sampled populations except perhaps that they are continuous. It will be assumed, however, that sampling is random. The assumption that the sampled populations are continuously distributed, when made, is often more of a semantic convenience than a statistical necessity. The tests can be legitimately applied in many cases when population distributions are discrete, provided that certain procedures are followed; however, these subjects will not be dealt with here. What we shall say about the tests will be accurate but simplified. For a full appreciation of their versatility in applications, their statistical characteristics (such as efficiency), and the proper procedure for dealing with tied observations or difference scores having a value of zero, as well as the presentation of many more tests, the reader is referred to the writer's comprehensive text on the subject.[1]

---

[1] *Distribution-Free Statistical Tests* (Englewood Cliffs, N.J.: Prentice-Hall, Inc., 1968).

A useful convention, when we wish to tell what a test tests, *in general*, without specifying details about where the rejection region lies, is to say simply that H_0 is that $\theta = \theta_0$. This is admissible for a variety of reasons. Irrespective of whether the test is to be left-tailed, right-tailed, or two-tailed, θ_0 is the most important θ-value in the null set. When the alternative hypothesis is specified, it tells us the θ-values in the alternative set which automatically tells us what θ-values are in the complementary null set. The alternative hypothesis, together with the probability distribution of S when $\theta = \theta_0$, tells us the tail or tails at which the rejection region is located; and the maximum probability α that the test statistic will fall in it if H_0 is true applies to the case where $\theta = \theta_0$. So knowing θ_0, the only additional information we need in order to apply the test is H_a, which depends upon the *specific* applicational situation and can therefore be omitted as long as we are speaking in generalities. The convention of stating that H_0 is that $\theta = \theta_0$ will be followed in the remainder of the book.

12-1   Binomial Tests

A number of excellent and useful tests are concerned with the proportion p of successes in an inexhaustible population and have a binomially distributed test statistic. The case where H_0 explicitly *states* what values p may have has already been covered in the preceding chapter. However, a far richer variety of tests have H_0's that are *immediately* concerned with other matters but that logically *imply* what values p may have. A few of them are outlined below. Once the equivalence is established between their nominal hypotheses and surrogate hypotheses about p, the tests are conducted in the same way as outlined in Chapter 11.

12-1-1   Test for the Median

The null hypothesis is that the median M of a continuously distributed population has the value M_0. Since exactly half of a continuously distributed population lies below the median, this implies that (and is equivalent to the hypothesis that) the proportion p of units in the population having values less than M_0 is .5. The test statistic is the number r of observations having values smaller than M_0 in a random sample of n observations from the population in question. The probability of exactly r such observations is

$$P(r) = \binom{n}{r} p^r (1 - p)^{n-r}$$

If H_0 is true and $M = M_0$, then $p = p_0 = .5$, and the formula becomes

$$P(r) = \binom{n}{r} .5^n.$$

The latter formula tells us the probability distribution of the test statistic when H_0 is true and $M = M_0$—from which we can determine the rejection region when we have decided upon α and H_a.

With this test one must be careful in selecting the proper rejection region. If actually M lies below M_0, then since half the population lies below M and an additional portion of it lies between M and M_0, the proportion p that lies below M_0 must be greater than .5, i.e., if $M < M_0$ then $p > .5$ (see Figure 12-1). And if $p > .5$, large values of r (e.g., values $> np$) will have greater probability than they would if $p = .5$. On the other hand, if M lies above M_0, so that M_0 lies below M, less than half the population must lie below M_0 and p must be less than .5. And if $p < .5$, small values of r (e.g., values $< np$) will have greater probability than they would if $p = .5$. Therefore, for testing

$$H_0 : M \leq M_0 \text{ against } H_a : M > M_0, \text{ use lower-tail test}$$

$$H_0 : M \geq M_0 \text{ against } H_a : M < M_0, \text{ use upper-tail test}$$

$$H_0 : M = M_0 \text{ against } H_a : M \neq M_0, \text{ use two-tail test}$$

Knowing the tail (or tails) at which the rejection region is to lie, one simply cumulates probabilities (from the tip of that tail inward) in the probability distribution of r when $p = .5$ until the cumulative probability is an acceptable α value.

P = proportion of population below M_0 = $P(X < M_0)$ = shaded area

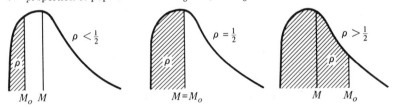

FIG. 12-1.   Relationships between M, M_0, and p in test for the median.

Example: A highway-safety engineer wishes to test at about the .05 level of significance the H_0 that the median time delay in applying the brake after observing a dangerous situation ahead is $\leq .75$ seconds against the H_a that it is $> .75$. After considering power functions, he decides to base the test upon the outcome of a random sample of $n = 13$ observations. The H_a that $M > .75$ is equivalent to the H_a that less that 50% of the population of time scores lies below .75—so that r should tend to take small values under H_a and the test should be left-tailed. From the probability distribution of r when $n = 13$ and $p = .5$ he finds that $P(r \leq 3) = .04614$; so he decides to use as rejection region the r-values 0, 1, 2, and 3, corresponding to an α of .04614. He now takes a random sample of 13 motorists, subjects them to the situation, and obtains the following time scores, rearranged to appear in ascending order: .61, 68, .76, .80, .83, .85, .86, .89, .90, .99, 1.21, 1.55, 2.17. Two of these time scores are $< .75$, so $r = 2$, falling in the rejection region. The engineer therefore rejects the $H_0 : M \leq .75$ in favor of the $H_a : M > .75$. The test attained the .04614 level of significance and its probability level is $P(r \leq 2 | n = 13, p = .5) = .01123$.

Note: This test is a special case of the Test for a Percentile. Let Q be the $100q$th percentile of a continuously distributed population, i.e., the value below which exactly a proportion q of the population lies. The null hypothesis is that Q has a value of Q_0. The test statistic is the number r of observations in a random sample of n observations that have values less than Q_0. Let p be the proportion of the population that lies below Q_0. Then the probability distribution of the test statistic is given by

$$P(r) = \binom{n}{r} p^r (1 - p)^{n-r}$$

which, if $Q = Q_0$, becomes

$$P(r) = \binom{n}{r} q^r (1 - q)^{n-r}$$

whereas if $Q < Q_0$, then $p > q$ and large values of r will have greater probability, while if $Q > Q_0$, then $p < q$ and small values of r will have greater probability, than they would if $Q = Q_0$. The test then consists of determining whether or not r falls in a rejection region based on its distribution

$$P(r) = \binom{n}{r} q^r (1 - q)^{n-r}$$

when $Q = Q_0$.

12-1-2 Test for the Median Difference

This test is simply the preceding test for the median in the special case where the population in question is a continuously distributed population of difference scores. H_0 is that the median M of a potential population of difference scores has the value M_0. The test statistic is the number r of difference scores in a random sample of n difference scores that have values less than M_0. And if H_0 is true, the distribution of r is given by the formula

$$P(r) = \binom{n}{r} .5^n$$

Rejection regions are the lower tail of this distribution when H_a is that $M > M_0$, the upper tail when H_a is that $M < M_0$, and both tails when H_a is that $M \neq M_0$.

The most common application of the test is in testing the influence of experimental treatments upon matched pairs of units. Imagine an infinite population of matched pairs of units. Suppose that one member of each matched pair is randomly assigned to an X treatment, the other member being assigned to a Y treatment. A measurement sensitive to the influence of treatment is made, after each treatment, upon the member assigned to it. Each pair's Y-score is subtracted from its X-score yielding a difference score $D = X - Y$. We now have a hypothetical population of difference scores and H_0 is that its median is M_0. But since the assignment of pair members to treatments was random and since the hypothetical difference-score population is infinite (so that no inequities due to random assignment remain), the H_0 that the median of the D population is M_0 is equivalent to the H_0 that the median difference in *treatment effects*, upon the units in question, is M_0. It is the latter hypothesis, and the accompanying H_a, that experimenters generally wish to test. They may do so by creating a sample of *actual* difference scores in the same way that we created the fictitious population of hypothetical difference scores and then applying the test for the median to the sample.

Had we applied both treatments to the same unit, the score measured under the later treatment might have been influenced by the lingering effects of the earlier treatment. However, if there are no such carry-over effects, single units can be used in place of matched pairs of units, the X-score and Y-score being the measured effects of the two different treatments upon the same unit at two different times. When it is legitimate, this is a very effective procedure since "matching" is often optimal when a unit is "matched" with itself.

Example: The director of an orphanage suspects that intellectual growth is greater for adopted orphans than for those raised in his institution. To test his theory, he randomly selects one member from each of the 18 pairs of identical twin infants in his orphanage and puts it up for adoption, keeping the other member in the institution. Ten years later he measures the IQ of all 36 of the individuals involved in the study. Let X stand for the IQ of an adopted twin (under treatment X—adoption) and Y stand for that of his matched orphanage twin (under treatment Y—institutionalization), and let $D = X - Y$ be the superiority of the adopted twin's IQ. Before viewing the data, the director states as his null hypothesis that the median M of the D population from which his sample was "drawn" is ≤ 0, and he states as his alternative hypothesis that the median is > 0. His test statistic will be the number r of difference scores whose value is less than 0. If the median of the D population is 0, r will have the binomial distribution

$$P(r \mid n = 18, p = .5) = \binom{18}{r}.5^{18}$$

Whereas if H_a is true, $M > 0$ so that $p < .5$ calling for a left-tail rejection region. Examining the left tail of the distribution of r when $n = 18$ and $p = .5$, the director decides to use as rejection region the r-values ≤ 5 corresponding to $\alpha = .04813$. When the 18 pairs of IQ's are obtained, the adopted twin's IQ is higher than that of his sibling in 15 of the 18 pairs, and 3 of the difference scores are less than 0. Since $r = 3$ falls in the rejection region, H_0 is rejected at the .04813 level of significance, so the director accepts H_a and concludes that foster homes are more conducive to intellectual development than is his orphanage. The probability level of the test is

$$P(r \leq 3 \mid n = 18, p = .5) = .00377.$$

12-1-3 Test for Trend

This test concerns a continuously distributed variable X whose values are generated sequentially in time or appear sequentially in space. The null hypothesis is that the sequence is a random one, so that there is no trend in the values of X. The alternative hypothesis, therefore, is that the sequence is nonrandom. However, the test is maximally sensitive to nonrandom sequences characterized by a unidirectional (i.e., nonreversing upward or downward) trend, and it may be greatly (or completely) lacking in sensitivity to other forms of nonrandomness, including cyclical trends. The test should therefore be regarded primarily as a test for *unidirectional* trend or, at least, the unidirectional *component* of whatever trend exists.

Consider a future random sample of $2n$ observations on the variable arranged in order of temporal (or spatial) appearance of their values: X_1, X_2, $\ldots$, X_i, $\ldots$, X_n, X_{n+1}, X_{n+2}, $\ldots$, X_{n+i}, $\ldots$, X_{2n}. If the null hypothesis is true, then for every possible value of i from 1 to n, X_{n+i} is just as likely to be less than X_i as to exceed it, and since X is continuously distributed

$$P(X_{n+i} = X_i) = 0$$

Therefore, if H_0 is true, $P(X_{n+i} > X_i) = p = .5$. The test statistic is the number r of X_{n+i}'s that are larger than the X_i with which they are paired, i.e., the X-value that occurred n observations earlier. If H_0 is true, the probability distribution of r is given by

$$P(r) = \binom{n}{r}.5^n$$

and r-values at the extremes of the distribution will occur infrequently. On the other hand, if there is a nonreversing upward trend, the later X's should tend to be larger than the earlier ones so that r, the number of times an X_{n+i} exceeds its paired X_i, should tend to take the large values in the upper tail of its range of possible values. And if there is a nonreversing downward trend, the X_{n+i}'s should tend to be smaller than their X_i counterparts so that $X_{n+i} > X_i$ in few of the n comparisons, causing r to assume the small values at the lower end of its range. Therefore, *roughly speaking*, to test the H_0 of no trend or downward trend (implying $p \leq .5$) against the H_a of upward trend (implying that $p > .5$), use as rejection region the upper tail of the distribution of r when $p = .5$ and n is half the number of observations, i.e., the number of comparisons of a later with an earlier observation. To test the H_0 of no trend or upward trend against the H_a of downward trend, use lower-tail values of r as rejection region. And to test the H_0 of no trend against the H_a of unspecified unidirectional trend, use a two-tail rejection region.

Example: A quality control engineer wants to test the H_0 that there is no trend (actually that there is randomness) in the manufacturing process producing ball bearings of a certain diameter. Since he wants to detect any systematic tendency to deviate from the nominal, intended, diameter, his H_a is that there is a trend of some type, upward or downward. This calls for a two-tail test. After examining power functions, he decides to let n, the number of pairs, be 10 and use the r-values 0, 1, 9, and 10 as rejection region, corresponding to an α of .02148. He then randomly selects one ball bearing every 30 seconds, measures its diameter, and records it, until he has taken a sample of $2n = 20$ diameters. The diameters, arranged in the same order in which the ball bearings to which they refer were produced, are: .3750, .3752, .3749, .3747, .3751, .3749, .3753, .3751, .3750, .3752, .3751, .3754, .3750, .3750, .3749, .3750, .3751, .3752, .3751, .3753. "Breaking the series in half" and comparing the 11th score .3751 with the 1st .3750, the 12th .3754 with the 2nd .3752, etc., he finds that in eight of the ten comparisons the later score X_{n+i} was larger than the earlier one X_i so that $r = 8$. Since $r = 8$ is not in the rejection region, the engineer accepts the H_0 of no trend. The probability level of the test is $P(r \geq 8) + P(r \leq 2) = .10938$.

Note: This test is Cox and Stuart's S_2 Sign Test for Trend in Location. It is highly effective in detecting nonreversing trends when sample sizes are reasonably large. It is far less sensitive to cyclical trends, especially when the cycles are rapid.

12-1-4 Test for Concentration

This test concerns the unknown proportion p of units in an infinite population that have values lying between two specified constants C and K, the latter being the larger constant. The null hypothesis H_0 is that p has a specified value p_0, i.e., that $100p_0\%$ of the population is concentrated between the values C and K. The test statistic is the number r of observatons in a random sample of n observations, drawn from the population in question, whose values lie between C and K.

The probability that the value X of a single randomly drawn observation will lie between C and K is $P(C < X < K) = p$. And the probability that in a random sample of n observations r of the observations will have values between C and K is

$$P(r) = \binom{n}{r} p^r (1 - p)^{n-r}$$

Therefore, if H_0 is true, the distribution of r is given by the formula

$$P(r) = \binom{n}{r} p_0^r (1 - p_0)^{n-r}$$

If $p < p_0$, r will be more likely to take small values, and if $p > p_0$, large values of r will be more probable, than is the case if $p = p_0$ as specified by H_0. Therefore, if H_a is that $p < p_0$, use a lower-tail rejection region; if H_a is that $p > p_0$, use an upper-tail rejection region; and if H_a is that $p \neq p_0$, use a two-tail rejection region.

Example: A political scientist believes that more than 40% of registered voters are in the age range from 25 to 45, inclusive. To test the validity of his belief, he states as his null hypothesis that the proportion p of registered voters whose ages fall in the above range (i.e., whose age A is such that $24 < A < 46$) is $\leq .40$, his alternative hypothesis being that the proportion is $> .40$. He decides to take a random sample of $n = 20$ registered voters and to use an α of .05653 for his upper-tail test, corresponding to a rejection region of r-values ≥ 12. He takes the sample and finds that six of the voters have ages > 24 but < 46; so since $r = 6$ is not in the rejection region, he accepts the H_0 that $p \leq .40$. The probability level of the test is

$$P(r \geq 6 \mid n = 20, p = .40) = .87440$$

12-2 Hypergeometric Tests

Several useful tests have a hypergeometrically distributed test statistic. One of the best known and most useful is due to R. A. Fisher. It is given below.

12-2-1 Test for Dependence

This test concerns an infinite population each of whose units is either a success S or a failure F and is also either an A or a B. The null hypothesis is that whether a unit is an S or an F is *independent* of whether it is an A or a B. If H_0 is true, then by the definition of independence, the proportion p of

S's in the entire population, the proportion p_A of S's among units that are A's, and the proportion p_B of S's among units that are B's, all have the same value, i.e., $p_A = p_B = p$. Now suppose that we draw a random sample of n observations from the infinite subpopulation of units that are A's and a random sample of $N - n$ observations from the infinite subpopulation of units that are B's; and suppose that r of the units in the A sample and $R - r$ of the units in the B sample are successes. Combining these two samples, we obtain a pooled sample of N units, R of which are successes. (See Table 12-1.)

Table 12-1

Sampling Situation in Test for Dependence

	Successes	Failures	Total
Sample of A's	r	$n - r$	n
Sample of B's	$R - r$	$(N - R) - (n - r)$	$N - n$
Combined sample	R	$N - R$	N

If H_0 is true, the A and B subpopulations are homogeneous so far as the proportion of successes is concerned. In that event the A sample, B sample, and the pooled sample are also "homogeneous" in the sense that whether a sample unit was an A or a B had no influence upon whether it was an S or an F—the A-B categorization was irrelevant. But this means that "A sample" and "B sample" might just as well have been arbitrary labels, the former being randomly assigned to n observations, and the latter being assigned to the remainder $N - n$, of N observations randomly selected from the overall population. And that means that the "A sample" might just as well have been randomly selected from the N observations in the pooled sample. If it had been, the probability that it would contain exactly r successes is given by the hypergeometric probability law

$$P(r) = \frac{\binom{R}{r}\binom{N - R}{n - r}}{\binom{N}{n}}$$

which therefore gives the probability distribution of the test statistic r when the null hypothesis is true.

On the other hand, if being an S or an F is not independent of being an A or a B, we would expect small values of r to be more likely if $p_A < p_B$, and large values of r to be more likely if $p_A > p_B$, than would be the case under independence where $p_A = p_B$. Therefore, to test the H_0 that $p_A \geq p_B$ against the H_a that $p_A < p_B$, use a lower-tail rejection region. To test the H_0 that

$p_A \leq p_B$ against the H_a that $p_A > p_B$, use an upper-tail rejection region. And to test the H_0 that $p_A = p_B$, i.e., the H_0 of independence, against the H_a that $p_A \neq p_B$, i.e., the H_a of nonindependence, use a two-tail rejection region.

Example: A doctor suspects that the survival rate for a certain rare disease depends upon the sex of the patient, but he is uncertain which sex has the better chance of survival. Therefore, he decides to test the null hypothesis that the survival rates are the same, so that survival is independent of sex, against the alternative hypothesis that they are different, so that survival rate depends upon sex. His H_a calls for a two-tail test and he decides to use the largest possible α-value that is less than .05, using as rejection region the r-values corresponding to the greatest absolute differences between male and female survival rates within the intended sample. He finds that there are only 16 known cases of the disease and that their outcomes are as shown in the upper portion of Table 12-2. Using the formula

$$P(r) = \frac{\binom{R}{r}\binom{N-R}{n-r}}{\binom{N}{n}} = \frac{\binom{10}{r}\binom{6}{9-r}}{\binom{16}{9}}$$

Table 12-2

Data for Doctor's Test for Dependence

Actually obtained figures

	Survived	Died	Total		Survival rate	Death rate	
Males	8	1	9		.889	.111	1.00
Females	2	5	7		.286	.714	1.00
Total	10	6	16		.625	.375	1.00

Possible figures, for fixed marginal totals, and their probabilities

Males survived r	Females survived $10-r$	Male survival rate $\dfrac{r}{9}$	Female survival rate $\dfrac{10-r}{7}$	Absolute difference $\left\lvert\dfrac{r}{9} - \dfrac{10-r}{7}\right\rvert$	$P(r)$
9	1	1.000	.143	.857	.0009
8	2	.889	.286	.603	.0236
7	3	.778	.429	.349	.1574
.					
.					
.					
4	6	.444	.857	.413	.1101
3	7	.333	1.000	.667	.0105

he obtains the figures shown in the lower portion of Table 12-2. Cumulating $P(r)$ over those r-values corresponding to the greatest absolute differences in sample survival rate, he finds that

$$P(r = 9 \text{ or } 3 \text{ or } 8) = .0009 + .0105 + .0236 = .0350$$

which is the largest such cumulative probability less than .05. Therefore, his two-tail rejection region consists of the r-values 9, 8, and 3 and his α level is .0350. The r-value for the actually obtained table is 8 which lies in the rejection region, so he rejects the H_0 of independence and concludes that survival rate depends upon sex.

Note: In this test the marginal frequencies of Table 12-1 are taken as given, i.e., fixed. This means that, in effect, there is only a single variable in the cells of the table since once a value is assigned to one cell, all other cell entries can be obtained by subtractions from marginal frequencies. If we take that variable to be r, then $P(r)$ gives us the probability for the entire set of four cell entries, in the case where the marginal frequencies $(R, N - R, n, N - n)$ are fixed and the row and column classifications are independent.

12-3   Rank Tests

A large number of excellent tests rank the sample observations in order of increasing size and use sums of ranks (or rank differences) as the test statistic. The probability distribution of the test statistic is generally not a standard probability distribution; rather, it is idiosyncratic to the particular test involved so that each different type of rank test requires its own set of tables of probabilities or critical values.

12-3-1   Treatment Populations

Suppose that every unit in an infinite population is randomly assigned to one of C different groups. The C groups are therefore C different, infinite "subpopulations" having identical distributions. Suppose now that a different treatment is applied to the units in each group. We now have C different "treatment populations." If the C treatments all have no effect, or if they all have exactly the same effect, then the C different treatment populations will be identically distributed. However, if some of the C treatments have different effects than others, then some of the treatment populations will have distributions that differ in some way from those of other treatment populations. These treatment populations are, of course, hypothetical, or potential, rather than real. Suppose, however, that finite random samples of units from a common population are randomly assigned to the C different treatments. After treatment, we may regard them as random samples drawn from their respective treatment populations. And if we now apply to these samples a test of whether or not the C parent populations have identical distributions, this is equivalent to testing whether or not the C treatments all have the same effect. Thus, a test of the H_0 of identically distributed populations is also a test of identical treatment effects if sample units have been drawn from a common population and randomly assigned to treatments. Two of the tests to follow, the Wilcoxon Rank-Sum Test and the Kruskal-Wallis Test, can be regarded as either tests for identical populations or as tests for identical treatment effects.

Now suppose that from each of R different populations a random sample of C units is drawn and randomly assigned, one unit to each of C groups, and that this is done an infinite number of times. The C units coming from the same population are, in a sense, homogeneous, and therefore "matched." Again, the C groups are C different, infinite populations each containing the same "mixture" of units and therefore having identical distributions. And if a different treatment is applied to the units of each group, we have C different treatment populations whose distributions will be identical only if all C treatments have the same effect. Therefore, if a sample of C units is randomly drawn from each of R populations and the C units are randomly assigned to treatments, one unit to each treatment, then after the treatments are administered, we may regard the R units receiving a given treatment as a random sample from that treatment population.

Finally, suppose that from each of R different populations a single unit is randomly drawn and successively subjected to all C treatments and that this is done an infinite number of times. And suppose that the effect of one treatment does not influence the measured effect of another. Then the measurements taken immediately after treatment form C different treatment populations. If treatments all have the same effect, these C treatment populations will have identical distributions; otherwise, some of them will differ. Therefore, if each of R units, drawn randomly from R different populations, is exposed to each of the C treatments, the R measurements of the effect of a given treatment can be regarded as a random sample of R observations drawn from the treatment population in question.

Two of the tests to follow, the Wilcoxon Signed-Rank Test and the Friedman Test, use matched units in testing for unequal treatment effects. In both cases the matching can be accomplished by using under each treatment either a different one of the C units comprising a homogeneous set or the same unit, which is therefore matched with itself.

12-3-2 Special Features of the Tests

In the tests to follow we can deduce the *exact* probability distribution of the test statistic in only one case, that in which $\theta = \theta_0$. So we cannot calculate the power of the test in other cases. Furthermore, θ_0 is not (and does not imply) a numerical value; rather, it is a *condition*, such as that two variables X and Y are *independent* or that they have *identical* distributions. This raises no difficulties when H_0 is a simple hypothesis (as in the case of conventional two-tail tests) stating the *single* condition $\theta = \theta_0$, e.g., "X and Y are independent" or "X and Y have identical distributions," and H_a states its negation $\theta \neq \theta_0$, e.g., "X and Y are dependent" or "X and Y have differing distributions." However, when H_0 is a compound hypothesis (as in all one-tail tests covered so far), it is difficult to be precise about what *sets* of possibilities or conditions are being covered by H_0 and H_a, especially about what conditions other than θ_0 (e.g., independence) are being covered by H_0. H_a is no longer the negation of $\theta = \theta_0$; rather it is the negation of something analogous to $\theta \geq \theta_0$. But if θ_0 is "independence" or "identical distributions," what conditions correspond to the θ's that are $> \theta_0$? It would be *accurate*, but not sufficiently *meaningful* to say, for example, that H_0 is that "X and Y are either independent or dependent in such a way that the probability ϕ that the test statistic will fall in the rejection region is $\leq \alpha$" and that H_a is that "they are dependent in such a way that $\phi > \alpha$." The trouble is that this does not tell us what those ways are. We would like to be more specific about precisely what types of dependency cause ϕ to be $> \alpha$ or $< \alpha$. Unfortunately, if X and Y are continuously distributed variables, there are an infinite number of ways in which they can be dependent or nonidentically distributed. So it is seldom the case that *only* under a certain specifiable type of dependency (or nonidentity) will ϕ tend to exceed α. However, we can often be quite certain that it *strongly* tends to do so under one known and easily specifiable type of condition. If under ordinary circumstances, this type of condition is almost certain to be *present* when $\phi > \alpha$ and highly likely to be the principle cause of ϕ exceeding α, we say that the test is *sensitive* to this type of condition.

When first introducing a test, we shall state as H_0 the single condition θ_0 under which the probability distribution of the test statistic can be obtained. And we shall state as H_a the negation of θ_0. These hypotheses will be what the test actually tests when H_0 is a simple and H_a a composite hypothesis. At a later point we shall depart from complete accuracy and re-express H_0 and H_a, with as much accuracy as possible, in terms of the specifiable type of condition to which the test is sensitive.

All of the tests to follow require that there be zero probability that certain observations will have exactly the same value. This requirement will be met if (a) all of the sampled populations, or measured variables, are continuously distributed and (b) measurements of the sampled units' values are invariably made with sufficient precision to prevent any ties from occurring. However, to avoid excess verbage, we shall not state this requirement in its entirety for every test.

12-3-3   Hotelling and Pabst's Test for Correlation

Consider a continuously distributed bivariate population in which each unit has two values, an X-value and a Y-value. The X and Y variables are *independent* if knowing the value of one tells you nothing more about the probable value of the other, i.e., if $P(X|Y) = P(X)$ or $P(Y|X) = P(Y)$. They are *positively correlated* if units having the smaller [larger] X-values tend also to have the smaller [larger] Y-values. They are *negatively correlated* if units having the smaller [larger] X-values tend to have the larger [smaller] Y-values. So as X increases: if Y tends to increase, there is a positive correlation; if Y tends to decrease, there is a negative one.

The null hypothesis is that X and Y are independent. The alternative hypothesis, therefore, is that they are dependent. However, the test is highly sensitive to forms of dependency involving a positive or a negative correlation between X and Y, and it may be greatly (or completely) lacking in sensitivity to other forms of dependency. Hence, the test should be regarded as a test *of* independence that is *sensitive* to alternatives involving positive or negative correlation and therefore as a test *for* unidirectional correlation.

To test the null hypothesis that X and Y are independent, and therefore uncorrelated, a random sample of n units is drawn from the continuously distributed bivariate population. Each unit's X-value and Y-value are measured and recorded. The n X-values are than ranked in order of increasing size from 1 (smallest) to n (largest) as are the n Y-values. Let the rank of the ith unit's X-value be $R(X_i)$, let the rank of its Y-value be $R(Y_i)$, and let $d_i = R(X_i) - R(Y_i)$ be the difference between the X-rank and Y-rank of the ith unit. The test statistic is

$$D = \sum_{i=1}^{n} d_i^2 = \sum_{i=1}^{n} [R(X_i) - R(Y_i)]^2 \qquad (12\text{-}1)$$

Now consider the distribution of the test statistic. If the H_0 of independence is true, there is no relationship between X and Y in the population so that every possible association between X-ranks and Y-ranks in the sample was equally likely before the sample was actually drawn. The total number of possible associations between sample X-ranks and Y-ranks is the number of different ways of assigning Y-ranks to X-ranks, and this is simply $n!$, the number of permutations of Y-ranks within n "positions" represented by X-ranks. Therefore, if X and Y are independent, every posssible sample association between X-ranks and Y-ranks has the same probability $1/n!$ and the probability distribution of D can be easily calculated. In the case where $n = 3$, there are $3! = 6$ possibilities. They are shown in Table 12-3 along with the resulting probability distribution of D in the case where X and Y are independent. On the other hand, if X and Y were positively [negatively] correlated, D would tend to take values at the lower [upper] end of its range of possible values more frequently than if X and Y were independent.

Table 12-3

Possibilities and Probabilities for Hotelling and Pabst's
Test Statistic When $n = 3$

(a) The six possible associations between X-ranks and Y-ranks, and the resulting
values of d_i and D

X-rank	1	2	3	1	2	3	1	2	3
Y-rank	1	2	3	1	3	2	2	1	3
Difference d_i	0	0	0	0	-1	1	-1	1	0
$D = \Sigma\, d_i^2$		0			2			2	

X-rank	1	2	3	1	2	3	1	2	3
Y-rank	2	3	1	3	1	2	3	2	1
Difference d_i	-1	-1	2	-2	1	1	-2	0	2
$D = \Sigma\, d_i^2$		6			6			8	

(b) Probability distribution of D if H_0 of independence is true

D	$P(D)$
0	1/6
2	2/6
6	2/6
8	1/6

Therefore, *roughly speaking*, to test the H_0 of no correlation or negative correlation (or, more succinctly, of no positive correlation) against the H_a of positive correlation, use a lower-tail rejection region; to test the H_0 of no correlation or positive correlation (and therefore of no negative correlation) against the H_a of negative correlation, use an upper-tail rejection region; and to test the H_0 of no correlation against the H_a of unspecified unidirectional correlation, use a two-tail rejection region. A table of critical values of D is given in Appendix B. (see Table B-7).

Example: A psychologist believes that among adults conservatism is positively correlated with age. Therefore, he states as his H_0 that age and conservatism are either independent (and therefore uncorrelated) or nega- tively correlated, and as his H_a that they are positively correlated, calling for a lower-tail rejection region. He decides to base his test upon a sample of 10 observations and the largest possible α less than or equal to .05. From Table B-7 in Appendix B he finds that for these conditions his rejection region con- sists of values of $D \leq 72$. He now draws a random sample of adults, asks their ages, and administers a test for conservatism. His sample data and the appropriate calculations are shown in Table 12-4. Since his obtained value of D, 22, falls in the rejection region, he rejects H_0 and accepts the H_a that age is positively correlated with conservatism. From Table B-7 in Appendix B he finds that if H_0 is true $P(D \leq 20) \leq .001$ and $P(D \leq 34) \leq .005$, where 20 and 34 are critical values for the largest possible α level $\leq .001$ and

Table 12-4

Data and Calculations for Psychologist's Test for
Correlation Between Age and Conservatism

Subject	S_1	S_2	S_3	S_4	S_5	S_6	S_7	S_8	S_9	S_{10}
Age	21	26	28	30	37	44	51	59	66	75
Conservatism score	30	35	42	16	39	55	68	81	62	94
Age rank	1	2	3	4	5	6	7	8	9	1·0
Conservatism rank	2	3	5	1	4	6	8	9	7	10
Rank difference d_i	-1	-1	-2	3	1	0	-1	-1	2	0
d_i^2	1	1	4	9	1	0	1	1	4	0

$$D = \sum_{i=1}^{10} d_i^2 = 22$$

$\leq .005$, respectively. Therefore, the probability level of the test $P(D \leq 22)$ must be some value between .001 and .005, presumably lying close to the former.

Note: If one lets the X variable be the time at which a unit is produced, this test becomes an excellent test for trend in the values of the Y variable.

12-3-4 Wilcoxon's Signed-Rank Test for Differences in Treatment Effects

This is a test for unequal treatment effects upon matched units. From each of n matched pairs of units one member is randomly assigned to receive an X treatment, and the remaining member is assigned to receive a Y treatment. After treatment, using a continuous scale of measurement, a precise measurement sensitive to treatment effects is made upon each member. Let X_i and Y_i be the measurements upon the members of the ith matched pair of units receiving the X and Y treatments, respectively. It is required that the measurements invariably be made with sufficient precision to prevent an X_i from ever having exactly the same value as the Y_i with which it is paired.

The null hypothesis is that the two treatments are exactly equivalent and therefore would cause originally identical units to have the same value following treatment. The matched units, of course, were not identical. But they were randomly assigned to treatments, so the "superior" unit was as likely to be assigned to treatment X as to treatment Y. It follows, therefore, (since we have required that $P(X_i = Y_i) = 0$) that if the null hypothesis of identical effects is true, $P(X_i > Y_i) = P(X_i < Y_i) = .5$, and if $P(X_i > Y_i) \neq .5$, the null hypothesis is false. (Unfortunately, it is possible in certain special and highly unlikely cases for H_0 to be false when $P(X_i > Y_i) = .5$; this will occur if the X and Y treatment populations have nonidentical but symmetric distributions with a common median.) As the negation of the null hypothesis, the alternative hypothesis is simply that the two treatments are not exactly equivalent. However, as might be expected from the above, the test is highly sensitive to differences in treatment effects such that X-observations tend to be larger, or tend to be smaller, than the Y-observations with which they are matched, i.e., to differences such that $P(X_i > Y_i) > .5$ or $P(X_i > Y_i) < .5$. Therefore, the test should be regarded as a test for unequal treatment effects that is *sensitive* to those forms of inequality that cause $P(X_i > Y_i)$ to differ unidirectionally from .5.

The test statistic is obtained as follows. Let $d_i = X_i - Y_i$, let S_i be the algebraic sign of d_i, and let $|d_i|$ be the absolute value of d_i (i.e., its magnitude with algebraic sign omitted). Finally, let R_i be the rank of $|d_i|$ among the n absolute difference scores when they are arranged in order of increasing size (the smallest receiving a rank of 1, the largest a rank of n). The test statistic is the sum of the positive signed ranks,

$$W_+ = \sum_{S_i = +} S_i R_i \tag{12-2}$$

Now consider the distribution of the test statistic. Suppose that *no* treatments had been administered but that the X_i and Y_i measurements had been made just the same. Since the assignment of members of matched pairs of units to treatments was *random*, the assignment of member A to X treatment and B to Y treatment had exactly the same probability as the assignment of member B to X treatment and member A to Y treatment. In the former case the difference score will be "A's score minus B's score"; in the latter it will be "B's score minus A's score"; so if the former difference score is K, the latter will be $-K$. Thus whatever a difference score's absolute magnitude, it was just as likely to be positive as to be negative, i.e., $P(d_i = -K) = P(d_i = +K)$ irrespective of the value of K. Furthermore, this will also be the case if the two treatments *are* administered but have either no effect or the same effect.

Therefore, if H_0 is true, every one of the n d_i's was just as likely to have been positive as to have been negative. So if H_0 is true, every R_i is as likely to be accompanied by a $+$ sign as by a $-$ sign, and the distribution of

$$W_+ = \sum_{S_i = +} S_i R_i$$

is obtained by assigning algebraic signs in each of the 2^n different possible ways to the n different R_i's and calculating W_+ for each way, each "way" having the same probability $1/2^n$. Thus, for the case where $n = 3$ we have the $2^3 = 8$ possibilities and the corresponding probability distribution of W_+ given in Table 12-5.

On the other hand, if treatments tend to make X-scores larger [smaller] than Y-scores, then d_i will be more likely to be $+$ than $-$ [$-$ than $+$] and W_+ will be more likely to take large [small] values than it would if H_0 were true and d_i were equally likely to be $+$ or $-$. Therefore, *roughly speaking*, to test the H_0 that X treatment tends to make scores the same as or larger than does Y treatment against the H_a that X treatment tends to make them smaller,

Table 12-5

Possibilities and Probabilities for Wilcoxon's
Signed-Rank Test Statistic When n = 3

(a) The eight possible sets of signed ranks, and the resulting values of W_+

	S_iR_i	S_iR_i	S_iR_i	S_iR_i	S_iR_i	S_iR_i	S_iR_i	S_iR_i
	+ 1	− 1	+ 1	+ 1	− 1	− 1	+ 1	− 1
	+ 2	+ 2	− 2	+ 2	− 2	+ 2	− 2	− 2
	+ 3	+ 3	+ 3	− 3	+ 3	− 3	− 3	− 3
W_+	6	5	4	3	3	2	1	0

(b) Probability distribution of W_+ if H_0 of identical treatment effects is true

W_+	$P(W_+)$
0	1/8
1	1/8
2	1/8
3	2/8
4	1/8
5	1/8
6	1/8

use a lower-tail rejection region. To test the H_0 that X-scores tend to be the same as or smaller than Y-scores against the H_a that they tend to be larger, use an upper-tail rejection region. And to test the H_0 that X-scores tend to have the same population distribution as Y-scores against the H_a that they do not, use a two-tail test.

If the effects of a previous treatment do not interact with those of a subsequent one, then instead of using matched pairs of units, a single unit may be subjected first to the X treatment and then to the Y treatment. An X-measurement is made upon the unit after the X treatment and a Y-measurement is obtained after the Y treatment. From that point on the conduct of the test is the same as already described.

The test as described so far tests the effect of treatments upon the particular n pairs of units (or n single units) used. If the experimenter wishes to be able to generalize his test results to an entire population of units, then the n pairs of units (or single units) must be a random sample from the population of pairs of units (or single units) to which he wishes to extend inference.

Example: A physiologist believes that muscular strength is increased by a certain drug. To test whether it is or not he states as H_0 that strength is either unaffected by the drug or weakened by it and states as H_a that strength is increased by the drug. He decides to test his H_0 using a Wilcoxon Signed-Rank Test upon a sample of $n = 10$ pairs of matched observations, conducted at the largest possible significance level $\alpha \leq .05$.

He draws a random sample of ten people and tests their strength of grip after they are given the drug (Treatment X); on another day he tests their strength of grip without the drug (Treatment Y). (Just to be safe, drug treatment is made the first treatment for a randomly selected five people and is made the second treatment for the remaining five.) The obtained data and the necessary mathematical operations upon them are given in Table 12-6.

Table 12-6

Data and Calculations for Physiologist's Test for Increase
of Strength Under Drug Treatment

| Subject | Strength of grip under | | $d_i = X_i - Y_i$ | Algebraic sign of d_i S_i | Rank of $\lvert d_i \rvert$ R_i |
	Drug X	No drug Y			
I	20.2	19.6	.6	+	5
II	11.4	11.2	.2	+	1
III	8.5	9.0	− .5	−	4
IV	27.4	25.1	2.3	+	8
V	31.5	28.4	3.1	+	10
VI	17.6	17.9	− .3	−	2
VII	7.3	6.5	.8	+	6
VIII	35.3	32.4	2.9	+	9
IX	12.0	11.6	.4	+	3
X	22.9	24.0	− 1.1	−	7

$$W_+ = \sum_{S_i = +} S_i R_i = 5 + 1 + 8 + 10 + 6 + 9 + 3 = 42$$

The H_a calls for an upper-tail test. If H_0 is true, the distribution of W_+ is symmetric, extending from 0 to $n(n + 1)/2$. From Table B-8 of Appendix B we find that the critical value of W_+ for a *lower*-tail test at the largest available $\alpha \leq .05$ is 10. Therefore, the critical value of W_+ for an upper-tail test is

$$\frac{n(n+1)}{2} - 10 = \frac{10(11)}{2} - 10 = 55 - 10 = 45$$

and the upper-tail rejection region is therefore W_+ values from 45 to 55. Our obtained W_+ value of 42 does not fall in the rejection region, so the physiologist must accept his H_0 that the drug does not increase muscular strength (as measured by strength of grip).

12-3-5 Wilcoxon's Rank-Sum Test for Nonidentical Population Distributions (or Unequal Treatment Effects)

This is a test of the H_0 that two continuously distributed populations X and Y have identical distributions. If the populations are those associated with treatment effects, it is a test of the H_0 that the treatments have identical effects.

Unlike the preceding test, this test does not use matched units and does not require equal numbers of X-observations and Y-observations.

The alternative hypothesis is that the two distributions differ in some way. However, the test is highly sensitive to those differences in which X-observations tend to be larger, or tend to be smaller, than Y-observations, and it tends to be rather insensitive in other cases. Thus, if p is the probability that a randomly drawn X-observation will exceed a randomly drawn Y-observation, so that $p = P(X > Y)$, the test tends to be sensitive to differences for which $p > .5$ or $p < .5$ and much less sensitive to those for which $p = .5$. This means that, in general, the test should be far more sensitive to differences in location than to differences in dispersion. Thus, the test should be regarded as a test *of* identical population distributions that is *sensitive* to those cases of nonidentical distributions in which $P(X > Y) \neq .5$.

A random sample of n observations is drawn from the X population and a random sample of m observations (where $m \geq n$) is drawn from the Y population. The $n + m$ observations in the combined sample are then ranked in order of increasing size, the smallest receiving a rank of 1, the largest a rank of $n + m$. Let X_i be the ith observation in the X sample and let R_i be the rank of X_i in the *combined* sample. The test statistic is

$$W_n = \sum_{i=1}^{n} R_i \qquad\qquad (12\text{-}3)$$

That is, the test statistic is the sum of the ranks of the observations in the smaller sample (or X sample).

Now consider the probability distribution of the test statistic. If H_0 is true and the X and Y populations have *identical* distributions, we may regard the combined sample as a single sample of $n + m$ observations drawn from the X and Y populations' common distribution. And we may regard "X sample" and "Y sample" as arbitrary labels assigned randomly to n and m of these observations, respectively. Chance alone had determined which n of the $n + m$ observations in the combined sample are the *values* associated with the n units drawn from the X population. There are

$$\binom{n + m}{n}$$

different ways of dividing the $n + m$ values in the combined sample into n "X-observations" and m "Y-observations," and if H_0 is true, each of these "splits" had the same probability

$$\frac{1}{\binom{n + m}{n}}$$

of becoming the actually obtained pair of samples. For each of these

$$\binom{n + m}{n}$$

"splits" there is a corresponding value of the test statistic W_n. Therefore, if H_0 is true, the probability distribution of W_n is obtained by investigating every different possible combination of n integers that can be obtained from the integers 1 to $n + m$, calculating the sum W_n of these n integers for each different combination, and giving the sum obtained for each combination the same probability

$$\frac{1}{\binom{n + m}{n}}.$$

This is illustrated in Table 12-7 for the case where $n = 2$ and $m = 3$. The table gives the

$$\binom{2 + 3}{2} = 10$$

different possible samples of two ranks from the ranks 1, 2, 3, 4, 5, their sums W_n, and the probability distribution of W_n when H_0 is true.

On the other hand, if X's tend to be greater [smaller] than Y's, so that p, the $P(X > Y)$, is $> .5$ [$< .5$], we would expect W_n to take large [small] values more often than it would if the X and Y populations had identical distributions. Therefore, *roughly speaking*, to test $H_0: p \leq .5$ against $H_a: p > .5$, use an upper-tail rejection region; to test $H_0: p \geq .5$ against $H_a: p < .5$, use a lower-tail rejection region; to test $H_0: p = .5$ against $H_a: p \neq .5$, use a two-tail region.

Table 12-7

Possibilities and Probabilities for Wilcoxon's Rank-Sum
Test Statistic When $n = 2$ and $m = 3$

(a) The ten possible sets of two ranks for the X sample and the resulting values of W_n

	1	1	1	1	2	2	2	3	3	4
	2	3	4	5	3	4	5	4	5	5
W_n	3	4	5	6	5	6	7	7	8	9

(b) Probability distribution of W_n if H_0 of identical population distributions is true

W_n	$P(W_n)$
3	.1
4	.1
5	.2
6	.2
7	.2
8	.1
9	.1

If the X and Y populations have identical distributions, as hypothesized by H_0, they must also have identical medians so that $M_X = M_Y$. And, although exceptions are entirely possible, we would expect it to be highly likely in general that $M_X > M_Y$ when $p > .5$ and that $M_X < M_Y$ when $p < .5$, especially if p differs appreciably from .5. (These relationships are certain to be the case if the two population distributions have identical "shapes" and dispersions so that, if they differ at all, they can differ only in location.) Therefore, the test is sometimes regarded as a *rough* test for unequal medians.

Example: A psychologist believes that seventh-graders can learn the Greek alphabet faster under Teaching Method A than under Teaching Method B. He decides to test this at the largest possible $\alpha \le .01$. He states as H_0 that the distributions of learning times are the same under the two methods or are such that scores tend to be higher under method A than under method B; and his H_a is that learning times tend to be lower under method A than under method B. He obtains a random sample of 13 seventh-graders, randomly selects 6 of them and assigns them to method A, and assigns the remaining 7 to method B. The resulting data and the necessary operations upon them are given in Table 12-8. His H_a calls for a lower-tail test. From Table B-9 of Appendix B he finds that when $n = 6$ and $m = 7$, the lower-tail critical value of W_n for the largest possible $\alpha \le .01$ is 25. His obtained value of W_n is 24 which falls within the rejection region. So he rejects H_0 and accepts the H_a that learning times tend to be shorter under Method A than under Method B.

Note: Although the actual operations performed in conducting the two tests are quite different, the Wilcoxon Rank-Sum Test is mathematically equivalent to the well-known Mann-Whitney U-Test, i.e., if the two tests are

Table 12-8

Data and Calculations for Psychologist's Test for Faster
Learning Under a Certain Teaching Method

Teaching method A		Teaching method B	
Learning time X_i	Its rank in combined sample R_i	Learning time Y_j	Its rank in combined sample R_j
207	8	275	12
149	3	196	7
152	4	229	10
174	6	344	13
78	1	165	5
126	2	257	11
		213	9
$W_n = \Sigma R_i = 24$			

applied to the same data to test H_0 against the same H_a at the same α-level, they will call for the same action and their test statistics will have the same probability level.

12-3-6 Friedman's Test for Nonidentical Treatment Effects

This test uses matched observations and tests whether or not all of several different treatments have identical effects. H_0 is that every treatment has exactly the same effect so that all treatment populations have identical distributions. H_a is that treatment effects are not all the same so that at least one treatment population's distribution differs in some way from another's. However, the test is highly sensitive to forms of nonidentity such that observations obtained under one treatment (or drawn randomly from one treatment population) tend to be larger, or tend to be smaller, than those obtained under another. This means that, in general, the test should be considerably more sensitive to differences in location indices than to differences in dispersion indices of the treatment populations. The test should therefore be regarded as a test of identical treatment effects that is sensitive to those nonidentical effects in which observations under one treatment tend to exceed those under another.

From each of R different sets of C matched units (the units being matched with other units in the *same* set, not necessarily with units in other sets) the C units in a set are randomly assigned, one each, to C different treatments. The treatments are then administered and a fine-grained (continuous-scale) measurement sensitive to the influence of treatments is then made upon each unit. It is required that the measurements be fine enough so that no units in the same matched set have measurements that are exactly equal and therefore tied. Alternatively, if there are no carry-over effects (i.e., if exposure to one treatment does not influence the effect of a subsequent treatment) or if carry-over effects can be successfully counterbalanced, each of R single units may be given all C treatments, the effect of each treatment being measured immediately after the treatment is administered and before the next treatment is applied.

In either case, the data may be recorded in a table having C columns representing the C different treatments and R rows representing the R different sets of matched observations so that the C different cell entries in any given row are matched observations obtained under different treatments (see Table 12-9). The C matched observations in each row are then ranked from 1 (smallest) to C (largest) and replaced by their ranks (see Table 12-9, bottom). Let T_j be the sum of the ranks in the jth column and let $\bar{T}$ be the average of such column sums. Each row contains the integers from 1 to C, so the average rank in any row is $(C + 1)/2$; and there are R rows, so the average

Table 12-9

Tabulation of Observations (0_{ij}) and Their Ranks (R_{ij}) in Friedman Test

Unit or Set of Matched Units	Treatment 1	2		j		C
1	0_{11}	0_{12}	$\cdots$	0_{1j}	$\cdots$	0_{1C}
2	0_{21}	0_{22}	$\cdots$	0_{2j}	$\cdots$	0_{2C}
$\vdots$	$\vdots$	$\vdots$		$\vdots$		$\vdots$
i	0_{i1}	0_{i2}	$\cdots$	0_{ij}	$\cdots$	0_{iC}
$\vdots$	$\vdots$	$\vdots$		$\vdots$		$\vdots$
R	0_{R1}	0_{R2}	$\cdots$	0_{Rj}	$\cdots$	0_{RC}

Unit or Set of Matched Units	Treatment 1	2		j		C
1	R_{11}	R_{12}	$\cdots$	R_{1j}	$\cdots$	R_{1C}
2	R_{21}	R_{22}	$\cdots$	R_{2j}	$\cdots$	R_{2C}
$\vdots$	$\vdots$	$\vdots$		$\vdots$		$\vdots$
i	R_{i1}	R_{i2}	$\cdots$	R_{ij}	$\cdots$	R_{iC}
$\vdots$	$\vdots$	$\vdots$		$\vdots$		$\vdots$
R	R_{R1}	R_{R2}	$\cdots$	R_{Rj}	$\cdots$	R_{RC}

$$T_j = \sum_{i=1}^{R} R_{ij}$$

column sum $\bar{T}$ is $R(C + 1)/2$. The test statistic S is the sum of the squared deviations of column sums from their average value:

$$S = \sum_{j=1}^{C} (T_j - \bar{T})^2 = \sum_{j=1}^{C} \left[T_j - \frac{R(C + 1)}{2} \right]^2 \qquad (12\text{-}4)$$

Now consider the probability distribution of the test statistic. If no treatments had been administered but the measurements had been made just the same, the observations, and therefore the ranks, in each row would be randomly associated with "treatments." (This is so because the matched units were randomly assigned to treatments; or in the case where one unit receives all treatments, because if there are no treatment effects and no sequential effects, fluctuations in measurements must be presumed random.) Therefore, if treatments are administered and have no effect, or have identical effects such as increasing every measurement by the same amount, the observations, and their ranks, within each row would still be randomly associated with treatments. Therefore, if the H_0 of identical treatment effects is true, every one of the $C!$ different possible assignments of the C ranks in a single row to the C different treatments (i.e., every one of the $C!$ permutations of ranks within the cells of a single row) was equally likely to be the actually obtained set. This is true for each of the R different rows, so there are $(C!)^R$ different possible assignments of within-row ranks to treatments in the entire table, and if H_0 is true, each of them had exactly the same probability $1/(C!)^R$ of becoming the actually obtained table of ranks. Therefore, the probability distribution of S when H_0 is true is obtained by assigning within-row ranks to treatments in all $(C!)^R$ different possible ways, calculating S for each way, and giving the S-value associated with each way the same probability $1/(C!)^R$. Thus in the case where there are $C = 2$ columns and $R = 3$ rows, we have $(C!)^R = 2^3 = 8$ different possibilities. These possibilities and the resulting probability distribution of S are given in Table 12-10.

On the other hand, if one treatment tends to make observations larger than do other treatments, its T_j will tend to be larger than $\bar{T}$; if it tends to make observations smaller than do other treatments, its T_j will tend to be smaller than $\bar{T}$; and in *either* case, the squared difference $(T_j - \bar{T})^2$ will tend to be *large*, as will the test statistic S. Therefore, *roughly speaking*, to test the H_0 that treatment effects are identical against the H_a that they are not and that some treatments systematically tend to make observations smaller than others or to make them larger than others, use an upper-tail rejection region. No other rejection regions should ordinarily be used with this test. The upper-tail rejection region automatically "tests" for treatment-effect deviations in either or both directions. If the user wishes to generalize his results beyond the sample of R single units or beyond the R sets of C matched units to the population or populations from which they came, then they must have been random samples from those populations.

Table 12-10

Possibilities and Probabilities for Friedman's Test Statistic S When $C = 2$ and $R = 3$

(a) The eight possible tables in which each row contains the ranks 1 and 2, and the resulting values of T_j and S

Table 1		Table 2		Table 3		Table 4		Table 5		Table 6		Table 7		Table 8	
1	2	1	2	1	2	1	2	2	1	2	1	2	1	2	1
1	2	1	2	2	1	2	1	1	2	1	2	2	1	2	1
1	2	2	1	1	2	2	1	1	2	2	1	1	2	2	1

T_j															
3	6	4	5	4	5	5	4	4	5	5	4	5	4	6	3

$S = \Sigma(T_j - 4.5)^2$							
4.5	.5	.5	.5	.5	.5	.5	4.5

(b) Probability distribution of S if H_0 of identical treatment effects is true

S	$P(S)$
.5	6/8
4.5	2/8

When $C = 2$, the Friedman Test is equivalent to a two-tail binomial test of the H_0 that, for every row in (upper) Table 12-9,

$$P(O_{i1} < O_{i2}) = P(O_{i1} > O_{i2}) = .5,$$

using as the test statistic r the number of observation pairs in which $O_{i1} < O_{i2}$. If H_0 is true, r is binomially distributed with $p = .5$ and n equal to the number of observation pairs, i.e., to the number of rows in the table. So tables of probabilities for the Friedman test statistic generally do not include the case $C = 2$, and one must perform a two-tail binomial test instead. In other cases not included in the tables for the Friedman Test one must resort to an approximation. Multiplying the Friedman test statistic S by the constant $12/[RC(C + 1)]$ one obtains a statistic

$$\chi_r^2 = \frac{12S}{RC(C + 1)} \tag{12-5}$$

which has approximately a "chi-square with $C - 1$ degrees of freedom" distribution if H_0 is true. Therefore, one need only calculate χ_r^2, refer it to the distribution of chi-square with $C - 1$ degrees of freedom, and reject H_0 if χ_r^2 equals or exceeds the upper-tail critical value of chi-square with $C - 1$ degrees of freedom for an upper-tail test at the chosen level of significance α.

Example: An agricultural research worker wishes to test (at the largest possible $\alpha \leq .05$) whether or not there are any differences in the effects of three different fertilizers upon the yield of a certain strain of corn. He knows that climate and soil-type can be influential factors, and he does not want his results to be highly qualified by these conditions. Therefore, he randomly selects eight counties and plants the corn in three equal-sized adjacent plots in each county. He randomly assigns the three different fertilizers to the three plots in each county. When the corn is harvested, he obtains the yields, and makes the calculations, shown in Table 12-11. Beside each yield, in parentheses, is its within-row rank. From Table B-10 of Appendix B he finds that for $C = 3$ and $R = 8$ and the largest possible $\alpha \leq .05$, the critical value of S is approximately 49; so his rejection region consists of S-values ≥ 49. His obtained S-value 62 falls in the rejection region, so he rejects H_0 and concludes that some of the fertilizers have different effects than others.

Note: Had I, II, and III been different strains of corn rather than different fertilizer-treatments, then, roughly speaking, the test would have tested whether or not the three strain-*populations* have identical distributions when climate and soil-type are "held constant." Thus, the test can be used either to test for nonidentical population distributions (where "population" is rather loosely defined) or for nonidentical treatment effects. In either case, the test permits the values of the "controlled" conditions to vary from row to row and therefore permits highly representative sampling and reasonably convincing generalizations.

Table 12-11

Data and Calculations for Agriculturist's Test for
Differences in Fertilizer Effects

	Fertilizer					
	I		II		III	
County	Yield	(Rank)	Yield	(Rank)	Yield	(Rank)
A	49.7	(3)	44.5	(1)	46.1	(2)
B	65.3	(1)	70.2	(2)	71.4	(3)
C	19.4	(2)	16.7	(1)	25.5	(3)
D	41.5	(3)	30.6	(1)	37.2	(2)
E	25.9	(1)	26.0	(2)	28.7	(3)
F	27.3	(2)	26.0	(1)	31.4	(3)
G	55.9	(2)	54.6	(1)	62.3	(3)
H	29.0	(1)	38.5	(2)	51.9	(3)

T_j	15	11	22
$T_j - 16$	-1	-5	6
$(T_j - 16)^2$	1	25	36

$$S = \sum_{j=1}^{3} (T_j - 16)^2 = 1 + 25 + 36 = 62$$

12-3-7 The Kruskal-Wallis Test for Nonidentical
Population Distributions

This test uses unmatched units to test whether or not all of several different continuously distributed populations have identical distributions. If the
populations are those associated with treatment effects, it is a test of whether
or not all of the treatments have identical effects. In the case where there are
only two populations, the test is equivalent to the two-tail Wilcoxon Rank-
Sum Test; so, in a sense, it is a generalization of that test to the case where
there are several populations.

The null hypothesis is that each of C different continuously distributed
populations has exactly the same distribution of values. The alternative
hypothesis, therefore, is that the C distributions are not all exactly the same.
However, the test is highly sensitive to differences in distributions such that
units in one population tend to be higher, or tend to be lower, in value than
those in another. So if p_{jk} is the probability that a randomly drawn observation from the jth population will exceed a randomly drawn observation from
the kth population, the test is highly sensitive to differences iñ the distributions of the jth and kth populations that cause p_{jk} to differ from 1/2. This
means that, in general, the test should be considerably more sensitive to
differences in indices of location than of dispersion for the distributions
involved. The test should therefore be regarded as a test of identical distribu-

tions that is sensitive to differences causing observations from one population to tend to exceed those from another.

A random sample is drawn from each of the C different populations. Let n_j be the number of observations in the sample from the jth population, and let

$$n = \sum_{j=1}^{C} n_j$$

be the total number of observations in all C samples. The n observations in the combined sample are now ranked in order of increasing size from 1 (smallest) to n (largest). Let $\bar{R}_j$ be the average rank assigned to the n_j observations in the jth sample. (So if R_{ij} is the rank assigned to the ith observation in the jth sample,

$$\bar{R}_j = \frac{\sum_{i=1}^{n_j} R_{ij}}{n_j} \Big)$$

The average rank assigned to all n observations in the combined sample is $(n + 1)/2$. The test statistic is

$$H = \frac{12}{n(n + 1)} \sum_{j=1}^{C} n_j \left(\bar{R}_j - \frac{n + 1}{2}\right)^2 \tag{12-6}$$

So the test statistic is a constant times the sum of the weighted squared deviations of the average rank in an individual sample from the average rank in the combined sample, the weights being the number of observations in the individual samples.

Now consider the distribution of the test statistic. If H_0 is true, i.e., if all C populations have identical distributions, then any given one of the n sample observations was equally likely, before sampling, to be the value obtained from measuring any one of the n sample units. Therefore, every one of the $n!$ different possible assignments of the ranks from 1 to n to the n different units had the same probability $1/n!$ of becoming the actually obtained association of ranks with units. Or, more simply, (since we are not interested in the distribution of ranks to units *within* a sample) every one of the

$$\frac{n!}{n_1! n_2! \cdots n_j! \cdots n_C!}$$

different possible ways of apportioning the n ranks into n_1 ranks to go to the first sample, n_2 to go to the second, etc., had the same probability of becoming the actually obtained allocation of the n ranks to the C samples. Therefore, the distribution of H when H_0 is true is obtained by apportioning the integers from 1 to n into predesignated groups containing $n_1, n_2, \ldots, n_C$ integers in all of the $n!/(n_1! n_2! \cdots n_C!)$ different possible ways, calculating a value of the test statistic H for each way, and giving each way a probability of

$$\frac{1}{\dfrac{n!}{n_1! n_2! \cdots n_C!}}$$

Thus in the case where $C = 2$, $n_1 = 3$, and $n_2 = 2$, there are $5!/(3!\,2!) = 10$ different possible assignments of ranks to groups, each having the same probability $1/10$, under H_0. They are shown in Table 12-12 together with the corresponding probability distribution of H.

On the other hand, if observations from one population either tend to be larger or tend to be smaller than those from others, then for samples from that population

$$\left(\bar{R}_j - \frac{n+1}{2}\right)^2$$

will tend to be large. The extremity of this $\bar{R}_j$ will tend to induce a compensating and oppositely extreme $\bar{R}_j$ and therefore to increase the sizes of the

$$\left(\bar{R}_j - \frac{n+1}{2}\right)^2$$

for the samples from other populations; so H will also tend to be large. Thus, if j and k are two of the populations whose variables are X_j and X_k, respectively, we would expect H to take large values more often when $P(X_j > X_k) \neq .5$ than when the probability is exactly $.5$, which it must be if all populations are identically distributed. Therefore, *roughly speaking*, to test the H_0 that for all possible pairs of populations, i.e., for all possible pairs of values for the subscripts j and k, $P(X_j > X_k) = .5$ against the H_a that for at least one pair $P(X_j > X_k) \neq .5$, use an upper-tail rejection region. No other rejection region should ordinarily be used with this test. For reasons analogous to those given for Wilcoxon's Rank-Sum Test, this test is sometimes regarded as a rough test for unequal medians.

Unfortunately, adequate tables giving the distribution (or critical values) of H, when H_0 is true, have not been published. However, the cumulative distribution of H, when H_0 is true, is approximately the same as that of a statistic called "chi-square with $C - 1$ degrees of freedom," the approximation generally improving as sample sizes increase. A table of critical values of chi-square (χ^2) for various degrees of freedom ν is given by Table B-13 in Appendix B for an upper-tail test at significance level $\alpha = Q$. One need only refer H to this table, using the appropriate α and $\nu = C - 1$, and reject H_0 if H exceeds the critical value listed. In the case where $C = 2$, the Kruskal-Wallis Test is equivalent to a two-tail Wilcoxon Rank-Sum Test. In this case, therefore, one need only perform the latter test at the same α level that one would have used for the Kruskal-Wallis Test.

Example: A mechanical engineer wishes to test (at the $\alpha = .05$ level of significance) the H_0 that the breaking strengths of 1-inch cubes of three different types of steel are identically distributed against the H_a that they are not and that some types tend to have different breaking strengths from others. He obtains random samples of 5 cubes of Type A, 6 of Type B, and 7 of Type

(a) The ten possible tables corresponding to different combinations of three ranks in Column 1 and two ranks in Column 2, and the calculation of the associated values of H

	Table 1		Table 2		Table 3		Table 4		Table 5	
	1	4	1	3	1	3	1	2	1	2
	2	5	2	5	2	4	3	5	3	4
	3		4		5		4		5	
$\bar{R}_j$	2	$4\frac{1}{2}$	$2\frac{1}{3}$	4	$2\frac{2}{3}$	$3\frac{1}{2}$	$2\frac{2}{3}$	$3\frac{1}{2}$	3	3
$\bar{R}_j - 3$	-1	$1\frac{1}{2}$	$-\frac{2}{3}$	1	$-\frac{1}{3}$	$\frac{1}{2}$	$-\frac{1}{3}$	$\frac{1}{2}$	0	0
$(\bar{R}_j - 3)^2$	1	$2\frac{1}{4}$	$\frac{4}{9}$	1	$\frac{1}{9}$	$\frac{1}{4}$	$\frac{1}{9}$	$\frac{1}{4}$	0	0
$n_j(\bar{R}_j - 3)^2$	3	$4\frac{1}{2}$	$\frac{4}{3}$	2	$\frac{1}{3}$	$\frac{1}{2}$	$\frac{1}{3}$	$\frac{1}{2}$	0	0
$\Sigma\, n_j(\bar{R}_j - 3)^2$	$7\frac{1}{2}$		$3\frac{1}{3}$		$\frac{5}{6}$		$\frac{5}{6}$		0	
$H = .4\,\Sigma\, n_j(\bar{R}_j - 3)^2$	3		$\frac{4}{3}$		$\frac{1}{3}$		$\frac{1}{3}$		0	

	Table 6		Table 7		Table 8		Table 9		Table 10	
	1	2	2	1	2	1	2	1	3	1
	4	3	3	5	3	4	4	3	4	2
	5		4		5		5		5	
$\bar{R}_j$	$3\frac{1}{3}$	$2\frac{1}{2}$	3	3	$3\frac{1}{3}$	$2\frac{1}{2}$	$3\frac{2}{3}$	2	4	$1\frac{1}{2}$
$\bar{R}_j - 3$	$\frac{1}{3}$	$-\frac{1}{2}$	0	0	$\frac{1}{3}$	$-\frac{1}{2}$	$\frac{2}{3}$	-1	1	$-1\frac{1}{2}$
$(\bar{R}_j - 3)^2$	$\frac{1}{9}$	$\frac{1}{4}$	0	0	$\frac{1}{9}$	$\frac{1}{4}$	$\frac{4}{9}$	1	1	$2\frac{1}{4}$
$n_j(\bar{R}_j - 3)^2$	$\frac{1}{3}$	$\frac{1}{2}$	0	0	$\frac{1}{3}$	$\frac{4}{3}$	$\frac{1}{2}$	2	3	$4\frac{1}{2}$
$\Sigma\, n_j(\bar{R}_j - 3)^2$	$\frac{5}{6}$		0		$\frac{5}{6}$		$3\frac{1}{3}$		$7\frac{1}{2}$	
$H = .4\,\Sigma\, n_j(\bar{R}_j - 3)^2$	$\frac{1}{3}$		0		$\frac{1}{3}$		$\frac{4}{3}$		3	

(b) Probability distribution of H if null hypothesis of identical population distributions is true

H	$P(H)$
0	.2
1/3	.4
4/3	.2
3	.2

C, and measures the breaking strength of each of them. The strengths, and in parentheses their ranks among the 18 observations, are given in Table 12-13 along with the necessary calculations for the Kruskel-Wallis Test. The obtained value of H is 4.047. For $\alpha = .05$ and $C - 1 = 2$ degrees of freedom, the critical upper-tail value of chi-square is 5.991. The rejection region is therefore H values ≥ 5.991. Since the obtained value of H does not fall in this rejection region, the engineer accepts the H_0 that the distributions of breaking strengths for 1-inch cubes are the same for the three types of steel.

Table 12-13

Data and Calculations for Engineer's Test for Differences
in Breaking Strengths of Steels

Breaking strengths for steel cubes of

	Type A		Type B		Type C	
	Strength	(Rank)	Strength	(Rank)	Strength	(Rank)
	82,285	(13)	80,014	(5)	79,359	(3)
	81,904	(9)	78,490	(1)	82,091	(11)
	81,957	(10)	82,105	(12)	83,145	(16)
	81,112	(6)	79,692	(4)	81,664	(8)
	81,335	(7)	78,665	(2)	85,107	(18)
			82,880	(15)	82,714	(14)
					84,422	(17)

Rank total	45	39	87
$\bar{R}_j$	9	6.5	12.43
$\bar{R}_j - 9.5$	$-.5$	-3	2.93
$(\bar{R}_j - 9.5)^2$	.25	9	8.585
$n_j(\bar{R}_j - 9.5)^2$	1.25	54	60.095

$\Sigma\, n_j(\bar{R}_j - 9.5)^2 = 115.345$

$$H = \frac{12}{n(n+1)} \Sigma\, n_j(\bar{R}_j - 9.5)^2 = \frac{12}{18(19)} \times 115.345 = 4.047$$

12-3-8 Tests of Main Effects and Interactions When There Are Two Causal Variables

Suppose that an experimenter is interested in both the individual and combined effects of two different variables, a "column" variable having C different possible values, called **levels**, and a "block" variable having B different levels. Imagine an experiment in which R different observations, i.e., fine-grained (continuous-scale) measurements, were obtained under each of the BC different combinations of levels of the two variables. Let O_{ijk} be the ith observation obtained under the jth level of the column variable and the kth level of the block variable. And let each O_{ijk} be recorded in the ith row of the jth column of a table—called a block—corresponding to the kth level of the block variable. Now each level of the "column" variable is represented by a column and each level of the "block" variable is represented by a block.

Now consider the population means, i.e., the means that would prevail if an infinite number of observations had been obtained under each combination of levels. Let μ_{jk} be the population mean of all potential observations obtained under the jth level of the column variable and the kth level of the block variable. Let

$$\mu_{j.} = \frac{1}{B} \sum_{k=1}^{B} \mu_{jk}$$

be the grand mean of all μ_{jk}'s associated with the jth level of the column variable. Note that $\mu_{j.}$ is obtained by summing or "collapsing" the μ_{jk}'s over block levels, then dividing by B. If the values of the $\mu_{j.}$'s, i.e., the grand column means, vary with j, i.e., vary from one column to another, we say that the column variable has a **main effect** or that there is a "main effect of columns"; if the $\mu_{j.}$'s all have the same value so that $\mu_{1.} = \mu_{2.} = \cdots = \mu_{C.}$, we say that there is no main effect associated with the column variable. Let

$$\mu_{.k} = \frac{1}{C} \sum_{j=1}^{C} \mu_{jk}$$

be the grand mean of all μ_{jk}'s associated with the kth level of the block variable. If the $\mu_{.k}$'s vary with k, we say that the block variable has a main effect, or that there is a "main effect of blocks," otherwise not. And if for any given pair of values of k, such as $k = a$ and $k = b$, the difference $\mu_{ja} - \mu_{jb}$ has the same value no matter what j is, we say that there is **no interaction** between levels of the column and block variables; if this is not always the case, we say that there *is* an **interaction**.

These points can be illustrated graphically as follows. Suppose that the μ_{jk}'s were plotted on a graph whose vertical axis gave their values and whose horizontal axis gave the levels, or values, of the column variable, the points corresponding to a given level of the block variable being connected by straight lines. Then we would have a family of "curves," one curve for each level of the block variable. Now only if *all* of the curves are parallel to each other is there no interaction, only if the mean of the C plotted points on a curve is the same for all B curves is there no main effect of blocks, and only if the mean of the B plotted points lying one above the other has the same value in all C cases is there no main effect of columns. It should be remembered that this graph represents the sampled *population*, not the sample. Figure 12-2 illustrates the eight different general types of "population" graph, letting C, B, and I stand for "column variable has main effect," "block variable has main effect," and "column and block variables interact," and

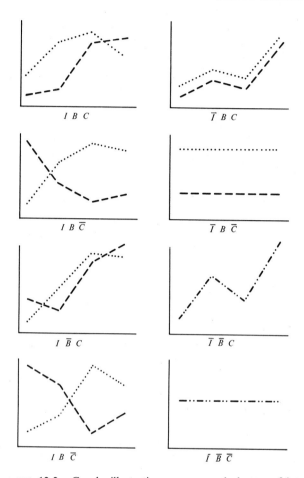

FIG. 12-2.   Graphs illustrating presence and absence of interactions,
main effect of columns, and main effect of blocks.

letting $\bar{C}$, $\bar{B}$, and $\bar{I}$ stand for the negation of these categories, i.e., no column
main effect, no block main effect, and no interaction.

Now suppose that a *finite* number R of observations O_{ijk} had been obtained, in an actual experiment, under each of the BC possible combinations
of levels of the two variables. Let

$$X_{ij} = \sum_{k=1}^{B} O_{ijk} \quad \text{and} \quad Y_{ik} = \sum_{j=1}^{C} O_{ijk}$$

so the former are obtained by collapsing or summing observations over
blocks and the latter, over columns. And let $Z_{ij} = O_{ija} - O_{ijb}$ in the case
where there are just two blocks, a and b. Then it follows from our previous
definitions that, using whichever test is appropriate, the Friedman Test if the
observations to which they are applied are matched and the Kruskal-Wallis

Test if they are not, (a) we can roughly test the $H_0: \mu_1. = \mu_2. = \cdots = \mu_C.$ that there is no main effect of columns by applying the test to the X_{ij}'s, (b) we can roughly test the $H_0: \mu_{.1} = \mu_{.2} = \cdots = \mu_{.B}$ that there is no main effect of blocks by applying the test to the Y_{ik}'s (c) in the case where there are just two blocks, we can roughly test the

$$H_0: \mu_{1a} - \mu_{1b} = \mu_{2a} - \mu_{2b} = \cdots = \mu_{Ca} - \mu_{Cb}$$

that there is no interaction by applying the test to the Z_{ij}'s. Of course, what we are actually testing is the H_0 that all of a set of populations are identically distributed, against the H_a that they are not, the populations being those whose means are: in case (a) the $\mu_j.$'s, in case (b) the $\mu_{.k}$'s, and in case (c) the $\mu_{ja} - \mu_{jb}$'s. If the H_0 of identical population distributions is true, the populations *must* have equal means; if it is false, they are highly *likely* to have *un*equal means. Whatever requirements the test makes about random sampling must, of course, have been met in designing the experiment.

These tests are inexact if we use the conventional definitions of main effects and interactions, which define them in terms of population means. They are exact tests if we substitute the following definitions. Let an infinite number of observations be taken under each of the BC conditions so that we are dealing with a population rather than a sample. Then if for each of the C different values of j the X_{ij}'s are identically distributed, there is no main effect of columns; otherwise there is a main effect. If for each of the B different values of k the Y_{ik}'s are identically distributed, there is no main effect of blocks; otherwise there is a main effect. And in the case where there are just two blocks, if for each of the C different values of j the Z_{ij}'s are identically distributed, there is no interaction; otherwise there is an interaction.

Example: A biologist is interested in testing (at the .05 level) the effects of time-of-weaning and subsequent diet upon the weight of 100-day-old rats. From each of ten litters of newborn rats he randomly selects four rats. He randomly divides the ten sets of four rats each into two groups of five sets each. One group is randomly assigned to Block 1 (early weaning) and the other group is assigned to Block 2 (late weaning). Within each group and each set, the four littermates are randomly assigned, one to each of four columns I, II, III, and IV, representing four different diets. The experiment is conducted, and the weight-data obtained when the rats reached the age of 100 days are given in Table 12-14, as are the necessary computational steps for testing the effects. The data collapsed over blocks are matched within rows, as are the Z_{ij}'s, so the Friedman Test is appropriate. The data collapsed over columns are not matched within rows, so the Kruskal-Wallis Test is called for, or, actually, the Wilcoxon Rank-Sum Test, since there are only two categories. For the Friedman Test on data collapsed over blocks we obtain $S = 53$ which does not fall in the rejection region for $\alpha = .05$ when $R = 5$ and $C = 4$; so we cannot conclude that there is a main effect for the

Table 12-14

Data and Calculations for Biologist's Test of Effects of Time-of-Weaning and Diet Upon Weight of Rats

	Block 1 (Early weaning) Diet					Block 2 (Late weaning) Diet			
Litter	I	II	III	IV	Litter	I	II	III	IV
1	72	67	71	69	6	54	60	63	60
2	43	43	51	41	7	38	45	50	48
3	50	47	52	40	8	52	51	58	49
4	52	60	50	50	9	37	42	40	43
5	61	59	58	56	10	43	46	48	45

Data collapsed over blocks

$$\Sigma_{k=1}^{2} O_{ijk} = X_{ij}$$

Friedman test on data collapsed over blocks

Ranks of X_{ij}'s

Litter pairs	Diet I	II	III	IV		I	II	III	IV
1, 6	126	127	134	129		1	2	4	3
2, 7	81	88	101	89		1	2	4	3
3, 8	102	98	110	89		3	2	4	1
4, 9	89	102	90	93		1	4	2	3
5, 10	104	105	106	101		2	3	4	1

$$\begin{array}{lcccc}
T_j & 8 & 13 & 18 & 11 \\
T_j - 12.5 & -4.5 & .5 & 5.5 & -1.5 \\
(T_j - 12.5)^2 & 20.25 & .25 & 30.25 & 2.25 \\
\end{array}$$

$$S = \Sigma(T_j - 12.5)^2 = 53$$

Data collapsed over columns

$$\Sigma_{j=1}^{C} O_{ijk} = Y_{ik}$$

Kruskal-Wallis (Wilcoxon) test on data collapsed over columns

Ranks of Y_{ik}'s

Litter	Block 1	Litter	Block 2		Block 1	Block 2
1	279	6	237		10	9
2	178	7	181		2	3
3	189	8	210		5	6
4	212	9	162		7	1
5	234	10	182		8	4

Rank-Sum   32   23 $W_n = 23$

Block 1 minus block 2 difference scores

$$Z_{ij} = O_{ij1} - O_{ij2}$$

Friedman test on Z_{ij}'s

Ranks of Z_{ij}'s

Litter pairs	I	II	III	IV		I	II	III	IV
1, 6	18	7	8	9		4	1	2	3
2, 7	5	-2	1	-7		4	2	3	1
3, 8	-2	-4	-6	-9		4	3	2	1
4, 9	15	18	10	7		3	4	2	1
5, 10	18	13	10	11		4	3	1	2

$$\begin{array}{lcccc}
T_j & 19 & 13 & 10 & 8 \\
T_j - 12.5 & 6.5 & .5 & -2.5 & -4.5 \\
(T_j - 12.5)^2 & 42.25 & .25 & 6.25 & 20.25 \\
\end{array}$$

$$S = 69$$

column variable, diet. For the Wilcoxon Rank-Sum Test on data collapsed over columns we obtain $W_n = 23$ which does not fall in the rejection region for a two-tail test at $\alpha = .05$ when $n = 5, m = 5$; so we cannot conclude that there is a main effect for the block variable, time-of-weaning. For the Friedman Test on the difference scores between corresponding observations in the two blocks we obtain $S = 69$ which falls in the rejection region for $\alpha = .05$ when $R = 5$ and $C = 4$; so we reject H_0 in favor of the H_a that there is an interaction between the block and column variables.

Note: The test for interaction is not limited to the case in which there are only two blocks. If there are $B > 2$ blocks, one forms $B - 1$ tables of difference scores, one for the $O_{ij2} - O_{ij1}$ difference scores, one for the

$$O_{ij3} - \frac{O_{ij1} + O_{ij2}}{2}$$

difference scores, one (if $B \geq 4$) for the

$$O_{ij4} - \frac{O_{ij1} + O_{ij2} + O_{ij3}}{3}$$

difference scores, and, in general, a table of

$$O_{ijk} - \frac{\sum_{m=1}^{k-1} O_{ijm}}{k - 1}$$

difference scores for each possible value of $k \geq 2$. Thus if there were three blocks, four columns, and two rows as shown in the top tier of matrices in Table 12-15, one would compute the three matrices in the second tier retaining only the first and last. One performs the appropriate test (Friedman or Kruskal-Wallis) upon each of the $B - 1$ tables of difference scores, obtaining a value of chi-square for each such table (by converting S to χ_r^2 if the test used was a Friedman Test). These $B - 1$ individual chi-square values are then summed. This sum is the test statistic. It has approximately a "chi-square with $(B - 1)(C - 1)$ degrees of freedom" distribution if H_0 is true.

12-3-9   Treatment of Tied Values

In assigning ranks to observations a problem may arise when two or more observations have the same value and therefore appear to be entitled to the same rank. Similarly, in conducting one of the binomial tests in which each observation must be categorized as "above" or "below" a certain value, a problem arises when the observation is neither above nor below the comparison-value but rather is equal to it. Such ambiguous data will occasionally occur even if the sampled populations are continuously distributed because observations are measurements and, since measurements can be made only to a finite number of decimal places, measurements are always discretely distributed. If the measurements of continuously distributed values are

Table 12-15

Illustration of Test for Interaction When $B = 3$

Block 1   (O_{ij1}'s)   Block 2   (O_{ij2}'s)   Block 3   (O_{ij3}'s)

2	3	7	5
0	2	4	8

7	5	6	5
4	8	1	0

6	0	5	2
3	1	7	9

$O_{ij2} - O_{ij1}$ $\dfrac{O_{ij2} + O_{ij1}}{2}$ $O_{ij3} - \dfrac{O_{ij2} + O_{ij1}}{2}$

5	2	-1	0
4	6	-3	-8

4.5	4	6.5	5
2	5	2.5	4

1.5	-4	-1.5	-3
1	-4	4.5	5

Within-Row ranks of
$O_{ij2} - O_{ij1}$

4	3	1	2
3	4	2	1

Within-Row ranks of
$O_{ij3} - \dfrac{O_{ij2} + O_{ij1}}{2}$

4	1	3	2
2	1	3	4

T_j	7	7	3	3	6	2	6	6
$(T_j - 5)^2$	4	4	4	4	1	9	1	1
$S = \sum_{j=1}^{4} (T_j - 5)^2$	16				12			
$\chi_r^2 = \dfrac{12S}{RC(C+1)}$	4.8				3.6			

$$\Sigma \chi_r^2 = 8.4$$

Degrees of freedom = $(B - 1)(C - 1) = (2)(3) = 6$
$P(\chi^2$ with 6 degrees of freedom $\geqslant 12.6) = .05$
$P(\chi^2$ with 6 degrees of freedom $\geqslant 8.4) > .05$
So, accept H_0 of no interaction

sufficiently fine-grained, ambiguous data will be rare; if the measurements are gross, the ambiguities are likely to be frequent.

The mere existence of equal observations in the collected data does not constitute a problem. There is no problem if the tied observations are not compared with each other in the binomial tests or if in rank tests they belong to sets of observations that are ranked separately. And even if they belong to a pooled set receiving a single ranking, there is no problem if arbitrarily assigning the tied-for ranks to the tied observations cannot affect the value of the test statistic. For example, if in conducting a Wilcoxon Rank-Sum Test, we obtain the data in Table 12-16a, there are tied observations, but all arbitrary assignments of the tied-for ranks to them produce the same value of W_n, and there is therefore no real difficulty.

Ties that create a problem will be called critical ties. The safest method of

Table 12-16

Illustrative Data Relevant to Treatment of Ties

(a) Innocuous ties in Wilcoxon Rank-Sum test

Data		Ranks (Assignment 1)		Ranks (Assignment 2)	
X	Y	X	Y	X	Y
14.2	11.6	4	3	4	2
17.4	11.6	5	2	5	3
	9.4		1		1
		$W_n = 9$		$W_n = 9$	

(b) Treatment of critical ties in Wilcoxon Rank-Sum test

Data		Approved method				Midrank method	
		First ranking (maximizing W_n)		Second ranking (minimizing W_n)			
X	Y	X	Y	X	Y	X	Y
1.1	1.1	2	1	1	3	2	2
1.1	2.0	3	4	2	4	2	4
3.7	3.7	6	5	5	6	5.5	5.5
4.4	4.1	8	7	8	7	8.5	7
4.4	4.5	9	10	9	10	8.5	10
	5.2		11		11		11
		$W_n = 28$		$W_n = 25$		$W_n = 26.5$	

(c) Treatment of critical ties in test for median difference

Data		Difference scores	"Signs" of difference scores	
X	Y	$D = X - Y$	maximizing r	minimizing r
11.2	11.2	0	−	+
12.7	11.5	1.2	+	+
13.4	12.2	1.2	+	+
17.9	15.6	2.3	+	+
21.5	22.4	− .9	−	−
26.7	26.7	0	−	+
28.3	25.9	2.4	+	+
			$r = 3$	$r = 1$

dealing with the problem of critical ties is to perform the test twice, once assigning tied-for ranks to tied observations in such a way as to maximize the value of the test statistic, once in such a way as to minimize it. If H_0 is accepted in both cases or rejected in both cases, there is no problem. If H_0 is accepted in one case and rejected in the other, the test outcome is ambiguous, and the best solution is simply to take more data until the ambiguity is resolved, i.e., until both tests call for the same action. Thus, if the data for the Rank-Sum Test were as given in Table 12-16b, one would rank the observations twice, as shown in that table, obtaining a different value of W_n for each ranking. Both values of W_n lie outside the rejection region corresponding to a lower-tail test at $\alpha = .05$, so if such a test had been intended, there would be no problem—H_0 would be accepted.

A less safe method that is distinctly makeshift but often more convenient is to give all observations having the same value the average of the set of ranks for which they are collectively tied (i.e., the average of the ranks they would receive if measured to enough decimal places to reveal differences). Following this *midrank* method, the previous data would have been ranked as shown at the extreme right in Table 12-16b, after which the test is conducted in the usual way. The method has a number of flaws. Perhaps the easiest to see is that if the ties are due to imprecision of measurement the obtained value of the test statistic is unlikely to coincide with the value that would have been obtained if measurement had been precise enough to eliminate ties. So the value of the test statistic obtained by the midrank method is likely to be the wrong value, and if it is the test is only approximate.

Most of the binomial tests are based upon the fact that in a continuously distributed population half the units have values less than the population median. However, while this is true not only for continuously distributed populations but indeed for any population in which the proportion of units having values equal to the median is zero, it is not true in many other cases. For example, in a population in which 70% of the units have the value 15, 20% have the value 16, and 10% have the value 17, the median is 15, but the proportion of units having values below it is zero rather than .5. And a large-sample binomial test of the true H_0 that $M \leq 15$ against the false H_a that $M > 15$ would be certain to reject the true H_0. It is important therefore that such binomial tests be confined to populations in which a negligible proportion of units "lie on the median." Another way of saying this is that the values of population units should belong to one of two categories "above the median" or "below the median" and that there should be no third category "on the median."

Occasionally, however, even if the population is continuous, some observations will fall on the hypothesized median M_0, or some difference scores will equal the hypothesized difference-score population median of zero, due to imprecision of measurement. In such cases, the safest procedure is again to perform the test twice, once counting all observations that are equal to M_0 as $< M_0$, once counting them as $> M_0$ (or, if the hypothesized median of the difference-score population is zero, once counting all zero difference scores as < 0, once as > 0), thus maximizing the test statistic in one case and minimizing it in the other. If H_0 is accepted in both cases or rejected in both cases, there is no problem. If it is accepted in one case and rejected in the other, the best procedure is to take more data until both tests call for the same action. Thus, if the data for testing at $\alpha = .05$ the H_0 that the median difference is zero against the H_a that it is not were as given in Table 12-16c, one would perform the test twice. First counting the two zero differences as < 0, i.e., negative, the number r of differences smaller than zero is 3. Then counting them as positive, r is 1. By calculation, or from binomial tables $P(r \leq 1 \text{ or } \geq 6) = .125$ whereas $P(r \leq 3 \text{ or } \geq 4) = 1.00$. Therefore, since both probability levels exceed .05, H_0 is accepted.

Other methods are considerably less safe and can be quite erroneous. One, analogous to the midrank method of dealing with ties, is to count half the ambiguous scores as "above," half as "below," resolving any odd ambiguous score by a flip of a coin. Using this method on the data of Table 12-16c, it turns out that $r = 2$ and $n = 7$. Another method is to drop the ambiguous data from the sample and perform the binomial test on the remaining data. Using this method on the same data, it turns out that $r = 1$ and $n = 5$, and the two-tail probability level of the test statistic is $P(r \leq 1 \text{ or } \geq 4) = .375$.

12-4 Estimation of Population Indices

12-4-1 Estimation Versus Hypothesis Testing

There are two approaches to determining the state of nature when there are many possible states. In hypothesis testing one guesses or *hypothesizes* that θ has a certain value or lies within a certain range of values and then uses sample data to test the tenability of the hypothesis. In *estimation* one takes the more direct approach of simply letting the sample data suggest the value of θ (point estimate) or the range within which its value lies (interval estimate). When samples are small, it is seldom possible to estimate convincingly and within narrow limits what the value of θ *is*; but it is often possible to convincingly reject a badly false H_0. And if the decision maker is interested not so much in the value of θ as in the *consequences* of *actions* called for by different θ-values, "knowing" that θ is very likely not to lie in the range of values associated with a certain action (and its consequence) may be all the information he needs. Thus, when samples are small, hypothesis testing is often a more satisfactory approach than estimation.

To a lesser extent the same thing can be said when samples are of moderate or large size. If interest is centered exclusively upon the consequences of actions, and if the actions fall into a natural dichotomy, which determine H_0 and H_a, then hypothesis testing may often be the superior approach. However, as sample size increases, convincing estimates of the value of θ can be made within narrower and narrower limits. Therefore, if there is any interest in knowing the value of θ for its own sake, or if more than two actions are associated with the different possible θ-values, estimation becomes an increasingly desirable alternative to hypothesis testing as sample size increases. And if interest resides exclusively in the value of θ, rather than in its consequences, estimation is the appropriate approach at any sample size, although it is not likely to be very effective when n is small.

When n is huge, estimation can generally pinpoint the value of θ, whereas $H_0: \theta = \theta_0$ may be a very wrong guess. The two-tail test of such an erroneous H_0 will almost certainly reject it, but knowing that θ is *not* θ_0 does not tell one what θ *is*. Furthermore, if the possible θ-values form a continuum, then there are an infinite number of them. Therefore, if H_0 and H_a are that $\theta = \theta_0$ and $\theta \neq \theta_0$, respectively, where θ_0 is a single one of the possible θ-values guessed to be the true value, H_0 is virtually certain to be false and if n is huge, the test is virtually certain to reject it (indeed since we "know" that it is false we may reject it without bothering to collect data and perform the test). A point estimate of θ is also virtually certain not to be the true value of θ, but point estimates are not claimed to be "true" or "false," as are hypotheses. Rather they are regarded as approximations which may be either "good" (i.e., close) or "bad." Thus hypothesis testing is an inadequate method of determining the point value of θ to a close approximation, whereas estimation is a highly appropriate method of doing so.

12-4-2   Interval Estimation

Suppose that one intends to (a) draw a random sample of n observations from a certain population, then (b) use these observations in a predesignated way to construct an interval of values, then (c) allege that the interval contains the value of a certain population index, such as the median. Before the sample is drawn, the interval is a variable, since the observations in a future random sample are variables. For certain population indices it is possible to calculate the probability P that this variable future interval will contain the constant population index, and, therefore, that the allegation will be true.

Now suppose that the sample is actually drawn, the interval is actually constructed, and the allegation that it contains the index is actually made. The obtained observations in the particular sample drawn are constants, and the upper and lower endpoints of the interval constructed from them are also constants. Therefore, the interval either does or does not contain the constant population index and the allegation is either true or false.

Nevertheless, one's degree of conviction (subjective probability), or confidence, that the value of the population index is contained within the interval, and therefore that the allegation is correct, should be well represented by the prior probability P of the event. From a slightly different point of view, suppose that one drew an infinite number of such samples, calculated the interval from the data in each one, and in each case alleged that the population index was contained within the interval. A proportion P of these allegations would be true and therefore P might well represent his confidence in the truth of any one particular allegation.

The interval obtained in this predesignated way from a particular already-drawn sample is called a ***confidence interval***, its endpoints are called ***confidence limits***, and P (or in percentage terms $100P\%$) is called the ***confidence level*** with which one can state that the constant population index I lies within the obtained interval. The confidence level P is numerically equal to the prior probability that I would lie in the interval constructed from a future sample in the predesignated way. In most books, confidence level is denoted by $1 - \alpha$ rather than by P.

In order to avoid bias, one must decide *before* viewing the observations of the actually obtained sample precisely how the confidence interval is to be constructed—which involves determining what the confidence level is to be.

To summarize, a confidence interval is an interval of values, constructed in a predesignated way from an obtained sample, within which one alleges that a certain population index lies. The allegation is made at a level of confidence that is numerically equal to the probability that the interval constructed in the prescribed way from a *future* sample would contain the index.

12-4-3 Confidence Intervals for Population Percentiles

We already know (see Section 12-1-1) that if Q_p is the $100p$th percentile of a continuously distributed population (so that a proportion p of the population units have values less than Q_p and a proportion $1 - p$ have values greater than Q_p), the probability that exactly r observations in a random sample of n observations will have values less than Q_p is

$$\binom{n}{r}p^r(1 - p)^{n-r}$$

Let $X_1, X_2, \ldots, X_r, X_{r+1}, \ldots, X_n$ represent the values of the n observations in a future random sample arranged in order of increasing size—so that X_r is the value of the rth smallest observation. And let $x_1, x_2, \ldots, x_r, x_{r+1}, \ldots, x_n$ represent the values of the n observations in an actually obtained random sample, arranged in order of increasing size. So the X_i's are variables and the

x_i's are the numerical, constant, values that they take in an actual sample. Now, if the number of sample observations whose values are less than Q_p is exactly r, then it follows that the rth smallest sample observation is $< Q_p$ and that the $r + 1$th smallest is $> Q_p$. So the probability that exactly r observations in a sample of n observations will have values $< Q_p$ is the same as the probability that the value X_r will be $< Q_p$ and the value X_{r+1} will be $> Q_p$, i.e.,

$$P(r \mid n, p) = P(X_r < Q_p < X_{r+1}) = \binom{n}{r} p^r (1 - p)^{n-r}$$

And in general, for any value of i,

$$P(X_i < Q_p < X_{i+1}) = \binom{n}{i} p^i (1 - p)^{n-i}$$

Therefore,

$$P(X_r < Q_p < X_{r+1}) = \binom{n}{r} p^r (1 - p)^{n-r}$$

$$P(X_{r+1} < Q_p < X_{r+2}) = \binom{n}{r+1} p^{r+1} (1 - p)^{n-(r+1)}$$

$$P(X_{r+2} < Q_p < X_{r+3}) = \binom{n}{r+2} p^{r+2} (1 - p)^{n-(r+2)}$$

$$\vdots \qquad\qquad \vdots$$

$$P(X_{s-1} < Q_p < X_s) = \binom{n}{s-1} p^{s-1} (1 - p)^{n-(s-1)}$$

And, since the population is continuous, the probability that Q_p will *equal* one of the X_i's is zero; so we can obtain $P(X_r < Q_p < X_s)$ by summing the above probabilities,

$$P(X_r < Q_p < X_s) = \sum_{i=r}^{s-1} P(X_i < Q_p < X_{i+1}) = \sum_{i=r}^{s-1} \binom{n}{i} p^i (1 - p)^{n-i}$$

(12-7)

Since

$$\sum_{i=r}^{s-1} \binom{n}{i} p^i (1 - p)^{n-i}$$

is the probability that in a future sample of n observations it will be true that $X_r < Q_p < X_s$, it is also the confidence level at which one can allege that $x_r < Q_p < x_s$ in an actually obtained sample (provided that it was decided prior to sampling that the confidence interval was to be based upon the values taken by X_r and X_s in the actually obtained sample).

If the population is continuous, it makes no difference whether one alleges that $x_r < Q_p < x_s$ or that $x_r \leq Q_p \leq x_s$ (i.e., whether or not one includes x_r and x_s in the interval); the confidence level for both allegations is the same.

If the population is discrete, the confidence level for the allegation that $x_r \leq Q_p \leq x_s$ is *at least*

$$\sum_{i=r}^{s-1} \binom{n}{i} p^i (1-p)^{n-i}$$

however, this will not be proved here.

A very important special case is that in which Q_p is the 50th percentile $Q_{.50}$ which, of course, is the median M. The median is a very common index of location, and it is highly convenient to be able to establish confidence limits for its value.

Example: The Air Force is interested in the time elapsed from the onset of a danger signal until a pilot can operate an ejection mechanism. It wishes to obtain an interval estimate for the 75th percentile of the potential population of such time scores; it wishes to have at least the .95 level of confidence that the interval contains the 75th percentile; and it wishes to obtain the interval from a sample of 12 observations. So it has already specified that

$$P(X_r < Q_{.75} < X_s) = \sum_{i=r}^{s-1} P(X_i < Q_{.75} < X_{i+1})$$
$$= \sum_{i=r}^{s-1} \binom{12}{i} .75^i .25^{12-i} \geq .95$$

but before drawing the sample it must also decide what values of r and s to use, among those pairs of r- and s-values that satisfy the above equation and inequality. From binomial tables, or by calculation, it obtains the following data

$$P(-\infty < Q_{.75} < X_1) = P(r = 0 \,|\, n = 12, p = .75) = .0000001$$
$$P(X_1 < Q_{.75} < X_2) = P(r = 1 \,|\, n = 12, p = .75) = .0000021$$
$$P(X_2 < Q_{.75} < X_3) = P(r = 2 \,|\, n = 12, p = .75) = .0000354$$
$$P(X_3 < Q_{.75} < X_4) = P(r = 3 \,|\, n = 12, p = .75) = .0003541$$
$$P(X_4 < Q_{.75} < X_5) = P(r = 4 \,|\, n = 12, p = .75) = .0023898$$
$$P(X_5 < Q_{.75} < X_6) = P(r = 5 \,|\, n = 12, p = .75) = .0114713$$
$$P(X_6 < Q_{.75} < X_7) = P(r = 6 \,|\, n = 12, p = .75) = .0401494$$
$$P(X_7 < Q_{.75} < X_8) = P(r = 7 \,|\, n = 12, p = .75) = .1032415$$
$$P(X_8 < Q_{.75} < X_9) = P(r = 8 \,|\, n = 12, p = .75) = .1935777$$
$$P(X_9 < Q_{.75} < X_{10}) = P(r = 9 \,|\, n = 12, p = .75) = .2581036$$
$$P(X_{10} < Q_{.75} < X_{11}) = P(r = 10 \,|\, n = 12, p = .75) = .2322932$$
$$P(X_{11} < Q_{.75} < X_{12}) = P(r = 11 \,|\, n = 12, p = .75) = .1267054$$
$$P(X_{12} < Q_{.75} < \infty) = P(r = 12 \,|\, n = 12, p = .75) = .0316764$$

In this case the selection of r and s is fairly easy. One cannot use an s of *less* than 12 while meeting the required confidence level, and the upper endpoint of the interval must be either X_{12} or infinity. The latter is undesirable, so one takes $s = 12$. In doing so one determines r as the largest value that satisfies the inequality

$$\sum_{i=r}^{s-1=11} \binom{12}{i}.75^i.25^{12-i} \geq .95$$

By trial and error, that value is found to be $r = 6$. So the confidence interval will be from just above x_6 to just below x_{12} and the exact confidence level will be

$$P(X_6 < Q_{.75} < X_{12}) = \sum_{i=6}^{11} \binom{12}{i}.75^i.25^{12-i} = .9540708$$

A random sample of 12 pilots is now drawn and their "ejection times" are measured. Arranged in order from smallest to largest, they are:

x_1	x_2	x_3	x_4	x_5	x_6	x_7	x_8	x_9	x_{10}	x_{11}	x_{12}
.251	.279	.282	.290	.299	.307	.311	.312	.319	.330	.345	.369

The value of x_6 is .307 and of x_{12} is .369, so one can allege at the .954 level of confidence that the 75th percentile of the population of ejection-time scores lies in the interval between .307 and .369, i.e., that $.307 < Q_{.75} < .369$.

12-4-4 Point Estimation

Generally speaking, the larger the intended sample size, the smaller the expected confidence interval for a fixed level of confidence. Therefore, even if one were to insist upon a confidence level of .99999, as sample size increased the expected values of the upper and lower confidence limits would approach each other until, eventually, at some huge value of n, the interval would have become virtually indistinguishable from a point. And this quasi-point interval would almost certainly contain the population index. This raises the question of whether or not at large n-values a "point estimate" might serve as well as an interval estimate of the population index.

A *point estimate* of a population index I is simply a statistic, calculated from a random sample from the population, whose value is used to represent or estimate the value of I. Often the statistic is the sample analogue of the population index, as when the sample median is used to estimate the population median. Since the estimator is a sample statistic, it has a sampling distribution, and since the sample is random, the statistic's sampling distribution is also its probability distribution. It is highly desirable that the estimator's probability distribution (a) be "centered" in some way upon the estimated population index and (b) have a variance that diminishes fairly rapidly as sample size n increases. That is, it is desirable that the sample estimates "converge," in some reasonable way, upon I as n increases. If the probability distribution of the estimator is "centered" upon I as its *mean* (i.e., if the mean of the probability distribution of the estimator equals I), the estimator is said to be a ***mean-unbiased estimator*** of I. If the probability distribution of the estimator is "centered" upon I as its *median* (i.e., if the median of the probability distribution of the estimator equals I), the estimator is said to be a ***median-unbiased estimator*** of I. If the estimator tends to fall closer and closer to I as n increases, and to converge upon it, it is said to be a ***consistent estimator***.

A number of common sample statistics have desirable properties as estimators. The sample proportion r/n is a mean-unbiased and consistent estimator of the population proportion p. The sample mean $\bar{X}$ is a mean-unbiased and consistent estimator of the population mean μ. And the sample median m is a median-unbiased and consistent estimator of the population median M, provided that the number of units n in the sample is odd and that the value of M equals the value of some unit in the population. (When n is small and even, the median of the probability distribution of the sample median may differ considerably from the population median. In most other cases the qualification is not very important.)

The mean-unbiasedness of r/n and $\bar{X}$ as estimators of p and μ will be proved in a later chapter. We shall now prove the median-unbiasedness of the sample median as an estimator of the population median when n is odd and M corresponds to a unit value.

Proof: Consider a population consisting of N units, whose values when arranged in order of increasing size are $u_1, u_2, \ldots, u_i, \ldots, u_{N-1}, u_N$. And suppose that at least one of these units has the same value as the population median. (This requirement is automatically met when N is odd, in which case

$$u_{(N+1)/2} = M$$

by definition. And if the requirement is met when N is even, it necessitates that

$$u_{N/2} = u_{(N/2)+1} = M)$$

Now suppose that we draw an odd number n of these population units (either with or without replacements), identify them as *sample* units, and arrange *them* in order of increasing size. The middle sample unit (whose value will be the sample median) can come from the left half of the sequence of population units, or from the right half, or, if N is odd, from the exact middle. For every sample in which it comes from the left half, there is a mirror-image sample in which it comes from the right half. (See Table 12-17.) In the former case the sample median will be $\leq M$ and in the latter it will be $\geq M$. And if the middle sample unit comes from the exactly middle population unit, then the sample median is equal to M.

Therefore, *at least* half the possible samples have a median that is $\leq M$ and *at least* half of them have a median that is $\geq M$. Furthermore, *some of*

Table 12-17

Mirror-Image and Non-Mirror Image Samples of Three Units
That Can Be Drawn Without Replacements From a Population of Five Units

(Values of population units are listed in each case. Circled values are those comprising the sample
Boldface value is median of sample.)

Symmetric samples (which have no mirror-image counterpart)	(1)	2	(4)	7	(9)	Middle sample unit *must* correspond to middle population unit
	1	(2)	(4)	(7)	9	
	(1)	2	(4)	(7)	9	Middle sample unit happens to correspond to middle population unit
	1	(2)	(4)	7	(9)	
Asymmetric samples arranged in mirror-image pairs	(1)	(2)	(4)	7	9	
	1	2	(4)	(7)	(9)	
						If middle units of pair members do not correspond to middle population unit, they *must* fall on opposite sides of it
	(1)	(2)	4	(7)	9	
	1	(2)	4	(7)	(9)	
	(1)	(2)	4	7	(9)	
	(1)	2	4	(7)	(9)	

them (at least one) have a median *exactly* equal to M since we have specified that at least one population unit has the value M and since that population unit must become the middle sample unit for *some* one or more of the possible samples. It follows, therefore, that if the medians of all possible samples were arranged in order of increasing size, the middle median (or medians) would have the value M. That is, the median of the distribution of sample medians is the population median.

12-5 Problems

1. Twelve randomly selected manufactured articles are taken from the as-
    sembly line and placed on life test. They expire after the following
    numbers of hours: 19, 31, 55, 43, 29, 39, 47, 49, 33, 51, 50, 32. Using
    the largest possible $\alpha \le .05$, test the H_0 that the (population) median
    life of the articles is 30 or less against the H_a that it is > 30. What are (a)
    the obtained value of the test statistic, (b) the critical value, (c) the

statistical decision, if the test statistic is the number of negative $X - M_0$ difference scores?

2. The following ten measurements were taken of a single subject's reaction time: .187, .279, .254, .331, .222, .290, .263, .277, .314, .255. Using the largest possible $\alpha \le .05$, test the H_0 that his (population) median reaction time is .250 or less against the H_a that it is $> .250$. What are (a) the obtained value of the test statistic, (b) its probability level, (c) the critical value, (d) the statistical decision?

3. What is the power of the above test if actually half the reaction time population lies at or below .300 and one-third lies at or below .250?

4. In Problem 2 suppose that half the reaction time population lies at or below .500 and one-tenth lies at or below .250. In that case, what would be the power of the test?

5. What would be the answer to the above problem if we used the largest possible $\alpha \le .01$?

6. A test for the median rejects H_0 if no observations in a random sample of three observations have values less than M_0. If actually the probability that in a random sample of one observation the one observation will have a value less than M_0 is .2, then, in performing the test: (a) What is the probability of committing a Type I Error? (b) What is the probability of committing a Type II Error? (c) What is the power of the test? (d) What is the alternative hypothesis, $M < M_0$, $M \ne M_0$, or $M > M_0$?

7. Using the largest possible $\alpha \le .02$ and the data given in Problem 2, test the H_0 that the 40th percentile of the continuously distributed reaction time population is .300 against the H_a that it is not. What are (a) the obtained value of the test statistic, (b) the critical value, (c) the statistical decision?

8. An experimenter wishes to test the H_0 that the 60th percentile of a certain time-score population is .200 or greater against the H_a that it is less, using the largest possible α less than or equal to .25. He draws a random sample of six scores from the population: .212, .187, .255, .194, .279, .306. What are (a) the obtained value of the test statistic, (b) its probability level, (c) the critical value, (d) the decision?

9. Given the following pairs of X and Y measurements on each of six randomly selected units

X	55	51	50	53	59	57
Y	13	17	9	97	88	2

test at the largest available $\alpha \leq .15$ the H_0 that the population of $X - Y$ difference scores has a median of zero against the H_a that the median is greater than zero. The difference-score population is continuously distributed. List all the values of the test statistic that lie in the rejection region and state their cumulative probability. What is the probability level for the obtained value of the test statistic? Should H_0 be rejected?

10. Below are reaction time scores for a group of subjects taken before (BB) and after (AB) the subject drank a bottle of beer. Using the largest $\alpha \leq .01$, test the H_0 that the median reaction time difference produced by drinking a bottle of beer is 0 against the H_a that it is not. What is (a) the obtained value of the test statistic, (b) its probability level, (c) the critical value, (d) the statistical decision?

Subject	BB	AB	Subject	BB	AB	Subject	BB	AB	Subject	BB	AB
A	.250	.244	F	.263	.255	K	.356	.257	O	.261	.232
B	.222	.210	G	.271	.235	L	.401	.333	P	.195	.288
C	.231	.230	H	.301	.279	M	.227	.201	Q	.289	.275
D	.452	.461	I	.199	.188	N	.284	.381	R	.344	.296
E	.187	.185	J	.342	.274						

11. A manufacturer wishes to test the H_0 that a new X method of production is equal or inferior to an old Y method against the H_a that the X method is superior. He assigns ten randomly selected machine operators to use method X on one day and method Y on the following day and another ten randomly selected operators to use method Y first, then method X. Five of the 20 operators turn out fewer units under the new than under the old method, and the other 15 turn out more. Can the manufacturer reject H_0 in favor of H_a if he uses the largest available $\alpha \leq .05$? What is the probability level of the obtained value of the test statistic?

12. Given the following 30 measurements, recorded in the sequence in which they were taken,

−1.04	1.49	−.15	.62	1.92	1.93	−.87	.81	1.23	.60
.47	−.51	.61	.62	.15	1.56	1.83	−.13	.94	−.75
1.98	1.90	.64	2.46	2.16	.07	.87	1.60	1.51	3.10

using the largest possible $\alpha \leq .05$, test the H_0 of no trend or downward trend against the H_a of upward trend. What is (a) the obtained value of the test statistic, (b) the critical value, (c) the decision?

13. A ball bearing manufacturing process is sampled every 15 minutes for 7.5 hours, producing a sample of 30 ball bearings whose diameters are

given below. Using the largest available $\alpha \le .05$, test the H_0 that the process is random against the H_a that it has a unidirectional trend. What are (a) the obtained value of the test statistic, (b) its probability level, (c) the rejection-region values, (d) the statistical decision?

```
.3751  .3755  .3750  .3752  .3757  .3749  .3747  .3751  .3754  .3750
.3749  .3752  .3747  .3753  .3753  .3750  .3754  .3754  .3752  .3751
.3748  .3746  .3750  .3751  .3747  .3748  .3748  .3747  .3750  .3751
```

14. Using the largest possible $\alpha \le .05$, and the following sequence of data, test the H_0 of no trend or downward trend against the H_a of upward trend. What are (a) the obtained value of the test statistic, (b) its probability level, (c) the critical value, (d) the decision?

```
81  84  81  83  83  85  84  83  81  85  83  86  85  86  89
88  89  82  89  84  86  82  86  86  84  85  84  87  87  88
```

15. The following sequence of data are lowest recorded temperatures for successive days of a given month at a southern New Mexico town. Using the largest possible $\alpha \le .025$, test the H_0 of no trend or upward trend against the H_a of downward trend. What are (a) the obtained value of the test statistic, (b) its probability level, (c) the critical value, (d) the statistical decision?

```
14  12  16  11  16  13  17  15  15  18  19  18  14  15  17
20  18  17  19  15  17  15  12  22  19  21  22  17  23  21
```

16. A quality control engineer wishes to test, at the largest available $\alpha \le .05$, the H_0 that 90% or less of the "100-ohm" resistors manufactured by his company have resistances between 99 and 101 ohms against the H_a that more than 90% do. To do so he draws a random sample of 50 resistors, of which 48 have resistances between 99 and 101 ohms. What is the probability level of the test? Can the engineer reject H_0?

17. An experimenter wishes to test the hypothesis that blue-eyedness is independent of sex by drawing a random sample of seven men and six women and observing the number of each sex that are blue-eyed. He draws the sample and finds that four of the men and none of the women are blue-eyed. Can he reject the hypothesis of indepedence at the largest available $\alpha \le .05$? What is the probability level of the test?

18. Give the complete probability distribution under H_0 of r, the number of blue-eyed men in the above situation where of 13 people, 7 are men, 6 are women, 4 are blue-eyed, and 9 are not.

19. A dean wishes to test whether or not the probability of a graduate student making A on a course is the same as that for an undergraduate. He randomly selects nine graduate students and six undergraduates and for each student he randomly selects one of the courses on his transcript and records the grade. A grade of A was made on the selected course by seven of the graduate students and by one of the undergraduate students. Perform the proper test to answer the dean's question at the largest possible α less than or equal to .05. What are (a) the probability level of the test, (b) the decision?

20. Obtain the probability distribution of Hotelling and Pabst's test statistic D for the case where there are $n = 4$ pairs of measurements.

21. Using Hotelling and Pabst's Test, test the following data for correlation at the largest available $\alpha \leq .01$. What is (a) the obtained value of the test statistic, (b) the rejection region, (c) the statistical decision?

Student	A	B	C	D	E	F	G	H	I
Homework Grade	68	87	59	52	75	64	82	61	47
Test Grade	51	63	41	39	80	67	59	44	33

22. Using the largest $\alpha \leq 01$, test whether or not there is any correlation between time scores and errors in the population from which the following random sample of subjects was drawn. What are (a) the obtained value of the test statistic, (b) the critical value, (c) the decision?

Subject	His Time Score	His Error Score
A	.446	17
B	1.614	9
C	.267	31
D	.201	36
E	.504	13
F	8.041	3
G	.311	25
H	.585	4
I	.255	33

23. Given the following X and Y measurements on five randomly drawn units from a bivariate population

X	0	1	4	6	9
Y	5	7	11	36	1,000

test the null hypothesis that the X and Y variables are independent against the alternative hypothesis that they are positively correlated. What is the probability level for the test?

24. Prepare a table giving the cumulative probability for each of the possible values of Wilcoxon's Signed Rank statistic W_+ for the case where $n = 4$. Cumulate from smallest to largest values of W_+.

25. Perform an upper-tail Wilcoxon Signed-Rank Test at the largest possible $\alpha \leq .05$ upon the following 15 pairs of measurements. What are (a) the obtained value of the test statistic, (b) the critical value, (c) the statistical decision?

X	.35	2.25	.89	1.23	4.78	3.39	.20	1.22	1.46	1.06	.99	.33	1.01	.98	.73
Y	.44	.64	−.59	.02	−2.72	.59	.48	−.22	.99	.57	−.81	−.20	.11	.05	2.15

26. Perform a Signed-Rank Test upon the data in Problem 10 to test the H_0 that the effect of beer upon reaction times is to reduce them or leave them unaffected against the H_a that its effect is to increase them. Use the largest available $\alpha \leq .01$. What are (a) the obtained value of W_+, (b) the critical value, (c) the statistical decision.

27. What is the smallest exact significance level at which the Wilcoxon Rank-Sum Test can be conducted when (a) $n = 2$, $m = 6$, lower-tail test, (b) $n = 5$, $m = 5$, two-tail test, (c) $n = 3$, $m = 7$, upper-tail test, (d) $n = 5$, $m = 10$, two-tail test, (e) $n = 2$, $m = 100$, lower-tail test?

28. Obtain the complete probability distribution of W_n for the case $n = 2$, $m = 4$. What are (a) the critical value for an upper-tail test at the largest possible $\alpha \leq .75$, (b) the corresponding exact significance level, (c) the critical value for a two-tail test at the largest possible $\alpha \leq .30$, (d) the corresponding exact value of α, (e) the critical value for a lower-tail test at the largest possible $\alpha \leq .25$, (f) the corresponding exact value of α.

29. Suppose that the X and Y observations in Problem 25 were independent. Perform a two-tail Wilcoxon Rank-Sum Test at the largest available $\alpha \leq .05$. What are (a) the obtained value of the test statistic, (b) the critical value, (c) the statistical decision?

30. Five artillery shell casings are manufactured by a new process and have diameters Ys of 7.650, 7.652, 7.657, 7.655, and 7.653. They are compared with two casings with diameters Xs of 7.651 and 7.648 left over from the old process, which was adequate. If the new casings are slightly too small, little damage will be done, but if they are slightly too large, they won't fit into the chamber; so a one-tail test is required. Use the Wilcoxon Rank-Sum Test, at the largest possible $\alpha \leq .05$, to test the null hypothesis of identical populations against the alternative hypothe-

sis that the new casings tend to be larger than the old ones. What are (a) the obtained value of the test statistic, (b) its probability level, (c) the critical value, (d) the statistical decision?

31. A warehouse owner suspects that he will lose more money in pilferage when Joe is night watchman than when Bill is. To test this, he randomly assigns Joe to 6 nights of duty and Bill to 7 and takes data on his losses during these 13 nights. The data are given below. Apply the Wilcoxon Rank-Sum Test to these data, at the largest possible $\alpha \leq .05$, to decide whether or not Joe is a bad risk. What are (a) the obtained value of the test statistic, (b) the critical value, (c) the statistical decision?

Joe	27.55	75.50	10.14	18.26	7.51	9.95		loss in dollars
Bill	31.40	97.06	81.32	99.60	0	97.35	79.95	loss in dollars

32. Tabulate the probability distribution of Friedman's S for the case where $C = 3$ and $R = 3$, showing the appropriate point and cumulative probabilities.

33. An experimenter wishes to test the H_0 that a person's reaction-time distribution is the same under conditions of noise, music, and quiet. He randomly selects three subjects, randomly determines the sequence of condition-presentation to each subject, and obtains the data given below. Using the largest $\alpha \leq .20$ and performing a Friedman Test, what are (a) the obtained value of S, (b) the critical value of S, (c) the statistical decision, (d) the probability level of the obtained value of S (using the results obtained in the previous problem)?

		Reaction Times Under		
		Noise	*Music*	*Quiet*
	A	.510	.447	.186
Subject	B	.333	.215	.312
	C	.302	.175	.227

34. Tabulate the probability distribution of the Kruskal-Wallis H statistic for the case where there are three samples, one containing one observation and the other two samples each containing two observations, so that $N = 5$, giving both point and cumulative probabilities.

35. An experimenter wishes to test the H_0 that the reaction times of Abe, Bill, and Charley are identically distributed against the H_a that they are not, using the largest possible $\alpha \leq .15$. He obtains the following reaction times: Abe, .112; Bill, .201 and .517; Charley, .224 and .263. Applying the Kruskal-Wallis Test to these data (and using the results of the

previous problem), (a) what is the obtained value of H, (b) what is its probability level, (c) what is the critical value of H, (d) what is the appropriate statistical decision?

36. Below are maze-running scores for 17 individual rats as recorded by 5 different experimenters, to whom the rats were randomly assigned. Using the Kruskal-Wallis Test at the nominal $\alpha = .05$ level of significance, test the H_0 that maze-running scores are independent of what experimenter does the recording against the H_a that some experimenters influence the data. What are (a) the obtained value of the test statistic, (b) the critical value, (c) the statistical decision?

E_1	E_2	E_3	E_4	E_5
67	285	318	501	409
	128	243	275	409
	141	96	275	
		222	275	
		184	433	
			302	

37. The hostility of four randomly selected subjects was measured numerically while the subject's antagonist was playing each of five roles. This was done both when the subject was rested and when he was fatigued. The data are given below. Using the Friedman Test and the largest possible $\alpha \le .01$, test for (a) the main effect of antagonist's role, (b) the main effect of subject's condition, and (c) the interaction between antagonist's role and subject's condition. In each case give the obtained value of the test statistic, the critical value, and the statistical decision.

Subject's Condition	Subject	*Subject's Hostility Score When Antagonist's Role Is To Be:*				
		Dogmatic	Aggressive	Sanctimonious	Condescending	Sarcastic
Rested	Abe	35	33	22	90	98
	Bill	62	33	17	21	68
	Chuck	7	2	6	20	18
	Don	13	33	21	47	43
Fatigued	Abe	84	10	24	32	35
	Bill	71	78	12	10	27
	Chuck	22	30	9	8	5
	Don	50	41	17	15	13

38. A sample of eight observations is to be drawn from a continuously distributed population whose median is M. What is the probability

that the second smallest observation will be less than M and the second largest observation will be greater than M (a single probability is being asked for)? The random sample is now drawn yielding the following values:

.617   .235   .884   .397   .104   2.775   .031   .928

Had we decided before drawing the sample to use its second smallest and second largest observations as limits of a confidence interval for M, we could now allege at the _____ level of confidence that _____ $< M <$ _____.

39. In the previous problem, had we decided prior to sampling to base our confidence interval on the first and seventh smallest sample observations, we could allege at the _____ level of confidence that _____ $< M <$ _____.

40. In Problem 38 had we made the proper announcements of our intentions prior to sampling, we could allege at the _____ level of confidence that the 10th percentile of the population lies between .031 and .397.

41. If a random sample from a continuous population is to contain ten observations, we can allege at the _____ level of confidence that the population median M will lie between the smallest and largest obtained sample observations, at the _____ level that it will lie between the second smallest and second largest, at the _____ level that it will lie between the third smallest and third largest, and at the _____ level that it will lie between the second and fifth smallest sample observations.

42. If the random sample is to contain ten observations, what is the largest value of k for which one can allege at the .90 level of confidence or better that M will lie between the kth smallest and kth largest sample observations? At the .95 level of confidence? At the .99 level? At the .999 level?

43. What is the smallest sample size n at which one can allege at at least the .95 level of confidence that M lies between the smallest and largest sample observations? Between the second smallest and second largest?

44. What is the smallest sample size n at which one can allege at at least the .99 level of confidence that M lies between the smallest and largest sample observations? Between the second smallest and second largest?

45. An experimenter draws 12 random observations from a continuously distributed population whose median is M. They are 44, 46, 47, 51, 55,

58, 58, 58, 59, 61, 68, 93. Had he announced his intentions prior to sampling, at what level of confidence could he now allege that $47 < M < 61$?

46. If an experimenter had decided upon his exact course of action before drawing the following sample of observations

$$1.311 \qquad 1.579 \qquad 1.832 \qquad 2.658 \qquad 27.591$$

he could state at exactly the _____ level of confidence that $1.311 < M < 2.658$, where M is the median of the continuous population from which the sample was randomly drawn.

47. 100,000 random samples of 5 observations each are to be drawn from a continuously distributed population whose 75th percentile is 100. About how many of the samples may we expect to be such that the two smallest observations have values smaller than 100 and the largest observation (in the same sample) has a value greater than 100?

48. If a sample of n observations is to be randomly drawn from a continuously distributed population with median M, what is the probability that M will lie between (a) the second smallest and second largest observations if $n = 5$, (b) the fourth smallest and fourth largest observations if $n = 10$, (c) the eighth smallest and eighth largest observations in a sample of size $n = 20$?

49. A statistician who had decided upon all of his actions in advance draws a random sample of five observations from a population of reaction times. The observation values are .112, .301, .335, .157, .226. The statistician announces that the first quartile Q_1 of the population lies in the interval between .226 and .335. At what level of confidence can he make this statement?

50. Given the following random sample of time scores, .115, .257, .289, .301, .360, .411, .572, .719, 1.487, (a) using the largest possible $\alpha \leq .05$, test the H_0 that $M = .400$ against the H_a that $M \neq .400$, (b) using the smallest possible confidence level $\geq .95$, set up a confidence interval for M, (c) in the light of the preceding answers, should accepting H_0 imply that it is exactly true?

51. A psychologist wishes to test the H_0 that the 75th percentile of a reaction-time population is $\leq .307$ seconds against the H_a that the 75th percentile is $> .307$ seconds using a sample of size $n = 5$ and the largest possible α-value $\leq .10$. If actually .307 is the 90th percentile, (a) what is the power of the test, (b) what is the exact probability of a Type I Error?

52. In a random sample of 14 peas, 5 were round and green, 1 was round and yellow, 2 were wrinkled and green, and 6 were wrinkled and yellow.

Test at the largest possible α-value $\leq .10$ the H_0 that the population proportion of green peas that are wrinkled is the same as the population proportion of yellow peas that are wrinkled against the H_a that the proportions are unequal. What is the exact value of α? What is the probability level of the test? What is the statistical decision?

53. A statistician (who had decided upon his course of action in advance) draws the following random sample of seven observations from a continuously distributed population: 13.67, 22.41, 6.59, 41.40, 18.96, 34.22, 29.03. He then anounces that the population median lies in the interval between 18.96 and 34.22. At exactly what level of confidence can he make this statement?

expected values and moments

The basic concept of expected value has already been introduced in Chapter 8 for a simple random variable X. In this chapter we shall be con-·cerned with expected values of slightly more complicated mathematical expressions involving X.

13-1 Laws of Expectation

13-1-1 Laws Involving a Single Variable or Function of It

As stated or implied in Section 8-1-2, if X is a variable whose possible values are $x_1, x_2, \ldots, x_i, \ldots, x_k$ and if it takes these values proportions $p_1, p_2, \ldots, p_i, \ldots, p_k$ of the time, respectively, so that $p_i = P(X = x_i)$, then, by definition, the *expected value* of X is

$$E(X) = x_1 p_1 + x_2 p_2 + \cdots + x_i p_i + \cdots + x_k p_k = \sum_{i=1}^{k} x_i p_i \qquad (13\text{-}1)$$

and this is the *average* of the values that X takes on all occasions. Each separate occurrence of X may be regarded as a unit in the X population; so it follows from Section 8-1-2 that if the mean of the X population is μ,

$$E(X) = \mu \qquad (13\text{-}2)$$

Now consider another variable V where V is a mathematical expression which may contain constants and X's but contains no variable other than X—so that V is a function solely of X. Then for each of the k different

329

possible values $x_1, x_2, \ldots, x_i, \ldots, x_k$ that X can take, V will have a corresponding value $v_1, v_2, \ldots, v_i, \ldots, v_k$ (some of which may be the same value). Whenever $X = x_i$ it will also be true that $V = v_i$. So if $X = x_i$ a proportion p_i of the time, during that same proportion of the time V must equal v_i. It follows, therefore, that the expected value of V must be

$$E(V) = \sum_{i=1}^{k} v_i P(V = v_i) = \sum_{i=1}^{k} v_i P(X = x_i) = \sum_{i=1}^{k} v_i p_i \qquad (13\text{-}3)$$

Thus if X takes the value $x_1 = 1$ a proportion $p_1 = .3$ of the time and the value $x_2 = 4$ a proportion $p_2 = .7$ of the time, the variable

$$V = X^2 - 2X + 1$$

must take the value $v_1 = x_1^2 - 2x_1 + 1 = 1^2 - 2(1) + 1 = 0$ a proportion $p_1 = .3$ of the time and the value $v_2 = x_2^2 - 2x_2 + 1 = 4^2 - 2(4) + 1 = 9$ a proportion $p_2 = .7$ of the time, so that $E(V) = \sum_{i=1}^{2} v_i p_i = 0(.3) + 9(.7) = 6.3$.

Using this relationship, we can obtain the expected values of some commonly occurring general types of expression. They are given below. In all cases, X is any random variable and C and K are constants.

$$E(C) = C \qquad (13\text{-}4)$$

Proof: Let $V = C$. Then $E(V) = \sum_i v_i p_i$. But since V is a constant, there is only one v_i, which equals C, and one p_i, which equals 1. So $E(V) = v_i p_i = C(1) = C$.

Example: $E(-2) = -2$

$$E(X + C) = E(X) + C \qquad (13\text{-}5)$$

Proof: Let $V = X + C$. Then

$$E(V) = \sum_i v_i p_i = \sum_i (x_i + C) p_i = \sum_i x_i p_i + C \sum_i p_i$$

But

$$\sum_i x_i p_i = E(X)$$

and

$$\sum_i p_i = 1$$

So $E(V) = E(X) + C$.

Example: $E(X - 3) = E(X) - 3$

$$E(CX) = CE(X) \qquad (13\text{-}6)$$

Proof: Let $V = CX$. Then

$$E(V) = \sum_i v_i p_i = \sum_i Cx_i p_i = C \sum_i x_i p_i = CE(X)$$

Example: $E(-7X) = -7E(X)$

$$E(CX + K) = CE(X) + K \qquad (13\text{-}7)$$

Proof: Let $V = CX + K$. Then

$$E(V) = \sum_i v_i p_i = \sum_i (Cx_i + K)p_i = C \sum_i x_i p_i + K \sum_i p_i = CE(X) + K \quad (1)$$

Example: $E(5X - 1) = 5E(X) - 1$.

Now suppose that $R, S, \ldots, V$ are several *different* functions of a single variable X and that A, B, C, and K are constants. Since $R, S, \ldots, V$ are all functions of the *same* variable X, it follows that the same proportion of the time that X takes the value x_i, R must take a corresponding value r_i, S a value $s_i, \ldots$, and V a value v_i. Consequently, a variable $\phi = AR + BS + \cdots + CV + K$ must take a corresponding value $\phi_i = Ar_i + Bs_i + \cdots + Cv_i + K$. This implies that

$$E(AR + BS + \cdots + CV + K) = AE(R) + BE(S) + \cdots + CE(V) + K \tag{13-8}$$

Proof: Let $\phi = AR + BS + \cdots + CV + K$. Then

$$E(\phi) = \sum_i \phi_i p_i = \sum_i (Ar_i + Bs_i + \cdots + Cv_i + K)p_i$$

$$= A \sum_i r_i p_i + B \sum_i s_i p_i + \cdots + C \sum_i v_i p_i + K \sum_i p_i$$

$$= AE(R) + BE(S) + \cdots + CE(V) + K \quad (1)$$

And as special cases of the above general formula, obtained by eliminating all but one of the functions of X, we have

$$E(CV) = CE(V) \tag{13-9}$$

$$E(V + K) = E(V) + K \tag{13-10}$$

$$E(CV + K) = CE(V) + K \tag{13-11}$$

Example: Let $R = 2X - 3$, $S = -X + 5$, and $T = X^3 + 1$. Then $E(2R - 3S + T - 7) = 2E(R) - 3E(S) + E(T) - 7$. We could continue by substituting for R, S, and T the corresponding expressions involving X, obtaining

$$E(2R - 3S + T - 7) = 2E(2X - 3) - 3E(-X + 5) + E(X^3 + 1) - 7$$

$$= 2[2E(X) - 3] - 3[-1E(X) + 5] + [E(X^3) + 1] - 7$$

$$= 7E(X) + E(X^3) - 27$$

$$= 7 \sum_i x_i p_i + \sum_i x_i^3 p_i - 27$$

It is important to notice that we have no law or formula expressing the expected value of a *power* or *root* of X, such as $E(X^r)$ or $E(X^{-1/r})$, in terms of $E(X)$. Therefore, if we want to obtain $E(X^r)$, we must do so by laboriously taking the summation

$$\sum_i x_i^r p_i$$

as indicated in the example.

Illustrative Problem: The daily earnings of an independent taxicab driver take one of three values. On 10% of the days he earns \$50, on 30% he earns \$60, and on 60% he earns \$70. Each night he gives half of the money he earned that day to his wife for household expenses plus a constant \$4 allowance for herself. If this is her only source of income, what is his wife's average daily income?

ANSWER: The problem tells us that a variable X takes the values 50, 60, and 70 proportions .1, .3, and .6 of the time, respectively, and asks for the expected value of $.5X + 4$. The expected value of X is

$$E(X) = \sum_i x_i p_i = 50(.1) + 60(.3) + 70(.6) = 5 + 18 + 42 = 65$$

So the answer is

$$E(.5X + 4) = .5E(X) + 4 = .5(65) + 4 = 32.5 + 4 = 36.5$$

13-1-2 Laws Involving Several Basic Variables

Now consider the case in which there is more than one basic variable. When there are two such variables X and Y,

$$E(X + Y) = E(X) + E(Y) \tag{13-12}$$

Proof: If the X variable can take k different values x_i's and the Y variable m different values y_j's, then there are km different possible pairings of an x_i with a y_j. So the variable $X + Y$ can take its possible values $x_i + y_j$ in km different possible ways, and the probability of occurrence of a particular one of these ways is simply $P(X = x_i$ and $Y = y_j)$, which we shall abbreviate to $P(x_i$ and $y_j)$. We shall also abbreviate $P(X = x_i)$ to $P(x_i)$ and $P(Y = y_j)$ to $P(y_j)$. So,

$$E(X + Y) = \sum_{ij} (x_i + y_j) P(x_i \text{ and } y_j)$$
$$= \sum_{i=1}^{k} \sum_{j=1}^{m} [x_i P(x_i \text{ and } y_j) + y_j P(x_i \text{ and } y_j)]$$
$$= \sum_{i=1}^{k} x_i \sum_{j=1}^{m} P(x_i \text{ and } y_j) + \sum_{j=1}^{m} y_j \sum_{i=1}^{k} P(x_i \text{ and } y_j)$$

Now, as can be seen from Table 13-1, the sum of $P(x_i$ and $y_j)$ over all values of j is $P(x_i)$ and over all values of i is $P(y_j)$. Substituting these equivalents and continuing the derivation,

$$E(X + Y) = \sum_{i=1}^{k} x_i P(x_i) + \sum_{j=1}^{m} y_j P(y_j)$$
$$= E(X) + E(Y)$$

And if A, B, and C are constants,

$$E(AX + BY + C) = AE(X) + BE(Y) + C \tag{13-13}$$

Proof: Let $U = AX$ and $V = BY + C$. Then U and V may be regarded

Table 13-1

Illustration for Derivation of Formula 13-12

	y_1	y_2	y_3	y_4	$\sum_j P(x_i \text{ and } y_j)$
x_1	$P(x_1 \text{ and } y_1)$	$P(x_1 \text{ and } y_2)$	$P(x_1 \text{ and } y_3)$	$P(x_1 \text{ and } y_4)$	$P(x_1)$
x_2	$P(x_2 \text{ and } y_1)$	$P(x_2 \text{ and } y_2)$	$P(x_2 \text{ and } y_3)$	$P(x_2 \text{ and } y_4)$	$P(x_2)$
x_3	$P(x_3 \text{ and } y_1)$	$P(x_3 \text{ and } y_2)$	$P(x_3 \text{ and } y_3)$	$P(x_3 \text{ and } y_4)$	$P(x_3)$
$\sum P(x_i \text{ and } y_j)$	$P(y_1)$	$P(y_2)$	$P(y_3)$	$P(y_4)$	1.00

as basic variables so that Formula 13-12 holds and

$$E(U + V) = E(U) + E(V) = E(AX) + E(BY + C)$$
$$= AE(X) + BE(Y) + C$$

Now, taking the most general case, if $X_1, X_2, \ldots, X_n$ are n different variables and $C_1, C_2, \ldots, C_n$ and K are constants, then

$$E(C_1X_1 + C_2X_2 + \cdots + C_nX_n + K)$$
$$= C_1E(X_1) + C_2E(X_2) + \cdots + C_nE(X_n) + K$$
$$= K + \sum_{i=1}^{n} C_iE(X_i) \tag{13-14}$$

Proof: We can generalize the two-variable formula (13-13) to any number of variables by the method of induction. If we know the formula for n variables and require the formula for $n + 1$ variables, we break the expression involving $n + 1$ variables into two expressions, one involving one variable and one involving n variables. We then treat the latter as a single variable so that we are back to the two-variable case. Its solution will give us two expected values, one involving one variable and the other involving n variables, both of which we can solve. Thus, if we wish to obtain the formula for $E(AX + BY + CZ + D)$, where X, Y, and Z are variables and A, B, C, and D are constants, we let $U = AX$ and $V = BY + CZ + D$. Formula 13-12 tells us that

$$E(AX + BY + CZ + D) = E(U + V) = E(U) + E(V)$$
$$= E(AX) + E(BY + CZ + D)$$

Formula 13-6 tells us that $E(AX) = AE(X)$ and Formula 13-13 says that $E(BY + CZ + D) = BE(Y) + CE(Z) + D$. So $E(AX + BY + CZ + D) = AE(X) + BE(Y) + CE(Z) + D$. Armed with this result we can obtain $E(AW + BX + CY + DZ + K)$ by letting $U = AW$ and $V = BX + CY + DZ + K$, and so on.

If X and Y are *independent* variables

$$E(XY) - E(X)E(Y) \tag{13-15}$$

Proof: If the X and Y variables take k and m different values, respectively, then the variable XY can take its possible values $x_i y_j$ in km different possible ways, and the probability of occurrence of a particular one of these ways is $P(X = x_i$ and $Y = y_j)$. And since X and Y are independent, $P(X = x_i$ and $Y = y_j) = P(X = x_i)P(Y = y_j)$. So using the abbreviated forms $P(x_i$ and $y_j)$, $P(x_i)$ and $P(y_j)$ for the last three probabilities, we have

$$E(XY) = \sum_{ij} x_i y_j P(x_i \text{ and } y_j)$$
$$= \sum_{i=1}^{k} \sum_{j=1}^{m} x_i y_j P(x_i)P(y_j)$$
$$= \sum_{i=1}^{k} x_i P(x_i) \sum_{j=1}^{m} y_j P(y_j)$$
$$= E(X)E(Y)$$

If $E(XY) = E(X)E(Y)$, then X and Y are **uncorrelated**; otherwise they are correlated. So if X and Y are independent, they are also uncorrelated. But they can be uncorrelated without also being independent. That is, independence is a sufficient, but not a necessary, condition for uncorrelatedness; and, by definition, X and Y are **correlated** if $E(XY) \neq E(X)E(Y)$.

Now, generalizing this result, *if X_1, X_2, ..., X_n are n mutually independent variables*

$$E[(X_1)(X_2)\cdots(X_n)] = [E(X_1)][E(X_2)]\cdots[E(X_n)]$$
$$E\left(\prod_{i=1}^{n} X_i\right) = \prod_{i=1}^{n} E(X_i) \tag{13-16}$$

So the expected value of the product of n independent variables is the product of the individual expected values of each variable.

Proof: Again, we generalize the two-variable formula by the method of induction. To obtain $E(X_1 X_2 X_3)$, we let $U = X_1 X_2$ and $V = X_3$

$$E(X_1 X_2 X_3) = E(UV) = E(U)E(V) = E(X_1 X_2)E(X_3) = E(X_1)E(X_2)E(X_3)$$

To obtain $E(X_1 X_2 X_3 X_4)$, we let $U = X_1 X_2 X_3$, whose expected value we now know, and let $V = X_4$, etc.

Finally, taking the most general case, if X_1, X_2, ..., X_n are n mutually independent random variables and C_1, C_2, ..., C_n and K are constants,

$$E[(C_1 X_1)(C_2 X_2)\cdots(C_n X_n) + K] = [C_1 E(X_1)][C_2 E(X_2)]\cdots[C_n E(X_n)] + K$$

or, putting it more succinctly,

$$E\left(K + \prod_{i=1}^{n} C_i X_i\right) = K + \prod_{i=1}^{n} C_i E(X_i) \tag{13-17}$$

Proof: Let $\prod_{i=1}^{n} X_i = V$ and let the constant $\prod_{i=1}^{n} C_i = A$. Then

$$E\left(K + \prod_{i=1}^{n} C_i X_i\right) = E[K + (C_1 C_2 \cdots C_n)(X_1 X_2 \cdots X_n)]$$
$$= E\left[K + \prod_{i=1}^{n} C_i \prod_{i=1}^{n} X_i\right] = E[K + AV]$$
$$= K + AE(V) = K + \left(\prod_{i=1}^{n} C_i\right)E\left(\prod_{i=1}^{n} X_i\right)$$
$$= K + \left(\prod_{i=1}^{n} C_i\right)\left(\prod_{i=1}^{n} E(X_i)\right)$$
$$= K + \prod_{i=1}^{n} C_i E(X_i)$$

13-2   Moments

13-2-1   General Case

Certain special classes of expected values are called *moments*. They are generally used to describe the characteristics of a population (whose units may be the occurrences of a variable X). Consider a population with mean μ and variance σ^2, consisting of units whose values are denoted by the variable X (a proportion p_i of the units having the value x_i), and let C be some arbitrarily chosen constant. Then

$$E\{(X - C)^r\} \tag{13-18}$$

is called the **rth moment of X about the arbitrary point** C, and, of course,

$$E\{(X - C)^r\} = \sum_{i=1}^{k}(x_i - C)^r p_i$$

13-2-2   Moments about the Origin

If we set C equal to zero we obtain

$$E(X^r) \tag{13-19}$$

which is called the **rth moment of X about the origin**. The most important moment of this type is obtained when $r = 1$ in which case $E(X^1)$ becomes simply

$$E(X) = \sum_{i=1}^{k} x_i p_i = \sum_{j=1}^{N} u_j \left(\frac{1}{N}\right) = \mu \tag{13-20}$$

the mean of the population. So the *first moment about zero* is the average of the values of all units in the population, which is the *mean of the population*.

13-2-3   Central Moments

If we set C equal to $E(X)$, that is to μ, we obtain

$$E\{(X - \mu)^r\} \tag{13-21}$$

which is called the **rth central moment of X** and is denoted μ_r. When $r = 1$, $E\{(X - \mu)^1\} = E(X) - E(\mu) = \mu - \mu = 0$. So the expected or average deviation of a unit's value from the population mean is zero. The most important central moment is obtained when $r = 2$, in which case

$$E\{(X - \mu)^2\} = \sum_{i=1}^{k}(x_i - \mu)^2 p_i = \sum_{i=1}^{k}(x_i - \mu)^2 \frac{f_i}{N} = \sigma^2 \tag{13-22}$$

So the *second central moment* is the average squared deviation of a unit's value from the mean of its population, i.e., is the *variance of the population*.

The population mean μ has the characteristic of being that point about which the average squared deviation is smallest. And the population variance σ^2 has the characteristic of being the smallest second moment, i.e., average squared deviation, about anything.

Proof:

$$
\begin{aligned}
E\{(X - C)^2\} &= E\{[(X - \mu) + (\mu - C)]^2\} \\
&= E\{(X - \mu)^2 + 2(X - \mu)(\mu - C) + (\mu - C)^2\} \\
&= E\{(X - \mu)^2\} + 2(\mu - C)E(X - \mu) + (\mu - C)^2 \\
&= \sigma^2 + 2(\mu - C)(\mu - \mu) + (\mu - C)^2 \\
&= \sigma^2 + (\mu - C)^2
\end{aligned}
$$

Since both terms on the right are squared values, both terms must be positive. Therefore, the value of C that minimizes the right side is $C = \mu$. So,

$$\text{minimum } E\{(X - C)^2\} = E\{(X - \mu)^2\} = \sigma^2 + (\mu - \mu)^2 = \sigma^2 \quad (13\text{-}23)$$

The equation proved above

$$E\{(X - C)^2\} = \sigma^2 + (\mu - C)^2 \quad (13\text{-}24)$$

is known as the *Parallel Axis Theorem*. It tells us that the second moment of X about an arbitrary axis C is equal to the variance of X plus the square of the distance between C and the mean of X. With the aid of this theorem, knowing μ and the second moment about one known and specified axis (μ or C), we can easily obtain the second moment about any other.

13-2-4 Standardized Values

When the value X is measured, it is expressed in terms of so many measuring units such as inches. So μ will be expressed in inches, σ^2 in (inches)^2 and the rth moment in (inches)^r. But had X been measured on a centimeter scale instead, the numerical values of X, σ^2, etc., would have been entirely different numbers of centimeters (raised to some power). Thus, all our figures are peculiar to the scale of measurement that is adopted. However, we can circumvent this difficulty by measuring X in terms of its distance in σ-units from μ. If the values of X, μ, and σ were all originally expressed in terms of the same measuring units, the deviation of X from μ in σ-*units* is the same number regardless of the common scale on which the values of X, μ, and σ were originally measured. The resulting *standardized value* of X is called Z, where

$$Z = \frac{X - \mu}{\sigma} \quad (13\text{-}25)$$

Thus, suppose that a man's height X is 70 inches and he belongs to a population in which $\mu = 68$ inches and $\sigma = 3$ inches. If the measurements had been made in centimeters instead of inches, X would be 177.80 cm, μ would be 172.72 cm, and σ would be 7.62 cm. But although the X's differ numerically, the Z scores do not:

$$Z = \frac{X - \mu}{\sigma} = \frac{70 \text{ in.} - 68 \text{ in.}}{3 \text{ in.}} = \frac{2 \text{ in.}}{3 \text{ in.}} = .667$$

$$Z = \frac{X - \mu}{\sigma} = \frac{177.80 \text{ cm.} - 172.72 \text{ cm.}}{7.62 \text{ cm.}} = \frac{5.08 \text{ cm.}}{7.62 \text{ cm.}} = .667$$

Notice that the units of measurement (inches and centimeters) appearing in both numerator and denominator cancel out, leaving a pure number.

13-2-5 Standardized Central Moments

In standardizing X, by converting it to Z, we also standardize its moments and make them independent of the scale of measurement. The ***r*th standardized central moment of X** is

$$E\{Z^r\} = E\left\{\left(\frac{X - \mu}{\sigma}\right)^r\right\} = \frac{E\{(X - \mu)^r\}}{\sigma^r} \tag{13-26}$$

When $r = 1$,

$$E\{Z^1\} = \frac{E\{X - \mu\}}{\sigma} = \frac{\mu - \mu}{\sigma} = 0$$

so *the mean of a population of Z-scores is zero*. When $r = 2$,

$$E\{Z^2\} = \frac{E\{(X - \mu)^2\}}{\sigma^2} = \frac{\sigma^2}{\sigma^2} = 1$$

and

$$\text{Var}\,(Z) = E\{[Z - E(Z)]^2\} = E\{[Z - 0]^2\} = E\{Z^2\} = 1$$

so *the variance of a population of Z-scores is 1*. Thus, for $r = 1$ and $r = 2$ we learn something about the distribution of Z, but nothing worthwhile about the distribution of X, since every population of X's has the same first two standardized central moments. However, when $r = 3$ we obtain

$$\frac{E\{(X - \mu)^3\}}{\sigma^3} \tag{13-27}$$

a very important index of the *asymmetry* of the X (or Z) population, which is called the coefficient of **skewness**. If the population is perfectly symmetric, for every unit whose value is a distance d above μ, there is another unit whose value is a distance d below it, so for every $X - \mu = d$ there is an $X - \mu = -d$. And since the $(X - \mu)$ deviations are cubed, the algebraic signs remain and the cubed values cancel each other out, leaving a coefficient of skewness of zero. But if the population is asymmetric, the coefficient is very sensitive to both the size and direction of unbalanced deviations of X from

μ, tending to be positive for unbalanced right-tailedness and negative for unbalanced left-tailedness. When $r = 4$, we obtain

$$\frac{E\{(X - \mu)^4\}}{\sigma^4} \qquad (13\text{-}28)$$

called the **kurtosis**, a less important index of the shape of the X (or Z) population's distribution, supposed to be sensitive to peakedness, i.e., concentration of X's about the mean or Z's about zero. Its limited meaningfulness is greatest in the case of symmetric distributions.

13-2-6 Example

To illustrate the various types of moments, let X be the number of errors committed in performing a task in which up to three errors can be made, and suppose that of those who attempt the task 20% make no errors, 10% make one, 20% make two, and 50% make three. Then the first moment of X about the origin, and therefore the mean of the X population, is

$$E(X^1) = \sum_{i=1}^{4} x_i^1 p_i = 0(.2) + 1(.1) + 2(.2) + 3(.5) = 2$$

The second central moment of X, and therefore the variance of the X population, is

$$
\begin{aligned}
E\{(X - \mu)^2\} &= \sum_{i=1}^{4} (x_i - 2)^2 p_i \\
&= (0 - 2)^2(.2) + (1 - 2)^2(.1) + (2 - 2)^2(.2) + (3 - 2)^2(.5) \\
&= 4(.2) + 1(.1) + 0(.2) + 1(.5) = 1.4
\end{aligned}
$$

The four possible values x_i's of X when converted to standardized values Z's become z_i's where

$$z_i = \frac{x_i - \mu}{\sigma} = \frac{x_i - 2}{\sqrt{1.4}}$$

Thus

$$z_1 = \frac{0 - 2}{1.1832} = -1.6903, \qquad z_2 = \frac{1 - 2}{1.1832} = -.8452,$$

$$z_3 = \frac{2 - 2}{1.1832} = 0, \qquad z_4 = \frac{3 - 2}{1.1832} = .8452$$

The third standardized central moment of X, and therefore the coefficient of skewness of the X population, is

$$E(Z^3) = \sum_{i=1}^{4} z_i^3 p_i = \sum_{i=1}^{4} \left(\frac{x_i - 2}{1.1832}\right)^3 p_i = \frac{1}{(1.1832)^3} \sum_{i=1}^{4} (x_i - 2)^3 p_i$$

$$= \frac{1}{1.6565}[(-2)^3(.2) + (-1)^3(.1) + 0(.2) + 1^3(.5)]$$

$$= .6037[-1.6 - .1 + 0 + .5] = .6037[-1.2] = -.7244$$

which could also have been obtained by direct substitution of the values of z_i into

$$\sum_{i=1}^{4} z_i^3 p_i$$

13-3   Laws of Variance

Perhaps the most important central moment is the variance. It is designated σ^2 or $\text{Var}(X)$ depending upon whether emphasis is being placed upon it as a population index or as a measure of the variability of a random variable. In addition to the formulas given for σ^2 in Chapter 2, the variance can be expressed in various equivalent ways:

$$\text{Var}(X) = E\{[X - E(X)]^2\} \tag{13-29}$$
$$= E\{(X - \mu)^2\} \tag{13-30}$$
$$= E(X^2) - [E(X)]^2 \tag{13-31}$$

The first formula is actually the definition of the variance of X. The second follows from the first because $E(X) = \mu$. And the third is proved as follows

$$\text{Var}(X) = E\{(X - \mu)^2\} = E\{X^2 - 2X\mu + \mu^2\} = E(X^2) - 2\mu E(X) + \mu^2$$
$$= E(X^2) - 2\mu(\mu) + \mu^2 = E(X^2) - \mu^2 = E(X^2) - [E(X)]^2$$

When arithmetical operations are performed upon X, the variance of the resulting variable follows certain laws. They are given below. The symbols C and K are constants and X, Y, and Z stand for variables.

$$\text{Var}(X + C) = \text{Var}(X) \tag{13-32}$$

Proof:

$$\text{Var}(X + C) = E\{[(X + C) - E(X + C)]^2\}$$
$$= E\{[X + C - E(X) - C]^2\}$$
$$= E\{[X - E(X)]^2\}$$
$$= \text{Var}(X)$$

$$\text{Var}(CX) = C^2\,\text{Var}(X) \tag{13-33}$$

Proof:

$$\text{Var}(CX) = E\{[CX - E(CX)]^2\}$$
$$= E\{[CX - CE(X)]^2\}$$
$$= E\{C^2[X - E(X)]^2\}$$
$$= C^2 E\{[X - E(X)]^2\}$$
$$= C^2\,\text{Var}(X)$$

$$\text{Var}(CX + K) = C^2\,\text{Var}(X) \tag{13-34}$$

Proof:

$$\begin{aligned}
\text{Var}\,(CX + K) &= E\{[CX + K - E(CX + K)]^2\} \\
&= E\{[CX + K - CE(X) - K]^2\} \\
&= C^2 E\{[X - E(X)]^2\} \\
&= C^2\,\text{Var}\,(X)
\end{aligned}$$

$$\text{Var}\,(X + Y) = \text{Var}\,(X) + \text{Var}\,(Y) + 2[E\{XY\} - (E\{X\})(E\{Y\})] \quad (13\text{-}35)$$

Proof: To simply notation let $EX = E(X)$ and $EY = E(Y)$. Then

$$\begin{aligned}
\text{Var}\,(X + Y) &= E\{[X + Y - E(X + Y)]^2\} \\
&= E\{[X + Y - EX - EY]^2\} \\
&= E\{[(X - EX) + (Y - EY)]^2\} \\
&= E\{(X - EX)^2 + 2(X - EX)(Y - EY) + (Y - EY)^2\} \\
&= E\{(X - EX)^2\} + 2E\{XY - XEY - YEX + (EX)(EY)\} \\
&\quad + E\{(Y - EY)^2\} \\
&= \text{Var}\,(X) + 2[E(XY) - (EX)(EY) - (EY)(EX) + (EX)(EY)] \\
&\quad + \text{Var}\,(Y) \\
&= \text{Var}\,(X) + \text{Var}\,(Y) + 2[E(XY) - (EX)(EY)]
\end{aligned}$$

The expression $E(XY) - (EX)(EY)$ or $E\{(X - EX)(Y - EY)\}$ which it equals is called the **covariance** of X and Y. If X and Y are uncorrelated, $E(XY) = (EX)(EY)$ and the covariance is zero, making

$$\text{Var}\,(X + Y) = \text{Var}\,(X) + \text{Var}\,(Y).$$

These results for two variables can be generalized to any number of variables. For example, for three variables,

$$\begin{aligned}
\text{Var}\,(X + Y + Z) &= E\{[X + Y + Z - E(X + Y + Z)]^2\} \\
&= E\{[(X - EX) + (Y - EY) + (Z - EZ)]^2\} \\
&= E\{(X - EX)^2 + (Y - EY)^2 + (Z - EZ)^2 \\
&\quad + 2(X - EX)(Y - EY) + 2(X - EX)(Z - EZ) \\
&\quad + 2(Y - EY)(Z - EZ)\} \\
&= E\{(X - EX)^2\} + E\{(Y - EY)^2\} + E\{(Z - EZ)^2\} \\
&\quad + 2E\{(X - EX)(Y - EY)\} + 2E\{(X - EX)(Z - EZ)\} \\
&\quad + 2E\{(Y - EY)(Z - EZ)\} \\
&= \text{Var}\,(X) + \text{Var}\,(Y) + \text{Var}\,(Z) + 2\,\text{Cov}\,(XY) \\
&\quad + 2\,\text{Cov}\,(XZ) + 2\,\text{Cov}\,(YZ)
\end{aligned}$$

Therefore, the variance of the sum of any number of variables is simply the sum of their individual variances plus twice the sum of all the covariances:

$$\text{Var} \left(\sum_j X_j \right) = \sum_j \text{Var}(X_j) + 2 \sum_{j > k} \text{Cov}(X_j X_k) \qquad (13\text{-}36)$$

And *if the variables are all mutually uncorrelated* (in which case their covariances all become zero) the variance of their sum is simply the sum of their separate variances:

$$\text{Var} \left(\sum_j X_j \right) = \sum_j \text{Var}(X_j) \qquad (13\text{-}37)$$

To illustrate the use of some of these laws, suppose that a store owner is considering renovating the store at a fixed cost of $5,000 and hiring a full-time clerk, a three-quarter-time secretary, and a half-time bookkeeper, but he wants to have some idea of the total cost. He knows that at a certain employment agency current salaries for available clerks, secretaries, and bookkeepers in general average $7,500, $6,000, and $8,000 and have standard deviations of $500, $400, and $1,000, respectively. So if a clerk, a secretary, and a bookkeeper will be randomly selected for him by the agency, he can proceed as follows. Letting full-time salaries for available clerks, secretaries, and bookkeepers registered with the agency be X, Y, and Z, respectively, the expected total cost to the store owner for the first year will be

$$
\begin{aligned}
E(X + .75Y + .5Z + 5,000) &= E(X) + .75E(Y) + .5E(Z) + 5,000 \\
&= 7,500 + .75(6,000) + .5(8,000) + 5,000 \\
&= 7,500 + 4,500 + 4,000 + 5,000 \\
&= 21,000
\end{aligned}
$$

dollars. So 21,000 is the mean of the population of $X + .75Y + .5Z + 5,000$ sums from which his sum will, in effect, be randomly drawn. Because of the random selection the salaries will be independent; so the variance of the sum will be

$$
\begin{aligned}
\text{Var}\,(X + .75Y &+ .5Z + 5,000) \\
&= \text{Var}(X) + \text{Var}(.75Y) + \text{Var}(.5Z) + \text{Var}(5,000) \\
&= \text{Var}(X) + (.75)^2\,\text{Var}(Y) + (.5)^2\,\text{Var}(Z) + 0 \\
&= (500)^2 + .5625(400)^2 + .25(1,000)^2 \\
&= 250,000 + 90,000 + 250,000 = 590,000
\end{aligned}
$$

square dollars and the standard deviation is the square root of that, which is 768.11 dollars.

Now consider a set of n variables $X_1, X_2, \ldots, X_n$ the sum or total of whose values is T. We know that

$$E(T) = \sum_{i=1}^{n} E(X_i)$$

And if the X_i's are all independent, or even uncorrelated, then

$$\text{Var}(T) = \sum_{i=1}^{n} \text{Var}(X_i)$$

Now the average of the n X_i values will be T/n, that is, a constant $1/n$ times T. Therefore, if we designate the average by $\bar{X}$, we have

$$E(\bar{X}) = E\left(\frac{1}{n}T\right) = \frac{1}{n}E(T) = \left(\frac{1}{n}\right)\sum_{i=1}^{n}E(X_i)$$

and

$$\text{Var}(\bar{X}) = \text{Var}\left(\frac{1}{n}T\right) = \left(\frac{1}{n}\right)^2\text{Var}(T) = \left(\frac{1}{n^2}\right)\sum_{i=1}^{n}\text{Var}(X_i)$$

Suppose, for example, that a scientist needs to know the value of a certain quantity to a greater degree of accuracy than that to which it can be measured. He assigns each of n different assistants the task of making the measurement (each using a different apparatus to do so) and uses the average of the n measurements. Since different assistants made the measurements using different apparatuses, the measurements X_i can be considered independent and the variance of the average is therefore

$$\left(\frac{1}{n^2}\right)\sum_{i=1}^{n}\text{Var}(X_i)$$

or $1/n$th of the average variance. Thus, the variance of the average is far smaller than the average variance, and accuracy of measurement has therefore been greatly increased by taking the average (provided, of course, that there is no constant bias in the measurements).

13-4 Expected Value and Variance of the Sample Mean

13-4-1 Formulas

Suppose that a random sample of n observations is to be drawn from a population whose mean and variance are μ and σ^2, respectively. Let X_i stand for the ith drawn sample observation, considered as a variable, and let

$$\bar{X} = \sum_{i=1}^{n}\frac{X_i}{n}$$

stand for the sample mean, considered as a variable (since the sample is not yet drawn). Then (a) irrespective of whether the sampled population is finite or infinite and of whether sampling is with or without replacements,

$$E(\bar{X}) = \mu \tag{13-38}$$

(b) if the sampled population is infinite, or if sampling is with replacements,

$$\text{Var}\,(\bar{X}) = \frac{\sigma^2}{n} \tag{13-39}$$

(c) if the sampled population is finite, consisting of N units, and sampling is without replacements,

$$\text{Var}\,(\bar{X}) = \left(\frac{N-n}{N-1}\right)\frac{\sigma^2}{n} \tag{13-40}$$

Therefore, the mean and variance of the probability distribution of $\bar{X}$ are μ and a constant times σ^2/n. If the sampled population is undepleted by the sample, the constant is 1, whereas if it is depleted, the constant is

$$\frac{N-n}{N-1}$$

The latter constant becomes 1 when N is infinitely larger than n. The symbol $\text{Var}\,(\bar{X})$ is often replaced by an equivalent symbol $\sigma_{\bar{X}}^2$.

13-4-2 Proof of Formulas In Case Where Population Is Undepletable

If the sampled population is infinite or if sampling is with replacements, the $i-1$ sample observations preceding the ith observation have not detectably depleted the sampled population. Therefore, irrespective of the value of i, the ith sample observation X_i has exactly the same possible values, and exactly the same probabilities of taking them, as has the variable X, whose population is being sampled. It follows, therefore, that $E(X_i) = E(X) = \mu$ and that $\text{Var}\,(X_i) = \text{Var}\,(X) = \sigma^2$.

Because they are randomly selected and do not deplete the sampled population, the n sample observations X_i's cannot influence each other and are therefore mutually independent. Therefore, we may regard the n different X_i's as n different, mutually independent, random variables. Formula 13-14 tells us that the expected value of their sum is equal to the sum of their individual expected values, i.e., that

$$E(\textstyle\sum_{i=1}^{n} X_i) = \sum_{i=1}^{n} E(X_i)$$

and Formula 13-37 tells us that the variance of their sum is equal to the sum of their individual variances, i.e., that

$$\text{Var}\,(\textstyle\sum_{i=1}^{n} X_i) = \sum_{i=1}^{n} \text{Var}\,(X_i)$$

We shall make use of this information in the following proofs:

Proof that $E(\bar{X}) = \mu$:

$$E(\bar{X}) = E\left(\frac{\sum_{i=1}^{n} X_i}{n}\right) = \frac{1}{n} E(\textstyle\sum_{i=1}^{n} X_i) = \frac{1}{n} \sum_{i=1}^{n} E(X_i)$$

$$= \frac{1}{n} \sum_{i=1}^{n} E(X) = \frac{1}{n} \sum_{i=1}^{n} \mu = \frac{1}{n}(n\mu) = \mu$$

Proof that Var $(\bar{X}) = \sigma^2/n$:

$$\text{Var}\,(\bar{X}) = \text{Var}\left(\frac{\sum_{i=1}^{n} X_i}{n}\right) = \frac{1}{n^2}\,\text{Var}\,(\sum_{i=1}^{n} X_i) = \frac{1}{n^2}\sum_{i=1}^{n}\text{Var}\,(X_i)$$

$$= \frac{1}{n^2}\sum_{i=1}^{n}\text{Var}\,(X) = \frac{1}{n^2}[n\,\text{Var}\,(X)] = \frac{\sigma^2}{n}$$

13-4-3 Proof of Formulas in Case Where Population Is Depletable

If sampling is without replacements from a finite population containing N units, we must consider the depletion of the population as observations are drawn from it. When the ith sample observation is to be drawn, $i - 1$ units have already been removed from the population; so the population now contains $N - (i - 1)$ units and the probability of drawing a specified one of them (i.e., a particular unit *known* not to have been drawn earlier) is $1/(N - i + 1)$, whereas the probability of not drawing it is

$$\frac{(N - i)}{(N - i + 1)}$$

The probability that the ith sample observation in sequence, X_i, will be the population unit designated u_s is the probability of not drawing u_s in each of the first $i - 1$ draws, drawing it on the ith, and not drawing it on each of the remaining draws:

$$P(X_i = u_s) = \frac{N-1}{N} \times \frac{N-2}{N-1} \times \frac{N-3}{N-2} \times \cdots \times \frac{N-i+1}{N-i+2}$$

$$\times \frac{1}{N-i+1} \times \frac{N-i}{N-i} \times \frac{N-i-1}{N-i-1} \cdots = \frac{1}{N}$$

Numerators cancel with denominators except for the N in the first denominator; so $P(X_i = u_s)$ is $1/N$.

Analogously, the probability that the ith drawn sample observation will be u_s and a subsequent jth drawn sample observation will be u_t is the probability of not drawing them on each of the first $i - 1$ draws, drawing u_s on the ith, not drawing u_t on each of the subsequent draws prior to the jth, and drawing it on the jth:

$$P(X_i = u_s \quad \text{and} \quad X_j = u_t) = \left[\frac{N-2}{N} \times \frac{N-3}{N-1} \times \frac{N-4}{N-2} \times \frac{N-5}{N-3}\right.$$

$$\times \cdots \times \frac{N-i}{N-i+2} \times \frac{1}{N-i+1} \times \frac{N-i-1}{N-i} \times \frac{N-i-2}{N-i-1}$$

$$\left.\times \cdots \times \frac{N-j+1}{N-j+2} \times \frac{1}{N-j+1} \times \frac{N-j}{N-j} \times \frac{N-j-1}{N-j-1} \times \cdots\right]$$

$$= \frac{1}{N} \times \frac{1}{N-1} = \frac{1}{N(N-1)}$$

Again, numerators cancel with denominators, first two fractions distant, then one fraction away, then in the same fraction, leaving only the denominators of the first two fractions. We would obtain exactly the same result if j were $< i$ as we did for $j > i$, so the formula holds for all cases where $i \neq j$ and $s \neq t$.

Armed with these results we are now ready to present the proofs.

Proof that $E(\bar{X}) = \mu$:

$$E(\bar{X}) = E\left\{\frac{\sum_{i=1}^{n} X_i}{n}\right\} = \frac{1}{n}\sum_{i=1}^{n} E(X_i) = \frac{1}{n}\sum_{i=1}^{n}\sum_{s=1}^{N} u_s P(X_i = u_s)$$

$$= \frac{1}{n}\sum_{i=1}^{n}\sum_{s=1}^{N} u_s\left(\frac{1}{N}\right) = \frac{1}{n}\sum_{i=1}^{n}\frac{\sum_{s=1}^{N} u_s}{N} = \frac{1}{n}\sum_{i=1}^{n}\mu$$

$$= \frac{1}{n}(n\mu) = \mu$$

Proof that $Var(\bar{X}) = \left(\frac{N-n}{N-1}\right)\frac{\sigma^2}{n}$

$$Var(\bar{X}) = E\{(\bar{X} - \mu)^2\} = E\left\{\left(\frac{\sum_{i=1}^{n} X_i}{n} - \mu\right)^2\right\}$$

$$= E\left\{\left(\frac{\sum_{i=1}^{n} X_i - n\mu}{n}\right)^2\right\} = \frac{1}{n^2}E\{[\sum_{i=1}^{n}(X_i - \mu)]^2\}$$

$$= \frac{1}{n^2}E\{[(X_1 - \mu) + (X_2 - \mu) + \cdots + (X_n - \mu)]^2\}$$

$$= \frac{1}{n^2}E\{(X_1 - \mu)^2 + (X_2 - \mu)^2 + \cdots + (X_n - \mu)^2$$

$$+ (X_1 - \mu)(X_2 - \mu) + (X_1 - \mu)(X_3 - \mu) + \cdots\}$$

$$= \frac{1}{n^2}E\{\sum_{i=1}^{n}(X_i - \mu)^2 + \sum_{i \neq j}(X_i - \mu)(X_j - \mu)\}$$

$$= \frac{1}{n^2}\sum_{i=1}^{n} E\{(X_i - \mu)^2\} + \frac{1}{n^2}\sum_{i \neq j} E\{(X_i - \mu)(X_j - \mu)\}$$

$$= \frac{1}{n^2}\sum_{i=1}^{n}\sum_{s=1}^{N}(u_s - \mu)^2 P(X_i = u_s)$$

$$+ \frac{1}{n_2}\sum_{i \neq j}\sum_{s \neq t}(u_s - \mu)(u_t - \mu)P(X_i = u_s \text{ and } X_j = u_t)$$

$$= \frac{1}{n^2}\sum_{i=1}^{n}\sum_{s=1}^{N}(u_s - \mu)^2\left(\frac{1}{N}\right)$$

$$+ \frac{1}{n^2}\sum_{i \neq j}\sum_{s \neq t}(u_s - \mu)(u_t - \mu)\left(\frac{1}{N(N-1)}\right)$$

$$= \frac{1}{n^2}\sum_{i=1}^{n}\sum_{s=1}^{N}\frac{(u_s - \mu)^2}{N} + \frac{1}{n^2 N(N-1)}\sum_{i \neq j}\sum_{s \neq t}(u_s - \mu)(u_t - \mu)$$

Now

$$\frac{\sum_{s=1}^{N}(u_s - \mu)^2}{N}$$

is, by definition, the population variance σ^2. And in the same way that we showed, in the first few lines of the proof, that

$$[\Sigma_{i=1}^n (X_i - \mu)]^2 = \Sigma_{i=1}^n (X_i - \mu)^2 + \sum_{i \neq j} (X_i - \mu)(X_j - \mu)$$

we can show that

$$[\Sigma_{s=1}^N (u_s - \mu)]^2 = \Sigma_{s=1}^N (u_s - \mu)^2 + \sum_{s \neq t} (u_s - \mu)(u_t - \mu)$$

But

$$\Sigma_{s=1}^N (u_s - \mu) = N\mu - N\mu = 0$$

and since

$$\sigma^2 = \frac{\Sigma_{s=1}^N (u_s - \mu)^2}{N} \text{ it follows that } \Sigma_{s=1}^N (u_s - \mu)^2 = N\sigma^2$$

Substituting these equivalents, we obtain

$$[0]^2 = N\sigma^2 + \sum_{s \neq t} (u_s - \mu)(u_t - \mu)$$

$$\sum_{s \neq t} (u_s - \mu)(u_t - \mu) = -N\sigma^2$$

So making these substitutions in the last line of the interrupted derivation, we obtain

$$\text{Var}(\bar{X}) = \frac{1}{n^2} \Sigma_{i=1}^n \sigma^2 + \frac{1}{n^2 N(N-1)} \sum_{i \neq j} - N\sigma^2$$

$$= \frac{1}{n^2}(n\sigma^2) + \frac{-N}{n^2 N(N-1)} \sum_{i \neq j} \sigma^2$$

Now since the summation over all values of $i \neq j$ originally applied to values of X_i and X_j in a sample of n observations, in which the pairs of values of i and j can differ in $_nP_2 = n(n-1)$ ways, the summation amounts to adding $n(n-1)$ instances of the constant σ^2 and is therefore equal to $n(n-1)\sigma^2$. So

$$\text{Var}(\bar{X}) = \frac{\sigma^2}{n} - \frac{1}{n^2(N-1)}[n(n-1)\sigma^2]$$

$$= \frac{\sigma^2}{n} - \frac{(n-1)\sigma^2}{n(N-1)} = \frac{(N-1)\sigma^2 - (n-1)\sigma^2}{n(N-1)}$$

$$= \frac{(N-n)\sigma^2}{n(N-1)} = \left(\frac{N-n}{N-1}\right)\frac{\sigma^2}{n}$$

13-5 Expected Values and Variances of Other General Sample Statistics

Again, suppose that a random sample of n observations is to be drawn from a population with mean μ and variance σ^2, and let X_i stand for the ith drawn sample observation. Since the sample is not yet drawn, the X_i's are variables and so are any sample statistics constructed from them. The following sections give the expected values or variances of such future sample statistics, and therefore the means or variances of their probability distributions, under the conditions named.

13-5-1 Expected Value of the Sample Variance

If the random observations are to be drawn from an infinite population or with replacements from a finite one, the expected value of the variance

$$s^2 = \frac{\sum_{i=1}^{n} (X_i - \bar{X})^2}{n}$$

of the future sample is

$$E(s^2) = \frac{(n-1)}{n}\sigma^2 \qquad (13\text{-}41)$$

whereas if the random observations are to be drawn from a finite population without replacements

$$E(s^2) = \frac{N(n-1)}{n(N-1)}\sigma^2 \qquad (13\text{-}42)$$

Proof:

$$E(s^2) = E\left\{\frac{\sum_{i=1}^{n} (X_i - \bar{X})^2}{n}\right\} = \frac{1}{n}E\left\{\sum_{i=1}^{n} [(X_i - \mu) - (\bar{X} - \mu)]^2\right\}$$

$$= \frac{1}{n}E\left\{\sum_{i=1}^{n}[(X_i - \mu)^2 - 2(\bar{X} - \mu)(X_i - \mu) + (\bar{X} - \mu)^2]\right\}$$

$$= \frac{1}{n}E\left\{\sum_{i=1}^{n} (X_i - \mu)^2 - 2(\bar{X} - \mu) \sum_{i=1}^{n} (X_i - \mu) + n(\bar{X} - \mu)^2\right\}$$

$$= \frac{1}{n}E\left\{\sum_{i=1}^{n} (X_i - \mu)^2 - 2(\bar{X} - \mu)n(\bar{X} - \mu) + n(\bar{X} - \mu)^2\right\}$$

$$= \frac{1}{n}E\left\{\sum_{i=1}^{n} (X_i - \mu)^2 - n(\bar{X} - \mu)^2\right\}$$

$$= \frac{1}{n}\left[\sum_{i=1}^{n} E(X_i - \mu)^2 - nE(\bar{X} - \mu)^2\right]$$

$$= \frac{1}{n}\left[\sum_{i=1}^{n} \sum_{s=1}^{N} (u_s - \mu)^2 P(X_i = u_s) - n \operatorname{Var}(\bar{X})\right]$$

$$= \frac{1}{n}\left[\sum_{i=1}^{n} \sum_{s=1}^{N} (u_s - \mu)^2 \frac{1}{N} - n \operatorname{Var}(\bar{X})\right]$$

$$= \frac{1}{n}\left[\sum_{i=1}^{n} \operatorname{Var}(X) - n \operatorname{Var}(\bar{X})\right]$$

$$= \operatorname{Var}(X) - \operatorname{Var}(\bar{X})$$

So, if sampling is with replacements or from an infinite population, $\operatorname{Var}(X) = \sigma^2$ and $\operatorname{Var}(\bar{X}) = \sigma^2/n$ and

$$E(s^2) = \sigma^2 - \frac{\sigma^2}{n} = \frac{n\sigma^2 - \sigma^2}{n} = \frac{n-1}{n}\sigma^2$$

Whereas if sampling is from a finite population without replacements, Var $(X) = \sigma^2$ and

$$\text{Var}(\bar{X}) = \left(\frac{N-n}{N-1}\right)\frac{\sigma^2}{n}$$

and

$$E(s^2) = \sigma^2 - \left(\frac{N-n}{N-1}\right)\frac{\sigma^2}{n} = \frac{Nn-n-N+n}{(N-1)n}\sigma^2 = \frac{N(n-1)}{n(N-1)}\sigma^2$$

13-5-2 Expected Value and Variance of the Difference between Sample Means

If $\bar{X}$ is the mean of a sample of n_X observations made upon a random variable X with mean μ_X and variance σ_X^2 and $\bar{Y}$ is the mean of a sample of n_Y observations made upon a random variable Y with mean μ_Y and variance σ_Y^2, then

$$E(\bar{X} - \bar{Y}) = \mu_X - \mu_Y \tag{13-43}$$

$$\text{Var}(\bar{X} - \bar{Y}) = \text{Var}(\bar{X}) + \text{Var}(\bar{Y}) - 2\,\text{Cov}(\bar{X}\bar{Y}) \tag{13-44}$$

Proof:

$$E(\bar{X} - \bar{Y}) = E(\bar{X}) + (-1)E(\bar{Y}) = E(\bar{X}) - E(\bar{Y}) = \mu_X - \mu_Y$$

$$\text{Var}(\bar{X} - \bar{Y}) = E\{[(\bar{X} - \bar{Y}) - E(\bar{X} - \bar{Y})]^2\} = E\{[(\bar{X} - \bar{Y}) - (\mu_X - \mu_Y)]^2\}$$

$$= E\{[(\bar{X} - \mu_X) - (\bar{Y} - \mu_Y)]^2\}$$

$$= E\{(\bar{X} - \mu_X)^2 - 2(\bar{X} - \mu_X)(\bar{Y} - \mu_Y) + (\bar{Y} - \mu_Y)^2\}$$

$$= E\{(\bar{X} - \mu_X)^2\} + E\{(\bar{Y} - \mu_Y)^2\} - 2E\{(\bar{X} - \mu_X)(\bar{Y} - \mu_Y)\}$$

$$= \text{Var}(\bar{X}) + \text{Var}(\bar{Y}) - 2\,\text{Cov}(\bar{X}\bar{Y})$$

If $\bar{X}$ and $\bar{Y}$ are uncorrelated, as they will be if X and Y are independent or if the observations in *both* samples are truly random *selections* from their respective populations (so that Y-observations are in no way matched with X-observations, e.g., are not taken at the same moment or upon the same unit), then Cov $(\bar{X}\bar{Y})$ will equal zero and

$$\text{Var}(\bar{X} - \bar{Y}) = \text{Var}(\bar{X}) + \text{Var}(\bar{Y}) \tag{13-45}$$

13-5-3 Expected Value and Variance of the Sum of Sample Observations

If

$$T = \sum_{i=1}^{n} X_i$$

is the sum of all n observations in a future random sample from a population with mean μ and variance σ^2, then (a) irrespective of whether the sampled population is finite or infinite and of whether sampling is with or without replacements

$$E(T) = n\mu \tag{13-46}$$

(b) if the sampled population is infinite or sampling is with replacements

$$\text{Var}(T) = n\sigma^2 \tag{13-47}$$

(c) if the sampled population is finite, consisting of N units, and sampling is without replacements

$$\text{Var}(T) = \left(\frac{N-n}{N-1}\right)n\sigma^2 \tag{13-48}$$

Proof: Since

$$\bar{X} = \frac{1}{n}\sum_{i=1}^{n} X_i, \qquad T = \sum_{i=1}^{n} X_i = n\bar{X}$$

So

$$E(T) = E(n\bar{X}) = nE(\bar{X}) = n\mu$$
$$\text{Var}(T) = \text{Var}(n\bar{X}) = n^2 \text{Var}(\bar{X})$$

So if sampling is from an infinite population or is with replacements,

$$\text{Var}(T) = n^2\left(\frac{\sigma^2}{n}\right) = n\sigma^2$$

And if sampling is without replacements from a finite population,

$$\text{Var}(T) = n^2\left[\left(\frac{N-n}{N-1}\right)\frac{\sigma^2}{n}\right] = \left(\frac{N-n}{N-1}\right)n\sigma^2$$

13-6   Expected Values and Variances of Sample Statistics Following Specific Probability Laws

13-6-1   Formulas

In Chapter 7 we derived the probability laws for certain sample statistics under specified sampling and population conditions. Since we know the probability that the sample statistic will take each of its possible values, we can calculate the mean and variance of its probability distribution. Here we consider only the case where the sampled population is dichotomous, consisting of N units of which R are successes, and in which the proportion of successes is p (so that when N is finite, $p = R/N$).

If the population is finite and sampling is without replacements, the number of successes r in a random sample of n observations follows the hypergeometric probability law and therefore has a hypergeometric probability distribution. The *mean of the hypergeometric distribution* is

$$E(r) = \frac{nR}{N} \tag{13-49}$$

and *its variance* is

$$\text{Var}\ (r) = \frac{nR}{N}\left(\frac{N-R}{N}\right)\left(\frac{N-n}{N-1}\right) \tag{13-50}$$

If the population is infinite or sampling is with replacements, the number of successes r in a random sample of n observations follows the binomial probability law and has a binomial probability distribution. The *mean and variance of the binomial distribution* are

$$E(r) = np \tag{13-51}$$
$$\text{Var}\ (r) = np(1 - p) \tag{13-52}$$

The number of observations, or trials, n necessary to produce r successes, when sampling randomly from an undepletable population, follows the negative binomial probability law. The *mean and variance of the negative binomial distribution* are

$$E(n) = \frac{r}{p} \tag{13-53}$$

$$\text{Var}\ (n) = \frac{r(1-p)}{p^2} \tag{13-54}$$

The geometric probability law is a special case of the negative binomial that occurs when $r = 1$. Therefore, the *mean and variance of the geometric distribution* are

$$E(n) = \frac{1}{p} \tag{13-55}$$

$$\text{Var}\ (n) = \frac{1-p}{p^2} \tag{13-56}$$

If successes are randomly distributed over a continuum of time or space such that the average number of successes in portions of size c is m, the number x of successes in a sample portion of size c follows the Poisson probability law. The *mean and variance of the Poisson distribution* are

$$E(x) = m \tag{13-57}$$
$$\text{Var}\ (x) = m \tag{13-58}$$

And, in the case of the Poisson approximation to the binomial probability law, the mean and variance of the Poisson distribution are

$$E(r) = np \tag{13-59}$$
$$\text{Var}\ (r) = np \tag{13-60}$$

13-6-2 Proofs of Formulas

Every unit in the sampled population is either a success or a failure and the proportion of successes is p, which equals R/N when N is finite. Therefore, if we arbitrarily assign a value of 1 to each success and a value of 0 to each failure, the mean of the population of values will be

$$\mu = E(X) = \sum_i x_i p_i = x_1 p_1 + x_2 p_2 = 1p + 0(1 - p) = p \qquad (13\text{-}61)$$

and the variance of the population will be

$$\sigma^2 = E(X - \mu)^2 = \sum_i (x_i - \mu)^2 p_i = (x_1 - p)^2 p_1 + (x_2 - p)^2 p_2$$

$$= (1 - p)^2 p + (0 - p)^2 (1 - p) = (1 - p)^2 p + p^2 (1 - p)$$

$$= (1 - p)p[1 - p + p] = p(1 - p) \qquad (13\text{-}62)$$

Now suppose that a random sample of n observations is drawn from the population. If the sample contains exactly r successes, there will be r observations having a value of 1 and $n - r$ having a value of 0; so the sum T of the values of the observations in the sample will equal the number of successes r, i.e.,

$$T = \sum_{i=1}^{n} x_i = r(1) + (n - r)(0) = r$$

Therefore, the formulas for the expected value and variance of T that we obtained in Section 13-5-3 will also give us the expected value and variance of r under corresponding sampling conditions.

Proof for Hypergeometric Distribution: If the sampled population is finite and sampling is without replacements,

$$E(T) = n\mu$$

and

$$\text{Var}(T) = \left(\frac{N - n}{N - 1}\right) n\sigma^2$$

We already know that

$$\mu = p = \frac{R}{N}$$

and that

$$\sigma^2 = p(1 - p) = \frac{R}{N}\left(1 - \frac{R}{N}\right) = \frac{R}{N}\left(\frac{N - R}{N}\right)$$

So

$$E(r) = E(T) = n\mu = \frac{nR}{N}$$

$$\text{Var}(r) = \text{Var}(T) = \left(\frac{N - n}{N - 1}\right) n\sigma^2 = \left(\frac{N - n}{N - 1}\right)\left(\frac{nR}{N}\right)\left(\frac{N - R}{N}\right)$$

Proof for Binomial Distribution: If the sampled population is infinite or sampling is with replacements, $E(T) = n\mu$ and Var $(T) = n\sigma^2$. And since we know that $\mu = p$ and $\sigma^2 = p(1 - p)$, we have

$$E(r) = E(T) = n\mu = np$$

$$\text{Var}(r) = \text{Var}(T) = n\sigma^2 = np(1 - p)$$

To obtain the mean and variance of the negative binomial and Poisson distributions, we must resort to a trick. The sum of the probabilities of a variable taken over all of its possible values is 1. The trick consists of converting the expected-value expression representing the mean or variance into another expression containing only the sum mentioned above and some constants. The proof for the Poisson is the easier of the two, so we shall present it first.

Proof for Poisson Distribution:

$$E(x) = \sum_i x_i p_i = \sum_x xP(x) = \sum_{x=0}^{\infty} x\frac{m^x}{x!\,e^m}$$

$$= 0 + \sum_{x=1}^{\infty} x\frac{m^x}{x!\,e^m} = m\sum_{x=1}^{\infty}\frac{m^{x-1}}{(x-1)!\,e^m}$$

Let $x - 1 = y$. So when $x = 1$, $y = 0$. Then substituting these values,

$$E(x) = m\sum_{y=0}^{\infty}\frac{m^y}{y!\,e^m}$$

Now we know that

$$\sum_{x=0}^{\infty}\frac{m^x}{x!\,e^m} = 1$$

and since y takes the same set of integer-values from 0 to ∞ as x,

$$\sum_{y=0}^{\infty}\frac{m^y}{y!\,e^m}$$

must also equal 1. Therefore, $E(x) = m(1) = m$.

$$\text{Var}(x) = E(x^2) - [E(x)]^2 = E(x^2) - m^2$$

$$= -m^2 + \sum_{x=0}^{\infty} x^2\frac{m^x}{x!\,e^m}$$

$$= -m^2 + \sum_{x=0}^{\infty} [x(x-1) + x]\frac{m^x}{x!\,e^m}$$

$$= -m^2 + \sum_{x=0}^{\infty} x(x-1)\frac{m^x}{x!\,e^m} + \sum_{x=0}^{\infty} x\frac{m^x}{x!\,e^m}$$

$$= -m^2 + 0 + 0 + \sum_{x=2}^{\infty} x(x-1)\frac{m^x}{x!\,e^m} + 0 + \sum_{x=1}^{\infty} x\frac{m^x}{x!\,e^m}$$

$$= -m^2 + m^2\sum_{x=2}^{\infty}\frac{m^{x-2}}{(x-2)!\,e^m} + m\sum_{x=1}^{\infty}\frac{m^{x-1}}{(x-1)!\,e^m}$$

Now let $x - 2 = y$ and let $x - 1 = z$ so when $x = 2$, $y = 0$ and when $x = 1$, $z = 0$. Making the appropriate substitutions,

$$\text{Var}(x) = -m^2 + m^2 \sum_{y=0}^{\infty} \frac{m^y}{y! e^m} + m \sum_{z=0}^{\infty} \frac{m^z}{z! e^m}$$

$$= -m^2 + m^2(1) + m(1)$$

$$= m$$

Proof for Negative Binomial Distribution: Since n is the number of observations required to produce r successes, the possible values of n extend from r to infinity. Therefore,

$$E(n) = \sum_{n=r}^{\infty} nP(n) = \sum_{n=r}^{\infty} n\binom{n-1}{r-1} p^r(1-p)^{n-r}$$

$$= \sum_{n=r}^{\infty} \frac{n(n-1)!}{(r-1)!(n-r)!} p^r(1-p)^{n-r}$$

$$= \frac{r}{p} \sum_{n=r}^{\infty} \frac{n!}{r!(n-r)!} p^{r+1}(1-p)^{n-r}$$

Now let $n = y - 1$ and let $r = z - 1$. So when $n = r$, $y = r + 1 = z$. Making the appropriate substitutions,

$$E(n) = \frac{r}{p} \sum_{y=z}^{\infty} \frac{(y-1)!}{(z-1)!(y-z)!} p^z(1-p)^{y-z}$$

$$= \frac{r}{p} \sum_{y=z}^{\infty} \binom{y-1}{z-1} p^z(1-p)^{y-z} = \frac{r}{p}(1) = \frac{r}{p}$$

since the expression being summed is the negative binomial probability that the zth success will occur on the yth trial and the summation takes place over all possible values of y from z to infinity.

$$\text{Var}(n) = E(n^2) - [E(n)]^2 = E(n^2) - \left(\frac{r}{p}\right)^2$$

$$E(n^2) = \sum_{n=r}^{\infty} n^2 P(n) = \sum_{n=r}^{\infty} n^2 \frac{(n-1)!}{(r-1)!(n-r)!} p^r(1-p)^{n-r}$$

$$= \sum_{n=r}^{\infty} [(n+1)-1] \frac{n!}{(r-1)!(n-r)!} p^r(1-p)^{n-r}$$

$$= \sum_{n=r}^{\infty} \frac{(n+1)!}{(r-1)!(n-r)!} p^r(1-p)^{n-r}$$

$$- \sum_{n=r}^{\infty} \frac{n!}{(r-1)!(n-r)!} p^r(1-p)^{n-r}$$

$$= \frac{(r+1)r}{p^2} \sum_{n=r}^{\infty} \frac{(n+1)!}{(r+1)!(n-r)!} p^{r+2}(1-p)^{n-r}$$

$$- \frac{r}{p} \sum_{n=r}^{\infty} \frac{n!}{r!(n-r)!} p^{r+1}(1-p)^{n-r}$$

$$= \frac{(r+1)r}{p^2} \sum_{n=r}^{\infty} \binom{n+1}{r+1} p^{r+2}(1-p)^{(n+2)-(r+2)}$$

$$- \frac{r}{p} \sum_{n=r}^{\infty} \binom{n}{r} p^{r+1}(1-p)^{(n+1)-(r+1)}$$

Now let $w = n + 2$, $x = r + 2$, $y = n + 1$, and $z = r + 1$. Then when

$n = r, w = n + 2 = r + 2 = x$ and $y = n + 1 = r + 1 = z$. So

$$E(n^2) = \frac{(r+1)r}{p^2} \sum_{w=x}^{\infty} \binom{w-1}{x-1} p^x (1-p)^{w-x}$$

$$- \frac{r}{p} \sum_{y=z}^{\infty} \binom{y-1}{z-1} p^z (1-p)^{y-z}$$

$$= \frac{(r+1)r}{p^2}(1) - \frac{r}{p}(1) = \frac{r^2 + r - rp}{p^2} = \frac{r^2}{p^2} + \frac{r(1-p)}{p^2}$$

$$\text{Var}(n) = E(n^2) - \left(\frac{r}{p}\right)^2 = \frac{r^2}{p^2} + \frac{r(1-p)}{p^2} - \frac{r^2}{p^2} = \frac{r(1-p)}{p^2}$$

13-7   Mean-Unbiased Estimators

Suppose that a random sample is to be drawn from a population, the value of a certain sample statistic S is to be calculated, and it is then to be alleged that a certain index I of the population has the same value as that just obtained for S. The sample statistic S is then being used as an estimator of I and the actually obtained value of S is an estimate of I. If this procedure were repeated an infinite number of times, the estimates would give us the probability distribution of S. And if the average of the alleged values of I, i.e., the mean of the probability distribution of S, exactly equaled the true value of I (and did so no matter what the sample size n), S would be, by definition (see Section 12-4-4), a *mean-unbiased estimator* of I.

But the mean of the distribution of S is $E(S)$. So S is a mean-unbiased estimator of I if $E(S) = I$. Having obtained $E(S)$ for many sample statistics in the preceding sections, we can now state and prove the following.

The sample mean is a mean-unbiased estimator of the population mean. And this is the case provided only that sampling is random. The population may be finite or infinite and sampling may be with or without replacements.

Proof: The proof in Section 13-4 that $E(\bar{X}) = \mu$, both when samples deplete the population and when they do not, proves the statement.

The proportion r/n of successes in the sample is a mean-unbiased estimator of the proportion p of successes in the sampled population. This too is the case provided only that sampling is random, irrespective of whether or not sampling depletes the population.

Proof: It was proved in Section 13-6 that $E(r) = np$ when sampling from an undepletable population and that

$$E(r) = n\left(\frac{R}{N}\right) = np$$

when sampling from a depletable one. Therefore, in both cases,

$$E\left(\frac{r}{n}\right) = \frac{1}{n}E(r) = \frac{1}{n}(np) = p$$

The sample variance is *not* a mean-unbiased estimator of the population variance by itself, but the product of the sample variance and a constant is such an estimator.

When sampling randomly from an infinite population or with replacements

$$\left(\frac{n}{n-1}\right)s^2 = \hat{\sigma}^2 = \sum_{i=1}^{n} \frac{(X_i - \bar{X})^2}{n-1}$$

is a mean-unbiased estimator of the population variance.

Proof: It was proved in Section 13-5-1 that under the circumstances named

$$E(s^2) = \left(\frac{n-1}{n}\right)\sigma^2$$

Therefore,

$$E\left\{\left(\frac{n}{n-1}\right)s^2\right\} = \left(\frac{n}{n-1}\right)E(s^2) = \left(\frac{n}{n-1}\right)\left(\frac{n-1}{n}\right)\sigma^2 = \sigma^2$$

When sampling randomly without replacements from a finite population,

$$\left[\frac{n(N-1)}{N(n-1)}\right]s^2$$

is a mean-unbiased estimator of the population variance.

Proof: It was proved in Section 13-5-1 that under these sampling conditions

$$E(s^2) = \frac{N(n-1)}{n(N-1)}\sigma^2$$

So

$$E\left\{\left[\frac{n(N-1)}{N(n-1)}\right]s^2\right\} = \left[\frac{n(N-1)}{N(n-1)}\right]E(s^2) = \left[\frac{n(N-1)}{N(n-1)}\right]\frac{N(n-1)}{n(N-1)}\sigma^2 = \sigma^2$$

If the sampled population is a dichotomous one in which successes have a value of 1 and failures a value of 0, the mean-unbiased estimator of the population variance $\sigma^2 = p(1-p)$ is

$$\frac{r(n-r)}{n(n-1)}$$

if the population is infinite or if sampling is with replacements and is

$$\frac{r(n-r)(N-1)}{n(n-1)N}$$

if the population is finite and sampling is without replacements.

Proof: If successes have a value of 1 and failures a value of 0, then both the sum

$$\sum_{i=1}^{n} X_i$$

of the n sample observations and the sum

$$\sum_{i=1}^{n} X_i^2$$

of their squared values must be r, the number of successes in the sample, and the sample mean

$$\frac{1}{n}\sum_{i=1}^{n} X_i$$

must be r/n. Therefore,

$$s^2 = \frac{1}{n}\sum_{i=1}^{n}(X_i - \bar{X})^2 = \frac{1}{n}\sum_{i=1}^{n}(X_i^2 - 2X_i\bar{X} + \bar{X}^2)$$

$$= \frac{1}{n}\left[\sum_{i=1}^{n}X_i^2 - 2\bar{X}\sum_{i=1}^{n}X_i + n\bar{X}^2\right] = \frac{1}{n}\left[\sum_{i=1}^{n}X_i^2 - 2\bar{X}n\bar{X} + n\bar{X}^2\right]$$

$$= \frac{1}{n}\left[\sum_{i=1}^{n}X_i^2 - n\bar{X}^2\right] = \frac{1}{n}\left[r - n\left(\frac{r}{n}\right)^2\right]$$

$$= \frac{1}{n}\left[\frac{nr - r^2}{n}\right] = \frac{r(n-r)}{n^2}$$

Therefore, if sampling does not deplete the population, the mean-unbiased estimator of σ^2 is

$$\frac{n}{n-1}s^2 = \frac{n}{n-1} \times \frac{r(n-r)}{n^2} = \frac{r(n-r)}{n(n-1)}$$

and if sampling depletes the population, the mean-unbiased estimator of σ^2 is

$$\frac{n(N-1)}{N(n-1)}s^2 = \frac{n(N-1)}{N(n-1)} \times \frac{r(n-r)}{n^2} = \frac{r(n-r)(N-1)}{n(n-1)N}$$

Finally, suppose that $\bar{X}$ and $\bar{Y}$ are the means of samples to be drawn randomly from an X population with mean μ_X and a Y population with mean μ_Y, respectively. Then *the difference between sample means $\bar{X} - \bar{Y}$ is a mean-unbiased estimator of the difference between population means $\mu_X - \mu_Y$.* And this is the case irrespective of whether or not the populations are finite and sampling is with replacements.

   Proof:

$$E(\bar{X} - \bar{Y}) = E(\bar{X}) - E(\bar{Y}) = \mu_X - \mu_Y$$

13-8   Problems

1.   Given two independent variables, X, Y, and a third variable Z where Z equals $(X^2 - 2X + 5)/3$ and given that the X variable takes the values 0, 1, 4, and 9 with probabilities .4, .3, .2, and .1, respectively, and that the Y variable takes the values -1, 0, and 8, with probabilities .1, .6, and .3, respectively, what are the expected values of: (a) $[\sqrt{125}/(2\sqrt{5})] - 14$, (b) X, (c) Y, (d) $5X$, (e) $2X + Y$, (f) X^2, (g) $Y^{1/3}$, (h) $X + Y$, (i) $3X + 4Y - 7$, (j) XY, (k) $X^2 + 5X - 7$, (l) $\sqrt{X^2 + 5X - 7}$, (m) $14X^2 - 7XY^{1/3} + 2Y - 26$, (n) $(Y^2\sqrt{X})/2$, (o) $(5Z + 4)/7$?

2.   According to Laplace, the greater frequency of male over female babies

has sometimes been attributed to a general desire on the part of fathers to perpetuate the family name by continuing to have children until a son is produced. If $P(\text{son}) = P(\text{daughter}) = 1/2$ and if every married couple stopped having children as soon as the first son was born or after the tenth daughter, whichever came first, and had at least one child, what would be the expected number of sons and the expected number of daughters per family?

3. A variable X takes the values -3, 1, and 5, 70%, 10%, and 20% of the time, respectively. Give (a) the first moment of X about zero, (b) the second central moment of X, (c) the third standardized central moment of X, (d) the fourth moment of X about -3.

4. A variable X takes the values -4, -1, 7 with probabilities $.80$, $.15$, $.05$, respectively. Give the zeroth, first, second, third, and fourth moments of X about (a) 0, (b) 2, (c) the mean. Give the rth standardized central moment of X for $r = 0$, $r = 1$, $r = 2$, $r = 3$, and $r = 4$. Give the following indices for the X population: mode, median, mean, range, interquartile range, variance, standard deviation, average absolute deviation from the mean, skewness, and kurtosis.

5. A population consists of four units having the following values: 0, 0, 0, 4. What are the mean, median, variance, skewness, and kurtosis of the population?

6. Thirty percent of the units in a population have the value -2 and the remaining seventy percent have the value 8. What are the mean and variance of the population.

7. A game consists of throwing a fair die and then tossing a fair coin. If the die comes up six, $X = 1$, and if it comes up anything else, $X = 0$. If the coin comes up heads, $Y = 1$, and if it comes up tails, $Y = 0$. The score for the game is $Z = X - Y$. What are the expected value and variance of Z?

8. A random variable X is -1 twenty percent of the time, 0 thirty percent of the time, and $+1$ fifty percent of the time. Z is equal to $X^2 + 3X - 2$. What are the expected value and variance of Z?

9. X is a random variable whose mean and standard deviation are 7 and 3, respectively. Y is a random variable, independent of X, whose mean and standard deviation are 4 and 5, respectively. Let $Z = 2X - 3Y + 1.5$. What are the mean and standard deviation of Z?

10. If X is the number of heads that come up when two fair coins are tossed, (a) what are the expected value and variance of X, (b) what is the expected value of X^3, (c) what is the third moment of X about its mean, (d) what are the mean and variance of $3X$?

11. What are the formulas for $E(X - Y)$ and $\text{Var}\,(X - Y)$?

12. A variable X and another variable Y have the values listed as column and row headings respectively in the table below, and cell entries give the probabilities with which X = column heading and Y = row heading, corresponding to that cell. Calculate (a) $E(X)$, (b) $E(Y)$, (c) Var (X), (d) Var (Y), (e) Var $(X + Y)$, (f) $E(XY)$.

X

		-1	0	1
Y	0	.05	.10	.15
	1	.15	.10	.45

13. Given three variables, X which takes the value -1 eighty percent of the time and the value 4 the other twenty percent of the time, Y which is -2 thirty percent of the time and 8 the rest of the time, and Z which is equal to $X^2 - 2X - 3$, if X and Y are independent, what are: (a) $E(X)$, (b) $E(Y)$, (c) $E(Z)$, (d) $E(X + Y)$, (e) $E(X + Z)$, (f) $E(Y + Z)$, (g) $E(X + Y + Z)$, (h) Var(X), (i) Var(Y), (j) Var(Z), (k) Var$(X + Y)$, (l) Var $(X + Z)$, (m) Var $(Y + Z)$, (n) Var $(X + Y + Z)$?

14. A population consisting of 100 units has a mean of 45 and a standard deviation of 5. A sample consisting of 40 units is to be permanently withdrawn from the population. What are the expected value and standard deviation of $\bar{X}$, the mean of the sample not yet drawn (and therefore a variable quantity)?

15. A sample is randomly drawn from a large lot and is found to contain 2 defectives and 23 nondefectives. Using the sample results, estimate the proportion of defectives in the lot.

16. A random sample of five observations $-4, -1, 0, 6, 9$ is drawn from an infinite X population and a random sample of three observations, 21, 22, 26 is drawn from an infinite Y population. Estimate (a) the value of the constant $\mu_X - \mu_Y$, where μ_X and μ_Y are the means of the X and Y populations, respectively, (b) the variance of the variable $\bar{X} - \bar{Y}$, where $\bar{X}$ is the mean of a random sample of five X-observations, $\bar{Y}$ is the mean of a random sample of three Y-observations, and neither sample depletes its population.

17. A population consists of four units whose values are 0, 1, 1, 10. A random sample of n observations is to be drawn with replacements. What is the probability that the mean of the sample will have the same value as the mean of the population if (a) $n = 1$, (b) $n = 2$, (c) $n = 3$? What is the probability that the median of the sample will have the same value as the median of the population if (d) $n = 1$, (e) $n = 2$, (f) $n = 3$?

18. If a random sample of ten observations is to be drawn from an infinite dichotomous population in which the proportion of successes is $p = .27$, (a) what is the expected value of r, the number of successes in the sample, (b) what is the probability that r will equal its expected value?

19. A European scientist reports that the mean and standard deviation of some temperature data are 35 and 2 degrees Centigrade, respectively. The formula for converting C degrees Centigrade into F degrees Fahrenheit is $F = 1.8C + 32$. What are the mean and standard deviation of the temperature data in degrees Fahrenheit?

20. From past records the editor of a dictionary knows that a word and its definition consume an average of 6.21 lines with a standard deviation of .81 lines. He intends to include 100,000 words in the next edition of the dictionary. If each page contains 200 lines, what are the mean and standard deviation of the number of *pages* that the next edition could contain? What is the ratio of the standard deviation to the mean, for the number of pages?

21. A mountain climbing expedition will fly from A to B, take a bus from B to C, and hike from C to D. If the mean and standard deviation of the time to travel from A to B are 12 hours and 2 hours, from B to C are 2 days and .5 days, and from C to D are 3 weeks and 1.5 weeks, respectively, and if the travel times for the three legs of the journey are independent, what are the mean and standard deviation of the total time to travel from A to D?

22. A candle placed at A produces a light whose intensity at B has a mean of 1 and a standard deviation of .1 units. If 10,000 such candles are placed at A, and if between A and B a filtering screen is placed which reduces the intensity of the emerging light to 1/10,000th of that of the entering light, what are the mean and standard deviation of the light intensity at B?

23. The proper dosage of a certain drug is exactly 5 grains. A drug company can supply the dosage either in a single pill with a mean of 5 grains and a standard deviation of .5 grains or in a set of 5 pills where the pill has a mean of 1 grain and a standard deviation of .2 grains. Which is preferable and why?

24. The weight of an aspirin tablet has a mean of .005 ounces and a standard deviation of .0004 ounces. If the tablets are packed randomly 1,000 to a bottle, and if T and $\bar{X}$ are the total weight and mean weight of the 1,000, (a) what are the mean and standard deviation of T, (b) what are the mean and standard deviation of $\bar{X}$?

14

the distribution of
the sample mean

The mean of a future random sample is a variable, and, as such, has a probability distribution. The distribution of the sample mean is determined by three factors: the distribution of the sampled population, the method of sampling, and the size of the sample. Therefore, if we know all three factors, it is at least theoretically possible to derive a probability law for $\bar{X}$, giving the probability distribution of the sample mean. In a few cases it is easy to do so. However, in some cases the derivation is prohibitively laborious or complex, and in many others we cannot obtain the distribution of $\bar{X}$ because we do not know the population distribution of X. In such cases where complexity or ignorance of the population distribution prevents us from obtaining an exact probability law for $\bar{X}$, we must resort to probability formulas involving inequalities or approximations. We shall discuss all these approaches in the following sections.

14-1   The Exact Distribution of $\bar{X}$

14-1-1   Case Where Sampled Population Is Dichotomous

Suppose that a proportion p (or a number R) of units in the sampled population have the value C and the remaining units all have the value K, where $K \neq C$. Let us call the former units successes. Then either the binomial or hypergeometric probability law tells us the probability of r successes in a random sample of n observations, and this is also the probability that the sample mean will have the value $[rC + (n - r)K]/n$. (This is the case only because the latter never takes the same value for two different values of r. If it did take the same value for two different r's, say r' and r'', the following equation, and its solutions, would be true

$$\frac{r'C + (n - r')K}{n} = \frac{r''C + (n - r'')K}{n}$$

$$r'C + nK - r'K = r''C + nK - r''K$$

$$(r' - r'')C = (r' - r'')K$$

$$C = K$$

or

$$r' = r''$$

But both solutions are false statements contradicting our input information that $C \neq K$, which must be the case if the population is to be dichotomous, and that r' and r'' are two different values of r. So there are no two different r-values for which the corresponding $\overline{X}$'s are equal.)

Therefore, if the sampled population is infinite or sampling is with replacements,

$$P\left(\overline{X} = \frac{rC + (n - r)K}{n}\right) = \binom{n}{r}p^r(1 - p)^{n-r} \tag{14-1}$$

and if sampling is without replacements from a finite population containing N units,

$$P\left(\overline{X} = \frac{rC + (n - r)K}{n}\right) = \frac{\binom{R}{r}\binom{N - R}{n - r}}{\binom{N}{n}} \tag{14-2}$$

14-1-2 General Case Where Sampled Population Is Discretely Distributed

We can generalize the above approach to any discretely distributed population. Suppose that the units in the sampled population have k different values $x_1, x_2, \ldots, x_i, \ldots, x_k$, that the proportion of units having these values are $p_1, p_2, \ldots, p_i, \ldots, p_k$, respectively, and, if the population is finite consisting of N units, that the numbers of units having these values are $R_1, R_2, \ldots, R_i, \ldots, R_k$, respectively. Now we may regard the k different x_i's as *categories*. So if a random sample of n observations is to be drawn, the probability $P(r_1, r_2, \ldots, r_i, \ldots, r_k)$ that r_1 of the observations will have the value x_1, r_2 the value $x_2, \ldots, r_i$ the value $x_i, \ldots,$ and r_k the value x_k is given either by the multinomial probability law or by the multivariate hypergeometric probability law, depending upon whether or not sampling detectably depletes the population. Furthermore, if the stated numbers of observations have the stated values (i.e., if there are r_1 x_1's, r_2 x_2's, etc.), then the sample mean $\bar{X}$ must have the value

$$\bar{X} = \frac{r_1 x_1 + r_2 x_2 + \cdots + r_i x_i + \cdots + r_k x_k}{n} = \frac{\sum_{i=1}^{k} r_i x_i}{n}$$

whose numerical value we shall call Q. Therefore, if $\bar{X} = Q$ for this set of r_i's *only*, then $P(r_1, r_2, \ldots, r_i, \ldots, r_k)$ for this set of r_i's is also $P(\bar{X} = Q)$. However, different sets of r_i's sometimes yield the same value of $\bar{X}$. So, *in general, $P(\bar{X} = Q)$ is the sum of the* $P(r_1, r_2, \ldots, r_i, \ldots, r_k)$'s for all sets of r_i's for which the corresponding $\sum_{i=1}^{k} r_i x_i / n$ equals Q. Let $\sum_{\bar{X}=Q}$ stand for this summation.

Then, if the sampled population is infinite or sampling is with replacements

$$P(\bar{X} = Q) = \sum_{\bar{X}=Q} \frac{n!}{r_1! r_2! \cdots r_i! \cdots r_k!} p_1^{r_1} p_2^{r_2} \cdots p_i^{r_i} \cdots p_k^{r_k} \qquad (14\text{-}3)$$

and if sampling is without replacements from a finite population

$$P(\bar{X} = Q) = \sum_{\bar{X}=Q} \frac{\binom{R_1}{r_1}\binom{R_2}{r_2}\cdots\binom{R_i}{r_i}\cdots\binom{R_k}{r_k}}{\binom{N}{n}} \qquad (14\text{-}4)$$

In theory, therefore, if the sampled population is discrete and we know its distribution, we can always obtain the complete exact probability distribution of $\bar{X}$ by forming every possible set of r_i's, calculating both $P(r_1, r_2, \ldots, r_i, \ldots, r_k)$ and $\bar{X}$ for each of them, and summing the probabilities associated with the same value of $\bar{X}$.

Example: Suppose that a random sample of two observations is to be drawn from an infinite population in which units have one of three values $x_1 = 1, x_2 = 3,$ or $x_3 = 5,$ and in which the proportions of units having these values are $p_1 = .5, p_2 = .2,$ and $p_3 = .3,$ respectively. The sample will be composed of r_1 x_1's, r_2 x_2's, and r_3 x_3's. Table 14-1a gives for each of the six different possible sample compositions (i.e., sets of r_i's) the value

$$\frac{r_1 x_1 + r_2 x_2 + r_3 x_3}{n}$$

of $\bar{X}$ and the probability

$$P(r_1, r_2, r_3) = \frac{n!}{r_1! r_2! r_3!} p_1^{r_1} p_2^{r_2} p_3^{r_3}$$

that the sample will be composed in that way (i.e., the probability of the set). Table 14-1b sums the latter probabilities for samples having the same $\bar{X}$ and presents the resulting probability distribution of $\bar{X}$.

Table 14-1

Calculation of Probability Distribution of $\overline{X}$ for Random
Sample of Two Observations from Infinite Population
in Which 50% of Units Have the Value 1, 20% the
Value 3, and 30% the Value 5

(a) Values of $\overline{X}$ and $P(r_1, r_2, r_3)$ for samples consisting of r_1 1's, r_2 3's, and r_3 5's

r_1	r_2	r_3	$\overline{X} = \dfrac{r_1(1) + r_2(3) + r_3(5)}{2}$	$P(r_1, r_2, r_3) = \dfrac{2!}{r_1!\,r_2!\,r_3!}\,.5^{r_1}\,.2^{r_2}\,.3^{r_3}$
2	0	0	1	.25
1	1	0	2	.20
1	0	1	3	.30
0	2	0	3	.04
0	1	1	4	.12
0	0	2	5	.09

(b) Probability distribution of $\overline{X}$

$\overline{X}$	$P(\overline{X})$
1	.25
2	.20
3	.34
4	.12
5	.09

14-1-3 Critique

When k is greater than 2, the probability law (Formula 14-3 or 14-4) is
vague. It tells us to sum the probabilities of *all* sets of r_i's yielding the same
$\overline{X}$ as a single known set, but it does not tell us how to identify these other
sets. When k and n are both small, we can simply calculate the mean for every
possible set of r_i's, thereby determining, by brute force, which values of $\overline{X}$
result from more than one set. However, as either k or n increases, this
becomes prohibitively laborious and then impracticable. Other, more
insightful, approaches also fail us as k or n increases. Therefore, our exact
probability laws for the sample mean are of limited practical usefulness.

14-1-4 Information about $\overline{X}$ When Population Distribution Is Incompletely Known

Thus even if we know the population distribution of X, obtaining the
exact and complete probability distribution of $\overline{X}$ may be, and often is, impracticable. And if we do not know the population distribution, it is impossible.

However although we must know the complete and exact distribution of
the sampled population in order to learn the complete and exact probability

distribution of $\bar{X}$, we can derive much useful *partial* information about the distribution of $\bar{X}$ without knowing anything more about the distribution of the sampled population than its mean μ and variance σ^2. That partial information includes the following items: (a) the mean and variance of the distribution of $\bar{X}$, (b) certain qualities of $\bar{X}$ as an estimator of μ, (c) an upper limit for the probability that $\bar{X}$ will deviate from μ by more than a specified amount, (d) a lower limit for the proportion of the distribution of $\bar{X}$ that lies within a specified interval centered on μ, (e) the limiting value of the variance of $\bar{X}$ as n approaches infinity, (f) the limiting shape of the distribution of $\bar{X}$ as n increases, (g) an approximate probability law for $\bar{X}$ when sample size is quite large.

We shall take up these matters in the sections to follow. The development of these topics will require *at most* a knowledge of the mean and variance of the sampled population. And the results will be applicable to any population that could be encountered in dealing with the real world, i.e., to any population whose variance is not infinite.

14-2   Mean and Variance of the Distribution of $\bar{X}$ and Behavior of Var ($\bar{X}$) as n Increases

The mean $E(\bar{X})$ and variance $\text{Var}(\bar{X})$ of the distribution of $\bar{X}$ were derived in the preceding chapter which showed that for random samples

$$E(\bar{X}) = \mu \tag{14-5}$$

and

$$\text{Var}(\bar{X}) = \left(\frac{N-n}{N-1}\right)\frac{\sigma^2}{n} \tag{14-6}$$

if sampling is without replacements from a finite population or

$$\text{Var}(\bar{X}) = \frac{\sigma^2}{n} \tag{14-7}$$

otherwise.

As n increases Var $(\bar{X})$ must decrease, becoming zero when $n = N$ in the first variance formula and approaching zero as n approaches infinity in the second. These formulas imply, therefore, that as n increases the bulk of the distribution of $\bar{X}$ becomes more and more tightly concentrated in the immediate vicinity of its mean μ. (This qualitative implication will be partially quantified and proved in more elegant fashion in a later section.) Thus, not only is the sample mean a mean-unbiased estimator of the population mean, but at larger and larger sample sizes it becomes a better and better estimator.

14-3 Tchebycheff Inequalities

The formulas for the mean and variance of the distribution of $\bar{X}$ provide us with useful information. However, two distributions with the same variance and mean do not necessarily have the same concentration of units within the same intervals. And it is generally not possible to calculate "concentrations" from variances and means (although we shall see later that it can be done approximately for the distribution of $\bar{X}$ when n is very large). Therefore, in addition to information about the variance of the distribution of $\bar{X}$, we would like to know what *proportions* of the distribution lie within specifiable distances of its mean μ.

Unfortunately, we cannot derive the *exact* proportion lying within a specified distance of μ unless we know the entire distribution of the sampled population. However, we can obtain a *lower limit* for the proportion; the formula is expressed in terms of the mean μ and variance σ^2 of the sampled population, but presupposes no further information about the sampled population and holds for any population having finite mean and variance (as do all populations encountered in real-world applications).

But first we need to develop some very useful formulas concerning the proportion of *any* variable's distribution (not just that of the sample mean) that lies within, or outside, a specified interval centered upon the distribution's mean. If the variable is a random one, these proportion formulas are also probability formulas. That is, they concern the probability that the variable will take a value lying within, or outside, the specified interval. For the sake of conciseness, all formulas will be presented as applying to a random variable; so all formulas will give probabilities as well as proportions.

14-3-1 Tchebycheff's Inequality

If X is any random variable having mean μ and variance σ^2, the proportion of its distribution that lies a distance of more than K standard deviations from its mean, or the probability that X will take a value more than K standard deviations from μ, is less than $1/K^2$. That is,

$$P(|X - \mu| > K\sigma) < \frac{1}{K^2} \tag{14-8}$$

This is *Tchebycheff's Inequality*. Its derivation, given below, is a classic of sparkling simplicity and elegance. It is given below only for the case where X is discretely distributed. The proof for the case where X is continuously distributed is entirely analogous but requires calculus.

Proof: Let the values taken by X be designated by x_i's. Then, by definition

$$\sigma^2 = E[(X - \mu)^2] = \sum_i (x_i - \mu)^2 P(X = x_i)$$

the summation taking place over all values of x_i. Now we can divide the set of all values of x_i into three mutually exclusive and exhaustive subsets: one in which $x_i < \mu - K\sigma$, another in which $\mu - K\sigma \leq x_i \leq \mu + K\sigma$, and a third in which $x_i > \mu + K\sigma$, where K is some positive constant. We can therefore express the original summation, which covered all x_i values, as the sum of three separate summations covering the three separate cases, as follows:

$$\sigma^2 = \sum_{x_i < \mu - K\sigma} (x_i - \mu)^2 P(X = x_i) + \sum_{\mu - K\sigma \leq x_i \leq \mu + K\sigma} (x_i - \mu)^2 P(X = x_i)$$
$$+ \sum_{x_i > \mu + K\sigma} (x_i - \mu)^2 P(X = x_i)$$

Since the squared quantities and the probabilities involved in each summation must, by nature, be positive, each of the three summations must be positive. Therefore, if we remove the middle summation, what remains on the right must either be the same as or smaller than the sum σ^2 of the three original summations. So if we remove it, we must replace the $=$ sign with a $\geq$ sign. Doing so, we obtain

$$\sigma^2 \geq \sum_{x_i < \mu - K\sigma} (x_i - \mu)^2 P(X = x_i) + \sum_{x_i > \mu + K\sigma} (x_i - \mu)^2 P(X = x_i)$$

Now consider the x_i values over which the first summation takes place. Since K and σ are both positive, $\mu - K\sigma$ lies below μ. And the x_i values over which the summation takes place all lie below $\mu - K\sigma$. So all of these x_i values lie farther below μ than does $\mu - K\sigma$, and therefore for *every* one of the x_i values over which the first summation takes place $(x_i - \mu)^2$ is greater than $[(\mu - K\sigma) - \mu]^2$. Therefore, if we substitute $\mu - K\sigma$ for x_i in the squared expression, the resulting summation must be smaller than before so that not only does the $\geq$ sign still hold true, but it should now be replaced with a $>$ sign, which is also true. Likewise, in the second summation all of the x_i's over which the summation takes place are farther above μ than is $\mu + K\sigma$, so substituting $\mu + K\sigma$ for x_i in the squared expression must reduce the value of the summation, again justifying the replacement of the $\geq$ sign with a $>$ sign. Making these substitutions, we obtain

$$\sigma^2 > \sum_{x_i < \mu - K\sigma} (\mu - K\sigma - \mu)^2 P(X = x_i) + \sum_{x_i > \mu + K\sigma} (\mu + K\sigma - \mu)^2 P(X = x_i)$$
$$\sigma^2 > K^2 \sigma^2 [\sum_{x_i < \mu - K\sigma} P(X = x_i) + \sum_{x_i > \mu + K\sigma} P(X = x_i)]$$

Now the summation of $P(X = x_i)$ over all x_i values that are less than $\mu - K\sigma$ is simply the probability that X will take one of those values; hence it is the probability that X will be less than $\mu - K\sigma$. And the summation of

$P(X = x_i)$ over those x_i values that are greater than $\mu + K\sigma$ is simply the probability that X will exceed $\mu + K\sigma$. So making the appropriate substitution of terms

$$\sigma^2 > K^2\sigma^2[P(X < \mu - K\sigma) + P(X > \mu + K\sigma)]$$

$$\frac{1}{K^2} > P(X - \mu < -K\sigma) + P(X - \mu > +K\sigma)$$

$$\frac{1}{K^2} > P(|X - \mu| > K\sigma)$$

And turning the formula around, we have Tchebycheff's Inequality

$$P(|X - \mu| > K\sigma) < \frac{1}{K^2}$$

Example: If X is a random variable whose mean is 5 and whose variance is 9, the probability that on a particular occasion X will take a value more than 2 standard deviations from its mean is

$$P(|X - 5| > 2 \times 3) < \frac{1}{2^2}$$

or

$$P(|X - 5| > 6) < .25$$

Therefore, the probability that X will take a value that is < -1 or > 11 is less than .25.

Illustrative Problems: A certain random variable is known to have a mean of 50 and a variance of 25. Nothing more is known about its distribution. What can we say *for certain* about the probability that the variable, on a specified future occasion, will take a value that deviates from its mean by more than 10?

ANSWER: Since we know nothing about the variable's distribution except its mean and variance, Tchebycheff's Inequality is appropriate. First, let us write the formula for Tchebycheff's Inequality. Then, beneath it, let us cast the question asked by the problem into the form of a probability formula similar to Tchebycheff's Inequality.

$$P(|X - \mu| > K\sigma) < \frac{1}{K^2}$$

$$P(|X - \mu| > 10) < \; ?$$

We see that $K\sigma$ and 10 are in corresponding positions. So $10 = K\sigma = K(5)$ and $K = 2$. Now $1/K^2$ and ? are also in corresponding positions and we now know the value of K; so we can solve for ? Thus $? = 1/K^2 = 1/2^2 = 1/4$. The answer to the problem, therefore, is that we can say with certainty that $P(|X - \mu| > 10)$ lies somewhere in the region below .25.

For the same variable in the preceding problem, what deviation from its mean has a probability that is less than .04?

ANSWER: Proceeding as before

$$P(|X - \mu| > K\sigma) < \frac{1}{K^2}$$

$$P(|X - \mu| > \,?\,) < .04$$

Since $1/K^2$ and .04 are in corresponding positions, $1/K^2 = .04$; so $K^2 = 1/.04 = 25$ and $K = 5$. Knowing K and σ, we can now solve for ?, which must equal $K\sigma$ because they are in corresponding positions: ? $= K\sigma = 5(5) = 25$, which is the answer to the problem.

14-3-2 Complementary and Alternative Forms of Tchebycheff's Inequality

Instead of being interested in the probability that the deviation of X from μ will exceed a specified value we may be interested in the probability that it will stay within designated limits. In that event we need the *complementary form of Tchebycheff's Inequality* which is

$$P(|X - \mu| \leq K\sigma) > 1 - \frac{1}{K^2} \qquad (14\text{-}9)$$

Proof: Since $P(|X - \mu| > K\sigma)$ equals $1 - P(|X - \mu| \leq K\sigma)$ we can substitute the latter for the former in the original inequality, obtaining

$$1 - P(|X - \mu| \leq K\sigma) < \frac{1}{K^2}$$

Adding

$$P(|X - \mu| \leq K\sigma) - \frac{1}{K^2}$$

to both sides of the above inequality, we obtain

$$1 - \frac{1}{K^2} < P(|X - \mu| \leq K\sigma)$$

And turning this formula around we obtain Formula 14-9.

There is an *alternative form of Tchebycheff's Inequality* which differs from the original in that the $>$ and $<$ signs are replaced by $\geq$ and $\leq$ signs. This alternative form and *its* complementary form are given below:

$$P(|X - \mu| \geq K\sigma) \leq \frac{1}{K^2} \qquad (14\text{-}10)$$

$$P(|X - \mu| < K\sigma) \geq 1 - \frac{1}{K^2} \qquad (14\text{-}11)$$

Proof: In deriving the original inequality, we broke

$$\sum_i (x_i - \mu)^2 P(X = x_i)$$

into three summations, one for $x_i < \mu - K\sigma$, one for $\mu - K\sigma \leq x_i \leq \mu + K\sigma$ and one for $x_i > \mu + K\sigma$. Had we included the boundary values $\mu - K\sigma$ and $\mu + K\sigma$ in the first and third summations rather than in the second, i.e., had our three summations been for $x_i \leq \mu - K\sigma$, $\mu - K\sigma < x_i < \mu + K\sigma$, and $x_i \geq \mu + K\sigma$, respectively, then we would have derived Formula 14-10. Formula 14-11 is derived from it in a manner analogous to the derivation of Formula 14-9 from Formula 14-8.

14-3-3   Limitations of Tchebycheff's Inequality

Unless K is ≥ 1, Tchebycheff's Inequality gives us no useful information. We already know that probabilities cannot be less than zero or greater than one. Yet when K is < 1, the right side of Tchebycheff's Inequality $1/K^2$ becomes > 1, and the right side of its complementary form $1 - (1/K^2)$ becomes < 0.

Even for K's greater than 1, the information provided is rather gross. For example, values 10 standard deviations from the mean are extremely rare in many, if not most, distributions. Yet the inequality only tells us that the probability that a random variable will take a value more than 10 standard deviations from its mean is less than .01, a fairly large probability. This grossness is to be expected, of course, since for many distributions "dropping the middle summation" (see derivation) removes a large component of variance in the useful cases where $K \geq 1$. The degree of grossness is illustrated by Table 14-2 which for known distributions and a variety of K values gives the exact value of $P(|X - \mu| > K\sigma)$ and compares it with $1/K^2$.

When X is a simple basic variable, if one knows its mean and variance one generally *also* knows its entire distribution. But in that case one can easily obtain from the distribution of X the *exact* probability that X will deviate from its mean by more than a specified amount, thereby obviating Tchebycheff's Inequality. However, when X is a function of a more basic variable or set of variables, it may be easy to calculate the mean and variance of X but impossible or prohibitively tedious to calculate its entire distribution. It is in these cases that Tchebycheff's Inequality finds its greatest usefulness. The most important of them are those in which the variable X is actually the mean or the sum of a sample of n observations drawn from a population of values of some more basic variable. These cases are especially important because they have features that tend to make the inequality more manageable and to compensate for some of its shortcomings. It is to these cases that we now turn our attention.

Table 14-2

Tchebycheff Upper Limit Contrasted with Exact Value of
$P(|X - \mu| > K\sigma)$ for Several Known Distributions

R_V = Rectangular distribution consisting of V discrete, equally spaced, equally frequent values
U = Uniform distribution—continuous rectangular-shaped distribution of values from 0 to 1
P_m = Poisson distribution with mean of m
N = Normal distribution—a continuous bell-shaped distribution

| K | $\dfrac{1}{K^2}$ | \multicolumn{6}{c}{Exact $P(|X - \mu| > K\sigma)$ for} |
		R_2	R_3	U	$P_{.01}$	P_1	N
1	1.000	0	.667	.423	.010	.080	.317
1.1	.826	0	.667	.365	.010	.080	.271
1.2	.694	0	.667	.307	.010	.080	.230
1.3	.592	0	0	.249	.010	.080	.194
1.4	.510	0	0	.192	.010	.080	.162
1.5	.444	0	0	.134	.010	.080	.134
1.6	.391	0	0	.076	.010	.080	.110
1.7	.346	0	0	.019	.010	.080	.089
1.8	.309	0	0	0	.010	.080	.072
1.9	.277	0	0	0	.010	.080	.057
2	.250	0	0	0	.010	.019	.046
2.25	.198	0	0	0	.010	.019	.024
2.5	.160	0	0	0	.010	.019	.012
2.75	.132	0	0	0	.010	.019	.006
3	.111	0	0	0	.010	.004	.003
3.5	.082	0	0	0	.010	.004	.0005
4	.063	0	0	0	.010	.0006	.00006
5	.040	0	0	0	.010	.00008	$\sim 10^{-6}$
6	.028	0	0	0	.010	10^{-5}	$\sim 10^{-9}$
7	.020	0	0	0	.010	10^{-6}	$\sim 10^{-12}$
8	.016	0	0	0	.010	10^{-7}	$\sim 10^{-15}$
9	.012	0	0	0	.010	$\sim 10^{-8}$	$\sim 10^{-19}$
10	.010	0	0	0	.00005	$\sim 10^{-9}$	$\sim 10^{-23}$

14-3-4   Tchebycheff's Inequality for the Sample Mean

Tchebycheff's Inequality applies to any random variable V and tells us
that

$$P\left(|V - E(V)| > K\sqrt{\text{Var}(V)}\right) < \frac{1}{K^2} \qquad (14\text{-}12)$$

Now the mean $\bar{X}$ of a future random sample of n observations drawn from
a population with mean μ and variance σ^2 is a random variable. And we
know that $E(\bar{X}) = \mu$ and that Var $(\bar{X})$ is equal to σ^2/n or

$$\left(\frac{N - n}{N - 1}\right)\frac{\sigma^2}{n}$$

depending upon the method of sampling. Therefore, substituting these values
in Formula 14-12, we obtain a ***Tchebycheff's Inequality for the Sample Mean***

which tells us that if the population is infinite or if sampling is with replacements

$$P\left(|\bar{X} - \mu| > \frac{K\sigma}{\sqrt{n}}\right) < \frac{1}{K^2} \tag{14-13}$$

whereas if sampling is from a finite population and with replacements

$$P\left(|\bar{X} - \mu| > K\sigma\sqrt{\frac{N - n}{(N - 1)n}}\right) < \frac{1}{K^2} \tag{14-14}$$

The original Tchebycheff Inequality $P(|X - \mu| > K\sigma) < 1/K^2$ lacked flexibility. If we equated the probability boundary $1/K^2$ to a desired value, this rigidly determined the deviation $K\sigma$ to which it applied (since we cannot control σ which is a fixed constant). And if we chose a certain deviation, this completely determined the corresponding probability boundary. However, with the Tchebycheff Inequality for the mean

$$P\left(|\bar{X} - \mu| > \frac{K\sigma}{\sqrt{n}}\right) < \frac{1}{K^2}$$

we can choose K in such a way as to give us the desired probability boundary $1/K^2$ and then choose a value for n that will give the corresponding deviation $K\sigma/\sqrt{n}$ as small a value as we desire. Thus, in general, we can usually manipulate n in such a way as to compensate for any undesired effects of K.

Furthermore, we do not actually need to know the σ of the sampled population. If we substitute for σ any value that is $\geq \sigma$, Tchebycheff's Inequality still holds. Therefore, we can substitute for σ anything that we are certain cannot underestimate it, however vague or inaccurate it may be in other respects. For example, the range of the sampled population's distribution must be $\geq$ the maximum absolute deviation of X from μ, which must be $\geq \sigma$ (since σ^2 is by definition the average squared deviation of X from μ). So if we are ignorant of σ but not of the range, we can substitute the sampled population's known range for its unknown σ. Indeed, we may use whatever partial information is available to overestimate σ^2 as closely as possible. For example, if the population distribution is known to be symmetric, then μ must be in the center so that half the range is $\geq$ the maximum $|X - \mu|$ which is $\geq \sigma$, and therefore half the range can be substituted for σ.

The complementary forms of Formulas 14-13 and 14-14 provide us with a lower limit for the probability that $\bar{X}$ will deviate from μ by no more than a specified amount. Thus, if sampling does not detectably deplete the population,

$$P\left(|\bar{X} - \mu| \leq \frac{K\sigma}{\sqrt{n}}\right) > 1 - \frac{1}{K^2} \tag{14-15}$$

which can also be written in the logically equivalent form

$$P\left(\mu - \frac{K\sigma}{\sqrt{n}} \leq \bar{X} \leq \mu + \frac{K\sigma}{\sqrt{n}}\right) > 1 - \frac{1}{K^2} \tag{14-16}$$

And if sampling does detectably deplete the population,

$$P\left(|\bar{X} - \mu| \le K\sigma\sqrt{\frac{N-n}{(N-1)n}}\right) > 1 - \frac{1}{K^2} \qquad (14\text{-}17)$$

which can also be written as

$$P\left(\mu - K\sigma\sqrt{\frac{N-n}{(N-1)n}} \le \bar{X} \le \mu + K\sigma\sqrt{\frac{N-n}{(N-1)n}}\right) > 1 - \frac{1}{K^2} \quad (14\text{-}18)$$

Illustrative Problems: A proposed item for a new IQ test is an intellectual task whose score X will be the number of seconds required to perform it, except that the maximum score will be 10. The test constructor knows nothing more than this about the distribution of X, but he needs a good estimate of the average score μ that people will make on the task. To how many randomly selected people will he have to administer the task in order to be sure that there will be less than one chance in ten that their mean score $\bar{X}$ will differ from the population mean μ by more than 1 second?

ANSWER: The question can be restated as follows: How large should n be in order that $P(|\bar{X} - \mu| > 1) < .1$? Tchebycheff's Inequality for the sample mean tells us that

$$P\left(|\bar{X} - \mu| > \frac{K\sigma}{\sqrt{n}}\right) < \frac{1}{K^2}$$

We don't know σ, but we do know that the maximum X-value is 10 and that time scores cannot be less than zero, so that the range of possible X-values is < 10. And since σ is $\le$ the range, we can be sure that 10 is an overestimate of σ which can therefore be substituted for it in the inequality. Doing so, we obtain

$$P\left(|\bar{X} - \mu| > \frac{10K}{\sqrt{n}}\right) < \frac{1}{K^2}$$

Now the first inequality above is similar to this one with .1 in a position corresponding to $1/K^2$ and 1 in a position corresponding to $10K/\sqrt{n}$. Therefore, $.1 = 1/K^2$ and $K^2 = 1/.1 = 10$ so that $K = \sqrt{10}$. Also

$$1 = \frac{10K}{\sqrt{n}} = \frac{10\sqrt{10}}{\sqrt{n}}$$

so that $\sqrt{n} = 10\sqrt{10}$ and $n = 1,000$, which is the answer.

14-3-5 Tchebycheff's Inequality for the Sample Proportion and Other Sample Statistics

Any statistic S to be calculated from a future random sample is a random variable, and if we know $E(S)$ and Var (S) we can substitute these expressions into Formula 14-12 and obtain a Tchebycheff's Inequality for S. In Chapter 13 we derived the expected value and variance of the sample statistics T and r (see Formulas 13-46 through 13-52) and from the latter we can easily obtain

$E(r/n) = (1/n) E(r)$ and Var $(r/n) = (1/n^2)$ Var (r). So making use of this information we can state Tchebycheff Inequalities for $T, r,$ and r/n. The most important of these is the one for r/n which is analogous to the inequality for $\bar{X}$ since r/n, the proportion of successes in the sample, is also the mean number of successes per observation.

Consider a future random sample of n observations to be drawn from a population containing N units. If the population units have numerical values, let the mean and variance of the population be μ and σ^2 and let T be the sum of the values of the n sample observations. If the values of the population units are categorical labels, let R be the number and p be the proportion of units in the population that are successes, and let r be the number and r/n be the proportion of observations in the sample that are successes. Then, if the sampled population is infinite or if sampling is with replacements,

$$P\left(|T - n\mu| > K\sigma\sqrt{n}\right) < \frac{1}{K^2} \tag{14-19}$$

$$P\left(|r - np| > K\sqrt{np(1-p)}\right) < \frac{1}{K^2} \tag{14-20}$$

$$P\left(\left|\frac{r}{n} - p\right| > K\sqrt{\frac{p(1-p)}{n}}\right) < \frac{1}{K^2} \tag{14-21}$$

whereas if the sampled population is finite and sampling is without replacements,

$$P\left(|T - n\mu| > K\sigma\sqrt{\frac{n(N-n)}{N-1}}\right) < \frac{1}{K^2} \tag{14-22}$$

$$P\left(\left|r - \frac{nR}{N}\right| > K\sqrt{\frac{nR}{N}\left(\frac{N-R}{N}\right)\left(\frac{N-n}{N-1}\right)}\right) < \frac{1}{K^2} \tag{14-23}$$

$$P\left(\left|\frac{r}{n} - \frac{R}{N}\right| > K\sqrt{\frac{R}{nN}\left(\frac{N-R}{N}\right)\left(\frac{N-n}{N-1}\right)}\right) < \frac{1}{K^2} \tag{14-24}$$

Of course, all these inequalities have their complementary and alternative forms. And they all have the advantage of being more flexible than Tchebycheff's Inequality for a single sample observation X. The inequality for r/n may be regarded as a special case of the inequality for $\bar{X}$, since r/n equals $\bar{X}$ when successes are given a value of one and nonsuccesses are assigned a value of zero. For similar reasons, the inequality for r may be regarded as a special case of the inequality for T. And, of course, T is simply $n\bar{X}$. So all of the above inequalities are closely related to the one for $\bar{X}$ and tend to share in the latter's virtues.

Illustrative Problems: The weight X of taffy apples produced by a certain candy company has a mean of 5 ounces and a variance of 4 square ounces. The apples are packed a gross to a box. What is the upper limit to the probability that the net shipping weight of the apples in a box will deviate from its mean of 45 pounds by more than 5 pounds?

ANSWER: The question being asked can be put in the form

$$P(|T - n\mu| > 5 \text{ lb}) < ?$$

whereas Tchebycheff's Inequality for the sample sum tells us that

$$P(|T - n\mu| > K\sigma\sqrt{n}) < \frac{1}{K^2}$$

So if we let 5 lb equal $K\sigma\sqrt{n}$, solve for K, and then calculate $1/K^2$, we have our answer.

$$K\sigma\sqrt{n} = 5 \text{ lb} = 80 \text{ oz}$$

$$K = \frac{80 \text{ oz}}{\sigma\sqrt{n}} = \frac{80 \text{ oz}}{2 \text{ oz}\sqrt{144}} = \frac{40}{12} = \frac{10}{3}$$

$$\frac{1}{K^2} = \left(\frac{3}{10}\right)^2 = .3^2 = .09$$

So, $P(|T - n\mu| > 5 \text{ lb}) < .09$.

A company produces small items, a random 1% of which are defective, and sells them wholesale in lots containing 10,000 items. A customer asks if the company can guarantee that less than one lot in 25 contains more than 150 defectives. Can it?

ANSWER: The problem tells us that $p = .01$ and that $n = 10,000$ so that $np = 100$ and asks whether or not the following inequality is true

$$P(r > 150) < \frac{1}{25}$$

The inequality can also be written in the equivalent forms

$$P(r - 100 > 50) < \frac{1}{25}$$

and

$$P(r - np > 50) < \frac{1}{25}$$

Now $P(r - np > 50)$ can be no larger than $P(|r - np| > 50)$ since the latter includes both the former and $P(r - np < -50)$. Therefore, if we can prove that $P(|r - np| > 50) < 1/25$, the original inequality must be true. Now Formula 14-20 is in a form similar to the last inequality, and if we substitute $K = 5$ and $np(1 - p) = 10,000(.01)(.99) = 99$, so that $K\sqrt{np(1 - p)} = 5(9.95) = 49.75$, it tells us that

$$P(|r - np| > K\sqrt{np(1 - p)}) < \frac{1}{K^2}$$

$$P(|r - np| > 49.75) < \frac{1}{25}$$

Now since 50 is greater than 49.75, if the probability that $|r - np|$ will exceed the latter is less than 1/25, the probability that it will exceed the former must be also. So the company can truthfully make the guarantee.

A million school children are to be inoculated against a certain disease. It is not known what proportion of the infinite population of potential school children would suffer serious side effects, but the sample proportion will be used to estimate it. What can be said about the probability that the estimate will lie within $\pm.005$ of the population proportion?

ANSWER: If the million children were randomly selected, or if the sample consisted of all children in school, Formula 14-21 tells us that

$$P\left(\left|\frac{r}{n} - p\right| > K\sqrt{\frac{p(1-p)}{n}}\right) < \frac{1}{K^2}$$

We do not know the value of p, but it is easy to see that the p-value .5 maximizes $p(1-p)$ and therefore maximizes

$$K\sqrt{\frac{p(1-p)}{n}}$$

for fixed values of K and n. Therefore, if we substitute .5 for p under the square root sign, the inequality must still hold. Doing so and substituting 1,000,000 for n,

$$\sqrt{\frac{p(1-p)}{n}}$$

becomes

$$\sqrt{\frac{.25}{1,000,000}} = \frac{.5}{1,000} = .0005$$

and the inequality becomes

$$P\left(\left|\frac{r}{n} - p\right| > .0005\,K\right) < \frac{1}{K^2}$$

whereas the question asked can be put in the (complementary) form

$$P\left(\left|\frac{r}{n} - p\right| > .005\right) < ?$$

Therefore, letting $.0005K = .005$ we find that $K = 10$ so that $1/K^2 = .01$ and

$$P\left(\left|\frac{r}{n} - p\right| > .005\right) < .01$$

which means that

$$P\left(p - .005 \leq \frac{r}{n} \leq p + .005\right) > .99$$

So we can say that there is a probability greater than .99 that the sample proportion r/n will lie within $\pm.005$ of the population proportion p.

14-3-6 Critique

It has already been pointed out that the inequalities given for $\bar{X}$ are considerably more useful in applications than those given for a single sample observation X, and this is also true of derivatives of $\bar{X}$ such as T and special cases of $\bar{X}$ such as r/n. In the case of X we can choose either the deviation limit or the probability limit, but not both, whereas in the case of $\bar{X}$ we can make both of them whatever we wish (or, at least, whatever we are *likely* to wish) simply by choosing a sufficiently large value of n.

Nevertheless, despite this considerable advantage the method is still gross in the sense that it does not tell us the *exact* probability, or even a close approximation to it, but only an upper (or lower) limit, somewhere below (or above) which the true probability must lie. This upper limit is generally many times greater than the true probability. So the method is expensive in the sense that it often calls for n-values enormously greater than would actually be required to produce the desired probability for the specified deviation if we knew the distribution of $\bar{X}$ at each possible n-value. Furthermore, other methods (soon to be discussed) are available which generally produce much closer (although more ambiguous) approximations to the exact probability, especially at the large n-values so often demanded by the inequality. These factors tend to make the inequality method an inefficient one in practical applications unless a huge n is involved *anyway* (i.e., is part of the statement of the problem) and an exact probability, or a close estimate of it, is not required.

Tchebycheff's Inequality for $\bar{X}$ has the advantages of ease, simplicity, and clarity (i.e., lack of ambiguity), which sometimes make it preferable to otherwise more efficient methods. Its principle virtue, however, lies not in practical applications but rather in the "theoretical" information it provides about the distribution of the sample mean. In the complementary forms given by formulas such as 14-16

$$P\left(\mu - \frac{K\sigma}{\sqrt{n}} \leq \bar{X} \leq \mu + \frac{K\sigma}{\sqrt{n}}\right) > \frac{K^2 - 1}{K^2}$$

it gives us an absolute lower limit for the proportion of the distribution of $\bar{X}$ that lies within a specified interval centered on μ. It therefore gives us explicit *quantitative* information about the *concentration* of the distribution of $\bar{X}$ in the neighborhood of the population mean μ. Thus, it provides us with valuable information about the effectiveness of $\bar{X}$ as an estimator of μ.

14-4   The Law of Large Numbers

14-4-1   The General Case

In the preceding sections on Tchebycheff's Inequality it was proven that if $\bar{X}$ is the mean of a future random sample of n observations from an inexhaustible population with mean μ and variance σ^2 then,

$$P\left(|\bar{X} - \mu| > \frac{K\sigma}{\sqrt{n}}\right) < \frac{1}{K^2}$$

Now suppose that we wanted a deviation limit $K\sigma/\sqrt{n}$ such that the probability is negligible that the limit will be exceeded by the deviation of $\bar{X}$ from μ. We can make the probability negligible by letting $1/K^2$ be an extremely small value, e.g., $1/1{,}000{,}000$. This of course means that K must be very large, e.g., $1{,}000$ which makes the deviation limit $1{,}000\,\sigma/\sqrt{n}$. But we can still make the deviation limit as small a value > 0 as we like by making n sufficiently large, e.g., $10{,}000{,}000{,}000{,}000{,}000$ so that $\sqrt{n} = 100{,}000{,}000$ and $1{,}000\,\sigma/\sqrt{n} = \sigma/100{,}000$.

We can express this idea more formally as follows (where instead of fixing the probability limit and then adjusting the deviation limit, we fix the latter and then adjust the former). Let the deviation limit $K\sigma/\sqrt{n}$ be a constant ϵ, so that $K = \epsilon\sqrt{n}/\sigma$. Substituting these values into Tchebycheff's Inequality for $\bar{X}$ given above, we obtain

$$P(|\bar{X} - \mu| > \epsilon) < \frac{\sigma^2}{\epsilon^2 n} \tag{14-25}$$

Now, no matter how small ϵ is, we can always make the probability limit $\sigma^2/(\epsilon^2 n)$ as small a value > 0 as we wish simply by making n sufficiently large. Indeed, even if ϵ is an "infinitesimally" (i.e., a negligibly) small constant value, as n increases toward infinity, the probability upper limit $\sigma^2/(\epsilon^2 n)$ must diminish and must approach zero as a limit.

We could also write the above inequality in the complementary form

$$P(|\bar{X} - \mu| \leq \epsilon) > 1 - \frac{\sigma^2}{\epsilon^2 n} \tag{14-26}$$

or its equivalent

$$P(\mu - \epsilon \leq \bar{X} \leq \mu + \epsilon) > 1 - \frac{\sigma^2}{\epsilon^2 n} \tag{14-27}$$

In this case, no matter how small ϵ is, we can always make the probability lower limit $1 - \sigma^2/(\epsilon^2 n)$ as large a value < 1 as we like by making n sufficiently large, and as n approaches infinity, the probability lower limit must approach 1 as a limit.

This bit of mathematical rationale constitutes the proof of the **Law of Large Numbers**, which, in mathematical symbols, says that:
For every $\epsilon > 0$, as $n \longrightarrow \infty$,

$$P(|\bar{X} - \mu| > \epsilon) \longrightarrow 0 \tag{14-28}$$

or, in complementary form,

$$P(|\bar{X} - \mu| \leq \epsilon) \longrightarrow 1 \tag{14-29}$$

In either form the Law of Large Numbers tells us, in effect, that no matter how microscopically small a value we choose for ϵ, the probability that $\bar{X}$ will fall within the interval $\mu \pm \epsilon$ approaches a limiting value of 1 as sample size increases. Or, in simpler words, if sample size is sufficiently large, $\bar{X}$ is virtually certain to fall practically on top of μ and therefore to estimate μ extremely well. When the probability of a sample statistic deviating from a population index by less than any constant preassigned amount, however small, approaches a limiting value of 1 as n approaches infinity, the sample statistic is said to be a **consistent estimator** of the population index. So $\bar{X}$ is a consistent (as well as a mean-unbiased) estimator of μ. The Law of Large Numbers also tells us that, no matter how small ϵ is, as n increases toward infinity, the *proportion* of the distribution of $\bar{X}$ that lies within the interval from $\mu - \epsilon$ to $\mu + \epsilon$ approaches a limiting value of 1.

Our derivation of the Law of Large Numbers assumed that either the sampled population was infinite or sampling was with replacements. However, in essence, the law is also valid if the population is finite and sampling is without replacements. For in that case, the closest n can come to infinity is when it equals N, the number of units in the population, in which case $\bar{X} = \mu$ and the deviation of $\bar{X}$ from μ is zero. Thus, in the "finite case" the Law of Large Numbers would be completely accurate if we simply replaced the condition "as $n \rightarrow \infty$" by the condition "as $n \rightarrow N$." This can be formally proved by using Formula 14-14 rather than Formula 14-13 as the starting point for a derivation analogous to the one already presented.

14-4-2   Bernoulli's Theorem

Now if the sampled population is dichotomous, containing a proportion p of successes, each of which may be arbitrarily assigned a value of 1, and a proportion $1 - p$ of failures, each of which may be arbitrarily assigned a value of 0, then the population mean is p and the sample mean is r/n, which is also the proportion of successes in the sample. Substituting these values into the general form of the Law of Large Numbers, we obtain a *Law of Large Numbers for Proportions:*
For every $\epsilon > 0$, as $n \rightarrow \infty$,

$$P\left(\left|\frac{r}{n} - p\right| > \epsilon\right) \longrightarrow 0 \qquad (14\text{-}30)$$

or, in complementary form,

$$P\left(\left|\frac{r}{n} - p\right| \leq \epsilon\right) \longrightarrow 1 \qquad (14\text{-}31)$$

This special case of the Law of Large Numbers is known as **Bernoulli's Theorem**. It could have been derived directly from Tchebycheff's Inequality for the Sample Proportion r/n (Formula 14-21) in a manner analogous to that used in deriving the general form of the Law of Large Numbers. It tells us that no matter how small ϵ is, the probability that r/n will fall within the interval from $p - \epsilon$ to $p + \epsilon$ —and therefore the proportion of the distribution of r/n that does so—can be made as large a value < 1 as we like by making n sufficiently large. Consequently, it affirms that the sample proportion of successes r/n is a consistent estimator of the population proportion of successes p.

14-4-3 Illustrations

The implications of Bernoulli's Theorem are easily illustrated. Suppose that the sampled population is infinite. Then binomial tables (or repeated applications of the binomial probability law) give us, for fixed values of p and n, the probabilities that r will take each of the integral values from 0 to n. These probabilities are, of course, also the probabilities that r/n will take each of the above values divided by n. Therefore, we can easily obtain the complete probability distribution of r/n for fixed values of p and n; and we can do it for each of several increasing values of n, thereby illustrating the "drawing in" of the r/n distribution about p as n increases. This has been done for the case in which $p = .2$, and the results are shown graphically in Figures 14-1 and 14-2, the former showing the relative frequency, or probability, distribution of r/n and the latter showing the cumulative relative frequency, or cumulative probability, distribution for the cases where n equals 1, 10, 100, and 1,000. In each of the four graphs in each figure, r/n is represented on the horizontal axis—always to exactly the same scale.

Clearly, as n increases, r/n becomes more and more compactly distributed about its expected value p. And this is just as true when the increasing occurs at the very smallest possible values of n as when n is "approaching infinity." Consequently, as n increases, r/n becomes a better and better estimator of p in the sense that the r/n value for a particular random sample is more and more likely to fall close to p.

Bernoulli's Theorem tells us that as n approaches infinity, the entire probability distribution of r/n draws in closer and closer to p. However, in any *particular* sample it is *possible* for r to have any value from 0 to n and therefore for r/n to have any value from 0 to 1, no matter how large n is. So Bernoulli's Theorem does not *guarantee* that as $n \to \infty$ $r/n \to p$, although it does *suggest* that it becomes *overwhelmingly probable*.

If we were actually to draw a series of successive random observations and calculate r/n the sample proportion of successes after n draws, for each value of n from 1 up to some very large number, what might we expect? At very early n-values we would expect r/n to fluctuate widely due to the randomness of sampling. And we would expect *some* fluctuation due to randomness even at very large n-values. However, knowing that as n increases Var (r/n)

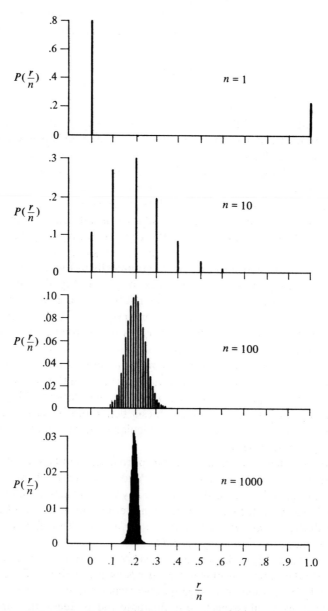

FIG. 14-1. Increasing concentration of distribution of sample proportion r/n about population proportion p as sample size n increases
(population is infinite and $p = .2$).

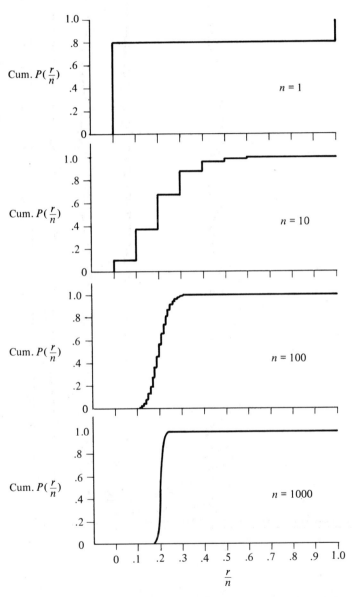

FIG. 14-2. Increasing concentration of cumulative distribution of r/n about p as n increases (population is infinite and $p = .2$).

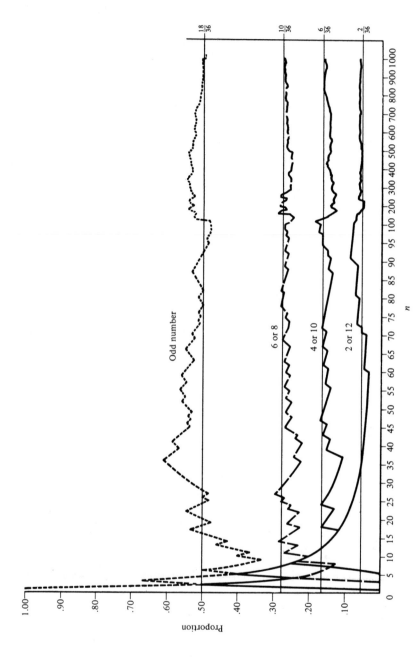

FIG. 14-3. Illustration of effect suggested by Bernoulli's theorem. Proportion of times, at the nth throw, that a pair of dice had come up as indicated. (For $100 < n \leqq 1000$, only the most devient outcomes in blocks of 10 throws are plotted, and scale is condensed.)

approaches a limiting value of zero and that the bulk of the distribution of r/n draws in closer and closer to p, we would expect the fluctuating values of r/n to *tend* to deviate less and less from their expected value p as n increases. Thus, as n increases we would expect r/n to migrate, somewhat erratically, toward p. And only in the most exceedingly rare exceptional cases should our expectation be disappointed.

This effect is illustrated by Figure 14-3 which gives the proportion r/n of outcomes of certain types that occurred in the first n throws of a pair of dice thrown 1,000 times. Thus, the proportion r/n of "successes" having been obtained by the nth throw is plotted against n, which goes from 1 to 1,000. Actually, this is done for each of four different mutually exclusive and exhaustive types of outcome: odd number comes up, number coming up is either a 6 or an 8, number is either 4 or 10, number is either 2 or 12. These outcomes were chosen because the biases due to "number holes" (see Section 5-4-4) tend to cancel each other out, so that we can calculate p almost exactly, and because the corresponding p-values, 18/36, 10/36, 6/36, 2/36 represent a reasonably diverse variety. Although the curves for the four different sample proportions intermingle chaotically for the first few throws, they soon become disentangled and begin to stabilize in the region of their corresponding p-value. Sometimes they approach p and sometimes they depart from it, but the former tendency eventually prevails over the latter with the result that by the time $n = 1,000$ all four curves are impressively close to a horizontal line through p.

14-5 The Central Limit Effect

The Law of Large Numbers told us that as n increases, the bulk of the distribution of $\bar{X}$ "draws in" toward μ, but it gave us no clue as to the shape of the distribution. In this section we shall consider the limiting shape of the distribution of $\bar{X}$ as n increases.

14-5-1 The Normal Distribution

The normal distribution is a *continuous, symmetric, bell-shaped* distribution extending from $-\infty$ to $+\infty$, although 99.99% of it is concentrated within ± 3.89 standard deviations σ's of its mean μ. Because of its symmetry and bell-shape, its mean, median, and mode all coincide, having the same value which "bisects the bell." Its standard deviation is the distance from the mean to the "point of inflection"—the point at which the curvature of one side of the bell changes from concave to convex.

If X is a normally distributed variable with mean μ and variance σ^2, its density function is given by the formula

$$f(X) = \frac{1}{\sigma\sqrt{2\pi}}e^{-\frac{1}{2}\left(\frac{X-\mu}{\sigma}\right)^2}$$

where e is the constant 2.71828 . . . and $\pi = 3.14159$ . . . . Thus, if we plotted a curve whose horizontal coordinates were values of X and whose vertical coordinates were the corresponding values of $f(X)$, the curve would describe the bell-shaped distribution of X. It should be clear from the above formula that the normal distribution is a "two-parameter" distribution, the parameters being μ and σ, and that two normal distributions differ if, and only if, they have different means or different variances. So if we converted *any* normally distributed variable X to a standardized variable $Z = (X - \mu)/\sigma$, *all* such standardized normal variables should have the *same standardized* normal distribution with zero mean and unit variance, whose density function is

$$f(Z) = \frac{1}{\sqrt{2\pi}}e^{-Z^2/2}$$

The formulas for $f(X)$ and $f(Z)$ describe the normal curve but are of little further use to the practitioner. Tables of cumulative probabilities for a standardized normal variable are easily obtainable (one can be found in Appendix B). And it is far easier to standardize a normally distributed variable and obtain its probabilities from tables than to obtain them by applying the methods of calculus to the above formulas.

The normal distribution is a mathematical fiction. It is *impossible*, not just unlikely, for any variable in the real world to have an *exactly* normal distribution, although for certain variables the *bulk* of their distribution is *approximately* normal. The impossibility is insured by the fact that the normal distribution extends from $-\infty$ to $+\infty$. It therefore treats all values in that infinite range as possible values that the variable could actually assume on some occasions, however rare they might be. But of course no real-world variable has an unlimited range of possible values, and for most the range is relatively short. Distributions of heights, or weights, of people are often cited as examples of a normal distribution. Within the interval $\mu \pm 2\sigma$ they are often approximately normal. But since neither heights nor weights can take values less than zero, they certainly cannot extend to $-\infty$ as required if their distributions are to be exactly normal.

Another feature of the normal distribution—its exact symmetry—although not impossible, is seldom encountered in real-world distributions. Often the most probable values of a variable lie very close to one end of its range of possible values, which acts as an absolute limit that cannot be transcended. In such cases the distribution is relatively free to spread out in one direction but not in the other. This tends to make the distribution asymmetric. This is often the case with error or time scores for the performance of an easy task. The error scores cannot be less than zero and the time scores cannot fall below some physiologically determined lower limit representing the fastest performance that is humanly possible. Yet because the task is easy, the most frequent scores are small ones lying close to the "barrier." So the distribution can, and does, spread far upward, but very little downward, from its mode, therefore presenting a markedly asymmetric appearance. (See Figures 14-4 and 14-5.)

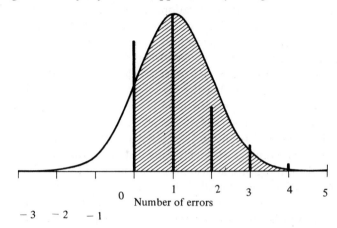

FIG. 14-4. Distribution of error scores for an easy task and "fitted" normal distribution with the same mean and variance. Unshaded area of normal distribution refers to impossible scores.

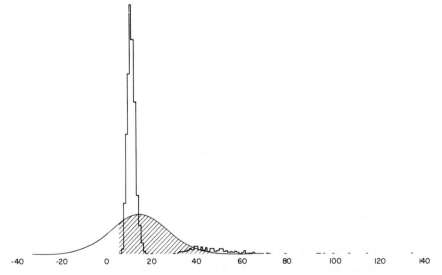

FIG. 14-5. Distribution of time scores for an easy task (histogram) and normal distribution (bell-curve) with the same mean, variance, and area. Unshaded area of normal distribution refers to impossible scores lying below the physiological limit.

A final characteristic of the normal distribution was its continuousness. Many real-world variables, such as time, have continuous distributions. But measurements made upon such variables are never made with infinite precision (to an infinite number of decimal places); they are always "rounded" (e.g., to the nearest thousandth of a second) and therefore have discrete distributions. Since we generally deal with observations, or measurements of values, rather than with the values themselves, we seldom deal directly with continuous distributions(although we *do* make inferrences about them based on statistics).

Thus, as an exact distribution, the normal distribution has characteristics some of which are impossible, and some of which are rarely encountered, in real-world distributions. Within the region $\mu \pm 2\sigma$, and sometimes a little farther, it serves as a fair approximation to the distributions of some, but by no means all, real-world variables. However, outside this interval—at the "tails" of the normal distribution—the goodness of the approximation often tends to deteriorate rapidly with increasing distances from μ. Unfortunately, it is often at the tail regions, at distances of about 1.5σ to 3.5σ from μ, that a good approximation is most needed. Thus, the virtues of the normal distribution as a sort of omnibus approximation to a variety of real-world distributions have been rather badly exaggerated. Its principle virtue is that it is the form of distribution *approached* by the distribution of certain sample statistics as sample size increases toward infinity.

14-5-2   The Central Limit Theorem

Let $\bar{X}$ be the mean of a future random sample of n observations to be drawn from an undepletable population with finite mean μ and finite variance σ^2. Then the **Central Limit Theorem** states that:

As n approaches infinity, the cumulative distribution of the standardized sample mean

$$Z = \frac{\bar{X} - \mu}{\dfrac{\sigma}{\sqrt{n}}}$$

(for finite values of Z) approaches the cumulative standardized normal distribution.

Consequently if we let ϕ stand for a normally distributed variable with mean of zero and variance of 1, the Central Limit Theorem tells us that (under the stated qualifications) for any finite value of K

$$As \quad n \longrightarrow \infty, \quad P(Z \leq K) \longrightarrow P(\phi \leq K)$$

or, equivalently, that

$$As \ n \longrightarrow \infty, \ P\left(\bar{X} \leq \mu + \frac{K\sigma}{\sqrt{n}}\right) \longrightarrow P(\phi \leq K)$$

The latter probability, $P(\phi \leq K)$, is extensively tabled and can therefore serve as an approximation to $P(Z \leq K)$ when n is very large.

The Central Limit Theorem is an exercise in pure mathematics and some of its qualifications need not trouble the practitioner. Since no variable in the real world has an unlimited range of values, all populations encountered in the real world must have finite means, variances, and ranges with the result that Z must necessarily be finite. So the "finite" qualification concerning μ, σ^2, and Z is automatically satisfied in practice. However, an important implication stems from the qualification that Z and K must remain finite while n approaches infinity. It is that at finite sample sizes n must be *extremely large relative to* $|K|$ in order for $P(Z \leq K)$ to be well approximated by $P(\phi \leq K)$ when K is negative or for $P(Z \geq K)$ to be well approximated by $P(\phi \geq K)$ when K is positive. And this in turn implies that the farther one goes out on either tail of the Z distribution to a point K, the worse the normal approximation to the cumulative tail probability (cumulated from tail inward to K) tends to become. This can be proved mathematically (see Appendix A). The "cumulative" qualification is included because whereas the normal distribution is continuous, if the sampled population is discrete the distribution of $\bar{X}$ will also be discrete. Only in their cumulative forms can a discrete distribution be properly regarded as approaching, or being approximated by, a continuous distribution. (However, the frequency polygon for the discrete relative frequency distribution of $\bar{X}$ generally approaches a normal curve, i.e., increasingly resembles the density function for a normally distributed variable, as n increases.) Finally, the theorem refers to the *approach* by the distribution of Z to a *standardized* normal distribution because the latter remains constant as n increases and it is easier to conceptualize a changing distribution approaching a constant distribution than another changing distribution. Actually, the theorem could just as validly have stated that the "*fit*" between the cumulative distribution of $\bar{X}$ and the cumulative distribution of a normally distributed variable with mean μ and *changing* variance σ^2/n improves as n approaches infinity.

Thus, the Central Limit Theorem offers us an approximation to the cumulative distribution of $\bar{X}$ that improves and approaches ultimate perfection with increasing sample sizes and does so for any population that could be encountered in actual practice. Furthermore, it does not require that we know anything about the sampled population other than its mean and variance— nothing is said in the theorem about the shape of the sampled population. And finally, by implication, it provides an easy method of obtaining appproximate probabilities for $\bar{X}$. Unfortunately, however, it does not tell us how good the approximation is nor how large n must be in order for us to feel justified in using it.

14-5-3   The Central Limit Effect

The Central Limit Theorem tells us the shape of the *approached* distribution, not the shape of the "approaching" distribution of $\bar{X}$ nor the rapidity of the approach. The latter are manifestations of what we shall call the Central Limit *Effect*. It is important to distinguish between the Central Limit Theorem and the Central Limit Effect. Whereas the *approach* to normality is a mathematically demonstrable theorem, the *degree* of *approximation* to normality, and the *rapidity* of the approach, at particular finite n-values (especially at small ones) are simply empirical facts about which the theorem guarantees nothing. And if we wish to determine these facts, we must abandon the Central Limit Theorem and investigate the distribution of $\bar{X}$ directly.

One way of determining the goodness of the approximation is to calculate the exact probability distribution of $\bar{X}$ for samples of specified size from known populations. This can be a tedious chore even when it is feasible; therefore, the first such attempts used populations for which the necessary calculations were easiest. The "discrete rectangular" distribution presents a particularly easy case since each of its K values has the same probability $1/K$ of being drawn; so it was investigated in many of the early studies of the Central Limit Effect. Figure 14-6 shows the distribution of $\bar{X}$ when sampling randomly and with replacements from a population consisting of four equally frequent equally spaced values and therefore having a discrete rectangular distribution. For $n = 1$ the distribution of $\bar{X}$ would be the same as that of X, i.e., the same as that of the sampled population. When $n = 2$, the distribution of $\bar{X}$ is "triangular"; at $n = 3$ it is already beginning to appear bell-shaped, and by the time $n = 5$ it is impressively so. (Here we are speaking of the imaginary "curve" connecting the tops of the vertical lines. We have already said that only the cumulative form of a discrete distribution can actually approach normality, but by disregarding the gaps between discrete values, the approach can be "seen" even in the uncumulated form.) The early investigators were greatly encouraged by such results and some very rash generalizations resulted from their enthusiasm.

Another way of investigating the shape of the distribution of $\bar{X}$ is by drawing a large number of random samples of specified size from the population in question, calculating the value of $\bar{X}$ for each sample, graphing the relative frequency distribution of these $\bar{X}$'s and thereby obtaining an *empirical* probability distribution of $\bar{X}$ for the case in question. This method does not yield the *exact* probability distribution of $\bar{X}$, but if the number of random samples drawn is quite large, the approximation can be very close at all but the rarest portions of the true distribution of $\bar{X}$. When the population

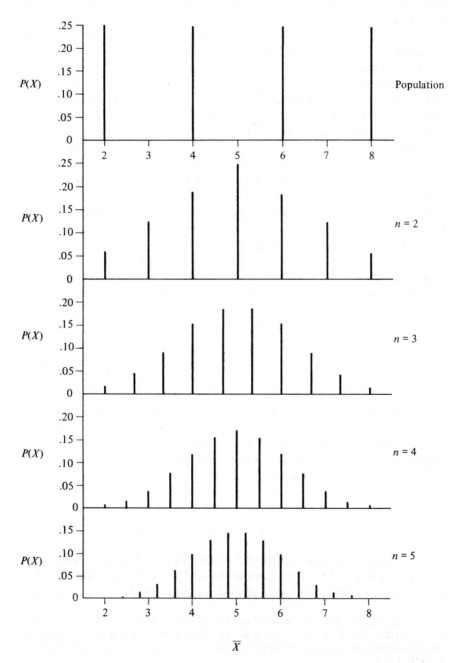

FIG. 14-6. Distribution of $\bar{X}$ for samples of size n from a discrete rectangular population.

contains many different values or when n is not small, this method may be the only practically feasible approach. It was used for this reason in some of the early studies, but this was before the advent of the high-speed electronic computer and consequently the number of samples drawn was almost invariably insufficient.

A more recent study using this method was conducted by the writer.[1] One of his experiments had used as the dependent variable the amount of time required for an instrument operator to reach up from a fixed hand-position and operate a push button. In order to obtain some idea of the shape of the distribution of time scores for this task, a single operator was given 2,520 trials at it yielding the distribution of time scores already shown in Figure 14-5. A very slightly "smoothed" and modified version of this distribution was then constructed and called the "X population"; another population was then constructed having exactly the same mean, variance, and number of units (and some other characteristics) as the X population but being as normal in shape as could be achieved under the limitations imposed. This "normal" population was called the "Y population." Both populations are shown in Figure 14-7.

For each of the "doubling" n values, 2, 4, 8, 16, 32, 64, 128, 256, 512, 1024, ten thousand random samples of n observations were drawn (with replacements) from the X population; for each sample $\bar{X}$ was calculated as well as

$$Z_X = \frac{\bar{X} - \mu}{\dfrac{\sigma}{\sqrt{n}}}$$

and the distribution of the 10,000 values of Z_X obtained at each n value was plotted. For each such sample drawn from the X population, a sample was drawn (with replacements) from the Y population using the same random numbers as had been used to draw the X sample. For each such sample the mean $\bar{Y}$ and standardized mean

$$Z_Y = \frac{\bar{Y} - \mu}{\dfrac{\sigma}{\sqrt{n}}}$$

were calculated and the distribution of 10,000 values of Z_Y at each n value was plotted just below that of the corresponding Z_X distribution. Actually, for both Z_X and Z_Y two graphs were plotted for each distribution, one a histo-

---

[1] J. V. Bradley, "A Large-Scale Sampling Study of the Central Limit Effect," *Journal of Quality Technology*, **3** (1971), 51–68, and "The Central Limit Effect for a Variety of Populations and the Influence of Population Moments," *Journal of Quality Technology*, **5** (1973), 171–177.

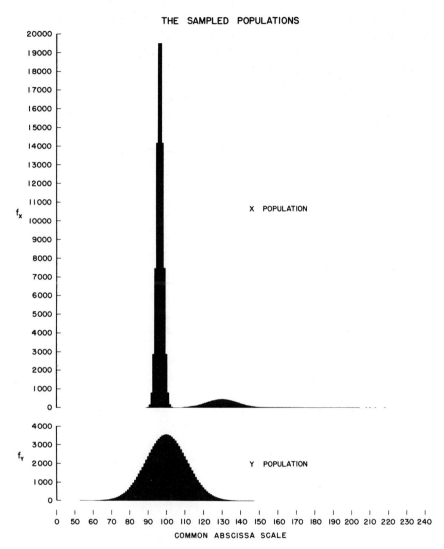

FIG. 14-7. Distributions of the *X* and *Y* populations, sampled in the
writer's study of the Central Limit Effect.

gram showing the relative frequency distribution, the other a step-curve
showing the cumulative relative frequency distribution. And on all graphs,
for comparison, the "approached" normal distribution was also plotted—a
bell-shaped curve to accompany the histograms and a sigmoid (∫-shaped)

curve to accompany the cumulative step-curves.[2] The purpose of the Z_Y distributions was to serve as "controls." The true distribution of means or standardized means of random samples from a normally distributed population is exactly normal. Therefore, if sampling was properly random and if the computer was properly programmed, the Z_Y distributions should conform closely to the "approached" normal distributions, discrepanies being convincingly attributable to chance (or to the fact that the sampled Y population is not *exactly* normal). The fact that they do so tends to confirm the validity of the results for Z_X.

The results are shown in Figures 14-8 through 14-18 in which sample size

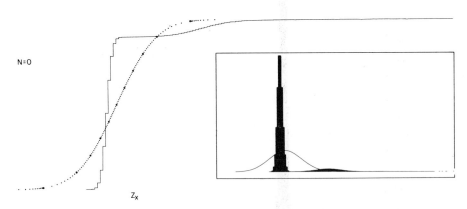

FIG. 14-8. Cumulative distribution (step-curve) and frequency distribution (histogram) of the X population converted to Z-scores, and corresponding normal distributions with the same mean and variance (dotted sigmoid curve and smooth bell-curve).

n is denoted by a capital N. Figure 14-8 labeled $N = 0$ simply shows the relative frequency distribution and cumulative relative frequency distribution of the sampled X population converted to Z-scores $Z = (X - \mu)/\sigma$, i.e., in the form most appropriate for comparison with the figures to follow. Discrepancies between the distribution of Z_X and the normal curve are both gross and obvious until $n = 32$, at which point the distribution of Z_X appears to have a sort of lopsided quasi-bellshape. However, at $n = 32$ the distribution of Z_X clearly falls short of the normal curve at its far left tail and exceeds it at

[2]Although they extend to infinity in both directions, the normal and cumulative normal curves are shown in Figures 14-8 through 14-18 only for Z-values having cumulative probabilities in the range from .001 to .999. The cumulative normal curve is plotted as a series of dots. The 11 enlarged dots represent cumulative probabilities of .01, .1, .2, .3, .4, .5, .6, .7, .8, .9, and .99. The first 10 and last 10 dots represent cumulative probabilities that are integral multiples of .001, and the 99 dots from the first enlarged dot to the last enlarged dot represent cumulative probabilities that are integral multiples of .01.

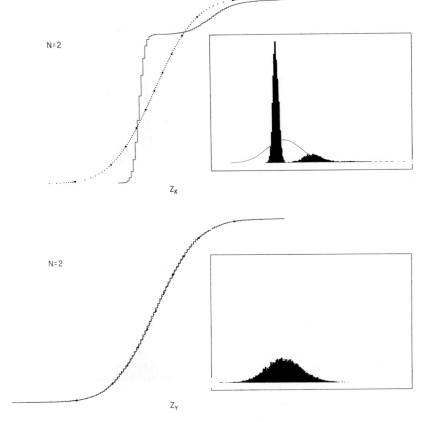

FIG. 14-9. Empirical cumulative distributions (step-curves) and frequency distributions (histograms) of the standardized sample means Z_X and Z_Y, at sample size $n = 2$, and corresponding normal distributions (dotted sigmoid curve and smooth bell-curve) that the Central Limit Theorem says they approach.

its far right tail. This tendency is present at *all* of the n-values investigated, although of course it diminishes with increasing n-values. Although it is not very apparent in the overall figures at the highest n-values, the effect is clearly present as one can see in Figures 14-19 and 14-20 which show the approach of the cumulative Z_X distribution to the cumulative standardized normal distribution at the far tails of the latter for all ten n-values.[3]

[3]In both figures the tail of the "approached" cumulative normal distribution is shown as a series of dots. The lowest and leftmost two dots have ordinates of .0001 and .0005. All other dots have ordinates that are integral multiples of .001. The dots having ordinates of .001, .01, .02, .03, .04 and .05 are enlarged.

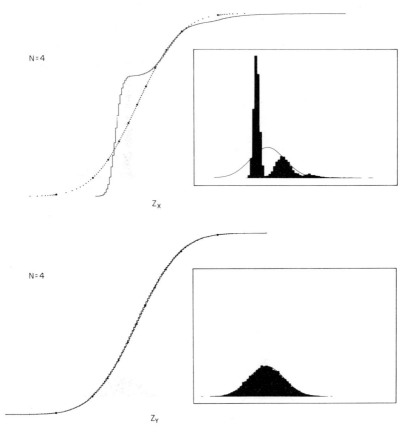

FIG. 14-10.   Central Limit Effect upon Z_X and Z_Y at $n = 4$.

These figures show an interesting characteristic of the Central Limit Effect that we mentioned earlier but that is seldom pointed out in statistics textbooks. Speaking very roughly, the farther we go out on the tail of the approached cumulative normal distribution, the greater tends to be the *relative* departure of the cumulative distribution of $\bar{X}$ (or Z_X) from it at a given *n*-value, and the greater *n* must be to bring this departure down to some "close fit." That is, if K is a positive number, ϕ is a normally distributed variable with mean of zero and variance of 1, and

$$Z = \frac{\bar{X} - \mu}{\frac{\sigma}{\sqrt{n}}}$$

is the standardized sample mean, then as K increases from some moderate

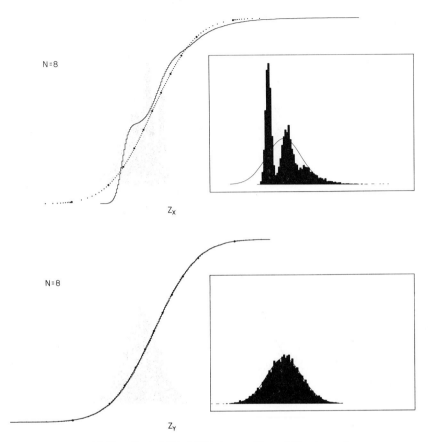

FIG. 14-11.   Central Limit Effect upon Z_X and Z_Y at $n = 8$.

value (like 1.5, the exact value depending upon the particular circumstances) toward larger values, the ratios

$$\frac{P(Z > K)}{P(\phi > K)}$$

and

$$\frac{P(Z < -K)}{P(\phi < -K)}$$

both tend to take values increasingly distant from 1 (unless "stopped" by becoming zero which will happen when K or $-K$ take values outside the range of the Z distribution), and the larger n must be to bring the ratio to a value within some reasonably close distance from 1, such as $1 \pm .2$. There may be "local" or temporary exceptions to this rule, but the general tendency

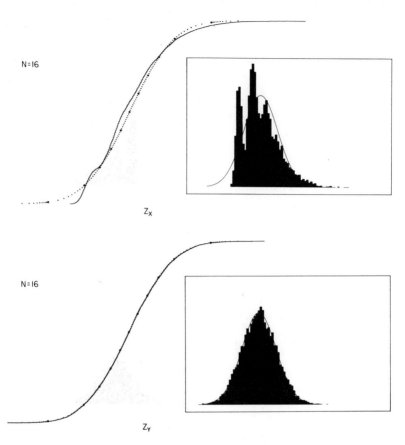

FIG. 14-12.   Central Limit Effect upon Z_X and Z_Y at $n = 16$.

is an empirical fact as well as a necessary implication of mathematical derivations (see Appendix A).

We can show this graphically as follows. Let ρ and α be the proportions of the distributions of Z and ϕ, respectively, that lie above K if "right tail" is specified, or that lie below $-K$ if "left tail" is specified. So ρ and α are the proportions of the "approaching" and "approached" (normal) distributions lying tailward of the same value. And ρ/α is the ratio referred to in the preceding paragraph: $P(Z > K)/P(\phi > K)$ or $P(Z < -K)/P(\phi < -K)$ in the right and left tail cases, respectively. Thus, it is a ratio between right- or left-tail cumulative probabilities, cumulated to a common abscissa-value K or $-K$. The ratio between ρ and α, then, is an excellent index of the departure of the empirical Z distribution from normality at a specified tail and at specified values of n and K (or α which implies K since $P(\phi > K) = \alpha$).

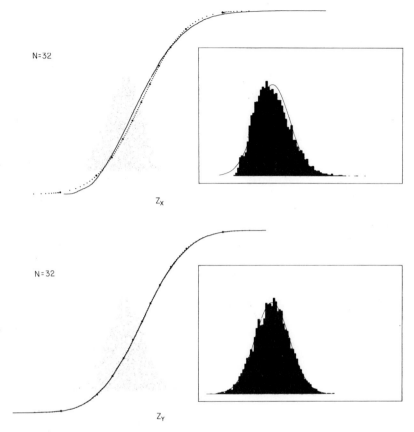

FIG. 14-13.   Central Limit Effect upon Z_X and Z_Y at $n = 32$.

For specified values of α and specified tails, Figure 14-21[4] presents curves showing how the ratio between p and α[5] changes with increasing sample size. For small values of α the ratio tends to progress steadily toward 1 as n increases from 2 to 1,024. And at any constant value of n, as α decreases

[4]Although the preceding figures are based upon 10,000 samples (and therefore Z's) at each n-value, Figure 14-21 is based upon 50,000 samples (the original 10,000 plus 40,000 more obtained in the same way) at each n-value.

[5]When p is less than α, the ratio p/α is plotted in the lower half of the graph on a scale rising from 0 to 1; when p exceeds α, the ratio α/p is plotted in the upper half of the graph on a scale rising physically and descending numerically from 1 to 0. The effect is the same as if p/α had been plotted in all cases, but on a linear scale when $p < \alpha$ and on a reciprocal scale when $p > \alpha$. Such a scale gives about equal prominence to p-values that are $< \alpha$ and $> \alpha$ which would not otherwise be the case since p can be many times greater than α but no more than one times smaller than α.

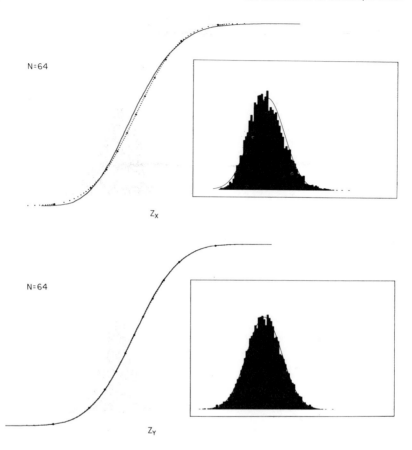

$N=64$

Z_X

$N=64$

Z_Y

FIG. 14-14.   Central Limit Effect upon Z_X and Z_Y at $n = 64$.

(i.e., as K increases) the ratio between ρ and α tends to depart farther and farther from 1, at least for small values of α, indicating greater and greater departures from normality at increasingly remote tail values.

Actually, the X population was only one of 24 populations investigated by the writer. Figures 14-22 through 14-45 show each of these sampled populations as a black histogram accompanied, for comparison, by a smooth curve depicting a normal distribution with the same mean and variance as the sampled population and the same area as the histogram. Accompanying each population is a graph, entirely analogous to Figure 14-21, showing the results of the sampling study conducted upon that population. That is, at each of the ten n-values representing the first ten powers of 2, at least 10,000 random samples of n observations were drawn (with replacements) from the population and an empirical distribution of at least 10,000 values of the standardized

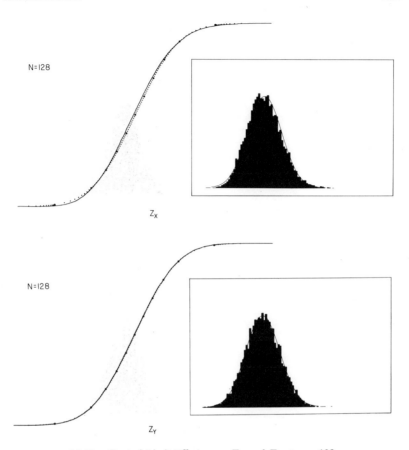

N=128

Z_X

N=128

Z_Y

FIG. 14-15.   Central Limit Effect upon Z_X and Z_Y at $n = 128$.

sample mean was obtained.[6] For each such distribution, the value of ρ corresponding to certain α-values was determined, and the ratio between ρ and α was plotted, for specified α-values and specified tail, against sample size n.

The 24 populations are numbered in order of increasing value of their third standardized central moment (their coefficient of skewness). And each figure gives δ_r, the difference between the rth standardized central moment of the population and that of a normal distribution. This is done for both $r = 3$ and $r = 4$. So δ_3 is the difference in skewness and δ_4 is the difference in

[6]Sampling distributions of Z obtained from a given population at different n-values are completely independent since different sets of random numbers were used to draw the samples. But those obtained from different populations at the same n-value are not independent since the same sets of random numbers were used.

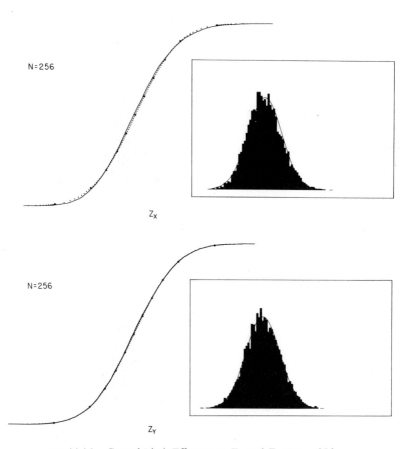

FIG. 14-16.   Central Limit Effect upon Z_X and Z_Y at $n = 256$.

kurtosis between the sampled population and a normally distributed population (whose skewness is 0 and whose kurtosis is 3).

The figures show that even when the "fit to normality" is excellent at most of the α-values, it tends to be poorer at the extreme values and to worsen rapidly as α becomes increasingly extreme. This is particularly apparent at small n-values. And, of course, the figures show that the fit to normality tends to improve as n increases, although there are sometimes temporary local exceptions. Perhaps the most important thing shown by the figures is the overwhelming influence of certain population moments upon the Central Limit Effect. Naively one might suppose that the critical population feature would be the extent of the population's departure from bell-shapedness (as measured, say, by the proportion of its histogram that lies outside the area enclosed by the accompanying normal curve). Instead, the critical population

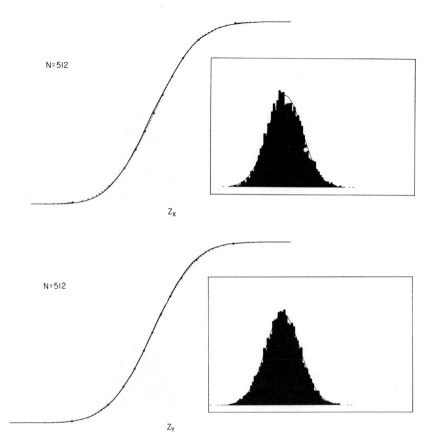

FIG. 14-17.   Central Limit Effect upon Z_X and Z_Y at $n = 512$.

feature appears to be its third, and to a much lesser extent its fourth, standardized central moment.

The third standardized central moment of a normal distribution, and of any symmetric distribution, is zero. In general, the farther the third standardized central moment of the sampled population is from this value, the more sluggish the Central Limit Effect tends to be, i.e., the worse the approach to normality by the distribution of $\bar{X}$. This can be seen in the tendency of the curves showing the ratio between ρ and α to depart farther and farther from the horizontal line through 1 as δ_3 increases. When δ_3 is very close to zero, δ_4 appears to have appreciable influence upon the Central Limit Effect; but as δ_3 increases in absolute value, it becomes increasingly dominant, especially at the larger n values.

This can be seen in Figures 14-46 and 14-47 each of which, for a different

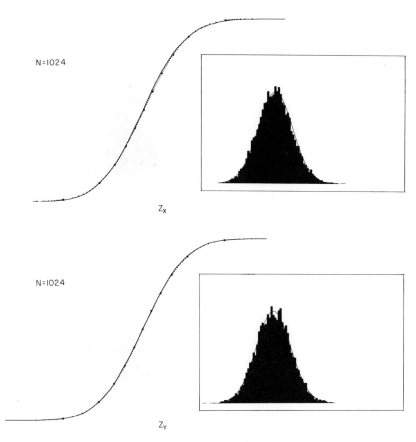

FIG. 14-18.   Central Limit Effect upon Z_X and Z_Y at $n = 1024$.

single n-value (see Figure 14-48 for the effect averaged accross all ten n-values) and for three small values of α, plots the ratio between ρ and α against the δ_3-values of the 24 sampled populations. Below this, δ_4, δ_5, δ_6, and δ_7 are plotted against δ_3. As δ_3 increases from 0 upward, the ratio between ρ and α tends to depart increasingly from 1. There are some perturbations in this departure, the more prominent of which are associated with perturbations in the curve plotting δ_4 against δ_3. The extent of these perturbations tends to diminish as either δ_3, n, or α increase. Thus, the sluggishness of the Central Limit Effect appears to be highly correlated with $|\delta_3|$ and, when n or $|\delta_3|$ is small, appreciably correlated with δ_4.

The principle population characteristic contributing to δ_3 (which is also the third standardized central moment) is unbalanced tailedness, i.e.,

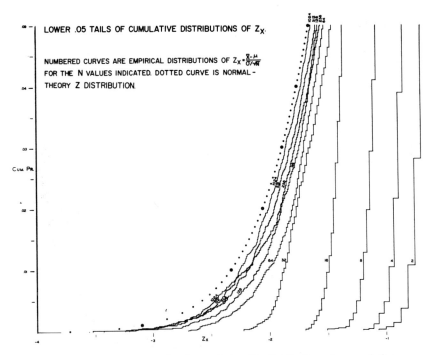

FIG. 14-19. Approach of cumulative Z_X distributions to cumulative standardized normal distribution at the left tail.

unequally long or unequally thick tails.[7] So asymmetry at the tails of the sampled population appears to be the population characteristic most responsible for pronounced nonnormality in the distribution of $\bar{X}$.

We have seen that the distribution of $\bar{X}$ became impressively bell-shaped at a very small n-value when the sampled population had a rectangular distribution. And we have seen that a much larger n-value was required to make the distribution of $\bar{X}$ impressively bell-shaped when the sampled population

[7]A unit whose value lies a distance d (measured in σ-units) from the population mean contributes an amount d^3 to the third standardized central moment. The greater the distance d from the mean, the greater the contribution; so the farther out on the tail a value lies, the greater its potential contribution. However, its contribution is completely nullified if there is a corresponding mirror-image unit whose value lies a distance $-d$ from the mean (i.e., a distance d in the opposite direction) since the mirror-image unit will contribute an exactly opposite amount $-d^3$ which together with the first unit contributes a net amount of zero. Thus, only those portions of the population distribution that have no nullifying counterpart located an equal distance on the opposite side of the mean can contribute to the third standardized central moment. And of these, their contribution is proportional to the cube of their distance from μ.

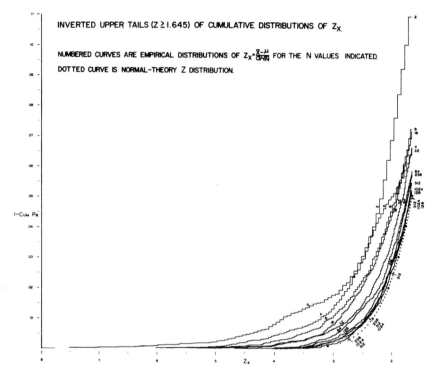

FIG. 14-20. Approach of cumulative Z_X distributions to cumulative standardized normal distribution at the right tail.

was the X population. Unfortunately, the X population is *not* the "worst" population in this respect that could be encountered in actual practice, nor is it anywhere close to it. The fact is that if we know nothing about the shape characteristics (e.g., the skewness) of the sampled population, we cannot specify a nonhuge n-value at which we can be certain that the distribution of $\bar{X}$ will be impressively bell-shaped, to say nothing of "approximately normal." Indeed, even to think of attempting it is to oversimplify since the "goodness-of-fit-to-normality" depends upon a variety of factors, not just n and the shape characteristics of the population distribution. We shall now summarize some of these influencing factors.

The *"shape" of the distribution of the sampled population* is extremely important. If the shape were exactly normal, the probability distribution of $\bar{X}$ would also be exactly normal at all n values from 1 to ∞. (We have already said that exactly normal populations are never encountered in practice, so exactly normal distributions of $\bar{X}$ will never be encountered either.) But even if the shape is only symmetric (e.g., rectangular), this is highly conducive to early bell-shapedness in the distribution of $\bar{X}$. As the asymmetry (especially

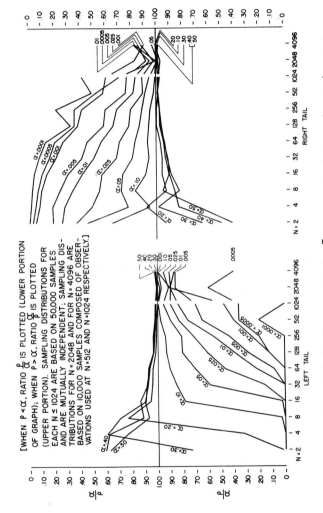

CENTRAL LIMIT EFFECT UPON MEANS OF N OBSERVATIONS
DRAWN FROM THE X POPULATION

RATIO BETWEEN EMPIRICAL, ρ, AND NORMAL-THEORY, α,
ONE-TAILED CUMULATIVE PROBABILITIES FOR $\bar{X}$'S HAVING
NORMAL-THEORY CUMULATIVE PROBABILITY OF α

[WHEN $P < \alpha$, RATIO $\frac{\rho}{\alpha}$ IS PLOTTED (LOWER PORTION
OF GRAPH); WHEN $P > \alpha$, RATIO $\frac{\alpha}{\rho}$ IS PLOTTED
(UPPER PORTION). SAMPLING DISTRIBUTIONS FOR
EACH $N \leq 1024$ ARE BASED ON 50,000 SAMPLES
AND ARE MUTUALLY INDEPENDENT; SAMPLING DIS-
TRIBUTIONS FOR N=2048 AND FOR N=4096 ARE
BASED ON 10,000 SAMPLES COMPOSED OF OBSER-
VATIONS USED AT N=512 AND N=1024 RESPECTIVELY.]

FIG. 14-21. Central Limit Effect upon Z_X (or $\bar{X}$) — approach of ρ toward α as n increases.

405

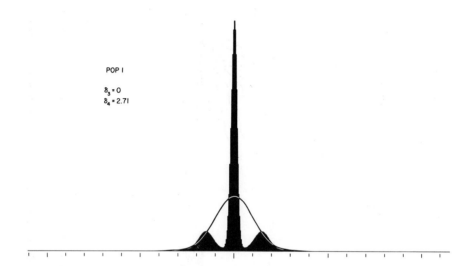

POP I

$\delta_3 = 0$
$\delta_4 = 2.71$

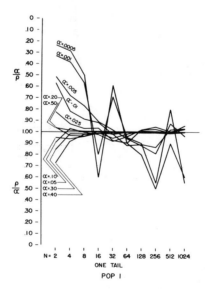

FIG. 14-22. Population 1 and Central Limit Effect upon means of samples drawn from it — approach of p toward in α as n increases.

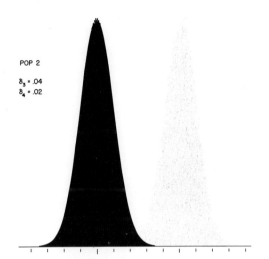

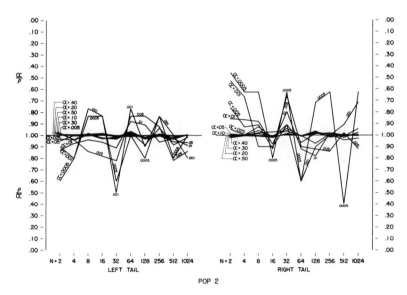

POP 2

FIG. 14-23. Population 2 and Central Limit Effect upon means of samples drawn from it.

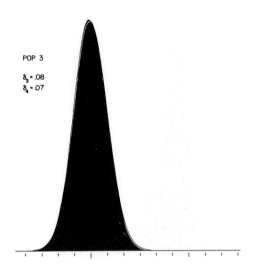

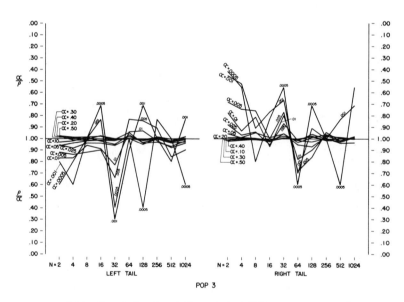

FIG. 14-24.   Population 3 and Central Limit Effect upon means of samples drawn from it.

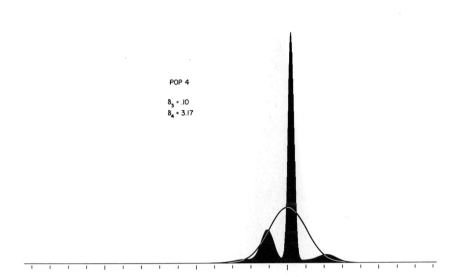

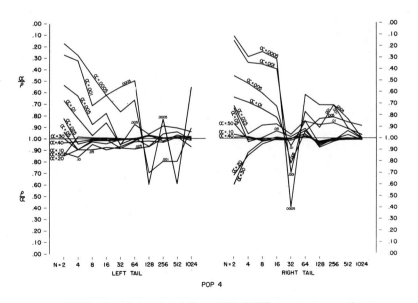

FIG. 14-25. Population 4 and Central Limit Effect upon means of samples drawn from it.

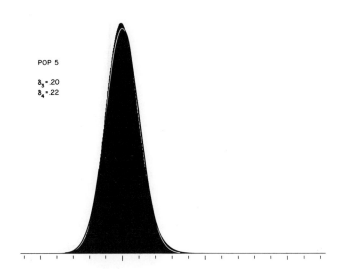

POP 5

$\delta_3 = .20$
$\delta_4 = .22$

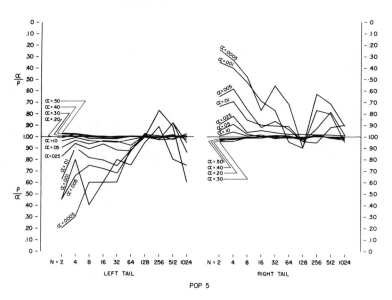

FIG. 14-26.   Population 5 and Central Limit Effect upon means of samples drawn from it.

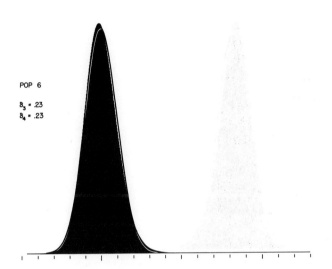

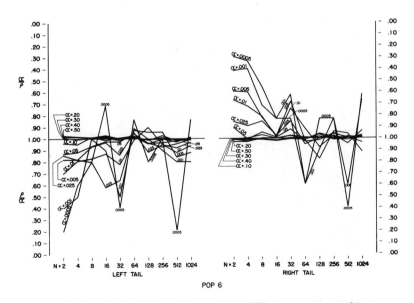

FIG. 14-27. Population 6 and Central Limit Effect upon means of samples drawn from it.

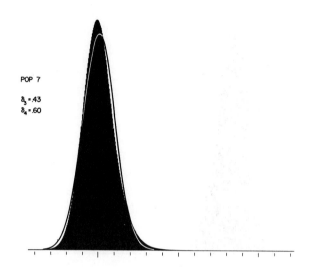

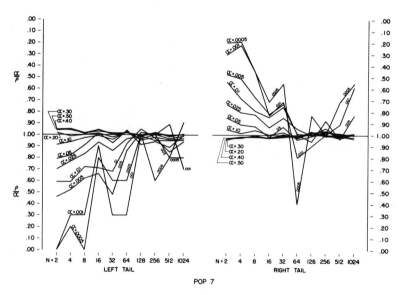

FIG. 14-28.   Population 7 and Central Limit Effect upon means of samples drawn from it.

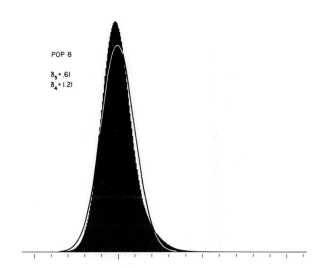

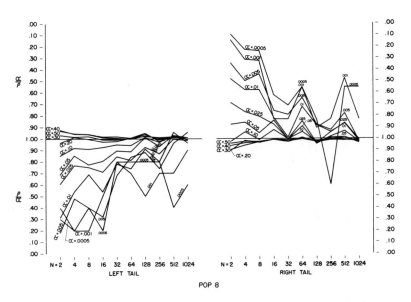

FIG. 14-29. Population 8 and Central Limit Effect upon means of samples drawn from it.

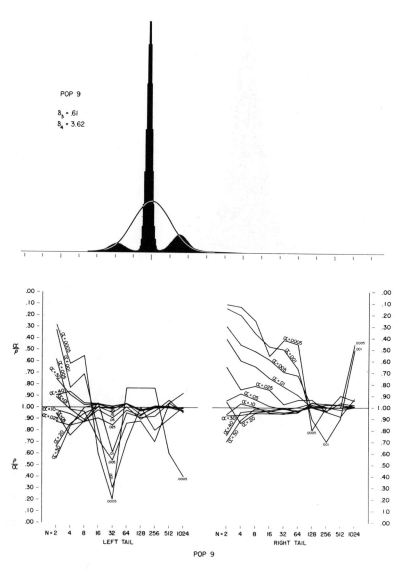

FIG. 14-30. Population 9 and Central Limit Effect upon means of samples drawn from it.

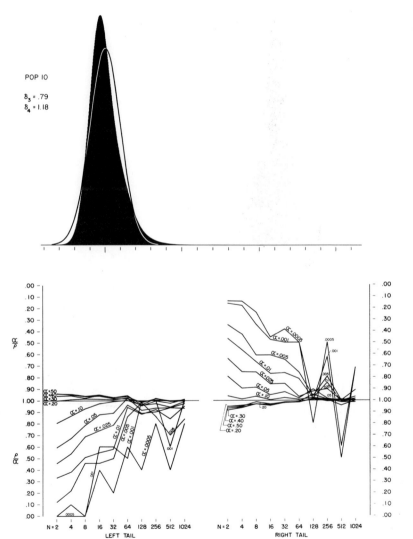

FIG. 14-31.   Population 10 and Central Limit Effect upon means of samples drawn from it.

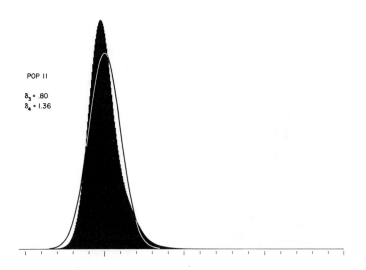

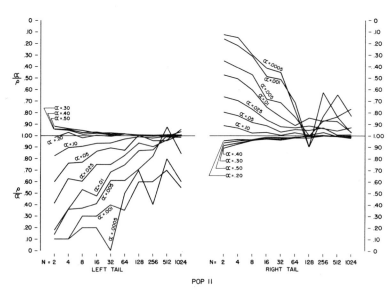

FIG. 14-32.   Population 11 and Central Limit Effect upon means of samples drawn from it.

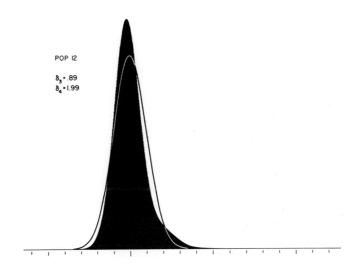

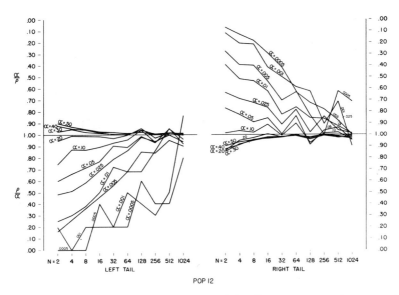

FIG. 14-33.   Population 12 and Central Limit Effect upon means of samples drawn from it.

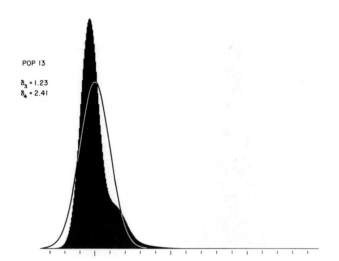

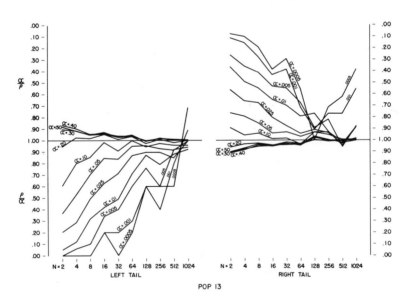

FIG. 14-34.   Population 13 and Central Limit Effect upon means of samples drawn from it.

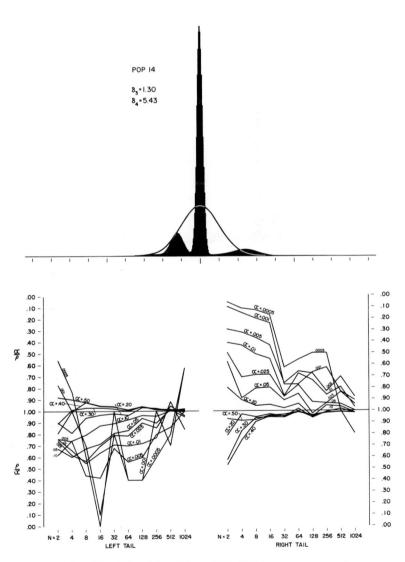

FIG. 14-35. Population 14 and Central Limit Effect upon means of samples drawn from it.

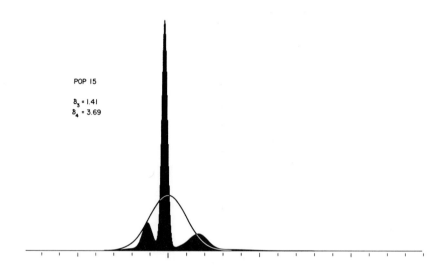

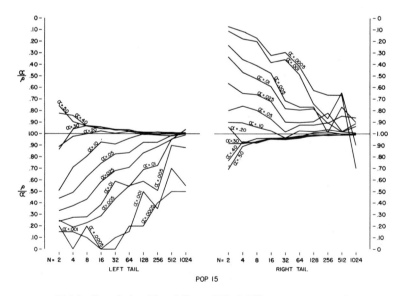

FIG. 14-36.   Population 15 and Central Limit Effect upon means of samples drawn from it.

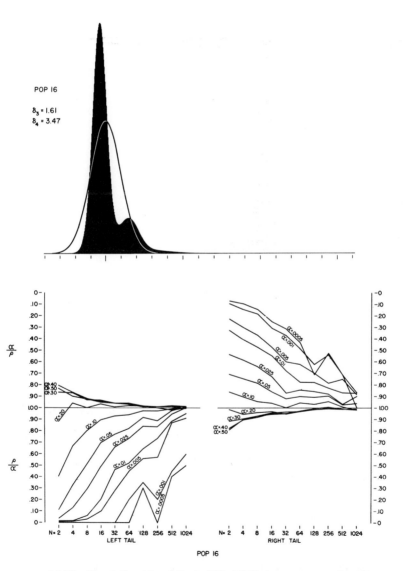

FIG. 14-37.   Population 16 and Central Limit Effect upon means of samples drawn from it.

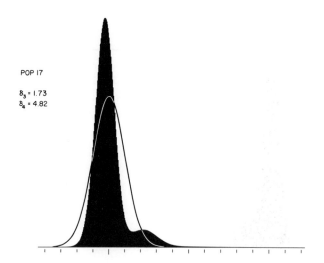

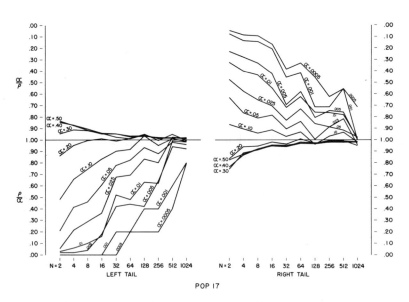

FIG. 14-38.   Population 17 and Central Limit Effect upon means of samples drawn from it.

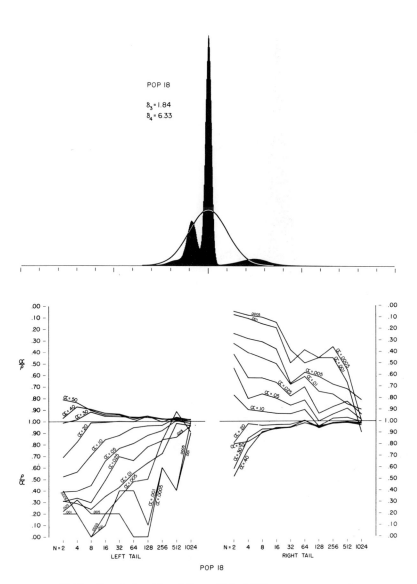

FIG. 14-39.   Population 18 and Central Limit Effect upon means of samples drawn from it.

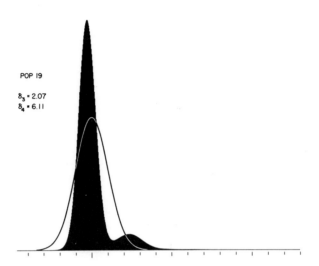

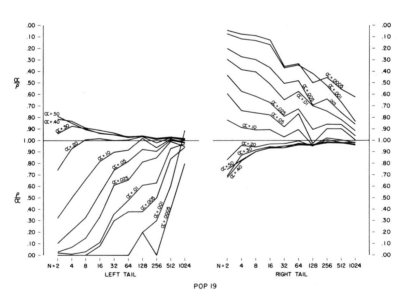

FIG. 14-40. Population 19 and Central Limit Effect upon means of samples drawn from it.

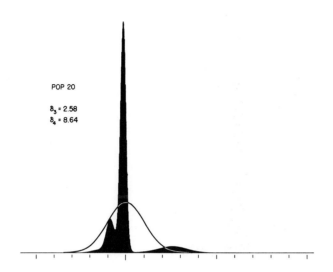

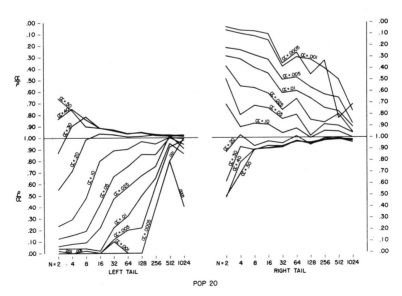

FIG. 14-41. Population 20 and Central Limit Effect upon means of samples drawn from it.

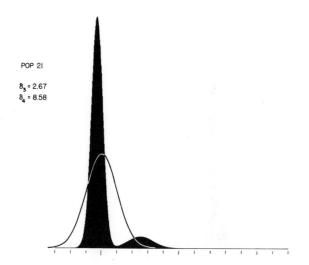

POP 21

$\delta_3 = 2.67$

$\delta_4 = 8.58$

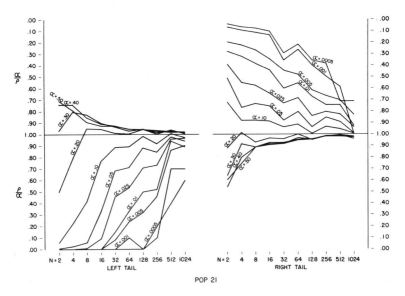

POP 21

FIG. 14-42.   Population 21 and Central Limit Effect upon means of samples drawn from it.

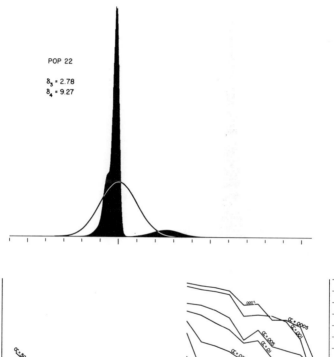

FIG. 14-43. Population 22 and Central Limit Effect upon means of samples drawn from it.

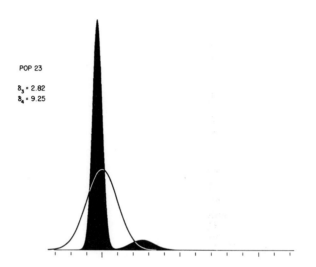

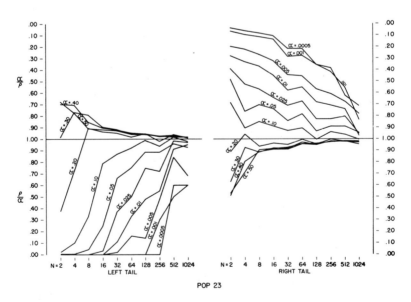

FIG. 14-44.   Population 23 and Central Limit Effect upon means of samples drawn from it.

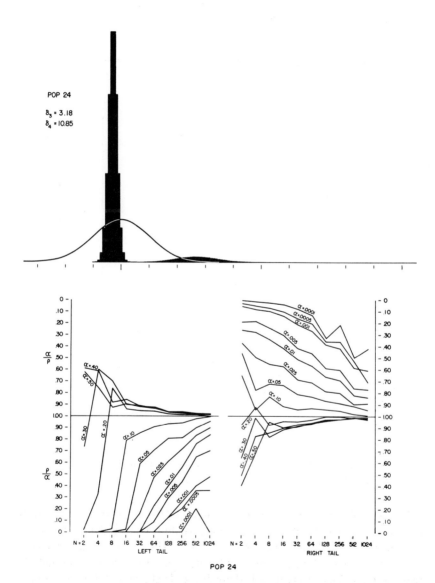

FIG. 14-45. Population 24 and Central Limit Effect upon means of samples drawn from it.

RELATIONSHIP BETWEEN THE ROBUSTNESS OF THE SAMPLE MEAN AT N = 2
AND THE EXTENT TO WHICH THE STANDARDIZED MOMENTS OF THE SAMPLED
POPULATION DIFFER FROM THOSE OF A NORMAL DISTRIBUTION (FOR 24
POPULATIONS WHOSE FIRST SEVEN STANDARDIZED CENTRAL MOMENTS EQUAL
OR EXCEED THOSE OF A NORMAL DISTRIBUTION)

ρ = ONE - TAILED CUMULATIVE PROBABILITY (IN AN EMPIRICAL SAMPLING DISTRIBUTION
CONTAINING AT LEAST 10,000 VALUES OF THE SAMPLE MEAN) OF THAT VALUE OF
THE SAMPLE MEAN WHOSE NORMAL - THEORY ONE - TAILED CUMULATIVE PROBABILITY
IS α

$\delta_i = \dfrac{\mu_i}{\sigma^i}$ MINUS NORMAL - THEORY VALUE OF $\dfrac{\mu_i}{\sigma^i}$, WHERE μ_i IS THE iTH CENTRAL
MOMENT AND σ IS THE STANDARD DEVIATION OF THE SAMPLED POPULATION

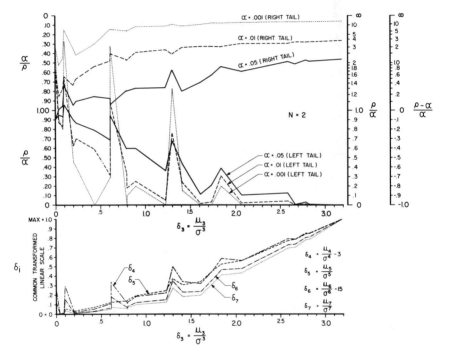

FIG. 14-46.   Influence of population moments upon the Central Limit
Effect when $n = 2$.

in the form of "unbalanced" long-tailedness) of the population distribution
increases, essential bell-shapedness in the distribution of $\bar{X}$ tends to make its
first appearance at larger and larger n-values. The *size n of the sample,* of
course, is another extremely important factor since the larger the sample the
greater the conduciveness to bell-shapedness. These two factors, of course,
are obvious. Another factor just as obvious but less frequently mentioned is

RELATIONSHIP BETWEEN THE ROBUSTNESS OF THE SAMPLE MEAN AT N = 8
AND THE EXTENT TO WHICH THE STANDARDIZED MOMENTS OF THE SAMPLED
POPULATION DIFFER FROM THOSE OF A NORMAL DISTRIBUTION (FOR 24
POPULATIONS WHOSE FIRST SEVEN STANDARDIZED CENTRAL MOMENTS EQUAL
OR EXCEED THOSE OF A NORMAL DISTRIBUTION)

ρ = ONE-TAILED CUMULATIVE PROBABILITY (IN AN EMPIRICAL SAMPLING DISTRIBUTION
CONTAINING AT LEAST 10,000 VALUES OF THE SAMPLE MEAN) OF THAT VALUE OF
THE SAMPLE MEAN WHOSE NORMAL-THEORY ONE-TAILED CUMULATIVE PROBABILITY
IS α

$\delta_i = \dfrac{\mu_i}{\sigma^i}$ MINUS NORMAL-THEORY VALUE OF $\dfrac{\mu_i}{\sigma^i}$, WHERE μ_i IS THE iTH CENTRAL
MOMENT AND σ IS THE STANDARD DEVIATION OF THE SAMPLED POPULATION

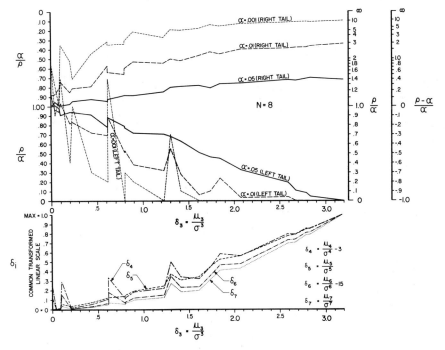

FIG. 14-47. Influence of population moments upon the Central Limit
Effect when $n = 8$.

one's *criterion of acceptable fit to normality*. Since, in practice, the distribution
of $\bar{X}$ *never* becomes *exactly* normal, the size n required to make the distribution of $\bar{X}$ approximately normal must depend upon one's definition of, or
criteria for, "approximate normality." A final and extremely important
factor that is seldom mentioned is the *portion of the distribution of $\bar{X}$* (or the
point on the *cumulative* distribution of $\bar{X}$) *in which one is interested*, i.e., over

RELATIONSHIP BETWEEN THE ROBUSTNESS OF THE SAMPLE MEAN AND THE
EXTENT TO WHICH THE STANDARDIZED MOMENTS OF THE SAMPLED POPULATION
DIFFER FROM THOSE OF A NORMAL DISTRIBUTION (FOR 24 POPULATIONS
WHOSE FIRST SEVEN STANDARDIZED CENTRAL MOMENTS EQUAL OR EXCEED
THOSE OF A NORMAL DISTRIBUTION)

$\bar{p}$ = AVERAGE VALUE OF ρ_N, WHERE ρ_N IS THE ONE-TAILED CUMULATIVE PROBABILITY
(IN AN EMPIRICAL SAMPLING DISTRIBUTION CONTAINING AT LEAST 10,000 VALUES
OF THE SAMPLE MEAN OF N OBSERVATIONS) OF THAT VALUE OF THE SAMPLE
MEAN WHOSE NORMAL-THEORY ONE-TAILED CUMULATIVE PROBABILITY IS α,
AND THE N VALUES UPON WHICH THE AVERAGE IS BASED ARE 2, 4, 8, 16, 32,
64, 128, 256, 512, AND 1024

$\delta_i = \frac{\mu_i}{\sigma^i}$ MINUS NORMAL-THEORY VALUE OF $\frac{\mu_i}{\sigma^i}$, WHERE μ_i IS THE iTH CENTRAL
MOMENT AND σ IS THE STANDARD DEVIATION OF THE SAMPLED POPULATION

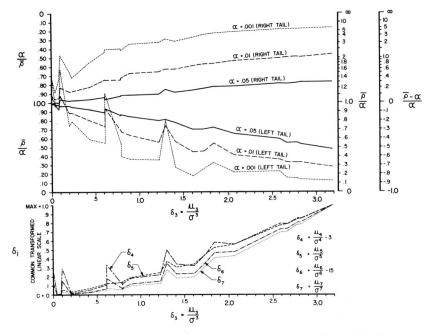

FIG. 14-48.   Influence of population moments upon the Central Limit
Effect — averaged over all *n*-values investigated.

which (or at which) goodness-of-fit to normality is to be measured. The
farther we go from μ, the larger n generally has to be in order to meet
whatever criterion of approximate normality we have adopted. (There are
localized exceptions to this, but nevertheless it is a strong general effect.)

Thus, while the Central Limit Effect is one of the most important and most
useful phenomena in statistics, it does *not* justify the simplistic belief that the

"approximate normality" of the distribution of the sample mean can be taken for granted at all but the tiniest of sample sizes. It is sensible to say that, in an overall sort of way, the distribution of $\bar{X}$ becomes more nearly normal as n increases. It is not sensible to say that the distribution of $\bar{X}$ ever becomes normal—or even approximately normal unless one is willing to specify the situation and define his terms. It is sometimes sensible to use one's statistical intuition, judgment, and educated guesses in deciding whether or not to assume that $\bar{X}$ is "approximately normally distributed" (sufficient to one's criterion) in specific cases, even in the absence of sufficient factual information. It is not reasonable, however, to suppose that the validity of the assumption is somehow guaranteed.

14-6 Problems

1. A fourth of the units in an infinite population have the value 0; the rest have the value 1. Give (a) the mean and standard deviation of the population, (b) the complete probability distribution for the mean of a random sample of two observations drawn from the population, (c) the mean and standard deviation of the distribution obtained in (b).

2. A random sample of two observations is to be drawn from a population consisting of four units having the values $0, -2, 0, 4$. Give the probability distribution for the sample mean and also for the sample median in the case of sampling (a) without replacements, (b) with replacements.

3. A random sample of three observations is to be drawn with replacements from a population consisting of three units having the values 0, 1, 9. What is the exact probability that there will be a difference of ≤ 1 between (a) the sample mean and the population mean, (b) the sample median and the population median?

4. Family income in the monarchy Petrolovia has a mean of \$3,000 and a variance of 100,000,000 square dollars. A sociologist desires to study a sample large enough for its mean to be a close estimate of the population mean. How large a random sample must the sociologist take in order for the standard deviation of the probability distribution of the mean to be $\leq$ \$100.

5. Ninety-nine percent of the units in a population have a value of zero and 1% have a value of 100. How large a sample must be taken in order for it to be *possible* for the sample mean to equal the population mean, i.e., what is the smallest n value at which this can happen?

6. What is the average number of times that a fair coin will come up heads

in 2 tosses, in 4, in 8? What is the probability that the coin will come up the average number of times in 2 tosses, in 4, in 8? As the size of the sample increases, what happens to $P(X = E\{X\})$?

7. A random variable X has an expected value of 100 and a variance of 25. Using Tchebycheff's Inequality, obtain the upper bound to the probability that X will deviate from its mean by more than 10.

8. A random sample of n observations is to be drawn from an infinite population with mean 100 and variance 25. How large must n be in order that there be a probability smaller than .01 that $\bar{X}$ will deviate from 100 by more than 10 units?

9. In the above problem, how large must n be so that there is a probability greater than .999 that $\bar{X}$ will lie in the interval between 99 and 101?

10. In Problem 8, how large must n be in order for there to be a probability of more than .9995 that $\bar{X}^2 - 7$ will lie in the interval between 9,597 and 10,602?

11. A random sample of n observations is to be drawn from an infinite population with a mean of 100 and a variance of 25. Using Tchebycheff's Inequality, obtain the upper bound for the probability that $\bar{X}$ will deviate from its expected value by more than 10 if (a) $n = 9$, (b) $n = 36$, (c) $n = 10,000$, (d) $n = 1,000,000$.

12. Prove that the following inequalities are simply modifications of Tchebycheff's Inequality:

(a) $P(|Z| > K) < 1/K^2$ where Z is any standardized random variable
(b) $P(|X - \mu| > C) < \sigma^2/C^2$
(c) $P(|X - \mu| > \sigma/\sqrt{C}) < C$

13. A population consists of five units having the values $-1, 1, 3, 3, 4$. A random sample of n units is to be drawn with replacements from this population. How large must the sample be to insure that the probability is greater than .99 that the sample mean will lie within the interval from $\mu - .01$ to $\mu + .01$?

14. A new IQ test has just been constructed. It consists of 75 items, each of which can receive a score of 0, 1, or 2; and the total score X on the entire test is the sum of the scores on the 75 individual items. Since the distribution of X is completely unknown, the test is to be administered to a standardization group of 90,000 adults randomly selected from the population to which the test is to apply. If μ is the mean value of X for the population for which it is intended, and $\bar{X}$ is the mean value of X for the standardization group, and p is the probability that $\bar{X}$ will differ from μ by more than 1, within what range of values can we be certain that p lies?

15. A random sample is to be drawn from an infinite dichotomous population in which the proportion of successes is p. How large must the sample size n be in order that: (a) there be a probability greater than .9 that the sample proportion of successes lie within .1 of the population proportion, (b) there be a probability greater than .99 that the sample proportion of successes lie within .01 of the population proportion?

16. In a certain community the weights of people have a mean of 150 and a standard deviation of 50. Twenty-five randomly selected people from this community board an elevator. The elevator cable will break if their combined weight exceeds 5,000 pounds. Within what range of values can we be certain that P(cable breaks on this occasion) lies?

17. A population consists of six units having the values 0, 0, 1, 1, 2, 2. For each value of n from 1 to 6, obtain the exact probability distribution of the mean $\bar{X}$ of a random sample of n observations drawn from the population without replacements. On the basis of these exact probability distributions answer the following questions: What is the exact $P(\bar{X} \geq 1.4)$ when (a) $n = 1$, (b) $n = 2$, (c) $n = 3$, (d) $n = 4$, (e) $n = 5$, (f) $n = 6$? What is the smallest value of K for which

$$P(|X - \mu| \leq K) \geq .90$$

when (g) $n = 1$, (h) $n = 2$, (i) $n = 3$, (j) $n = 4$, (k) $n = 5$, (l) $n = 6$?

18. What would be the answers to the above problem had sampling been *with* replacements? Give answers for n-values from 1 to 5 only.

19. A sample of 10 observations is drawn from a population whose mean is 100 and whose variance is 25. The mean of the sample is 90. Give the value of the sample mean in standardized form, i.e., as a Z score.

20. If X is a normally distributed variable and if $P(X \leq -K) = C$, then what is $P(-K < X < K)$?

21. If numerical grades X's are normally distributed with mean μ and variance σ^2 and alphabetical grades are assigned according to the rule

    A if $\mu + 2\sigma \leq X$

    B if $\mu + \sigma \leq X < \mu + 2\sigma$

    C if $\mu - \sigma \leq X < \mu + \sigma$

    D if $\mu - 2\sigma \leq X < \mu - \sigma$

    F if $X < \mu - 2\sigma$

what percentages of grades will be A's, B's, C's, D's, and F's?

22. If standardized numerical grades Z's are normally distributed and it is decided to assign alphabetical grades according to the equal interval

$(2K)$ scheme

$$
\begin{array}{lll}
\text{A} & \text{if} & 3K \le Z \\
\text{B} & \text{if} & K \le Z < 3K \\
\text{C} & \text{if} & -K \le Z < K \\
\text{D} & \text{if} & -3K \le Z < -K \\
\text{F} & \text{if} & Z < -3K
\end{array}
$$

and it is decided that 50% of the grades should be C's, what is the value of K and what percentages of grades will be A's, B's, D's and F's?

23. A random sample of 9 observations is to be drawn from a normally distributed population with a mean of 30 and a variance of 36. What is the probability that $\bar{X}$ will fall below 26?

24. Let $\bar{X}$ be the mean of a random sample of four observations to be drawn from an infinite population one-third of whose units have the value 0, one-third the value 1, and one-third the value 2. For each of the C-values, .1, .2, .3, .4, .5, .6, .7, .8, .9, 1, obtain and give $P(|\bar{X} - \mu| < C)$ based on (a) the exact probability distribution of $\bar{X}$, (b) the assumption that $\bar{X}$ and $Z = (\bar{X} - \mu)/(\sigma/\sqrt{n})$ are approximately normally distributed due to the Central Limit Effect, obtaining the required probabilities from standardized normal tables, (c) Tchebycheff's Inequality (in which case the probability obtained will be a lower bound).

25. If a fair coin is to be tossed n times, what is the probability that the number of heads will differ from $n/2$ by 1 or less if (a) $n = 4$, (b) $n = 8$? What is the probability that the proportion of heads will differ from .5 by .25 or less if (c) $n = 4$, (d) $n = 8$?

26. A proportion p of the units in an infinite dichotomous population have the value 1 and a proportion $1 - p$ have the value 0. For each of the n-values 1, 2, 3, 4, 5, 10, obtain the exact probability distribution of the mean $\bar{X}$ of a random sample of n observations drawn from the population, and give the exact $P(\bar{X} > .7)$ when (a) $p = .5$, (b) $p = .1$, (c) $p = .01$.

27. A proportion $p = .5$ of the units in an infinite dichotomous population have the value 1 and a proportion $1 - p$ have the value 0. For each of the n-values 1, 2, 3, 4, 5, 10, what is the smallest value of C for which $P(|\bar{X} - \mu| \le C) > .90$ if your answers are based upon (a) the exact probability distribution of $\bar{X}$, (b) Tchebycheff's Inequality, (c) the false assumption that $\bar{X}$ is normally distributed?

28. Work Problem 27 substituting $p = .1$ for $p = .5$.

29. Work Problem 27 substituting $p = .01$ for $p = .5$.

15

normal-approximation statistics

STATISTICAL METHODS BASED UPON THE NORMAL APPROXIMATION TO THE DISTRIBUTION OF $\bar{X}$

In the preceding chapter we learned that as $n \longrightarrow \infty$ the cumulative probability distribution of $\bar{X}$ (and therefore of r/n) approaches a cumulative normal distribution. In this chapter we shall discuss statistical methods that make use of this fact. When we speak of the "normality" or quasi-normality of the distribution of $\bar{X}$, or of the standardized mean Z, we shall always mean the *cumulative* probability distribution. The adjective will be omitted to avoid repetitiousness.

All of the methods presented in this chapter are approximate. So confidence levels (or the intervals they refer to), significance levels, probability levels, and power, all will be approximate. This will be made clear by using approximation signs in the derivation of their formulas. However, in order to avoid repetitousness, we shall often omit the qualification "approximate" when refering to them after their approximateness has been established.

15-1   The One-Sample Z Statistic

Consider an infinite population of X's (or a random variable X) with mean μ and variance σ^2. We already know that if $\bar{X}$ is the mean of a random sample of n X-observations, then the probability distribution of $\bar{X}$ has a mean of μ and a variance of σ^2/n and the probability distribution of the standardized mean

$$Z = \frac{\bar{X} - E\{\bar{X}\}}{\sqrt{\text{Var}(\bar{X})}} = \frac{\bar{X} - \mu}{\frac{\sigma}{\sqrt{n}}} \tag{15-1}$$

has a mean of zero and a variance of 1.

It can be shown that if the original distribution of X's is exactly normal, then the distributions of $\bar{X}$ and Z are also exactly normal, with the means and variances already given. And we have seen in the preceding chapter that no matter what the shape of the original distribution of X's the probability distributions of $\bar{X}$ and of Z become more and more nearly normal as n increases.

Therefore, if we regard the statistic

$$Z = \frac{\bar{X} - \mu}{\frac{\sigma}{\sqrt{n}}}$$

as having a standardized normal distribution, i.e., a normal distribution with mean 0 and variance 1, we shall be exactly right when X has an exactly normal distribution, and in other cases we shall be approximately right when the Central Limit Effect has caused the probability distribution of $\bar{X}$ to become approximately normal. Therefore, if we are correct in making the required "assumption" that the distribution of $\bar{X}$ is essentially normal, we can use tables of cumulative probabilities for the standardized normal distribution to give us the analogous cumulative probabilities for Z, i.e., we can treat Z as approximately a standardized normal variable.

Specifically, let ϕ stand for a normally distributed variable whose mean is 0 and whose variance is 1, and let $\Phi(K)$ stand for the cumulative probability of all values of ϕ less than or equal to K. Then, if the normality assumption for $\bar{X}$ is justified,

$$P(Z \leq K) \cong P(\phi \leq K) = \Phi(K) \tag{15-2}$$

and, letting C be a constant that is smaller than the constant K,

$$P(C \leq Z \leq K) \cong P(C \leq \phi \leq K)$$
$$\cong P(\phi \leq K) - P(\phi < C)$$
$$\cong \Phi(K) - \Phi(C) \tag{15-3}$$

since for a normal (and therefore continuous) distribution $P(\phi = C) = 0$ so that $P(\phi \leq C) = P(\phi < C)$.

15-1-1 Confidence Intervals for μ, Based on the Z Statistic

By simply performing some algebraic manipulations upon one of the above formulas, we obtain a formula giving all the information necessary to establish a confidence interval for μ.

$$\Phi(K) - \Phi(C) \cong P(C \le Z \le K)$$

$$\cong P\left(C \le \frac{\bar{X} - \mu}{\frac{\sigma}{\sqrt{n}}} \le K\right)$$

$$\cong P\left(\frac{C\sigma}{\sqrt{n}} \le \bar{X} - \mu \le \frac{K\sigma}{\sqrt{n}}\right)$$

$$\cong P\left(-\frac{C\sigma}{\sqrt{n}} \ge -(\bar{X} - \mu) \ge -\frac{K\sigma}{\sqrt{n}}\right)$$

$$\cong P\left(\bar{X} - \frac{C\sigma}{\sqrt{n}} \ge \mu \ge \bar{X} - \frac{K\sigma}{\sqrt{n}}\right)$$

$$\cong P\left(\bar{X} - \frac{K\sigma}{\sqrt{n}} \le \mu \le \bar{X} - \frac{C\sigma}{\sqrt{n}}\right) \tag{15-4}$$

Therefore, if we knew the value of the population standard deviation σ, if we chose the values of K and C before sampling, and if we then drew a random sample of n observations from the infinite population, calculated the numerical value $\bar{x}$ of its mean, and substituted these numbers for the symbols K, C, σ, and $\bar{X}$ in the formula, we could state at approximately the $\Phi(K) - \Phi(C)$ level of confidence that the population mean μ lies at or between the numerical value of $\bar{x} - K\sigma/\sqrt{n}$ and that of $\bar{x} - C\sigma/\sqrt{n}$, i.e., that

$$\bar{x} - \frac{K\sigma}{\sqrt{n}} \le \mu \le \bar{x} - \frac{C\sigma}{\sqrt{n}} \tag{15-5}$$

It may help conceptualization to make a small change in notation. Let ϕ_p and ϕ_q be constants such that $P(\phi \le \phi_p) = p$ and $P(\phi \le \phi_q) = q$. That is, roughly speaking, let ϕ_p and ϕ_q be the $100p$ th and $100q$ th percentiles in the distribution of ϕ. Then, if we let $C = \phi_p$ and $K = \phi_q$, so that $\Phi(C) = \Phi(\phi_p) = p$ and $\Phi(K) = \Phi(\phi_q) = q$, Formula 15-4 can be written in the equivalent form

$$q - p \cong P\left(\bar{X} - \frac{\phi_q\sigma}{\sqrt{n}} \le \mu \le \bar{X} - \frac{\phi_p\sigma}{\sqrt{n}}\right) \tag{15-6}$$

So if we chose p and q (or ϕ_p and ϕ_q) before actually drawing a random sample whose mean turned out to be $\bar{x}$, we could state at approximately the $q - p$ level of confidence that

$$\bar{x} - \frac{\phi_q\sigma}{\sqrt{n}} \le \mu \le \bar{x} - \frac{\phi_p\sigma}{\sqrt{n}} \tag{15-7}$$

provided that we substituted obtained or known numerical values for all symbols except μ in the above inequality.

Examples: To illustrate, suppose that X is the amount of coffee poured into a cup by an automatic vending machine, and X is known to have a quasi-normal distribution. The mean μ of the X variable can be varied by adjusting the machine and is unknown, but the variance σ^2 of the X variable is known from past experience to be independent of μ, having the constant value .25 square ounces. Since customers have been complaining of being shorted, it is desired to obtain for the 90% level of confidence a symmetric confidence interval for μ based on a sample of nine cups of coffee. Thus, it is desired that $\Phi(K) - \Phi(-K) = .90$ and since the normal distribution is symmetric this implies that $\Phi(K) = .95$ and $\Phi(-K) = .05$. Consulting normal tables we find that $P(\phi < 1.6449) = .95$ and $P(\phi < -1.6449) = .05$, so $K = 1.6449$ and $C = -1.6449$. The sample of nine cups of coffee is now taken and yields the following values of X: 4.8, 5.7, 6.2, 4.9, 5.5, 5.2, 5.2, 5.1, 6.0 from which we calculate $\bar{x} = 5.4$. Now the formula tells us that we can have confidence at approximately the $\Phi(K) - \Phi(C)$ level that

$$\bar{x} - \frac{K\sigma}{\sqrt{n}} \le \mu \le \bar{x} - \frac{C\sigma}{\sqrt{n}}$$

We have already found that, for about the 90% level of confidence and a symmetric confidence interval, $K = 1.645$ and $C = -1.645$, and we know that $\sigma = .5$, $n = 9$, and $\bar{x} = 5.4$. Substituting these numerical values for their symbols, the desired confidence interval is

$$5.4 - 1.645\left(\frac{.5}{\sqrt{9}}\right) \le \mu \le 5.4 - (-1.645)\left(\frac{.5}{\sqrt{9}}\right)$$

$$5.4 - \left(\frac{.8225}{3}\right) \le \mu \le 5.4 + \left(\frac{.8225}{3}\right)$$

$$5.4 - .274 \le \mu \le 5.4 + .274$$

$$5.126 \le \mu \le 5.674$$

So we can be confident at about the 90% level that μ lies in the interval from 5.126 to 5.674 ounces.

Suppose that, instead of desiring the interval corresponding to a specified confidence level and sample size, we had wanted to know how large a sample we would have to take in order to be confident at about the .99 level that $\bar{x} - .1 \le \mu \le \bar{x} + .1$. This is a symmetric interval, so we know that $\Phi(K) = .995$ and $\Phi(C) = \Phi(-K) = .005$, and consulting normal tables we find that $K = 2.5758$. So we have the equation

$$.99 \cong P\left(\bar{X} - \frac{K\sigma}{\sqrt{n}} \le \mu \le \bar{X} - \frac{(-K)\sigma}{\sqrt{n}}\right) = P(\bar{X} - .1 \le \mu \le \bar{X} + .1)$$

So

$$\frac{K\sigma}{\sqrt{n}} = .1$$

and

$$n = \left(\frac{K\sigma}{.1}\right)^2 = [10(2.5758)(.5)]^2 = (12.879)^2 = 165.87$$

and the sample size would therefore have to be at least 166.

15-1-2   Test of an Hypothesized Value for μ, Using the One-Sample Z Statistic

We know that if the X population is infinite, if the sample is random, and if the distribution of $\bar{X}$ is essentially normal, then

$$Z = \frac{\bar{X} - \mu}{\dfrac{\sigma}{\sqrt{n}}}$$

is essentially normally distributed with zero mean and unit variance. Therefore, if under these conditions we *correctly* hypothesize $H_0: \mu = \mu_0$, then, since $\mu = \mu_0$, the test statistic

$$Z_0 = \frac{\bar{X} - \mu_0}{\dfrac{\sigma}{\sqrt{n}}} \tag{15-8}$$

must be essentially normally distributed with zero mean and unit variance. Furthermore, we know that, since μ is the mean of the distribution of $\bar{X}$, if actually $\mu < \mu_0$, $\bar{X}$ will tend also to be less than μ_0 so that Z_0 will tend to take values less than zero, whereas if actually $\mu > \mu_0$, $\bar{X}$ will tend to be $> \mu_0$ and Z_0 will tend to be > 0. Therefore, lower-tail, upper-tail, and both-tail values of the distribution of the test statistic Z_0 when $\mu = \mu_0$, and therefore of the distribution of ϕ, are appropriate rejection regions, respectively, for testing $H_0: \mu \geq \mu_0$ against $H_a: \mu < \mu_0$, for testing $H_0: \mu \leq \mu_0$ against $H_a: \mu > \mu_0$, and for testing $H_0: \mu = \mu_0$ against $H_a: \mu \neq \mu_0$.

If we take as our rejection region all values of Z_0 less than C or greater than K, the probability of committing a Type I Error is

$$P(Z_0 < C) + P(Z_0 > K) \cong \Phi(C) + 1 - \Phi(K) \tag{15-9}$$

Again let ϕ_p be that value of ϕ for which $P(\phi \leq \phi_p) = p$ and let ϕ_q be that value of ϕ for which $P(\phi \leq \phi_q) = q$. Then if we let $C = \phi_p$ and $K = \phi_q$, i.e., if we take as our rejection region all values of Z less than ϕ_p or greater than ϕ_q, the probability of a Type I Error can alternatively be written as

$$P(Z_0 < \phi_p) + P(Z_0 > \phi_q) \cong p + (1 - q) \tag{15-10}$$

So if we let $p = \alpha$ and $q = 1$ (so that $\phi_q = +\infty$), we have a left-tail test at nominal significance level α; if we let $p = 0$ (so that $\phi_p = -\infty$) and $q = 1 - \alpha$, we have a right-tail test at nominal significance level α; and if we let $p = 1 - q = \alpha/2$ so that $p + (1 - q) = \alpha$, we have a symmetric two-tail test at nominal significance level α.

If we *incorrectly* hypothesize $H_0: \mu = \mu_0$ when actually $\mu = \mu_a$, then we still reject if Z_0 is less than ϕ_p or greater than ϕ_q, but the probability of one of these events happening is no longer $p + (1 - q)$; rather, it is the power of the test. The power of the test, then, is the probability that the test statistic

Z_0 will fall in the rejection region, which is:

$$\text{Power} = P(Z_0 < \phi_p \,|\, \mu = \mu_a) + P(Z_0 > \phi_q \,|\, \mu = \mu_a)$$

$$= P\left(\frac{\bar{X} - \mu_0}{\frac{\sigma}{\sqrt{n}}} < \phi_p \,\bigg|\, \mu = \mu_a\right) + P\left(\frac{\bar{X} - \mu_0}{\frac{\sigma}{\sqrt{n}}} > \phi_q \,\bigg|\, \mu = \mu_a\right)$$

$$= P\left(\frac{\bar{X} - \mu_0 + \mu_a - \mu_a}{\frac{\sigma}{\sqrt{n}}} < \phi_p \,\bigg|\, \mu = \mu_a\right)$$

$$\qquad\qquad + P\left(\frac{\bar{X} - \mu_0 + \mu_a - \mu_a}{\frac{\sigma}{\sqrt{n}}} > \phi_q \,\bigg|\, \mu = \mu_a\right)$$

$$= P\left(\frac{\bar{X} - \mu_a}{\frac{\sigma}{\sqrt{n}}} < \phi_p + \frac{\mu_0 - \mu_a}{\frac{\sigma}{\sqrt{n}}} \,\bigg|\, \mu = \mu_a\right)$$

$$\qquad\qquad + P\left(\frac{\bar{X} - \mu_a}{\frac{\sigma}{\sqrt{n}}} > \phi_q + \frac{\mu_0 - \mu_a}{\frac{\sigma}{\sqrt{n}}} \,\bigg|\, \mu = \mu_a\right)$$

$$= P\left(Z_a < \phi_p + \frac{\mu_0 - \mu_a}{\frac{\sigma}{\sqrt{n}}} \,\bigg|\, \mu = \mu_a\right)$$

$$\qquad\qquad + P\left(Z_a > \phi_q + \frac{\mu_0 - \mu_a}{\frac{\sigma}{\sqrt{n}}} \,\bigg|\, \mu = \mu_a\right)$$

But since μ_a is the true value of μ, Z_a is essentially a normally distributed variable ϕ with zero mean and unit variance. So we can replace Z_a by ϕ in the formula as follows:

$$\text{Power} \cong P\left(\phi < \phi_p + \frac{\mu_0 - \mu_a}{\frac{\sigma}{\sqrt{n}}}\right) + P\left(\phi > \phi_q + \frac{\mu_0 - \mu_a}{\frac{\sigma}{\sqrt{n}}}\right)$$

$$\cong \Phi\left(\phi_p + \frac{\mu_0 - \mu_a}{\frac{\sigma}{\sqrt{n}}}\right) + 1 - \Phi\left(\phi_q + \frac{\mu_0 - \mu_a}{\frac{\sigma}{\sqrt{n}}}\right) \qquad (15\text{-}11)$$

and, since we know the values of ϕ_p, ϕ_q, μ_0, σ, and n, we can look up the values of the Φ's in normal tables for specific alternative values μ_a of μ, thereby obtaining the approximate power of the test when $\mu = \mu_a$.

Example: To illustrate, suppose in the coffee-machine example that it had been desired to test at the nominal .05 level of significance the $H_0: \mu \leq 5$ against the $H_a: \mu > 5$. Then, a right-tail test is called for so that in the formula giving the probability of a Type I Error,

$$P(Z_0 < \phi_p | \mu = \mu_0) + P(Z_0 > \phi_q | \mu = \mu_0) \cong p + 1 - q$$

we let $\phi_p = -\infty$ so that $P(Z_0 < \phi_p | \mu = \mu_0) = p = 0$ and $1 - q$ must therefore be .05

$$P(Z_0 > \phi_q | \mu = \mu_0) \cong P(\phi > \phi_q) = .05$$

From normal tables we find that $P(\phi > 1.6449) = .05$ so we know that $\phi_q = 1.6449$ and therefore that our rejection region is values of Z_0 greater than 1.6449. Now

$$Z_0 = \frac{\bar{X} - \mu_0}{\frac{\sigma}{\sqrt{n}}} = \frac{5.4 - 5}{\frac{.5}{\sqrt{9}}} = \frac{.4}{\frac{.5}{3}} = \frac{1.2}{.5} = 2.4$$

which lies in the rejection region; so we reject the $H_0 : \mu \leq 5$ in favor of the $H_a : \mu > 5$. The probability level of the obtained value of the test statistic is $P(Z_0 > 2.4 | \mu = \mu_0) \cong P(\phi > 2.4) = .0082$. Now, if the value of the population mean were actually $\mu = \mu_a = 5.5$, then the power of the test to reject the false null hypothesis $H_0 : \mu \leq 5$ in favor of the true alternative hypothesis $H_a : \mu = 5.5$ would be

$$\text{Power} \cong \Phi\left(\phi_p + \frac{\mu_0 - \mu_a}{\frac{\sigma}{\sqrt{n}}}\right) + 1 - \Phi\left(\phi_q + \frac{\mu_0 - \mu_a}{\frac{\sigma}{\sqrt{n}}}\right)$$

$$\cong \Phi\left(-\infty + \frac{5 - 5.5}{\frac{.5}{\sqrt{9}}}\right) + 1 - \Phi\left(1.6449 + \frac{5 - 5.5}{\frac{.5}{\sqrt{9}}}\right)$$

$$\cong \Phi(-\infty) + 1 - \Phi\left(1.6449 - \frac{.5}{\frac{.5}{3}}\right)$$

$$\cong 0 + 1 - \Phi(1.6449 - 3) = 1 - \Phi(-1.3551)$$

From normal tables we find that $P(\phi < -1.3551) = .0877$; so

$$\text{Power} \cong 1 - .0877 = .9123$$

15-1-3 The Difference-Score *Z* Statistic

A special case of the one-sample Z statistic is that in which the n observations are n difference scores $D_i = X_i - Y_i$, where X_i and Y_i are measurements made under an X treatment and a Y treatment, respectively. Both treatments may be administered to a single unit, the ith unit in a sample of n units. Or one member of a matched pair of units may be randomly assigned to the X treatment, the other being assigned to the Y treatment, so that D_i is the difference score for the ith of n matched pairs of units. In either case there is only a *single* sample of n difference scores, which may be regarded as having been drawn from an infinite population of potential difference scores with mean μ_D and variance σ_D^2. The mean of the difference-score population is

$$\mu_D = E(D_i) = E(X_i - Y_i) = E(X_i) - E(Y_i) = \mu_X - \mu_Y$$

So the mean of the difference-score population is also the difference between the means of the X and Y treatment populations. (See Sections 12-3-1 and 12-1-2.) Likewise if $\bar{D}$, $\bar{X}$, and $\bar{Y}$ are, respectively, the mean of the sample of n difference scores, the mean of the n X-measurements, and the mean of the n Y-measurements, then

$$\bar{D} = \frac{\sum_{i=1}^{n} D_i}{n} = \frac{\sum_{i=1}^{n} (X_i - Y_i)}{n} = \frac{\sum_{i=1}^{n} X_i}{n} - \frac{\sum_{i=1}^{n} Y_i}{n} = \bar{X} - \bar{Y}$$

Therefore, if the sample can be regarded as having been drawn randomly, and if n is large enough (or the difference-score population normal enough) to cause $\bar{D}$ to be nearly normally distributed, then

$$Z = \frac{\bar{D} - \mu_D}{\sqrt{\dfrac{\sigma_D^2}{n}}} = \frac{\bar{X} - \bar{Y} - (\mu_X - \mu_Y)}{\sqrt{\dfrac{\sigma_D^2}{n}}} \qquad (15\text{-}12)$$

is distributed approximately as ϕ the standardized normal variable. So the difference-score Z statistic is simply a special case of the one-sample Z statistic and can be used, in exactly the same way as previously described, to establish confidence intervals or test hypotheses about the mean μ_D of the difference-score population or about the difference between means of the treatment populations $\mu_X - \mu_Y$.

For reasons given in Section 12-3-4, if an X population and a Y population are identically distributed, then the $X - Y$ difference-score population will be symmetrically distributed about zero. Obviously, if we add Δ to the value of every unit in the X population, and therefore to the value of every unit in the $X - Y$ difference-score population, the resulting X and Y populations will be identical except for location and the difference-score population will be symmetrically distributed about Δ. Therefore, returning to the difference-score Z statistic, if treatments have no effect, or if they influence primarily the means, rather than the variances and shapes of the treatment populations, then the difference-score population will tend to be symmetric. As we have already seen in Chapter 14, this symmetry is conducive to early quasi-normality in the distribution of the sample mean, $\bar{D}$ in this case, so that the difference-score Z statistic should be relatively well approximated by ϕ at relatively small values of n.

The difference-score Z statistic is rarely used because it requires that we know the true variance of the difference-score population, whereas we generally have only the sample estimate of it. We have discussed it, however, because a modification of it, which is a common and useful statistic, will be taken up later.

Example: A shoe manufacturer wished to determine whether or not a new leather treating process, the X process, was superior to the old Y process. He randomly selected 26 subjects and randomly divided them into two groups of 13 subjects each. He then manufactured a pair of shoes for each subject. For subjects in the first group, the leather in the left shoe was treated by the X process and that in the right shoe by the Y process. For the subjects in the second group, the left shoe was made of Y process leather and the right shoe of X process leather. The 26 subjects then wore these shoes exclusively for exactly six months. At the end of this period the amount of wear on the outside portion of the heel of each shoe was measured (by measuring the thickness of the heel and subtracting it from the original thickness). Table 15-1 gives, for each subject, the amount of wear on the X-treated shoe, the amount of wear on the Y-treated shoe, and the difference score, as well as some calculations appropriate to this and subsequent examples. The table does not state whether a shoe was a right shoe or a left shoe; the "sidedness" variable has been controlled by having each type of leather worn in half of the cases on the right and in half on the left, so, in effect, it has been eliminated from the immediate problem. The table shows that the mean $\bar{D}$ of the $X_i - Y_i$ difference scores is -1.5. And it will be assumed that the manufacturer knows that the standard deviation σ_D of the sampled population of difference scores has the value 5.00. (We shall not speculate about how he knows it.)

If the manufacturer wanted a symmetric confidence interval for μ_D corresponding to the 95% level of confidence, he would procede as follows. Since 95% of the standardized normal distribution lies between -1.96 and $+1.96$, he lets $K = 1.96$ and $C = -1.96 = -K$, so that $\Phi(K) - \Phi(C) = .95$. Substituting $\bar{D}$ for $\bar{X}$, μ_D for μ, and σ_D for σ in Formula 15-4, he obtains

$$\Phi(K) - \Phi(C) \cong P\left(\bar{D} - \frac{K\sigma_D}{\sqrt{n}} \le \mu_D \le \bar{D} - \frac{C\sigma_D}{\sqrt{n}}\right)$$

which tells him that the desired confidence interval is given by the expression inside the large parentheses, when the appropriate numerical values are substituted. Doing so he obtains:

$$-1.5 - \frac{1.96(5)}{\sqrt{26}} \le \mu_D \le -1.5 - \frac{-1.96(5)}{\sqrt{26}}$$

$$-1.5 - \frac{9.8}{5.099} \le \mu_D \le -1.5 + \frac{9.8}{5.099}$$

$$-1.5 - 1.922 \le \mu_D \le -1.5 + 1.922$$

$$-3.422 \le \mu_D \le .422$$

So the manufacturer could be confident at approximately the 95% level that μ_D was in the interval from -3.422 to $.422$.

Table 15-1

Data and Calculations for Shoe Manufacturer's Problem

	Amount of wear on			
	X treated shoe	Y treated shoe	Difference	
Subject	X_i	Y_i	$D_i = X_i - Y_i$	D_i^2
A	113	117	-4	16
B	91	97	-6	36
C	107	103	4	16
D	138	131	7	49
E	120	115	5	25
F	97	94	3	9
G	99	107	-8	64
H	86	89	-3	9
I	112	105	7	49
J	111	114	-3	9
K	96	100	-4	16
L	86	92	-6	36
M	131	129	2	4
N	127	126	1	1
O	115	115	0	0
P	104	106	-2	4
Q	88	90	-2	4
R	90	96	-6	36
S	121	123	-2	4
T	102	109	-7	49
U	115	119	-4	16
V	125	133	-8	64
W	101	101	0	0
X	96	99	-3	9
Y	88	90	-2	4
Z	126	124	2	4
Σ	2,785	2,824	-39	533

$$\bar{X} = \frac{2,785}{26} = 107.115 \qquad \bar{Y} = \frac{2,824}{26} = 108.615 \qquad \bar{D} = \frac{-39}{26} = -1.500$$

$$s_D^2 = \frac{\Sigma D_i^2}{n} - \bar{D}^2 = \frac{533}{26} - 2.25 = 20.50 - 2.25 = 18.25$$

$$\hat{\sigma}_D^2 = \frac{n}{n-1}s_D^2 = \frac{26}{25}(18.25) = 1.04(18.25) = 18.98 \qquad \hat{\sigma}_D = 4.357$$

If the manufacturer wanted to test the $H_0 : \mu_D \geq 0$ against the $H_a : \mu_D < 0$ at about $\alpha = .01$, he would take as his test statistic

$$Z_0 = \frac{\bar{D} - (\mu_D)_0}{\dfrac{\sigma_D}{\sqrt{n}}} = \frac{-1.5 - 0}{\dfrac{5}{\sqrt{26}}} = -.3(5.099) = -1.5297$$

His H_a tells him that the test is left-tailed, and since $\Phi(-2.3263) = .01 \cong \alpha$, the critical value is -2.3263, and the rejection region consists of all values ≤ -2.3263. The obtained value of the test statistic -1.5297 does not fall in the rejection region, so H_0 is accepted. The approximate probability level of the test is $\Phi(-1.5297) = .063$. If the true value of μ_D had been -1.00, the power of the test (see Formula 15-11) would have been approximately

$$\Phi\left(\phi_p + \frac{(\mu_D)_0 - (\mu_D)_a}{\frac{\sigma_D}{\sqrt{n}}}\right) = \Phi\left(-2.3263 + \frac{0 - (-1)}{\frac{5}{\sqrt{26}}}\right)$$

$$= \Phi\left(-2.3263 + \frac{5.099}{5}\right) = \Phi(-2.3263 + 1.0198)$$

$$= \Phi(-1.3065) = .096$$

15-2   The Two-Sample Z Statistic

Now let $\bar{X}$ be the mean of a random sample of n_X observations drawn from an infinite population of X's with mean μ_X and variance σ_X^2 and let $\bar{Y}$ be the mean of an *independent* (this is a very important qualification) random sample of n_Y observations drawn from an infinite population of Y's with mean μ_Y and variance σ_Y^2. Then the difference between sample means $\bar{X} - \bar{Y}$ is a random variable whose mean is

$$E(\bar{X} - \bar{Y}) = E(\bar{X}) - E(\bar{Y}) = \mu_X - \mu_Y$$

and whose variance (since $\bar{X}$ and $\bar{Y}$ are independent) is

$$\text{Var}\,(\bar{X} - \bar{Y}) = \text{Var}\,(\bar{X}) + (-1)^2\,\text{Var}\,(\bar{Y}) = \frac{\sigma_X^2}{n_X} + \frac{\sigma_Y^2}{n_Y}$$

Consequently, the standardized variable

$$Z = \frac{(\bar{X} - \bar{Y}) - E(\bar{X} - \bar{Y})}{\sqrt{\text{Var}\,(\bar{X} - \bar{Y})}} = \frac{(\bar{X} - \bar{Y}) - (\mu_X - \mu_Y)}{\sqrt{\frac{\sigma_X^2}{n_X} + \frac{\sigma_Y^2}{n_Y}}} \tag{15-13}$$

has a mean of zero and a variance of 1.

If the X and Y populations are normally distributed, then Z will also be normally distributed. But even if the sampled populations are not normally distributed, if $\bar{X}$ and $\bar{Y}$ are both essentially normally distributed due to the Central Limit Effect, then Z will be essentially normally distributed. That the Central Limit Effect applies in this situation can be seen as follows. Suppose that $n_X = Cn$ and $n_Y = Kn$ where C, K, and n are all integers. Imagine the Cn X-observations arranged in a line and divided into n successive blocks of C observations each. Let X_{gh} be the hth observation in the gth block, so

$$\bar{X}_g = \frac{\sum_{h=1}^C X_{gh}}{C}$$

is the mean of the gth block of C X-observations. Likewise, imagine the Kn Y-observations arranged in a line and divided into n successive blocks of K observations each. Let Y_{gh} be the hth observation in the gth block, so

$$\bar{Y}_g = \frac{\sum_{h=1}^{K} Y_{gh}}{K}$$

is the mean of the gth block of K Y-observations. Now consider a new variable $V_g = \bar{X}_g - \bar{Y}_g$. The mean of the n "observations" on the new variable is

$$\bar{V} = \frac{\sum_{g=1}^{n} V_g}{n} = \frac{\sum_{g=1}^{n} (\bar{X}_g - \bar{Y}_g)}{n} = \frac{\sum_{g=1}^{n} \bar{X}_g}{n} - \frac{\sum_{g=1}^{n} \bar{Y}_g}{n}$$

$$= \frac{1}{n} \sum_{g=1}^{n} \frac{\sum_{h=1}^{C} X_{gh}}{C} - \frac{1}{n} \sum_{g=1}^{n} \frac{\sum_{h=1}^{K} Y_{gh}}{K}$$

$$= \frac{\sum_{i=1}^{nC} X_i}{nC} - \frac{\sum_{j=1}^{nK} Y_j}{nK} = \frac{\sum_{i=1}^{n_X} X_i}{n_X} - \frac{\sum_{j=1}^{n_Y} Y_j}{n_Y} = \bar{X} - \bar{Y}$$

So we may regard $\bar{X} - \bar{Y}$ as the mean of n observations on a single random variable V, and we already know that the Central Limit Effect applies in the case of a single variable.

15-2-1   Confidence Intervals for $\mu_X - \mu_Y$ Based on the Two-Sample Z Statistic

If sampling is random, if the X and Y populations are infinite, and if the distributions of $\bar{X}$ and $\bar{Y}$ are essentially normal, then Z is approximately a standardized normal variable so that

$$P(C \leq Z \leq K) \cong \Phi(K) - \Phi(C) \qquad (15\text{-}14)$$

So,

$$\Phi(K) - \Phi(C) \cong P\left(C \leq \frac{(\bar{X} - \bar{Y}) - (\mu_X - \mu_Y)}{\sqrt{\dfrac{\sigma_X^2}{n_X} + \dfrac{\sigma_Y^2}{n_Y}}} \leq K \right)$$

$$\cong P\left(-(\bar{X} - \bar{Y}) + C\sqrt{\frac{\sigma_X^2}{n_X} + \frac{\sigma_Y^2}{n_Y}} \leq -(\mu_X - \mu_Y) \right.$$

$$\left. \leq -(\bar{X} - \bar{Y}) + K\sqrt{\frac{\sigma_X^2}{n_X} + \frac{\sigma_Y^2}{n_Y}} \right)$$

$$\cong P\left(\bar{X} - \bar{Y} - C\sqrt{\frac{\sigma_X^2}{n_X} + \frac{\sigma_Y^2}{n_Y}} \geq \mu_X - \mu_Y \geq \bar{X} - \bar{Y} - K\sqrt{\frac{\sigma_X^2}{n_X} + \frac{\sigma_Y^2}{n_Y}} \right)$$

$$\cong P\left(\bar{X} - \bar{Y} - K\sqrt{\frac{\sigma_X^2}{n_X} + \frac{\sigma_Y^2}{n_Y}} \leq \mu_X - \mu_Y \leq \bar{X} - \bar{Y} - C\sqrt{\frac{\sigma_X^2}{n_X} + \frac{\sigma_Y^2}{n_Y}} \right)$$

$$(15\text{-}15)$$

Therefore, if we select the values of K, C, n_X, and n_Y prior to sampling and if we know the numerical values of σ_X^2 and σ_Y^2, we may have confidence at approximately level $\Phi(K) - \Phi(C)$ that the difference between population means $\mu_X - \mu_Y$ lies in the interval from

$$\bar{x} - \bar{y} - K\sqrt{\frac{\sigma_X^2}{n_X} + \frac{\sigma_Y^2}{n_Y}} \text{ to } \bar{x} - \bar{y} - C\sqrt{\frac{\sigma_X^2}{n_X} + \frac{\sigma_Y^2}{n_Y}}$$

i.e., that

$$\bar{x} - \bar{y} - K\sqrt{\frac{\sigma_X^2}{n_X} + \frac{\sigma_Y^2}{n_Y}} \leq \mu_X - \mu_Y \leq \bar{x} - \bar{y} - C\sqrt{\frac{\sigma_X^2}{n_X} + \frac{\sigma_Y^2}{n_Y}} \quad (15\text{-}16)$$

where $\bar{x}$ and $\bar{y}$ are the numerical values of the means of the actually obtained X and Y samples. And if we choose $C = \phi_p$ and $K = \phi_q$, our approximate confidence level will be $q - p$.

Example: To illustrate, in "saturation bombing" raids the point of impact of bombs dropped by enemy planes, as measured along the direction of flight, is known to have a standard deviation of 200 ft for daytime flights and 500 ft for night flights. On one day and on the following night enemy planes followed the same line of flight, along which are two prime targets. As measured in feet (along the line of flight) from the first prime target, the mean values of the points of impact were 80 ft for the 100 bombs dropped in the daytime raid and 420 ft for the 500 bombs dropped in the night raid. Allied Intelligence desires a symmetric confidence interval for the distance between actual points of aim (treated as population means) in which it can have confidence at about the .90 level. They proceed as follows. Since the normal distribution is symmetric, the values of $\phi_{.05}$ and $\phi_{.95}$ will provide approximate endpoints for a symmetric 90% confidence interval for Z about its mean of zero. Normal tables tell them that $\phi_{.05} = -1.645$ and $\phi_{.95} = 1.645$. So they can be confident at approximately the .90 level that

$$\bar{x} - \bar{y} - \phi_{.95}\sqrt{\frac{\sigma_X^2}{n_X} + \frac{\sigma_Y^2}{n_Y}} \leq \mu_X - \mu_Y \leq \bar{x} - \bar{y} - \phi_{.05}\sqrt{\frac{\sigma_X^2}{n_X} + \frac{\sigma_Y^2}{n_Y}}$$

$$80 - 420 - 1.645\sqrt{\frac{200^2}{100} + \frac{500^2}{500}} \leq \mu_X - \mu_Y \leq 80 - 420 - (-1.645)\sqrt{\frac{200^2}{100} + \frac{500^2}{500}}$$

$$-340 - 1.645\sqrt{\frac{40,000}{100} + \frac{250,000}{500}} \leq \mu_X - \mu_Y \leq -340 + 1.645\sqrt{\frac{40,000}{100} + \frac{250,000}{500}}$$

$$-340 - 1.645\sqrt{900} \leq \mu_X - \mu_Y \leq -340 + 1.645\sqrt{900}$$

$$-340 - 49.35 \leq \mu_X - \mu_Y \leq -340 + 49.35$$

$$-389.35 \leq \mu_X - \mu_Y \leq -290.65$$

So Allied Intelligence estimates at a confidence level of about .90 that the night target was between 290.65 and 389.35 feet farther down the flight path from the first prime target than was the day target.

15-2-2 Test of an Hypothesized Value of $\mu_X - \mu_Y$, Using the Two-Sample Z Statistic

If both sampled populations are infinite, both samples are random, and the distributions of $\bar{X}$ and $\bar{Y}$ are essentially normal, we can use the two-sample Z statistic to test the hypothesis that the distance Δ between population means equals some specific value Δ_0. That is letting $\Delta = \mu_X - \mu_Y$, we can test the $H_0: \Delta = \Delta_0$ against H_a's that $\Delta < \Delta_0$, $\Delta \neq \Delta_0$, or $\Delta > \Delta_0$ calling for left-tail, two-tail, or right-tail tests, respectively. For if $H_0: \Delta = \Delta_0$ is true, then $\Delta_0 = \mu_X - \mu_Y$ and

$$Z_0 = \frac{\bar{X} - \bar{Y} - \Delta_0}{\sqrt{\dfrac{\sigma_X^2}{n_X} + \dfrac{\sigma_Y^2}{n_Y}}} \tag{15-17}$$

has an essentially normal distribution with zero mean and unit variance so that

$$P(Z_0 < \phi_p) + P(Z_0 > \phi_q) \cong p + (1 - q) \tag{15-18}$$

and we can reject H_0 at about the $p + (1 - q)$ level of significance if either

$$\frac{\bar{X} - \bar{Y} - \Delta_0}{\sqrt{\dfrac{\sigma_X^2}{n_X} + \dfrac{\sigma_Y^2}{n_Y}}} < \phi_p \quad \text{or} \quad \frac{\bar{X} - \bar{Y} - \Delta_0}{\sqrt{\dfrac{\sigma_X^2}{n_X} + \dfrac{\sigma_Y^2}{n_Y}}} > \phi_q$$

If actually $\mu_X - \mu_Y < \Delta_0$, $\bar{X} - \bar{Y}$ will tend to be smaller than Δ_0 and Z_0 will tend to be $< \phi_p$ a greater proportion of the time than p. On the other hand, if $\mu_X - \mu_Y > \Delta_0$, $\bar{X} - \bar{Y}$ will tend to exceed Δ_0 and Z_0 will tend to exceed ϕ_q a greater proportion of the time than $1 - q$. Therefore, if H_a is that $\Delta < \Delta_0$, we let $p = \alpha$ and $q = 1$, doing a left-tail test at the nominal α level of significance; if H_a is that $\Delta \neq \Delta_0$, we let $p = \alpha/2$ and $1 - q = \alpha/2$, doing a two-tail test at the nominal α level; and if H_a is that $\Delta > \Delta_0$, we let $p = 0$ and $q = 1 - \alpha$, which results in a right-tail test at a nominal significance level of α.

Example: To illustrate, in the previous example, suppose that Allied Intelligence knew that the two prime targets were located exactly 400 ft apart but was not sure whether or not the enemy knew it, and wished to find out. They therefore wanted to test the H_0 that $\mu_X - \mu_Y = -400$ against the H_a that $\mu_X - \mu_Y \neq -400$, using the nominal $\alpha = .05$ level of significance. The test statistic is

$$Z_0 = \frac{\bar{X} - \bar{Y} - \Delta_0}{\sqrt{\dfrac{\sigma_X^2}{n_X} + \dfrac{\sigma_Y^2}{n_Y}}} = \frac{80 - 420 - (-400)}{\sqrt{\dfrac{40{,}000}{100} + \dfrac{250{,}000}{500}}} = \frac{-340 + 400}{\sqrt{400 + 500}}$$

$$= \frac{60}{30} = 2$$

And for a two-tail test at nominal $\alpha = .05$ we take $\phi_p = \phi_{\alpha/2} = \phi_{.025} = -1.96$ and $\phi_q = \phi_{1-(\alpha/2)} = \phi_{.975} = 1.96$ and we reject H_0 in favor of H_a: $\Delta \neq -400$ since $Z_0 > \phi_q$.

The power of the test is the probability that the test statistic

$$Z_0 = \frac{\bar{X} - \bar{Y} - \Delta_0}{\sqrt{\dfrac{\sigma_X^2}{n_X} + \dfrac{\sigma_Y^2}{n_Y}}}$$

will fall in the rejection region if actually $\Delta = \mu_X - \mu_Y$ is equal to some other value Δ_a. Thus,

$$\text{Power} = P(Z_0 < \phi_p | \Delta = \Delta_a) + P(Z_0 > \phi_q | \Delta = \Delta_a)$$

$$= P\left(\frac{\bar{X} - \bar{Y} - \Delta_0}{\sqrt{\dfrac{\sigma_X^2}{n_X} + \dfrac{\sigma_Y^2}{n_Y}}} < \phi_p \middle| \Delta = \Delta_a\right) + P\left(\frac{\bar{X} - \bar{Y} - \Delta_0}{\sqrt{\dfrac{\sigma_X^2}{n_X} + \dfrac{\sigma_Y^2}{n_Y}}} > \phi_q \middle| \Delta = \Delta_a\right)$$

$$= P\left(\frac{\bar{X} - \bar{Y} - \Delta_0 + \Delta_a - \Delta_a}{\sqrt{\dfrac{\sigma_X^2}{n_X} + \dfrac{\sigma_Y^2}{n_Y}}} < \phi_p \middle| \Delta = \Delta_a\right)$$

$$+ P\left(\frac{\bar{X} - \bar{Y} - \Delta_0 + \Delta_a - \Delta_a}{\sqrt{\dfrac{\sigma_X^2}{n_X} + \dfrac{\sigma_Y^2}{n_Y}}} > \phi_q \middle| \Delta = \Delta_a\right)$$

$$= P\left(\frac{\bar{X} - \bar{Y} - \Delta_a}{\sqrt{\dfrac{\sigma_X^2}{n_X} + \dfrac{\sigma_Y^2}{n_Y}}} < \phi_p + \frac{\Delta_0 - \Delta_a}{\sqrt{\dfrac{\sigma_X^2}{n_X} + \dfrac{\sigma_Y^2}{n_Y}}} \middle| \Delta = \Delta_a\right)$$

$$+ P\left(\frac{\bar{X} - \bar{Y} - \Delta_a}{\sqrt{\dfrac{\sigma_X^2}{n_X} + \dfrac{\sigma_Y^2}{n_Y}}} > \phi_q + \frac{\Delta_0 - \Delta_a}{\sqrt{\dfrac{\sigma_X^2}{n_X} + \dfrac{\sigma_Y^2}{n_Y}}} \middle| \Delta = \Delta_a\right)$$

But since Δ_a is the true value of Δ and therefore of $\mu_X - \mu_Y$, the expression on the left of the inequality signs is essentially a normally distributed variable with mean of zero and variance of 1 and can therefore be replaced by ϕ and we have

$$\text{Power} \cong P\left(\phi < \phi_p + \frac{\Delta_0 - \Delta_a}{\sqrt{\dfrac{\sigma_X^2}{n_X} + \dfrac{\sigma_Y^2}{n_Y}}}\right) + P\left(\phi > \phi_q + \frac{\Delta_0 - \Delta_a}{\sqrt{\dfrac{\sigma_X^2}{n_X} + \dfrac{\sigma_Y^2}{n_Y}}}\right) \quad (15\text{-}19)$$

Since all of the ingredients of the expressions to the right of the inequalities are known constants except perhaps for Δ_a, we can obtain the power of the test for any specified value of Δ_a, i.e., for any specified true difference between population means.

Example: To illustrate, if the true value of $\mu_X - \mu_Y$ in the previous example had been $\Delta = \Delta_a = -300$ ft, then the power of the test would have been

$$\text{Power} \cong P\left(\phi < \phi_p + \frac{\Delta_0 - \Delta_a}{\sqrt{\frac{\sigma_X^2}{n_X} + \frac{\sigma_Y^2}{n_Y}}}\right) + P\left(\phi > \phi_q + \frac{\Delta_0 - \Delta_a}{\sqrt{\frac{\sigma_X^2}{n_X} + \frac{\sigma_Y^2}{n_Y}}}\right)$$

$$\cong P\left(\phi < -1.96 + \frac{-400 - (-300)}{\sqrt{\frac{40,000}{100} + \frac{250,000}{500}}}\right)$$

$$+ P\left(\phi > 1.96 + \frac{-400 - (-300)}{\sqrt{\frac{40,000}{100} + \frac{250,000}{500}}}\right)$$

$$\cong P\left(\phi < -1.96 - \frac{100}{30}\right) + P\left(\phi > 1.96 - \frac{100}{30}\right)$$

$$\cong P(\phi < -1.96 - 3.33) + P(\phi > 1.96 - 3.33)$$

$$\cong P(\phi < -5.29) + P(\phi > -1.37)$$

$$\cong .00000\,00612 + .91466$$

$$\cong .915$$

15-3   Critique of Z Statistics

All of the uses of the Z statistics that we have discussed so far assume:

1. that sampling is random and, if two (unmatched) samples are drawn, that the two samples are independent;
2. that either the sampled populations are infinite or that sampling is with replacements;
3. that the distribution of each sample mean used in the statistic ($\bar{X}$, $\bar{Y}$, or in the case of difference scores, $\bar{D}$) is a sufficiently close approximation to a normal distribution to produce the degree of approximation to normality required in the distribution of Z (and this implies, among other things, that either the distributions of the sampled populations are "sufficiently" quasi-normal or that sample sizes are large enough to produce "sufficient" quasi-normality in the distributions of the sample means);
4. that the exact values of the variances of the sampled populations are known and can therefore be substituted into the formula for Z.

Assumption 3 requires good judgment on the part of the practitioner and tends to make the use of the Z statistic somewhat of an art. However assumption 4 tends to make the Z statistic academic. It tells us that in order to be able to use the Z statistic to obtain information (through estimation or tests) about the means of the sampled populations we must first know their exact variances. But, in all but a very few practical situations, if we have enough information (either a priori or empirical) about a population to know its variance, we also know its mean. And, in such cases, there would therefore be no need to estimate, or test hypotheses about, the values of the population means since these would already be known. Therefore, the uses of the Z statistic, as outlined so far, tend to be restricted to those cases where the population in which we are interested is known to have the same variance, but not the same mean, as many other populations about whose distribution we have considerable information. An example of such a situation is provided by a thoroughly checked-out target rifle whose sights have been tampered with; one might then "know" in advance the true (population) variance of the distribution of shots along a horizontal line 100 yards distant without knowing in advance where the mean would lie, that is, one could accurately predict the variance but not the mean. Although such situations sometimes arise in applications, they tend to be confined primarily to certain areas of application (such as Quality Control) and to be rare in others.

15-4 $\hat{Z}$ Statistics

15-4-1 Rationale

An obvious way of making Z statistics more versatile is to obtain from each sample not only the sample mean but also the sample estimate of population variance $\hat{\sigma}^2$ and to substitute the latter for the actual population variance σ^2 in the denominator of Z. This is tantamount to substituting

$$\hat{\sigma}_{\bar{X}}^2 = \frac{\hat{\sigma}^2}{n}$$

for

$$\sigma_{\bar{X}}^2 = \frac{\sigma^2}{n}$$

in the denominator of the one-sample Z statistic and

$$\hat{\sigma}_{\bar{X}-\bar{Y}}^2 = \frac{\hat{\sigma}_X^2}{n_X} + \frac{\hat{\sigma}_Y^2}{n_Y}$$

for

$$\sigma_{\bar{X}-\bar{Y}}^2 = \frac{\sigma_X^2}{n_X} + \frac{\sigma_Y^2}{n_Y}$$

in the denominator of the two-sample Z statistic.

If we do this, we are substituting a variable for a constant which is the expected value of the substituted variable. The resulting statistic $\hat{Z}$ now has variable quantities both in its numerator and in its denominator and the variability of the latter is undesirable. It prevents $\hat{Z}$ from having the same probability distribution as Z. Thus the $\hat{Z}$ statistic will not have an exactly normal probability distribution even if all sampled populations are exactly normally distributed. However, this alone should not dissuade us since, in practice, exactly normal populations are impossible and so therefore is an exactly normally distributed Z statistic. What *should* make us hesitate is the fact that this solution tends to cause a *further* distortion in the actual distribution of the statistic relative to the normal distribution used to obtain probabilities for it.

When n is small, the variability of the denominator of $\hat{Z}$ is large enough, relative to that of the numerator, to cause an intolerable amount of distortion and renders the $\hat{Z}$ statistic unsatisfactory. However, as n increases, the variance of $\hat{\sigma}_{\bar{X}}^2$ about its expected value $\sigma_{\bar{X}}^2$ diminishes at a much faster rate than does the variance of $\bar{X}$ about its expected value of μ. So, relatively speaking, $\hat{\sigma}_{\bar{X}}^2$ begins to behave like its own constant expected value, which it was substituted to estimate, while $\bar{X}$ is still behaving like a variable and, indeed, becomes almost entirely responsible for the variability of $\hat{Z}$. And in the two-sample case, the variance of $\hat{\sigma}_{\bar{X}-\bar{Y}}^2$ about its expected value of $\sigma_{\bar{X}-\bar{Y}}^2$ diminishes much faster than that of $\bar{X} - \bar{Y}$ about its expected value of $\mu_X - \mu_Y$, with results entirely analogous to the one-sample case. More importantly in both the one and two sample cases, as n-values increase, the variance of the entire denominator of $\hat{Z}$ diminishes much faster than does that of the entire numerator. In fact the variance of the denominator *relative* to that of the numerator approaches zero. So the denominator of $\hat{Z}$ becomes more and more similar to the constant denominator of Z. Hence as $n \to \infty$ (or, in the two-sample case, as n_X and n_Y both $\to \infty$), $\hat{Z}$ behaves more and more like Z and becomes equivalent to it when n-values become infinite. Therefore, as n-values increase, $\hat{Z}$ becomes an increasingly satisfactory substitute for Z. Table 15-2 provides some data on how good a substitute the one-sample $\hat{Z}$ statistic is for the one-sample Z statistic when the sampled population has an exactly normal distribution.

Proof:　　Consider the statistic $\hat{\sigma}^2$. We already know that its expected value is σ^2. And although we shall not prove it, it can be shown that, except at the smallest sample sizes, $\text{Var}(\hat{\sigma}^2) \cong$ (a constant, C)$/n$ and also that $\text{Var}(\hat{\sigma}) \cong$ (a different constant, c)$/n$ (where the constant depends primarily upon the second and fourth central moments of the sampled population).

If we substitute $\hat{\sigma}^2$ for σ^2 in the one-sample Z statistic

$$Z = \frac{\bar{X} - \mu}{\sqrt{\dfrac{\sigma^2}{n}}} = \frac{\bar{X} - \mu}{\sqrt{\sigma_{\bar{X}}^2}}$$

Table 15-2

True and Normal-Approximation Probabilities for One-Sample $\hat{Z}$
Statistic When Sampled Population is Normally Distributed

True $P(Z \leqslant Z_\alpha) = \alpha$

Normal-approximation $P(Z \leqslant \hat{Z}_\alpha) = P(\phi \leqslant \hat{Z}_\alpha) = \Phi(\hat{Z}_\alpha)$

n	Normal-approximation probability corresponding to true probability of α		
	$\alpha = .05$	$\alpha = .01$	$\alpha = .001$
2	.0000 +	.00000 +	.000000 +
3	.0018	.00000 +	.000000 +
4	.0093	.00000 +	.000000 +
5	.0165	.00009	.000000 +
10	.0334	.00239	.000009
15	.0391	.00434	.000076
20	.0419	.00555	.000172
25	.0436	.00634	.000263
50	.0468	.00809	.000544
100	.0484	.00903	.000749
200	.0492	.00951	.000871
500	.0497	.00980	.000948
1,000	.0498	.00990	.000974
∞	.0500	.01000	.001000

we are, in effect, substituting $\hat{\sigma}_{\bar{X}}^2 = \hat{\sigma}^2/n$ for $\sigma_{\bar{X}}^2$. The expected value of $\hat{\sigma}_{\bar{X}}^2$ is

$$E(\hat{\sigma}_{\bar{X}}^2) = E\left(\frac{\hat{\sigma}^2}{n}\right) = \frac{1}{n}E(\hat{\sigma}^2) = \frac{\sigma^2}{n} = \sigma_{\bar{X}}^2$$

the constant that it replaced, and its variance is

$$\text{Var}\,(\hat{\sigma}_{\bar{X}}^2) = \text{Var}\left(\frac{\hat{\sigma}^2}{n}\right) = \left(\frac{1}{n}\right)^2 \text{Var}\,(\hat{\sigma}^2) = \left(\frac{1}{n}\right)^2 \frac{C}{n} = \frac{C}{n^3}$$

So as n increases, the variance of $\hat{\sigma}_{\bar{X}}^2$ about its expected value of $\sigma_{\bar{X}}^2$ diminishes at a rate proportional to n^3, whereas the variance σ^2/n of $\bar{X}$ about its expected value of μ diminishes only at a rate proportional to n.

Likewise, if we substitute $\hat{\sigma}_X^2$ for σ_X^2 and $\hat{\sigma}_Y^2$ for σ_Y^2 in the two-sample Z statistic

$$Z = \frac{(\bar{X} - \bar{Y}) - (\mu_X - \mu_Y)}{\sqrt{\dfrac{\sigma_X^2}{n_X} + \dfrac{\sigma_Y^2}{n_Y}}} = \frac{(\bar{X} - \bar{Y}) - (\mu_X - \mu_Y)}{\sqrt{\sigma_{\bar{X}}^2 + \sigma_{\bar{Y}}^2}}$$

$$= \frac{(\bar{X} - \bar{Y}) - (\mu_X - \mu_Y)}{\sqrt{\sigma_{\bar{X}-\bar{Y}}^2}}$$

we are, in effect, substituting

$$\hat{\sigma}_{\bar{X}-\bar{Y}}^2 = \hat{\sigma}_{\bar{X}}^2 + \hat{\sigma}_{\bar{Y}}^2 = \frac{\hat{\sigma}_X^2}{n_X} + \frac{\hat{\sigma}_Y^2}{n_Y}$$

for $\sigma^2_{\bar{X}-\bar{Y}}$. The expected value of $\hat{\sigma}^2_{\bar{X}-\bar{Y}}$ is

$$E(\hat{\sigma}^2_{\bar{X}-\bar{Y}}) = E\left(\frac{\hat{\sigma}^2_X}{n_X} + \frac{\hat{\sigma}^2_Y}{n_Y}\right) = \frac{1}{n_X}E(\hat{\sigma}^2_X) + \frac{1}{n_Y}E(\hat{\sigma}^2_Y) = \frac{\sigma^2_X}{n_X} + \frac{\sigma^2_Y}{n_Y} = \sigma^2_{\bar{X}-\bar{Y}}$$

which is the constant that it replaced, and its variance is

$$\text{Var}\,(\hat{\sigma}^2_{\bar{X}-\bar{Y}}) = \text{Var}\left(\frac{\hat{\sigma}^2_X}{n_X} + \frac{\hat{\sigma}^2_Y}{n_Y}\right) = \left(\frac{1}{n_X}\right)^2 \text{Var}\,(\hat{\sigma}^2_X) + \left(\frac{1}{n_Y}\right)^2 \text{Var}\,(\hat{\sigma}^2_Y)$$

$$= \left(\frac{1}{n_X}\right)^2 \frac{C_X}{n_X} + \left(\frac{1}{n_Y}\right)^2 \frac{C_Y}{n_Y} = \frac{C_X}{n_X^3} + \frac{C_Y}{n_Y^3}$$

Now if we let $n_X = n$ and $n_Y = Kn$, where K is a constant,

$$\text{Var}\,(\hat{\sigma}^2_{\bar{X}-\bar{Y}}) = \frac{C_X}{n^3} + \frac{C_Y}{(Kn)^3} = \frac{C_X + \dfrac{C_Y}{K^3}}{n^3} = \frac{\text{(a constant)}}{n^3}$$

and

$$\text{Var}\,(\bar{X} - \bar{Y}) = \frac{\sigma^2_X}{n} + \frac{\sigma^2_Y}{Kn} = \frac{\sigma^2_X + \dfrac{\sigma^2_Y}{K}}{n} = \frac{\text{(a different constant)}}{n}$$

So, again, as n increases, the variance of $\hat{\sigma}^2_{\bar{X}-\bar{Y}}$ about its expected value of $\sigma^2_{\bar{X}-\bar{Y}}$ diminishes at a rate proportional to n^3, whereas the variance of $\bar{X} - \bar{Y}$ about its expected value of $\mu_X - \mu_Y$ diminishes only at a rate proportional to n. Clearly, the two-sample case is analogous to the one-sample case.

The critical question however is the variability of the entire denominator relative to that of the numerator. This can be shown most easily in the one-sample case where

$$Z = \frac{\bar{X} - \mu}{\dfrac{\sigma}{\sqrt{n}}}$$

The variance of the numerator is

$$\text{Var}\,(\bar{X} - \mu) = \text{Var}\,(\bar{X}) + \text{Var}\,(\mu) = \text{Var}\,(\bar{X}) + 0 = \frac{\sigma^2}{n}$$

and if we replace σ with $\hat{\sigma}$, the variance of the denominator is

$$\text{Var}\left(\frac{\hat{\sigma}}{\sqrt{n}}\right) = \left(\frac{1}{\sqrt{n}}\right)^2 \text{Var}\,(\hat{\sigma}) \cong \left(\frac{1}{n}\right)\frac{c}{n} = \frac{c}{n^2}$$

So the variance of the denominator is (almost) inversely proportional to n^2 whereas that of the numerator is inversely proportional to n and the ratio of the variance of the denominator to that of the numerator is

$$\frac{\text{Var (denominator)}}{\text{Var (numerator)}} \cong \frac{\dfrac{c}{n^2}}{\dfrac{\sigma^2}{n}} = \frac{\dfrac{c}{\sigma^2}}{n} = \frac{\text{a constant}}{n}$$

which approaches zero as n approaches infinity.

15-4-2 The One-Sample $\hat{Z}$ Statistic

Substituting $\hat{\sigma}^2$ for σ^2 in the one-sample Z statistic, we obtain the one-sample $\hat{Z}$ statistic

$$\hat{Z} = \frac{\bar{X} - \mu}{\sqrt{\dfrac{\hat{\sigma}^2}{n}}} = \frac{\bar{X} - \mu}{\sqrt{\dfrac{\sum_{i=1}^{n} (X_i - \bar{X})^2}{n(n-1)}}} \qquad (15\text{-}20)$$

If n is large enough to have caused $\hat{\sigma}_{\bar{x}}$ to behave almost like a constant relative to the variability of $\bar{X}$, the $\hat{Z}$ statistic will behave almost like the Z statistic. And if n is also large enough to have caused $\bar{X}$ to be quasi-normally distributed, the $\hat{Z}$ statistic will be nearly normally distributed—at least within two or three standard deviations of its mean. In such cases, if one substitutes $\hat{\sigma}$ for σ, $\hat{Z}$ for Z, and $\hat{Z}_0$ for Z_0 in Formulas 15-1 through 15-11, the formulas will continue to be approximately true. Thus, we need only make these substitutions in order to obtain the necessary probability formulas for $\hat{Z}$.

It can be seen in Table 15-2 that for a given n-value the approximation worsens for increasingly remote tail values, e.g., as α goes from .05 to .01 to .001. Consequently, in order to meet a given criterion of approximation (i.e., ratio between true and alleged significance levels or confidence levels) one must use larger and larger n-values as significance level α goes from .05 to .01 to .001 or as confidence level goes from .95 to .99 to .999.

Examples: In the coffee-machine example for the one-sample Z statistic, the sample size $n = 9$ was large enough for that statistic since the population distribution was known to be quasi-normal. For comparative purposes we shall use the $\hat{Z}$ statistic upon the same data. However, at such a small sample size, even with the relatively innocuous α of .05 and confidence level of .90, we cannot expect the normal approximation to be excellent.

In the coffee-machine example, if we had not known the value of σ^2, we could have estimated it from the sample data as follows:

$$\hat{\sigma}^2 = \frac{\sum_{i=1}^{n} (x_i - \bar{x})^2}{n-1}$$

$$= [(4.8 - 5.4)^2 + (5.7 - 5.4)^2 + (6.2 - 5.4)^2 + (4.9 - 5.4)^2$$
$$+ (5.5 - 5.4)^2 + (5.2 - 5.4)^2 + (5.2 - 5.4)^2 + (5.1 - 5.4)^2$$
$$+ (6.0 - 5.4)^2]/8$$

$$= [.36 + .09 + .64 + .25 + .01 + .04 + .04 + .09 + .36]/8$$

$$= \frac{1.88}{8} = .235$$

$$\hat{\sigma} = \sqrt{.235} = .485$$

Substituting this value for $\hat{\sigma}$, as well as the appropriate values of $\bar{x}$, n, C, and K (the same values as used when dealing with Z), into the estimated confidence interval

$$\bar{x} - \frac{K\hat{\sigma}}{\sqrt{n}} \le \mu \le \bar{x} - \frac{C\hat{\sigma}}{\sqrt{n}}$$

we obtain

$$5.4 - 1.645\left(\frac{.485}{\sqrt{9}}\right) \le \mu \le 5.4 - (-1.645)\left(\frac{.485}{\sqrt{9}}\right)$$

$$5.4 - .266 \le \mu \le 5.4 + .266$$

$$5.134 \le \mu \le 5.666$$

as the estimated confidence interval corresponding to approximately the 90% level of confidence.

If we substitute $\hat{\sigma}$ for σ and $\hat{Z}_0$ for Z_0 in the previous test of the $H_0: \mu \le 5$ against the $H_a: \mu > 5$ at the nominal .05 level of significance, we obtain

$$\hat{Z}_0 = \frac{\bar{X} - \mu_0}{\frac{\hat{\sigma}}{\sqrt{n}}} = \frac{5.4 - 5}{\frac{.485}{\sqrt{9}}} = \frac{.4}{.162} = 2.47$$

which is greater than 1.645 and therefore lies in the rejection region, so H_0 is rejected in favor of H_a, just as before. (However, the $\hat{Z}_0$ of 2.47 is closer to the Z_0 of 2.40 than we have any right to expect at such a small n-value.)

Finally, if actually $\mu = \mu_a = 5.5$, the estimated power of the test would be

$$\text{Estimated power} = \Phi\left(\phi_p + \frac{\mu_0 - \mu_a}{\frac{\hat{\sigma}}{\sqrt{n}}}\right) + 1 - \Phi\left(\phi_q + \frac{\mu_0 - \mu_a}{\frac{\hat{\sigma}}{\sqrt{n}}}\right)$$

$$= \Phi\left(-\infty + \frac{5 - 5.5}{\frac{.485}{\sqrt{9}}}\right) + 1 - \Phi\left(1.645 + \frac{5 - 5.5}{\frac{.485}{\sqrt{9}}}\right)$$

$$= \Phi(-\infty) + 1 - \Phi(1.645 - 3.086)$$

$$= 0 + 1 - \Phi(-1.441)$$

$$= 1 - .075 = .925$$

Usually, we want to calculate power *before* the sample is drawn, and therefore, often, before $\hat{\sigma}^2$ is known. In such cases, the $\hat{Z}$ approach is ineffective since without a numerical value for $\hat{\sigma}^2$ we cannot calculate the power. Furthermore, the approximation will tend to be poor unless $\hat{\sigma}$ is a close estimator of σ, which requires that n be large. (For a given n-value, the approximation will tend to worsen as the absolute magnitude of $\mu_0 - \mu_a$ increases relative to that of $\sigma/\sqrt{n}$.)

15-4-3 The Difference-Score $\hat{Z}$ Statistic

If we substitute

$$\hat{\sigma}_D^2 = \frac{\sum_{i=1}^{n}(D_i - \bar{D})^2}{n - 1}$$

for σ_D^2 in the difference-score Z statistic, we obtain the difference-score $\hat{Z}$ statistic

$$\hat{Z} = \frac{\bar{D} - \mu_D}{\sqrt{\dfrac{\hat{\sigma}_D^2}{n}}} = \frac{\bar{D} - \mu_D}{\sqrt{\dfrac{\sum_{i=1}^{n}(D_i - \bar{D})^2}{n(n - 1)}}} = \frac{(\bar{X} - \bar{Y}) - (\mu_X - \mu_Y)}{\sqrt{\dfrac{\sum_{i=1}^{n}(D_i - \bar{D})^2}{n(n - 1)}}} \qquad (15\text{-}21)$$

If the number of difference scores n is sufficiently large (see Section 15-4-2), the difference-score $\hat{Z}$ statistic will be approximately normally distributed. In that event, if we substitute $\hat{\sigma}_D$ for σ, $\bar{D}$ for $\bar{X}$, μ_D for μ, $(\mu_D)_0$ for μ_0, $(\mu_D)_a$ for μ_a, $\hat{Z}$ for Z, and $\hat{Z}_0$ for Z_0 in Formulas 15-1 through 15-11, the formulas will apply to the case in which the data are difference scores drawn randomly from an infinite population of difference scores, and they will be approximately true.

Examples: In the shoe example for the difference-score Z statistic, if the manufacturer did not know the standard deviation σ_D of the difference-score population (as he almost certainly would not), he could use $\hat{\sigma}_D$ in place of it. The calculations in Table 15-1 yield $\hat{\sigma}_D = 4.357$. Substituting this value for σ_D, the estimated confidence interval for the nominal 95% level of confidence becomes

$$\bar{D} - \frac{K\hat{\sigma}_D}{\sqrt{n}} \leq \mu_D \leq \bar{D} - \frac{C\hat{\sigma}_D}{\sqrt{n}}$$

$$-1.5 - \frac{1.96(4.357)}{\sqrt{26}} \leq \mu_D \leq -1.5 - \frac{-1.96(4.357)}{\sqrt{26}}$$

$$-1.5 - 1.675 \leq \mu_D \leq -1.5 + 1.675$$

$$-3.175 \leq \mu_D \leq .175$$

In the test of the $H_0\colon \mu_D \geq 0$ against the $H_a\colon \mu_D < 0$ at the nominal .01 level of significance, the test statistic becomes

$$\hat{Z}_0 = \frac{\bar{D} - (\mu_D)_0}{\dfrac{\hat{\sigma}_D}{\sqrt{n}}} = \frac{-1.5 - 0}{\dfrac{4.357}{\sqrt{26}}} = -1.755$$

which is greater than -2.3263 and therefore does not fall in the rejection region, so H_0 is accepted as before. Finally, if actually $\mu_D = -1$, the very roughly estimated power of the test would be

$$\Phi\left(\phi_p + \frac{(\mu_D)_0 - (\mu_D)_a}{\frac{\hat{\sigma}_D}{\sqrt{n}}}\right) = \Phi\left(-2.3263 + \frac{0 - (-1)}{\frac{4.357}{\sqrt{26}}}\right)$$

$$= \Phi(-2.3263 + 1.1703) = \Phi(-1.156) = .124$$

15-4-4   The Two-Sample $\hat{Z}$ Statistic

If $\hat{\sigma}_X^2$ and $\hat{\sigma}_Y^2$ are substituted for σ_X^2 and σ_Y^2 in the two-sample Z statistic, the result is the two-sample $\hat{Z}$ statistic

$$\hat{Z} = \frac{(\bar{X} - \bar{Y}) - (\mu_X - \mu_Y)}{\sqrt{\frac{\hat{\sigma}_X^2}{n_X} + \frac{\hat{\sigma}_Y^2}{n_Y}}} \tag{15-22}$$

If n_X and n_Y are *both* large enough to make the relative variance of the denominator negligible, the probability distribution of the two-sample $\hat{Z}$ statistic will be almost the same as that of the two-sample Z statistic (at least within two or three standard deviations of its mean). And if both n_X and n_Y are also large enough to make the probability distribution of $\bar{X} - \bar{Y}$ quasi-normal, the probability distribution of the two-sample $\hat{Z}$ statistic will be approximately a standardized normal distribution (at least within two or three standard deviations of its mean). In such cases, if one substitutes $\hat{\sigma}_X$ and $\hat{\sigma}_Y$ for σ_X and σ_Y, $\hat{Z}$ for Z, and $\hat{Z}_0$ for Z_0, in Formulas 15-13 through 15-19, the formulas will continue to be approximately correct. Therefore, they can be used to obtain confidence intervals, test hypotheses, determine the power of the test, and obtain the probability level of the obtained value of the test statistic, always approximately. Other things being equal, if n_X and n_Y both increase while the ratio n_X/n_Y remains constant, the approximation will tend to improve. Other things being equal, for positive values of K, the approximation $P(\hat{Z} \le -K) \cong P(\phi \le -K)$ and $P(\hat{Z} \ge K) \cong P(\phi \ge K)$ will tend to worsen as K increases. That is, the ratio between the $\hat{Z}$ probability and the ϕ probability will tend to depart increasingly from 1.00 as K increases. So the approximation will tend to worsen as α goes from .05 to .01 to .001 and as confidence levels go from .95 to .99 to .999.

Examples: In the bombing example for the two-sample Z statistic, suppose that instead of knowing that $\sigma_X^2 = 40,000$ and $\sigma_Y^2 = 250,000$, they were estimated from the same sample data used to calculate $\bar{X}$ and $\bar{Y}$ and that $\hat{\sigma}_X^2 = 50,000$ and $\hat{\sigma}_Y^2 = 230,500$. Then

$$\sqrt{\frac{\hat{\sigma}_X^2}{n_X} + \frac{\hat{\sigma}_Y^2}{n_Y}} = \sqrt{\frac{50,000}{100} + \frac{230,500}{500}} = \sqrt{500 + 461}$$

$$= \sqrt{961} = 31$$

so that the estimated confidence interval for the nominal .90 level of confidence would be

$$\bar{x} - \bar{y} - \phi_{.95}\sqrt{\frac{\hat{\sigma}_X^2}{n_X} + \frac{\hat{\sigma}_Y^2}{n_Y}} \leq \mu_X - \mu_Y \leq \bar{x} - \bar{y} - \phi_{.05}\sqrt{\frac{\hat{\sigma}_X^2}{n_X} + \frac{\hat{\sigma}_Y^2}{n_Y}}$$

$$80 - 420 - 1.645(31) \leq \mu_X - \mu_Y \leq 80 - 420 - (-1.645)(31)$$

$$-391 \leq \mu_X - \mu_Y \leq -289$$

The test statistic for the test of the H_0: $\mu_X - \mu_Y = -400$ against the H_a: $\mu_X - \mu_Y \neq -400$ at the nominal .05 level of significance would be

$$\hat{Z}_0 = \frac{\bar{X} - \bar{Y} - \Delta_0}{\sqrt{\frac{\hat{\sigma}_X^2}{n_X} + \frac{\hat{\sigma}_Y^2}{n_Y}}} = \frac{80 - 420 - (-400)}{31} = \frac{60}{31} = 1.94$$

which does not fall in the rejection region consisting of values < -1.96 or > 1.96, so H_0 would be accepted. If actually the value of $\mu_X - \mu_Y$ had been $\Delta_a = -300$, the roughly estimated power of the test would have been

$$\text{Power} \cong P\left(\phi < \phi_p + \frac{\Delta_0 - \Delta_a}{\sqrt{\frac{\hat{\sigma}_X^2}{n_X} + \frac{\hat{\sigma}_Y^2}{n_Y}}}\right) + P\left(\phi > \phi_q + \frac{\Delta_0 - \Delta_a}{\sqrt{\frac{\hat{\sigma}_X^2}{n_X} + \frac{\hat{\sigma}_Y^2}{n_Y}}}\right)$$

$$\cong P\left(\phi < -1.96 + \frac{-400 - (-300)}{31}\right)$$

$$+ P\left(\phi > 1.96 + \frac{-400 - (-300)}{31}\right)$$

$$\cong P(\phi < -1.96 - 3.226) + P(\phi > 1.96 - 3.226)$$

$$\cong P(\phi < -5.186) + P(\phi > -1.266)$$

$$\cong .0000001074 + .89724$$

$$\cong .897$$

Again, it should be remembered that the power estimate will tend to be poor unless n_X and n_Y are large enough to cause $\hat{\sigma}_X^2$ and $\hat{\sigma}_Y^2$ to behave almost like constants relative to the variation of $\bar{X} - \bar{Y}$. And at given sample sizes, the approximation will tend to worsen as the absolute value of $\Delta_0 - \Delta_a$ increases relative to that of

$$\sqrt{\frac{\sigma_X^2}{n_X} + \frac{\sigma_Y^2}{n_Y}}$$

15-5 *t* Statistics

15-5-1 The *t* Distributions

There is a whole family of distributions known as t distributions, all of which are symmetric and bell-shaped but nonnormal—differing from a normal distribution by having thicker tails. The distribution with the thickest tails is identified as the distribution of t "with 1 degree of freedom." The one with the second thickest tails is the distribution of t "with 2 degrees of freedom." Thus, the individual t distributions are identified by their degrees of freedom. And as degrees of freedom increase from 1 to ∞ in steps of 1, the tails of the associated distribution become successively thinner until when degrees of freedom become infinite the corresponding t distribution is a normal distribution. The t distributions all have a mean of zero and a variance that diminishes rapidly toward 1 as degrees of freedom increase toward infinity. So as degrees of freedom $\longrightarrow \infty$ $t \rightarrow \phi$, that is, the t distribution approaches the standardized normal distribution.

When the number of degrees of freedom is small, the t distributions are poorly approximated by the standardized normal distribution. That is, if K is a positive number, $P(t < -K)$ is poorly approximated by $P(\phi < -K)$, and so is $P(t > K)$ by $P(\phi > K)$. Furthermore, the approximation worsens as K increases. We can quantify these statements as follows. Let t_α stand for the 100αth percentile (where $\alpha < .5$) in a t distribution, so that $P(t < t_\alpha) = \alpha$. And suppose that we regard the normal-approximation probability $P(\phi < t_\alpha)$ as a good approximation to $P(t < t_\alpha)$ only if the ratio of the former to the latter lies between .9 and 1, i.e., only if $.9\alpha < P(\phi < t_\alpha) < \alpha$. Then the smallest number of degrees of freedom required to make the approximation a good one is 32 when $\alpha = .05$, 97 when $\alpha = .01$, and about 300 when $\alpha = .001$.

15-5-2 The One-Sample t Statistic

If $\bar{X}$ is the mean, and $\hat{\sigma}^2$ the sample estimate of population variance, of a random sample of n observations to be drawn from an *exactly* normally distributed population, with mean μ and variance σ^2, then

$$t = \frac{\bar{X} - \mu}{\sqrt{\dfrac{\hat{\sigma}^2}{n}}} = \frac{\bar{X} - \mu}{\sqrt{\dfrac{\sum_{i=1}^{n} (X_i - \bar{X})^2}{n(n - 1)}}} \tag{15-23}$$

has exactly a t distribution with $n - 1$ degrees of freedom. The above statistic is the one-sample t statistic and it equals the one-sample $\hat{Z}$ statistic. So if the sampled population is exactly normally distributed, the one-sample $\hat{Z}$ statistic has *exactly* a t with $n - 1$ degrees of freedom distribution. And even if the population's distribution is only approximately normal, when n is small (e.g., < 25) the distribution of the one-sample $\hat{Z}$ statistic will probably be better approximated by the distribution of t with $n - 1$ degrees of freedom than by the standardized normal distribution. Thus, when n is small and the sampled population is quasi-normal, $P(\hat{Z} < K)$ tends to be much better approximated by $P(t < K)$ than by $P(\phi < K)$. [How *much* better (in the fictitious case where the sampled population is *exactly* normal) can be inferred from Table 15-2; let t_α be the 100αth percentile in the distribution of t with $n - 1$ degrees of freedom; then the cell entries are $P(\phi < t_\alpha)$ and the column headings are $\alpha = P(t < t_\alpha)$.] However, the variance of the distribution of $\hat{\sigma}^2$ (and therefore of the one-sample $\hat{Z}$ and t statistics when n is small) is very sensitive to (i.e., is greatly affected by) appreciable nonnormality in the sampled population. So when n is small and the sampled population does *not* have a very nearly normal distribution, the t approximation may be far from satisfactory. See Figure 15-1 which shows the distortion in significance levels (i.e., in the probability of a Type I Error) due to sampling from the nonnormal population shown in Figure 14-7. Figure 15-1 should be compared with Figure 14-21 which shows the same thing for the one-sample Z statistic.

Therefore, (unless n is quite large) if we know that the sampled population has an approximately normal distribution, we should treat the one-sample $\hat{Z}$ statistic as having a t with $n - 1$ degrees of freedom distribution rather than a standardized normal distribution. To do so, we proceed as follows. Let $P(t \leq K) = T(K)$ and $P(t \leq C) = T(C)$ where $C < K$. Then

$$P(C \leq t \leq K) = T(K) - T(C)$$

and by a derivation entirely analogous to that which produced Formula 15-4, we obtain

$$T(K) - T(C) \cong P\left(\bar{X} - \frac{K\hat{\sigma}}{\sqrt{n}} \leq \mu \leq \bar{X} - \frac{C\hat{\sigma}}{\sqrt{n}} \right) \qquad (15\text{-}24)$$

Now let t_p be that value of t for which $P(t \leq t_p) = p$, let t_q be that value of t for which $P(t \leq t_q) = q$, and let $t_p = C$ and $t_q = K$. Then

$$T(C) = P(t \leq t_p) = p \quad \text{and} \quad T(K) = P(t \leq t_q) = q$$

and Formula 15-24 can be written in the equivalent form

$$q - p \cong P\left(\bar{X} - \frac{t_q\hat{\sigma}}{\sqrt{n}} \leq \mu \leq \bar{X} - \frac{t_p\hat{\sigma}}{\sqrt{n}} \right) \qquad (15\text{-}25)$$

This formula tells us that if we chose p and q (or t_p and t_q) before drawing from a quasi-normal population a random sample whose mean turned out to be $\bar{x}$, the statement that

$$\bar{x} - \frac{t_q\hat{\sigma}}{\sqrt{n}} \leq \mu \leq \bar{x} - \frac{t_p\hat{\sigma}}{\sqrt{n}} \qquad (15\text{-}26)$$

could be made at approximately the $q - p$ level of confidence.

In *testing hypotheses* about μ, our test statistic is

$$t_0 = \frac{\bar{X} - \mu_0}{\frac{\hat{\sigma}}{\sqrt{n}}} = \frac{\bar{X} - \mu_0}{\sqrt{\frac{\sum_{i=1}^{n} (X_i - \bar{X})^2}{n(n - 1)}}} \qquad (15\text{-}27)$$

which will be distributed approximately as t with $n - 1$ degrees of freedom if actually $\mu = \mu_0$. If we take as rejection region all values of t_0 less than t_p

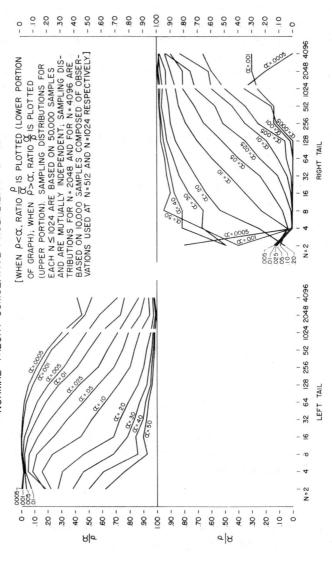

FIG. 15-1. Sensitivity of the *t* statistic to the nonnormality of the *X* population (see Figure 14-7), at various sample sizes and cumulative-probabilities. (Ratio between actual ρ and alleged α probability of a Type I Error when applying the one-sample *t* test to samples of *n* observations drawn randomly and with replacements from the *X* population.)

464

or greater than t_q, the probability of committing a Type I Error will be

$$P(t_0 < t_p | \mu = \mu_0) + P(t_0 > t_q | \mu = \mu_0)$$
$$\cong P(t < t_p) + P(t > t_q) = p + 1 - q \qquad (15\text{-}28)$$

So to test $H_0: \mu \geq \mu_0$ against $H_a: \mu < \mu_0$ at the nominal α level of significance, let $p = \alpha$ and $q = 1$ (so that $t_q = +\infty$ and we are therefore doing a left-tail test). To test $H_0: \mu \leq \mu_0$ against $H_a: \mu > \mu_0$ at a nominal significance level of α, let $p = 0$ and $q = 1 - \alpha$ (so that $t_p = -\infty$ and we are doing a right-tail test). And to perform the two-tail test of $H_0: \mu = \mu_0$ against $H_a: \mu \neq \mu_0$ at the nominal α level of significance, let $p = \alpha/2$ and $q = 1 - (\alpha/2)$.

The formula giving the approximate power of the test is obtained by a derivation entirely analogous to that which produced Formula 15-11. (We need only substitute t_0 and t_a for Z_0 and Z_a, t_p and t_q for ϕ_p and ϕ_q, $\hat{\sigma}$ for σ, t for ϕ and T for Φ, and recognize that if $\mu = \mu_a$ then t_a becomes t.) The formula is

$$\text{Power} \cong T\left(t_p + \frac{\mu_0 - \mu_a}{\frac{\hat{\sigma}}{\sqrt{n}}}\right) + 1 - T\left(t_q + \frac{\mu_0 - \mu_a}{\frac{\hat{\sigma}}{\sqrt{n}}}\right) \qquad (15\text{-}29)$$

Unless $\hat{\sigma}$ is based upon a large sample, this formula is likely to provide only a *very gross approximation* to the actual power of the test, and, at a given n-value, the approximation will tend to worsen as the absolute value of $\mu_0 - \mu_a$ increases relative to that of $\sigma/\sqrt{n}$.

Examples: In the coffee-machine example the population distribution was known to be quasi-normal but the sample size was too small for the $\hat{Z}$ statistic's distribution to be extremely well approximated by the standardized normal distribution. However, these are precisely the conditions under which the t distribution provides a much better approximation than the ϕ distribution as well as a very good approximation in the absolute sense (at least at nonextreme α and confidence levels). For this example, therefore, the t approach is clearly preferable.

In the example a random sample of $n = 9$ observations was drawn in which $\bar{x} = 5.4$ and $\hat{\sigma} = .485$. In order to set up a symmetric confidence interval for μ, corresponding to the 90% level of confidence, we need to know the 5th and 95th percentiles of the distribution of t with $n - 1 = 8$ degrees of freedom. From Table B-12 in Appendix B we find them to be $t_{.05} = -1.860$ and $t_{.95} = +1.860$. Formula 15-25 tells us that

$$q - p \cong P\left(\bar{X} - \frac{t_q \hat{\sigma}}{\sqrt{n}} \leq \mu \leq \bar{X} - \frac{t_p \hat{\sigma}}{\sqrt{n}}\right)$$

Therefore, we may have confidence at approximately the

$$q - p = .95 - .05 = .90$$

level that

$$5.4 - \frac{(1.86)(.485)}{\sqrt{9}} \leq \mu \leq 5.4 - \frac{(-1.86)(.485)}{\sqrt{9}}$$

which reduces to

$$5.0993 \leq \mu \leq 5.7007$$

To test the H_0: $\mu \leq 5$ against the H_a: $\mu > 5$ at the nominal .05 level of significance, the test statistic is

$$t_0 = \frac{\bar{X} - \mu_0}{\frac{\hat{\sigma}}{\sqrt{n}}} = \frac{5.4 - 5}{\frac{.485}{\sqrt{9}}} = 2.47$$

which, of course, is the same value as previously obtained for the test statistic $\hat{Z}_0$, since in the one-sample case $\hat{Z} = t$. The difference between the $\hat{Z}$ and t tests is that they have different rejection regions. For the t test, the rejection region consists of all values greater than the 95th percentile of the t distribution with $n - 1 = 8$ degrees of freedom. From t tables we find that, with 8 degrees of freedom, $t_{.95} = 1.860$. Since the obtained value of the test statistic 2.47 falls in the rejection region of values ≥ 1.860, H_0 is rejected.

If actually $\mu = \mu_a = 5.5$, the estimated power of the test is obtained by substituting this and other appropriate values into Formula 15-29.

$$\text{Power} \cong T\left(t_p + \frac{\mu_0 - \mu_a}{\frac{\hat{\sigma}}{\sqrt{n}}}\right) + 1 - T\left(t_q + \frac{\mu_0 - \mu_a}{\frac{\hat{\sigma}}{\sqrt{n}}}\right)$$

$$\cong T\left(-\infty + \frac{5 - 5.5}{\frac{.485}{\sqrt{9}}}\right) + 1 - T\left(1.860 + \frac{5 - 5.5}{\frac{.485}{\sqrt{9}}}\right)$$

$$\cong T(-\infty) + 1 - T(1.860 - 3.086)$$

$$\cong 0 + 1 - T(-1.226)$$

where $T(-1.226)$ is the proportion of the distribution of t with 8 degrees of freedom that lies below -1.226. This is found (by interpolation) from t tables to be about .14. So the estimated power is $1 - .14 = .86$. However, n is far too small (and $\mu_0 - \mu_a$ too large) for us to treat the .86 as anything more than a very gross estimate which may be far from the correct power.

15-5-3 The Difference-Score t Statistic

If $\bar{D}$ is the mean, and $\hat{\sigma}_D^2$ the sample estimate of population variance, of a random sample of n difference scores $D_i = X_i - Y_i$ to be drawn from an *exactly* normally distributed population of difference scores with mean μ_D and variance σ_D^2, the difference-score t statistic

$$t = \frac{\bar{D} - \mu_D}{\sqrt{\frac{\hat{\sigma}_D^2}{n}}} = \frac{\bar{D} - \mu_D}{\sqrt{\frac{\sum_{i=1}^{n}(D_i - \bar{D})^2}{n(n-1)}}} = \frac{(\bar{X} - \bar{Y}) - (\mu_X - \mu_Y)}{\sqrt{\frac{\sum_{i=1}^{n}(D_i - \bar{D})^2}{n(n-1)}}} \quad (15\text{-}30)$$

has exactly a t distribution with $n - 1$ degrees of freedom. And it will have approximately that distribution, even when n is small, if the distribution of the difference-score population is only quasi-normal. The difference-score t statistic has the same formula as the difference-score $\hat{Z}$ statistic. So when n is small and the population distribution is approximately normal, the difference-score $\hat{Z}$ statistic should be much better approximated by the appropriate t distribution than by the ϕ distribution. The difference-score t statistic is a special case of the one-sample t statistic, so if we substitute μ_D, $(\mu_D)_0$, $(\mu_D)_a$, $\hat{\sigma}_D$, D_i, and $\bar{D}$ for μ, μ_0, μ_a, $\hat{\sigma}$, X_i, and $\bar{X}$, respectively, in Formulas 15-23 through 15-29, the formulas will apply.

Difference-score populations are often quasi-normally distributed. However, there is no assurance that the population will be quasi-normal in a particular case, and one certainly cannot infer it from a small sample. Yet because difference-score populations are generally only *potential* or *conceptual*, rather than actually existing, populations, one's information about the population is often confined to the same small sample to be used in the test or interval-estimate of μ_D. Thus, we often do not know that the quasi-normality condition has been met until sample size is so large that the t approximation has lost most of its advantage over the ϕ approximation, even if the condition *has* been met. And if the condition has *not* been met, the ϕ and t approximations may *both* be unacceptably bad when n is small. Clearly, the use of the t approach when the population distribution is not known to be quasi-normal involves a dangerous gamble.

Examples: In the shoe example, if the manufacturer knew that the difference-score population had a quasi-normal distribution but did not know its variance, he could proceed as follows. The sample of $n = 26$ difference scores had $\bar{D} = -1.5$ and $\hat{\sigma}_D = 4.357$. To establish a symmetric confidence interval corresponding to the 95% level of confidence we need to know the 2.5th and 97.5th percentiles of the distribution of t with $n - 1 = 25$ degrees of freedom. From Table B-12 in Appendix B we find them to be $t_{.025} = -2.060$ and $t_{.975} = 2.060$. When properly modified for the difference-score situation, Formula 15-25 tells us that

$$q - p \cong P\left(\bar{D} - \frac{t_q \hat{\sigma}_D}{\sqrt{n}} \leq \mu_D \leq \bar{D} - \frac{t_p \hat{\sigma}_D}{\sqrt{n}}\right)$$

So we may have approximately the $q - p = .975 - .025 = .95$ level of confidence that

$$-1.5 - \frac{(2.060)(4.357)}{\sqrt{26}} \leq \mu_D \leq -1.5 - \frac{(-2.060)(4.357)}{\sqrt{26}}$$

which reduces to

$$-3.26 \le \mu_D \le .26$$

To test the $H_0: \mu_D \ge 0$ against the $H_a: \mu_D < 0$ at the nominal .01 level of significance, the test statistic, obtained by modifying Formula 15-27 to apply to difference scores, is

$$t_0 = \frac{\bar{D} - (\mu_D)_0}{\frac{\hat{\sigma}_D}{\sqrt{n}}} = \frac{-1.5 - 0}{\frac{4.357}{\sqrt{26}}} = -1.755$$

the same value as previously obtained for the $\hat{Z}_0$ test statistic. The test is left-tailed and the nominal α level is .01; so the rejection region consists of all values less than the 1st percentile of the distribution of t with $n - 1 = 25$ degrees of freedom. From t tables this is found to be $t_{.01} = -2.485$. Since the obtained value of t_0 is not less than -2.485, H_0 is accepted.

If actually $\mu_D = -1$, the estimated power of the test would be obtained by substituting this and other appropriate values into the modification of Formula 15-29 for difference scores:

$$\text{Power} \cong T\left(t_p + \frac{(\mu_D)_0 - (\mu_D)_a}{\frac{\hat{\sigma}_D}{\sqrt{n}}}\right) + 1 - T\left(t_q + \frac{(\mu_D)_0 - (\mu_D)_a}{\frac{\hat{\sigma}_D}{\sqrt{n}}}\right)$$

$$\cong T\left(-2.485 + \frac{0 - (-1)}{\frac{4.357}{\sqrt{26}}}\right) + 1 - T\left(+\infty + \frac{0 - (-1)}{\frac{4.357}{\sqrt{26}}}\right)$$

$$\cong T(-1.315) + 1 - 1 = T(-1.315)$$

From t tables -1.315 is found to be about the 10th percentile of the distribution of t with 25 degrees of freedom; so the power of the test is estimated to be approximately .10. Considering the sizes of n and α, and the relative sizes of $|(\mu_D)_0 - (\mu_D)_a|$ and $\hat{\sigma}_D/\sqrt{n}$, this estimate should be regarded as only grossly approximate.

15-5-4 The Two-Sample t Statistic

If $\bar{X}$ is the mean of a random sample of n_X observations, X_i's, to be drawn from a normally distributed X population with mean μ_X and variance σ_X^2 and if $\bar{Y}$ is the mean of an *independent* random sample of n_Y observations Y_j's to be drawn from a normally distributed Y population with mean μ_Y and variance σ_Y^2, then if $\sigma_X^2 = \sigma_Y^2$, i.e., if the two populations have the same variance,

$$t = \frac{(\bar{X} - \bar{Y}) - (\mu_X - \mu_Y)}{\sqrt{\frac{\sum_{i=1}^{n_X}(X_i - \bar{X})^2 + \sum_{j=1}^{n_Y}(Y_j - \bar{Y})^2}{n_X + n_Y - 2}\left(\frac{1}{n_X} + \frac{1}{n_Y}\right)}} \tag{15-31}$$

has exactly a t distribution with $n_X + n_Y - 2$ degrees of freedom. This statistic is the two-sample t statistic. Unlike the one-sample t statistic which was equal to the one-sample $\hat{Z}$ statistic, the two-sample t statistic is *not* equal to the two-sample $\hat{Z}$ statistic in the general case but only in the special case where $n_X = n_Y$.

The conditions necessary for a t statistic to have *exactly* a t distribution will never be encountered in practice. And when these requirements are not met, the goodness of the approximation of the true distribution of the t statistic by the theoretical t distribution is influenced by a multiplicity of interacting factors (entirely analogous to, but more numerous than, those mentioned in the penultimate paragraph of Section 14-5-3). *Any* concise description of the goodness of the approximation or of the conditions conducive to a good approximation therefore tends to be glib and overly simplified. The reader is warned that this is the case in what follows.

If the sampled populations have equal variances but are only quasi-normally distributed, the two-sample t statistic will have approximately a t with $n_X + n_Y - 2$ degrees of freedom distribution (and the approximation will become perfect when both sample sizes become infinite). So in that situation, when any sample size is small, the distribution of the two-sample t statistic tends to be much better approximated by the appropriate t distribution than is the distribution of the two-sample $\hat{Z}$ statistic by the standardized normal distribution. And when sample sizes are large, the former approximation tends to be slightly superior to the latter.

If the sampled populations are quasi-normal and have only slightly unequal variances, the t statistic's distribution still tends to be fairly well approximated by the appropriate t distribution, provided that the two sample sizes are nearly equal. And this approximation tends to be superior to that of the $\hat{Z}$ distribution by the ϕ distribution when any sample size is small, but not necessarily when both samples are large.

However, the goodness of the t approximation depends strongly upon the amount of departure from normality and from equality of variance when any sample size is small and upon the latter when both sample sizes are large and unequal.

If the two populations have different variances and the two samples have different sizes, the t approximation may be unacceptably bad if the differences are appreciable. And this will be the case even if both populations are normally distributed and both samples are "infinitely" large. The reason for this is that when $\sigma_X^2 \neq \sigma_Y^2$ and $n_X \neq n_Y$ the expected value of the denominator of the two-sample t statistic is not the same as when the condition $\sigma_X^2 = \sigma_Y^2$ is met.

Proof: The expression under the radical in the denominator of the two-sample t statistic can be written in the following equivalent form:

$$\frac{(n_X - 1)\hat{\sigma}_X^2 + (n_Y - 1)\hat{\sigma}_Y^2}{n_X + n_Y - 2}\left(\frac{1}{n_X} + \frac{1}{n_Y}\right)$$

and since the sample sizes are constants, its expected value is obtained by substituting σ^2's for $\hat\sigma^2$'s

$$\frac{(n_X - 1)\sigma_X^2 + (n_Y - 1)\sigma_Y^2}{n_X + n_Y - 2}\left(\frac{1}{n_X} + \frac{1}{n_Y}\right)$$

Now if the two populations have equal variances as required, $\sigma_X^2 = \sigma_Y^2$ and we can substitute σ^2 for each of them in which case the expected value reduces to

$$\frac{(n_X - 1 + n_Y - 1)\sigma^2}{n_X + n_Y - 2}\left(\frac{1}{n_X} + \frac{1}{n_Y}\right) = \sigma^2\left(\frac{1}{n_X} + \frac{1}{n_Y}\right) = \frac{\sigma_X^2}{n_X} + \frac{\sigma_Y^2}{n_Y}$$

which is the proper expected value since it was obtained under the required conditions. Likewise, if the two samples are of the same size, n_X equals n_Y and we can substitute n for each of them obtaining the same value as before

$$\frac{(n - 1)(\sigma_X^2 + \sigma_Y^2)}{2(n - 1)}\left(\frac{2}{n}\right) = \frac{\sigma_X^2 + \sigma_Y^2}{n} = \frac{\sigma_X^2}{n_X} + \frac{\sigma_Y^2}{n_Y}$$

The value obtained in both cases is the variance of the numerator of the two-sample t statistic, which is what it is supposed to be.

However, generally speaking, when $\sigma_X^2 \neq \sigma_Y^2$ and $n_X \neq n_Y$, the expected value is *not* the same as when $\sigma_X^2 = \sigma_Y^2$. As a result the denominator is biased, or displaced, from its "proper" value, and the t statistic does not have a t distribution. Futhermore, taking "infinite-sized" samples does not ameliorate the situation. This can be seen as follows. The t statistic can be written in the equivalent form

$$t = \frac{(\bar X - \bar Y) - (\mu_X - \mu_Y)}{\sqrt{\dfrac{(n_X - 1)\hat\sigma_X^2 + (n_Y - 1)\hat\sigma_Y^2}{n_X + n_Y - 2}\left(\dfrac{n_X + n_Y}{n_X n_Y}\right)}}$$

$$= \frac{(\bar X - \bar Y) - (\mu_X - \mu_Y)}{\sqrt{\left[\dfrac{n_X + n_Y}{n_X + n_Y - 2}\right]\left(\dfrac{n_X - 1}{n_X n_Y}\right)\hat\sigma_X^2 + \left[\dfrac{n_X + n_Y}{n_X + n_Y - 2}\right]\left(\dfrac{n_Y - 1}{n_X n_Y}\right)\hat\sigma_Y^2}}$$

Now if we let $n_X \longrightarrow \infty$ and $n_Y \longrightarrow \infty$, the expression inside square brackets approaches 1, the entire expression multiplying $\hat\sigma_X^2$ approaches $1/n_Y$ and that multiplying $\hat\sigma_Y^2$ approaches $1/n_X$, and for reasons already discussed, $\hat\sigma_X^2$ and $\hat\sigma_Y^2$ can be replaced by the variances they estimate σ_X^2 and σ_Y^2, respectively, so that

$$t \longrightarrow \frac{(\bar X - \bar Y) - (\mu_X - \mu_Y)}{\sqrt{\dfrac{\sigma_X^2}{n_Y} + \dfrac{\sigma_Y^2}{n_X}}}$$

which equals the two-sample Z statistic if either $\sigma_X^2 = \sigma_Y^2$ or $n_X = n_Y$. Now let $\sigma_X^2 = \sigma^2$ and $\sigma_Y^2 = R\sigma^2$. And let $n_X = n$ and $n_Y = \gamma n$. Then when n becomes "infinitely large,"

$$t = \frac{(\bar{X} - \bar{Y}) - (\mu_X - \mu_Y)}{\sqrt{\dfrac{\sigma_X^2}{n_Y} + \dfrac{\sigma_Y^2}{n_X}}} = \frac{(\bar{X} - \bar{Y}) - (\mu_X - \mu_Y)}{\sqrt{\dfrac{\sigma^2}{\gamma n} + \dfrac{R\sigma^2}{n}}}$$

$$= \frac{(\bar{X} - \bar{Y}) - (\mu_X - \mu_Y)}{\sigma \sqrt{\dfrac{1 + \gamma R}{\gamma n}}}$$

whereas

$$Z = \frac{(\bar{X} - \bar{Y}) - (\mu_X - \mu_Y)}{\sqrt{\dfrac{\sigma_X^2}{n_X} + \dfrac{\sigma_Y^2}{n_Y}}} = \frac{(\bar{X} - \bar{Y}) - (\mu_X - \mu_Y)}{\sqrt{\dfrac{\sigma^2}{n} + \dfrac{R\sigma^2}{\gamma n}}}$$

$$= \frac{(\bar{X} - \bar{Y}) - (\mu_X - \mu_Y)}{\sigma \sqrt{\dfrac{\gamma + R}{\gamma n}}}$$

and the ratio of t to Z, which is 1 when variances are equal (and sample sizes are infinite), is

$$\frac{t}{Z} = \frac{\dfrac{(\bar{X} - \bar{Y}) - (\mu_X - \mu_Y)}{\sigma \sqrt{\dfrac{1 + \gamma R}{\gamma n}}}}{\dfrac{(\bar{X} - \bar{Y}) - (\mu_X - \mu_Y)}{\sigma \sqrt{\dfrac{\gamma + R}{\gamma n}}}} = \frac{\sqrt{\dfrac{\gamma + R}{\gamma n}}}{\sqrt{\dfrac{1 + \gamma R}{\gamma n}}} = \sqrt{\frac{\gamma + R}{1 + \gamma R}}$$

so that

$$t = Z \sqrt{\frac{\gamma + R}{1 + \gamma R}}$$

Now because of the Central Limit Effect, the Z statistic has become a ϕ statistic, having a normal distribution with zero mean and unit variance. And because

$$\sqrt{\frac{\gamma + R}{1 + \gamma R}}$$

is a constant, the t statistic must also be normally distributed with the same mean

$$E(t) = E\left(Z\sqrt{\frac{\gamma + R}{1 + \gamma R}}\right) = \left(\sqrt{\frac{\gamma + R}{1 + \gamma R}}\right)E(Z) = \left(\sqrt{\frac{\gamma + R}{1 + \gamma R}}\right)0 = 0$$

but with variance

$$\text{Var}(t) = \text{Var}\left(Z\sqrt{\frac{\gamma + R}{1 + \gamma R}}\right) = \left(\frac{\gamma + R}{1 + \gamma R}\right)\text{Var}(Z) = \left(\frac{\gamma + R}{1 + \gamma R}\right)1 = \frac{\gamma + R}{1 + \gamma R}$$

So t is normally distributed with a mean of 0 and a standard deviation of

$$\sqrt{\frac{\gamma + R}{1 + \gamma R}}$$

whereas if its requirement that $\sigma_{\bar{X}}^2 = \sigma_{\bar{Y}}^2$ had been met it would have been normally distributed with a mean of zero and a standard deviation of 1 (as is Z), and it is the latter distribution upon which the t tables are based (for infinite degrees of freedom).

We can determine how much bias this introduces as follows. Since t is normally distributed with mean of 0 and standard deviation of

$$\sqrt{\frac{\gamma + R}{1 + \gamma R}}$$

we can standardize it by subtracting its mean and dividing the result by its standard deviation, obtaining a standardized normal variable

$$\frac{t}{\sqrt{\dfrac{\gamma + R}{1 + \gamma R}}}$$

which must therefore have a ϕ distribution. Therefore, if the t tables "allege" that for our two-sample t statistic $P(t < t_\alpha) = \alpha$, in actuality, the *true* probability is

$$P(t < t_\alpha) = P\left(\frac{t}{\sqrt{\dfrac{\gamma + R}{1 + \gamma R}}} < \frac{t_\alpha}{\sqrt{\dfrac{\gamma + R}{1 + \gamma R}}}\right) = P\left(\phi < \frac{t_\alpha}{\sqrt{\dfrac{\gamma + R}{1 + \gamma R}}}\right)$$

$$= \Phi\left(\frac{t_\alpha}{\sqrt{\dfrac{\gamma + R}{1 + \gamma R}}}\right) = p$$

which can be obtained (for a given α-value) simply by getting t_α from t tables (the theoretical t-value whose cumulative probability is α) and then looking up the cumulative probability of

$$\frac{t_\alpha}{\sqrt{\dfrac{\gamma + R}{1 + \gamma R}}}$$

in normal tables (tables giving the probability distribution of ϕ). This has been done, and the ratio between the true p and alleged α cumulative probabilities, for various values of R and γ, has been plotted in Figures 15-2 through 15-4 for several conventional α-values used in testing. (Although they may differ in variance, the true and alleged t distributions are symmetrically distributed about a common axis through zero. Therefore, the ratio between the true p and alleged α significance levels is the same for a left-tail test at level C, a right-tail test at level C, and a two-tail test at level $2C$.) Obviously, the t approach is unsatisfactory when population variances and sample sizes both differ considerably.

SENSITIVITY OF t TEST TO HETEROGENEITY OF VARIANCE WHEN BOTH SAMPLES ARE OF INFINITE SIZE

RATIO BETWEEN TRUE, P, AND NORMAL–THEORY, α, PROBABILITIES OF REJECTION WHEN TEST USES TWO–TAILED REJECTION REGION CORRESPONDING TO $\alpha = .05$ AND t STATISTIC IS BASED UPON SAMPLES OF SIZE N AND γ N, RESPECTIVELY, FROM NORMAL POPULATIONS WITH EQUAL MEANS BUT VARIANCES σ^2 AND $R\sigma^2$

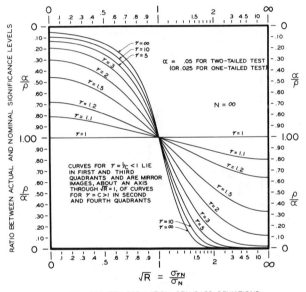

FIG. 15-2. Sensitivity of the two-sample *t* test to inequality of population variances (when sample sizes are infinite, but unequal, and $\alpha = .05$ for two-tailed test).

Formulas: The validity of the probability formulas to follow depends upon the samples being random and independent, the population variances being equal, and the population distributions being normal. If these conditions are met, the formulas for interval estimation and tests of hypotheses will be exactly (rather than approximately) true; and if, in addition, both samples are very large, the formula for power will often be a good approximation. The distributions cannot, of course, be exactly normal and this means that the formulas are approximate at best.

SENSITIVITY OF t TEST TO HETEROGENEITY OF VARIANCE WHEN BOTH SAMPLES ARE OF INFINITE SIZE

RATIO BETWEEN TRUE, P, AND NORMAL-THEORY, α, PROBABILITIES OF REJECTION WHEN TEST USES TWO-TAILED REJECTION REGION CORRESPONDING TO $\alpha = .01$ AND t STATISTIC IS BASED UPON SAMPLES OF SIZE N AND γ N, RESPECTIVELY, FROM NORMAL POPULATIONS WITH EQUAL MEANS BUT VARIANCES σ^2 AND $R\sigma^2$

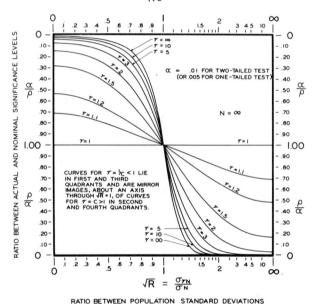

FIG. 15-3. Sensitivity of the two-sample *t* test to inequality of population variances (when sample sizes are infinite, but unequal, and $\alpha = .01$ for two-tailed test).

Let δ stand for the denominator of the two-sample *t* statistic so that

$$\delta = \sqrt{\frac{\sum_{i=1}^{n_X} (X_i - \bar{X})^2 + \sum_{j=1}^{n_Y} (Y_j - \bar{Y})^2}{n_X + n_Y - 2}\left(\frac{1}{n_X} + \frac{1}{n_Y}\right)} \qquad (15\text{-}32)$$

Then, by a derivation entirely analogous to that which produced Formula 15-4, it can be shown that if the required conditions are sufficiently well met

$$T(K) - T(C) \cong P(\bar{X} - \bar{Y} - K\delta \leq \mu_X - \mu_Y \leq \bar{X} - \bar{Y} - C\delta) \qquad (15\text{-}33)$$

SENSITIVITY OF t TEST TO HETEROGENEITY OF VARIANCE WHEN BOTH SAMPLES ARE OF INFINITE SIZE

RATIO BETWEEN TRUE, P, AND NORMAL-THEORY, α, PROBABILITIES OF REJECTION WHEN TEST USES TWO-TAILED REJECTION REGION CORRESPONDING TO $\alpha = .001$ AND t STATISTIC IS BASED UPON SAMPLES OF SIZE N AND γ N, RESPECTIVELY, FROM NORMAL POPULATIONS WITH EQUAL MEANS BUT VARIANCES σ^2 AND $R\sigma^2$

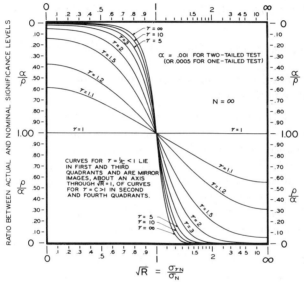

FIG. 15-4. Sensitivity of the two-sample t test to inequality of population variances (when sample sizes are infinite, but unequal, and $\alpha = .001$ for two-tailed test).

where $T(K) = P(t \leq K)$. Or, letting $C = t_p$ and $K = t_q$,

$$q - p \cong P(\bar{X} - \bar{Y} - t_q\,\delta \leq \mu_X - \mu_Y \leq \bar{X} - \bar{Y} - t_p\,\delta) \qquad (15\text{-}34)$$

where $P(t \leq t_p) = p$ and $P(t \leq t_q) = q$. So, if $\bar{x}$ and $\bar{y}$ are the means of the actually obtained X sample and Y sample, respectively, and d is the actually obtained value of δ, we can state at approximately the $q - p$ level of confidence that

$$\bar{x} - \bar{y} - t_q d \leq \mu_X - \mu_Y \leq \bar{x} - \bar{y} - t_p d \qquad (15\text{-}35)$$

In testing hypotheses about the value Δ of $\mu_X - \mu_Y$, the test statistic is

$$t_0 = \frac{\bar{X} - \bar{Y} - \Delta_0}{\delta} \tag{15-36}$$

which will be distributed approximately as t with $n_X + n_Y - 2$ degrees of freedom if actually $\mu_X - \mu_Y = \Delta_0$. So if we take as rejection region all values of t_0 less than t_p or greater than t_q, the probability of committing a Type I Error will be

$$P(t_0 < t_p \,|\, \mu_X - \mu_Y = \Delta_0) + P(t_0 > t_q \,|\, \mu_X - \mu_Y = \Delta_0)$$
$$\cong P(t < t_p) + P(t > t_q) = p + 1 - q \tag{15-37}$$

By a derivation analogous to those yielding Formulas 15-11 and 15-29, it can be shown that if actually $\mu_X - \mu_Y = \Delta_a$, then if both sample sizes are large,

$$\text{Power} \cong T\left(t_p + \frac{\Delta_0 - \Delta_a}{\delta}\right) + 1 - T\left(t_q + \frac{\Delta_0 - \Delta_a}{\delta}\right) \tag{15-38}$$

the approximation tending to worsen as the absolute value of $\Delta_0 - \Delta_a$ increases relative to that of d.

Examples: It would be improper to apply the two-sample t statistic in the bombing example used to illustrate the two-sample Z and $\hat{Z}$ statistics, since the variances of the populations and the sizes of the samples both differed appreciably. Therefore, consider the following example. It is known that final numerical semester grades made by students taught by an old established teaching method Y are quasi-normally distributed. And it is strongly suspected that under a new teaching method X the distribution of grades would also be quasi-normal and would have about the same variance so that if the new method has any appreciable effect it would be reflected primarily in the mean of the distribution. (Note that if the new teaching method has *no* differential effect upon the population distribution of grades, the X population's distribution *must* be quasi-normal and σ_X^2 *must* equal σ_Y^2, since, in that event, the two populations will have identical distributions.) In order to investigate the effect of the new method, 13 students are randomly divided into two classes, one containing 7 students and the other 6 students. The class of 7 is taught by the new method and the class of 6 by the old, both classes being taught by the same teacher. At the end of the semester the grades of the 7 students taught by the new method are 77, 79, 80, 84, 85, 88, 95 for which $\bar{x} = 84$ and

$$\sum_{i=1}^{7} (x_i - \bar{x})^2 = 228$$

and those of the 6 students taught by the old method are 68, 70, 72, 77, 79, 84 for which $\bar{y} = 75$ and

$$\sum_{j=1}^{6} (y_j - \bar{y})^2 = 184$$

Sample sizes are too small for the $\hat{Z}$ statistic to be trusted. But because sample sizes are almost equal and there is considerable previous evidence indicating that both populations are probably quasi-normally distributed, the teacher is willing to hazard the t approach. The actually obtained denominator d of the two-sample t statistic is

$$d = \sqrt{\frac{\sum_{i=1}^{n_X} (x_i - \bar{x})^2 + \sum_{j=1}^{n_Y} (y_j - \bar{y})^2}{n_X - n_Y - 2}\left(\frac{1}{n_X} + \frac{1}{n_Y}\right)}$$

$$= \sqrt{\frac{228 + 184}{7 + 6 - 2}\left(\frac{1}{7} + \frac{1}{6}\right)} = \sqrt{\frac{412}{11}\left(\frac{13}{42}\right)}$$

$$= \sqrt{\frac{5,356}{462}} = \sqrt{11.593} = 3.405$$

So if the teacher wanted a symmetric confidence interval for $\mu_X - \mu_Y$ corresponding to about the .90 level of confidence, he would let $q = .95$ and $p = .05$. And from tables of probabilities for t with $7 + 6 - 2 = 11$ degrees of freedom he would obtain $t_q = t_{.95} = 1.796$ and $t_p = t_{.05} = -1.796$. The desired confidence interval would be

$$\bar{x} - \bar{y} - t_q d \le \mu_X - \mu_Y \le \bar{x} - \bar{y} - t_p d$$
$$84 - 75 - 1.796(3.405) \le \mu_X - \mu_Y \le 84 - 75 - (-1.796)(3.405)$$
$$9 - 6.115 \le \mu_X - \mu_Y \le 9 + 6.115$$
$$2.885 \le \mu_X - \mu_Y \le 15.115$$

Or if he wanted to test, at about the $\alpha = .05$ level of significance, the $H_0: \mu_X - \mu_Y = 0$ against the $H_a: \mu_X - \mu_Y \ne 0$, he would calculate the value of the test statistic

$$t_0 = \frac{\bar{x} - \bar{y} - \Delta_0}{d} = \frac{84 - 75 - 0}{3.405} = 2.643$$

and take as his rejection region values of t_0 that are either less than $t_{.025}$ or greater than $t_{.975}$. From tables for t with 11 degrees of freedom he finds that $t_{.025} = -2.201$ and $t_{.975} = 2.201$ and since 2.643 falls in the upper rejection region, he would reject H_0. If he wanted to obtain some idea of the power of the test if actually $\mu_X - \mu_Y = 2$, he could obtain a gross estimate (which might be badly in error due to the small sample sizes) by using Formula 15-38 as follows:

$$\text{Power} \cong T\left(t_p + \frac{\Delta_0 - \Delta_a}{\delta}\right) + 1 - T\left(t_q + \frac{\Delta_0 - \Delta_a}{\delta}\right)$$

$$\cong T\left(-2.201 + \frac{0 - 2}{3.405}\right) + 1 - T\left(2.201 + \frac{0 - 2}{3.405}\right)$$

$$\cong T(-2.201 - .587) + 1 - T(2.201 - .587)$$

$$\cong T(-2.788) + 1 - T(1.614)$$

$$\cong .009 + 1 - .93$$

$$\cong .079$$

15-6   General Critique

We have discussed three general methods for estimating, or testing hypotheses about, population means: (a) treating Z statistics as having a ϕ distribution, (b) treating $\hat{Z}$ statistics as having a ϕ distribution, (c) treating t statistics as having one of the t distributions.

The flaws in the Z method are that (a) it requires that we know the values of the σ^2's, although we hardly ever do, and (b) when samples are small, it requires that we know that the population is quasi-normally distributed, although we often do not. The flaw in the $\hat{Z}$ method is that it requires that all sample sizes be large and doesn't work if they are very small. The flaws in the t approach are that (a) the two-sample tests and estimates require that we know that $\sigma_X^2 = \sigma_Y^2$ although we seldom really do, and (b) when samples are small, it requires that we know that the population is quasi-normally distributed, although we often do not, and it takes the variance of $\hat{\sigma}^2$ into account *properly* only when the population *is* quasi-normal.

If we actually knew that each method's requirements were met, the Z method would be the best choice (since a known σ^2 is better than an estimate of it), the t method would be the next best (since it is better to take the variability of $\hat{\sigma}^2$ into account than not to), and the $\hat{Z}$ method would be the worst. However, if we do not know that a method's population requirements are met, the method having the fewest such requirements is probably the best choice.

In a sense, the $\hat{Z}$ method makes no requirements of the population but only of the *sample*. The validity of the $\hat{Z}$ method depends upon sample size being large enough for two *known* and *reliable* effects to have sufficiently taken place, the Central Limit Effect and the shrinkage of the relative variance of the denominator to a negligible value, (i.e., large enough to have caused the distribution of $\bar{X}$ [or $\bar{X} - \bar{Y}$] to be essentially normal and to have made $\hat{\sigma}_{\bar{X}}^2$ [or $\hat{\sigma}_{\bar{X}-\bar{Y}}^2$] behave essentially like a constant relative to $\bar{X}$ [or $\bar{X} - \bar{Y}$]). It is true that these effects are influenced by population characteristics, but it is also true that if the samples involved are large *enough*, the $\hat{Z}$ method will meet any *practical* requirement of accuracy. The other methods (except for the one-sample t statistic when the sample is large) require that despite our ignorance of population means, we be well-informed about other characteristics of the population that are at least as hard to establish, such as distribution-shape and variance.

From where is such information to come? If it comes from a vast backlog of data or experience on similar problems, one should consider two questions: (a) How can one be sure that the backlog data applies to the present situation? (b) Why has the backlog not produced equally reliable information about population means, thereby obviating the present tests or estimates? On the other hand, if the information comes from the same sample used to test or estimate the mean, it will not be sufficiently reliable unless all sample sizes are so large that the $\hat{Z}$ method—which does not require the information (at least not to anything like the same degree)—would probably be quite satisfactory.

The safest approach, therefore, is always to take samples large enough to justify the use of the $\hat{Z}$ statistics, referred to the ϕ distribution. (And if one has no inkling whatever as to the shape of the population distribution, a *huge* sample is appropriate.) However, if the small sample sizes calling for other methods cannot be avoided, one should recognize that the accuracy of the method strongly depends upon how nearly the population distributions meet the requirements; and if the two-sample t statistic is to be used, one should strive for equal sample sizes.

15-7 The Normal Approximation to the Distribution of Successes

15-7-1 The Normal Approximation to the Binomial Distribution

If one wished to obtain the binomial probability of r successes in n trials, he could either compute it from the binomial formula (7-3) or look it up in binomial tables. However, if n is large, computation may be prohibitively laborious and the tables may not extend that far. In such cases, approximate probabilities may be obtained by making use of the Central Limit Effect.

Let X, the value of an observation, be 1 if the observation is a success and zero if it is not. Then

$$\bar{X} = \frac{\sum_{i=1}^{n} X_i}{n} = \frac{r(1) + (n - r)0}{n} = \frac{r}{n}$$

is the sample mean, and if n is sufficiently large,

$$Z = \frac{\bar{X} - E(\bar{X})}{\sqrt{\text{Var}(\bar{X})}} = \frac{\frac{r}{n} - E\left(\frac{r}{n}\right)}{\sqrt{\text{Var}\left(\frac{r}{n}\right)}} = \frac{\frac{r}{n} - \frac{1}{n}E(r)}{\sqrt{\frac{1}{n^2}\text{Var}(r)}} = \frac{r - E(r)}{\sqrt{\text{Var}(r)}}$$

$$= \frac{r - np}{\sqrt{np(1 - p)}} \tag{15-39}$$

will have an approximately normal distribution with zero mean and unit variance. And in that event, cumulative binomial probabilities for r will be approximately equivalent to the corresponding cumulative normal probabilities for Z obtained by looking Z up in standard normal tables. Specifically, if r

takes a particular value r', Z must take the particular value

$$Z' = \frac{r' - np}{\sqrt{np(1-p)}}$$

and if n is sufficiently large

$$P(r \leq r') = P(Z \leq Z')$$

$$= P\left(Z \leq \frac{r' - np}{\sqrt{np(1-p)}}\right)$$

$$\cong P\left(\phi \leq \frac{r' - np}{\sqrt{np(1-p)}}\right)$$

$$\cong \Phi\left(\frac{r' - np}{\sqrt{np(1-p)}}\right) \qquad (15\text{-}40)$$

Other things being equal, the approximation tends to improve as n increases, to worsen as p departs farther and farther from .5 in either direction, and to worsen as r/n departs farther and farther from p in either direction (see Figure 15-5, where the number of errors is r). If p and r are both extremely small (and if n is large and np is small), it is probably better to use the Poisson approximation.

For example, to obtain the probability of 10 or fewer successes in a random sample of 50 observations drawn from an infinite population in which the proportion of units that are successes is .30 we would proceed as follows, (see Formula 15-40):

$$P(r \leq 10) = P\left(Z \leq \frac{10 - 50(.30)}{\sqrt{50(.30)(.70)}}\right) = P\left(Z \leq \frac{10 - 15}{\sqrt{10.5}}\right)$$

$$= P\left(Z \leq \frac{-5}{3.24}\right) = P(Z \leq -1.543)$$

$$\cong P(\phi \leq -1.543) = \Phi(-1.543)$$

$$\cong .061$$

The exact binomial probability sought is .079. The approximation can be improved by introducing a "correction for continuity." Imagine the true probability distribution of r to be represented by a histogram (whose total area is 1.00) and the histogram to be approximated by a normal distribution. The true cumulative probability of r' is simply the sum of the areas of the histogram bars representing the r-values involved, and the "normal approximation" to it is the area in the normal distribution lying above the same set of abscissas, i.e., above the same base. The correction for continuity consists in making sure that the *entire* histogram bar representing each of the cumulated r's is counted, especially the ones on the ends. Thus, in the example, to make sure of counting all of the normal area corresponding to the histogram bar for $r = 10$, which extends from 9.5 to 10.5, we must let $r' = 10.5$ in which case

ERRORS ARE BINOMIALLY DISTRIBUTED.   THE NORMAL APPROXIMATION IS POOR IF THERE
IS AN APPRECIABLE PROBABILITY FOR EITHER ZERO ERRORS OR N ERRORS IN N TRIALS.

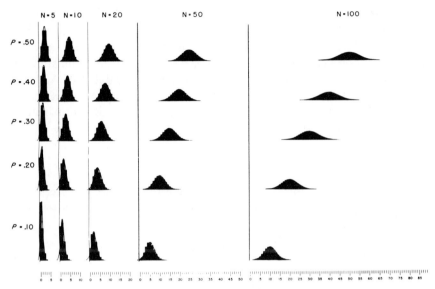

DISTRIBUTIONS FOR NUMBER OF ERRORS IN N TRIALS WHEN *P* = PROBABILITY OF AN ERROR
ON A SINGLE TRIAL. SMOOTH CURVES ARE $\bar{x} \pm 3\sigma$ OF THE NORMAL APPROXIMATION WITH
DOTTED PORTION OF CURVE COVERING IMPOSSIBLE, *i.e.* NEGATIVE, ERROR FREQUENCIES.

FIG. 15-5. Illustration of normal approximation to the binomial
distribution (where, to facilitate comparison, the binomial distributions are represented by histograms).

to get $P(r \leq 10)$ we calculate

$$P(r \leq 10.5) = P\left(Z \leq \frac{10.5 - 15}{\sqrt{10.5}}\right) = P\left(Z \leq \frac{-4.5}{3.24}\right)$$
$$\cong P(\phi \leq -1.389) = .082$$

which compares more favorably with the exact binomial probability of .079.

If the sampled population contains more than two categories (and is undepletable), the probabilities for completely specified sample compositions
have a multinomial distribution. However, if we are only interested in the
frequencies of observations that belong to a single category "success" and are
willing to lump all nonsuccesses into a single category, the probability of r
successes in a sample of size n is properly regarded as a binomial probability
and we can use the normal approximation under the same conditions as
before.

15-7-2   The Normal Approximation to the Poisson Distribution

The Poisson approximation to the binomial distribution applies when its mean np is small. However, when $m = np$ is large enough to permit the integral values of r to be distributed almost symmetrically about their expected value m, the normal approximation may be a satisfactory approximation for the Poisson distribution (in obtaining cumulative probabilities). And this should also be the case for the Poisson distribution in its own right. In the former case, for greatest accuracy one would use the normal approximation to the binomial referring

$$Z = \frac{r - np}{\sqrt{np(1 - p)}}$$

to the ϕ distribution. In the latter case, where the number of successes is denoted by the symbol x and the mean and variance of its distribution both have the value m, we would use

$$Z = \frac{x - E(x)}{\sqrt{\text{Var }(x)}} = \frac{x - m}{\sqrt{m}} \qquad (15\text{-}41)$$

as the approximately normally distributed variable, calculating

$$P(x \le x') = P\left(Z \le \frac{x' - m}{\sqrt{m}}\right)$$

$$\cong P\left(\phi \le \frac{x' - m}{\sqrt{m}}\right)$$

$$\cong \Phi\left(\frac{x' - m}{\sqrt{m}}\right) \qquad (15\text{-}42)$$

15-7-3   The Normal Approximation to the Hypergeometric Distribution

If the sampled population is finite, containing N units of which R are successes, and random sampling is without replacements, then the number r of successes in the sample has a hypergeometric distribution with mean nR/N and variance

$$\frac{nR}{N}\left(\frac{N - R}{N}\right)\left(\frac{N - n}{N - 1}\right)$$

When n is large and N is many times larger, we could treat r as an approximately binomially distributed variable and use the normal approximation to the binomial. However, a better approximation can usually be obtained by substituting the hypergeometric mean and variance into the penultimate term of Formula 15-39, obtaining

$$Z = \frac{r - E(r)}{\sqrt{\mathrm{Var}\,(r)}} = \frac{r - \dfrac{nR}{N}}{\sqrt{\dfrac{nR}{N}\left(\dfrac{N - R}{N}\right)\left(\dfrac{N - n}{N - 1}\right)}} \tag{15-43}$$

and treating it as having approximately a standardized (cumulative) normal distribution, so that

$$P(r \le r') = P\left(Z \le \frac{r' - \dfrac{nR}{N}}{\sqrt{\dfrac{nR}{N}\left(\dfrac{N - R}{N}\right)\left(\dfrac{N - n}{N - 1}\right)}} \right)$$

$$\cong P\left(\phi \le \frac{r' - \dfrac{nR}{N}}{\sqrt{\dfrac{nR}{N}\left(\dfrac{N - R}{N}\right)\left(\dfrac{N - n}{N - 1}\right)}} \right)$$

$$\cong \Phi\left(\frac{r' - \dfrac{nR}{N}}{\sqrt{\dfrac{nR}{N}\left(\dfrac{N - R}{N}\right)\left(\dfrac{N - n}{N - 1}\right)}} \right) \tag{15-44}$$

15-8   Chi-Square Statistics

15-8-1   The Chi-Square Distributions

Consider ν *normally* and *independently* distributed variables $X_1, X_2, \ldots, X_\nu$ with means, respectively, $\mu_1, \mu_2, \ldots, \mu_\nu$ and variances, respectively, $\sigma_1^2, \sigma_2^2, \ldots, \sigma_\nu^2$. We can standardize these variables by converting them to Z scores by means of the formula

$$Z_i = \frac{X_i - \mu_i}{\sigma_i}$$

The resulting variables $Z_1, Z_2, \ldots, Z_\nu$ are normally and independently distributed and each distribution has a mean of zero and a variance of 1 (so the Z_i's may be regarded as ϕ_i's).

Now the sum of the squared values of these ν individual Z scores is a variable known as chi-square with ν degrees of freedom, i.e.,

$$\chi_\nu^2 = \sum_{i=1}^{\nu} Z_i^2 \tag{15-45}$$

For each different value of ν, there is a different chi-square distribution. That is, "chi-square" is really a whole family of variables whose individual distributions depend upon the value of ν, the degrees of freedom. Cumulative probabilities for the chi-square family of distributions have been tabled, and the tables are entered with the appropriate number of degrees of freedom v as well as with the numerical value of either χ_v^2 or its cumulative probability.

A useful property of chi-square variables is that the sum of several mutually independent chi-square variables is itself a chi-square variable with degrees of freedom equal to the sum of those of the contributing chi-square variables. This follows from Formula 15-45, which tells us that

$$\chi_{a+b+c}^2 = \sum_{i=1}^{a+b+c} Z_i^2 = \sum_{i=1}^{a} Z_i^2 + \sum_{i=a+1}^{a+b} Z_i^2 + \sum_{i=a+b+1}^{a+b+c} Z_i^2$$
$$= \chi_a^2 + \chi_b^2 + \chi_c^2$$

It is important, however, that the components be independent. Although $Z_1^2 + Z_2^2 = \chi_2^2$ and $Z_1^2 + Z_2^2 + Z_3^2 = \chi_3^2$, in this case it is *not* true that $\chi_2^2 + \chi_3^2 = \chi_5^2$ since the component χ^2's contain some common Z_i's.

Since a chi-square variable presupposes normality of distribution for its contributing variables, no real-world variable has a chi-square distribution. The chi-square distributions are, however, useful as approximations to the distributions of certain real-world variables, especially those whose true distribution is multinomial in which case the computation of exact probabilities is often prohibitively laborious.

15-8-2 The Chi-Square Approximation to the Multinomial

The multinomial probability law applies when we sample randomly (and with replacements if the population is finite) from a population each of whose units belongs to a single one of K mutually exclusive and exhaustive categories. The law tells us that if p_i is the proportion of the population's units that belong to the ith category and r_i is the number of sample units drawn from the ith category in a random sample of n observations, then the probability of obtaining exactly r_1 units in the first category, r_2 in the second, . . . , and r_K units in the last, is

$$P(r_1, r_2, \ldots, r_K) = \frac{n!}{r_1! r_2! \cdots r_K!} p_1^{r_1} p_2^{r_2} \cdots p_K^{r_K} \qquad (15\text{-}46)$$

This is the probability for the entire pattern of r_i's obtained in the sample, i.e., the value of each and every one of the r_i's is specified. Had we been interested in only one of them, such as r_1, then we would have lumped all the rest into a complementary category of "not r_1's" and used the binomial probability law.

In inferential statistics the multinomial probability law is used mainly in hypothesis testing. We hypothesize that the p_i have certain values and reject the hypothesis if the sample results are too extreme a departure from it, i.e., if the cumulative probability under H_0 of the obtained sample and more extreme samples is too small. But what is a "more extreme" sample? If we were calculating the exact multinomial probabilities directly, we might define a more extreme sample as one having a smaller point probability under H_0 (although this approach does not always produce the intended result). But it is precisely because of the excessive labor involved in calculating these point probabilities that we are seeking an approximation, so this definition is ruled out. Since the expected value of an r_i is np_i, the deviation of r_i from np_i (under H_0) indicates extremity for r_i and we might use

$$\sum_{i=1}^{K} |r_i - np_i| \quad \text{or} \quad \sum_{i=1}^{K} (r_i - np_i)^2$$

as the measure of extremity. (The sum of the $r_i - np_i$ cannot be used because it is always zero.) This would be acceptable if the p_i were all equal. But if $p_1 = .60$, $p_2 = .39$, and $p_3 = .01$, either of these measures would have the same value for the two cases $r_1 = 55, r_2 = 44, r_3 = 1$ and $r_1 = 55, r_2 = 39$, $r_3 = 6$ although the latter case is less probable and strikes us as being less in harmony with the p-values. We could, however, overcome this difficulty by using *relative* departures from expectation, measuring "extremity" by

$$\sum_{i=1}^{K} \frac{|r_i - np_i|}{np_i} \quad \text{or} \quad \sum_{i=1}^{K} \frac{(r_i - np_i)^2}{np_i}$$

The latter approach is the one that we shall use because if n is infinitely large, the statistic

$$\sum_{i=1}^{K} \frac{(r_i - np_i)^2}{np_i} = \sum_{i=1}^{K} \frac{(o_i - e_i)^2}{e_i} \tag{15-47}$$

(where o_i and e_i are frequently encountered alternative symbols standing for the *observed* frequency r_i and the *expected* frequency np_i, respectively) has exactly a chi-square with $K - 1$ degrees of freedom distribution. And if n is less than infinite but sufficiently large, the above statistic is distributed approximately as chi-square with $K - 1$ degrees of freedom so that

$$P\left(\sum_{i=1}^{K} \frac{(r_i - np_i)^2}{np_i} \geq D\right) \cong P(\chi^2_{K-1} \geq D) \tag{15-48}$$

where χ^2_{K-1} is the theoretical chi-square variable with $K - 1$ degrees of freedom. The above formula gives us the *chi-square approximation to the multinomial distribution*. We have used the $>$ rather than the $<$ sign because we are almost always interested in the probability of large departures of the r_i's from their expected values np_i's rather than in the probability of very close correspondences.

In order for

$$\sum_{i=1}^{K} \frac{(r_i - np_i)^2}{np_i}$$

to be distributed *exactly* as chi-square with $K - 1$ degrees of freedom, each r_i (considered as a variable) must be normally distributed about its expected value np_i. And that occurs (in accordance with the Central Limit Theorem applied to the binomially distributed variable r_i) only when n becomes infinite. However, the above statistic will be *approximately* distributed as chi-square with $K - 1$ degrees of freedom provided that the distribution of each r_i about its expected frequency np_i is quasi-normal (at least over its more probable portions). Now since an r_i takes only integer values from 0 to n, its distribution about np_i cannot be bell-shaped if np_i is very close to zero (and cannot even be roughly symmetric, over its more probable portions, if, in addition, p_i is far from .5). Therefore, the quasi-normality requirement is often "replaced" by a "rule" stating some minimum acceptable expected frequency. Five is a favorite, although there is considerable variation. Such rules are oversimplifications, however, since the goodness-of-fit to normality by the distribution of r_i depends upon n, p_i, and the particular value of r_i involved, and this information is not adequately encompassed in the single product np_i. It cannot be denied, however, that for a method that requires infinite expected frequencies for exactitude, it works astonishingly well when they are small.

When $K = 2$, so that the multinomial distribution reduces to the binomial, the chi-square approximation to the multinomial is equivalent to the normal approximation to the binomial. This can be seen as follows. When n is large and $K = 2$, the statistic having approximately a χ^2_{K-1} distribution (see Formula 15-47) can be written

$$\chi^2_1 = \sum_{i=1}^{2} \frac{(r_i - np_i)^2}{np_i} = \frac{(r_1 - np_1)^2}{np_1} + \frac{(r_2 - np_2)^2}{np_2}$$

Now let $r_1 = r$ and $p_1 = p$. Then $r_2 = n - r_1 = n - r$ and $p_2 = 1 - p_1 = 1 - p$ and we have

$$\chi^2_1 = \frac{(r - np)^2}{np} + \frac{[(n - r) - n(1 - p)]^2}{n(1 - p)}$$

$$= \frac{(1 - p)(r - np)^2 + p[-1(r - np)]^2}{np(1 - p)}$$

$$= \frac{(1 - p + p)(r - np)^2}{np(1 - p)} = \frac{(r - np)^2}{np(1 - p)}$$

$$\sqrt{\chi^2_1} = \frac{r - np}{\sqrt{np(1 - p)}} = Z \tag{15-49}$$

which is approximately normally distributed when n is large (see Section 15-7-1). Since $\chi^2_1 = Z^2$, the chi-square statistic ignores the *direction* of the deviation of r from its expected value and since large deviations in either direction can only make χ^2 large, only the upper tail of a χ^2 distribution is used as a rejection region. Therefore, an *upper*-tail χ^2_1 test at level α is equivalent to a *two*-tail Z test at level α. Thus, the critical value of $|Z|$ for a two-tail test at $\alpha = .05$ is 1.96 and the critical value of χ^2_1 for an upper-tail test at $\alpha = .05$ is $1.96^2 = 3.84$, thereby showing their equivalence.

Although *all* of the r_i's are involved in the calculation of the chi-square approximation, those that are completely *dependent* do *not* contribute to its degrees of freedom. Thus, when $K = 2$, if we regard r_1 as completely free to vary, we must then regard r_2 as completely dependent since whatever value r_1 has, the value of r_2 is determined by the relationship $r_2 = n - r_1$, and knowing r_1 and n we must also know r_2. Similarly, when $K = 5$ if we know the values of r_1, r_2, r_3, r_4, and n, then we also know the value of r_5 since $r_5 = n - (r_1 + r_2 + r_3 + r_4)$; so one of the r_i's may be regarded as completely dependent. Thus, roughly speaking, the number of degrees of freedom in the chi-square approximation to the multinomial is equal to the number of relatively free r_i's, which is $K - 1$.

As we mentioned earlier, the principle application of the chi-square approximation to the multinomial is in testing hypotheses, using

$$\sum_{i-1}^{K} \frac{(o_i - e_i)^2}{e_i}$$

as the test statistic, where the e_i's are *hypothesized* expected frequencies specified by H_0. If H_0 were badly false in *any* way, one would expect some of the $|o_i - e_i|$ differences to be large. However, since large deviations of *any* one or more of the o_i's from its hypothesized e_i in *either* direction tend to give the test statistic large values (which therefore constitute the rejection region), the chi-square tests must test a simple H_0 (that specifies all of the e_i's or p_i's exactly as a single "state of nature") against a compound H_a that is simply the negation of H_0. If one had a *specific* alternative in mind, so that H_a also specified all of the p_i-values exactly, a more appropriate approach would be to do a likelihood ratio test applied to the hypothesized multinomial point probabilities. Partly because of this, and the other factors mentioned, chi-square tests, unlike most tests, are often used in an effort to *confirm* an H_0 rather than to refute it. For this reason, in describing chi-square tests emphasis will be placed almost entirely upon H_0.

15-8-3 The Chi-Square Test of Hypothesized Population Proportions

Suppose that every unit in a certain population belongs to one and only one of K categories and that one wishes to test a hypothesis that specifies the exact proportions $p_1, p_2, \ldots, p_K$ of units that belong to each category. After having decided upon the numerical values of the hypothesized p_i's and upon the significance level α, one draws a random sample of n *independent* observations (with replacements if the population is finite). Let o_i be the observed number of sample observations belonging to the ith category, and let $e_i = np_i$ be the expected value of o_i if H_0 is true. Then, if H_0 is true the set of o_i's (considered as variables) will be multinomially distributed, and the cumulative probability of sets of o_i's as extreme, or more so, (under H_0) as the obtained set will be approximately the same as the upper-tail cumulative probability of the test statistic

$$\sum_{i=1}^{K} \frac{(o_i - e_i)^2}{e_i} \tag{15-50}$$

in the distribution of chi-square with $K - 1$ degree of freedom. Therefore, one calculates

$$\sum_{i=1}^{K} \frac{(o_i - e_i)^2}{e_i}$$

enters the chi-square tables with the predetermined nominal value of α and with $K - 1$ degrees of freedom, and rejects H_0 if and only if

$$P\left(\chi_{K-1}^2 \geq \sum_{i=1}^{K} \frac{(o_i - e_i)^2}{e_i}\right) \leq \alpha$$

This approximate test is simply a direct and uncomplicated use of the near equivalence between multinomial and chi-square cumulative probabilities.

To illustrate, Gregor Mendel hypothesized, in effect, that in peas "round" and "yellow" were dominant characteristics, each having probability 3/4, that the alternative characteristics "wrinkled" and "green" were recessive, each having probability 1/4, and that shape and color were independent so that $P(R \& Y) = 9/16$, $P(R \& G) = 3/16$, $P(W \& Y) = 3/16$, and

$$P(W \& G) = 1/16.$$

(Let these probabilities, in the order named, be designated p_1, p_2, p_3, and p_4.) This hypothesis could be tested by drawing a random sample of 100 peas and performing a chi-square test at the nominal .05 level of significance. Before drawing the sample we can calculate the $e_i = np_i$: $e_1 = 100(9/16) = 56.25$, $e_2 = 100(3/16) = 18.75$, $e_3 = 100(3/16) = 18.75$, $e_4 = 100(1/16) = 6.25$; and entering the chi-square table with $K - 1 = 4 - 1 = 3$ degrees of freedom and with an upper-tail $\alpha = .05$, we find that the critical value of χ_3^2 for this test is 7.81. Suppose that we now draw the sample and find that the number of peas that are round and yellow is $o_1 = 65$, that are round and green is $o_2 = 12$, that are wrinkled and yellow is $o_3 = 20$, and that are wrinkled and green is $o_4 = 3$. We then calculate

$$\sum_{i=1}^{4} \frac{(o_i - e_i)^2}{e_i} = \frac{(65 - 56.25)^2}{56.25} + \frac{(12 - 18.75)^2}{18.75} + \frac{(20 - 18.75)^2}{18.75}$$
$$+ \frac{(3 - 6.25)^2}{6.25}$$
$$= \frac{8.75^2}{56.25} + \frac{(-6.75)^2}{18.75} + \frac{1.25^2}{18.75} + \frac{(-3.25)^2}{6.25}$$
$$= \frac{76.5625}{56.25} + \frac{45.5625 + 1.5625}{18.75} + \frac{10.5625}{6.25}$$
$$= 1.361 + 2.513 + 1.690 = 5.564$$

Since 5.564 is less than the critical value 7.81, we cannot reject H_0. The approximate probability level of the test is $P(\chi_3^2 \geq 5.564)$ which is less than .25 but greater than .10.

15-8-4   The Chi-Square Test of Independence

Sometimes every unit in a population belongs to one and only one categorical subdivision of one characteristic and also belongs to one and only one categorical subdivision of a second characteristic. For example, automobiles can be categorized according to color as well as according to body style. In such cases we may wish to test the hypothesis that the two characteristics are independent, i.e., that for each of the pools of units belonging to a category of the first characteristic the same proportion of the pool belongs to any specified category of the second characteristic (so that the proportion of convertibles that are yellow is the same as the proportion of sedans or station wagons that are yellow, etc.).

We can represent the composition of the population, or of a random sample drawn from it, by a table whose R rows represent the R different categories of the first characteristic and whose C columns represent the C different categories of the second. Let p_i, p_j, and p_{ij}, respectively, represent the unknown proportions of units in the population that belong to the ith row category, to the jth column category, and to *both* the ith row category and the jth column category. If the row and column characteristics are completely independent, then for every pair of i and j subscripts, $p_{ij} = p_i p_j$. Now suppose that a random sample of n observations is drawn from the population (with replacements if the population is finite). Let o_i, o_j, and o_{ij}, respectively, stand for the number of sample observations belonging to the ith row category, to the jth column category, and to both categories. Therefore, if we use the table to represent the composition of the sample, the o_{ij}'s will be cell entries corresponding to the ith row and jth column, and since $o_i = \sum_{j=1}^C o_{ij}$ and $o_j = \sum_{i=1}^R o_{ij}$, each o_i will be the marginal total for the ith row and each o_j will be the marginal total for the jth column.

Now if the H_0 of independence is true, $p_{ij} = p_i p_j$ and the expected frequencies corresponding to the o_{ij}'s would be $e_{ij} = n p_i p_j$. So the statistic

$$\sum_{i=1}^R \sum_{j=1}^C \frac{(o_{ij} - n p_i p_j)^2}{n p_i p_j}$$

would be distributed approximately as χ_{RC-1}^2 when H_0 is true but not otherwise and would therefore serve as a sensitive statistic with which to test H_0. (The test, however, would be superfluous, since if we knew the population proportions we could test H_0 by simply observing whether or not every $p_{ij} = p_i p_j$.)

We do not know the values of p_i and p_j, but we can estimate them from the sample: $\hat{p}_i = o_i/n$ and $\hat{p}_j = o_j/n$ are mean-unbiassed estimates of p_i and p_j, respectively. Therefore, a reasonable estimate of e_{ij} when H_0 is true would be

$$n\hat{p}_i\hat{p}_j = n\left(\frac{o_i}{n}\right)\left(\frac{o_j}{n}\right) = \frac{o_i o_j}{n}$$

and the test statistic

$$\sum_{i=1}^{R}\sum_{j=1}^{C}\frac{(o_{ij} - n\hat{p}_i\hat{p}_j)^2}{n\hat{p}_i\hat{p}_j} = \sum_{i=1}^{R}\sum_{j=1}^{C}\frac{\left(o_{ij} - \dfrac{o_i o_j}{n}\right)^2}{\dfrac{o_i o_j}{n}} \tag{15-51}$$

should be sensitive to the validity of H_0. It is, but because we have substituted the variable estimates $\hat{p}_i$ and $\hat{p}_j$ for the constant population proportions p_i and p_j, the statistic is no longer distributed approximately as χ^2_{RC-1}, but rather as $\chi^2_{(R-1)(C-1)}$, i.e., as *chi-square with* $(R-1)(C-1)$ *degrees of freedom.* Sample values are more inclined to conform to "expected values" estimated from the sample itself than to the true expected values. The reduction in degrees of freedom compensates for this artificially increased conformity.

Why exactly $(R-1)(C-1)$ degrees of freedom? The "rule" is that one additional degree of freedom is lost for every population parameter (index) that must be estimated from the sample. Had we used the actual p_i's and p_j's there would have been $RC-1$ degrees of freedom. Now there are R p_i's and C p_j's in the population, but if we obtain the values of $R-1$ of the p_i's, we can get the last one by subtracting their sum from 1, and likewise if we know $C-1$ of the p_j's, we can obtain the remaining one by subtraction. So we only "need" to estimate $R-1$ of the p_i's and $C-1$ of the p_j's, since we get the last ones "free." Our original $RC-1$ degrees of freedom is therefore reduced by $R-1 + C-1$ giving us

$$RC - 1 - (R + C - 2) = RC - R - C + 1 = (R-1)(C-1)$$

Another way of explaining it is that by using the o_i and o_j marginal frequencies to estimate the expected cell frequencies we were, in effect, treating the marginal frequencies as fixed. We were asking, in effect, "Given *these* marginal frequencies and independence, what are the expected cell frequencies?" And the resulting estimates correspond to appropriate expected cell proportions *for that table.* Thus, the "expected" frequencies are those expected for a restricted variety of tables, i.e., samples, having exactly those marginal frequencies. But when all marginal frequencies are fixed, one can fill in only $(R-1)(C-1)$ of the cell entries more or less arbitrarily (specifically those in the first $R-1$ rows and the first $C-1$ columns), the remaining cell entries being completely dependent in the sense that their values are the difference between the sum of those already entered in a row or column and its marginal frequency. Thus, degrees of freedom is the number of o_{ij} minus the number of them that are completely dependent. (See Table 15-3.)

Table 15-3

Illustration of Degrees of Freedom Involved in
Chi-Square Test of Independence

Having filled in $(R - 1)(C - 1)$ cells of an $R \times C$ table with fixed marginal frequencies,
all remaining cell entries are completely determined

2	10	25	(50 − 37) 13	50
5	4	3	(30 − 12) 18	30
3 (10 − 7)	6 (20 − 14)	2 (30 − 28)	(20 − 11) 9 (40 − 31)	20
10	20	30	40	100

In the test for independence, therefore, one selects a nominal value of α, draws a random sample of n *independent* observations from the population in question (with replacements if the population is finite), calculates the value of the test statistic

$$\sum_{i=1}^{R} \sum_{j=1}^{C} \frac{\left(0_{ij} - \dfrac{0_i 0_j}{n}\right)^2}{\dfrac{0_i 0_j}{n}}$$

and rejects the H_0 of independence if and only if

$$P\left(\chi^2_{(R-1)(C-1)} \geq \sum_{i=1}^{R} \sum_{j=1}^{C} \frac{\left(0_{ij} - \dfrac{0_i 0_j}{n}\right)^2}{\dfrac{0_i 0_j}{n}}\right) \leq \alpha$$

The H_0 is that the row and column characteristics are independent (so that for every i and j, $p_{ij} = p_i p_j$). The test requirements are that sampling is random, that the population units are independent of each other, and that every population unit belongs to one and only one of the R row categories and also to one and only one of the C column categories. The value of α should be chosen prior to sampling. And, of course, since the test is only approximate, the chosen value will only approximate the true value.

To illustrate, the personnel director of a large company wishes to test at nominal $\alpha = .05$ the hypothesis that for employees who resign within five years of being hired, the number of years of service is independent of the sex of the employee. Consulting his records, he obtains the observed frequencies

Table 15-4

Data for Personnel Director's Problem

Number of employees resigning within 5 years

Sex	Years of service					
	<1	1	2	3	4	
Male	34	12	8	6	0	60
Female	46	48	22	14	10	140
	80	60	30	20	10	200

shown in Table 15-4. Noting that in this table when $i = 1$, $o_i = 60$ and

$$\frac{o_i}{n} = \frac{60}{200} = .3$$

and when $i = 2$, $o_i = 140$ and $o_i/n = 140/200 = .7$, he then calculates

$$\sum_{i=1}^{2} \sum_{j=1}^{5} \frac{\left[o_{ij} - \frac{o_i o_j}{n}\right]^2}{\frac{o_i o_j}{n}} = \frac{[34 - .3(80)]^2}{.3(80)} + \frac{[12 - .3(60)]^2}{.3(60)}$$

$$+ \frac{[8 - .3(30)]^2}{.3(30)} + \frac{[6 - .3(20)]^2}{.3(20)} + \frac{[0 - .3(10)]^2}{.3(10)}$$

$$+ \frac{[46 - .7(80)]^2}{.7(.80)} + \frac{[48 - .7(60)]^2}{.7(60)} + \frac{[22 - .7(30)]^2}{.7(30)}$$

$$+ \frac{[14 - .7(20)]^2}{.7(20)} + \frac{[10 - .7(10)]^2}{.7(10)}$$

$$= \frac{10^2}{24} + \frac{(-6)^2}{18} + \frac{(-1)^2}{9} + \frac{0^2}{6} + \frac{(-3)^2}{3} + \frac{(-10)^2}{56}$$

$$+ \frac{6^2}{42} + \frac{1^2}{21} + \frac{0^2}{14} + \frac{3^2}{7}$$

$$= 4.167 + 2.000 + .111 + 0 + 3.000 + 1.786$$

$$+ .857 + .048 + 0 + 1.286$$

$$= 13.255$$

Entering the chi-square table, he finds that the critical value of $\chi^2_{(R-1)(C-1)} = \chi^2_4$ for nominal $\alpha = .05$ is 9.488. Since the obtained value 13.255 exceeds the critical value, the H_0 of independence is rejected. The approximate probability level of the test is $P(\chi^2_4 \geq 13.255)$ which is slightly greater than .01. So H_0 could not have been rejected if he had used a nominal α of .01.

In this example each of the 200 units in the table is a *different* employee, so the requirement that the population units be independent of each other was met. However, if the hypothesis had been that the distance (in hundred-mile categories) involved in long-distance telephone calls is independent of the sex of the employee making the call, the requirement of independence would be violated if the data used in the test included more than one call by the same person. Had one person been responsible for almost all of the 400–500 mile calls used as data in the test, this alone would spuriously tend to make it appear that that person's *sex* was associated with such calls whereas actually it is only that *person* who was associated with them.

15-8-5 The Chi-Square Test of Goodness-of-Fit

If we want to test the hypothesis that a set of data came from a discrete population whose probability distribution is completely specified, we can do so by means of the chi-square test of hypothesized population proportions; we need only treat the discrete values (or sets of them) as categories and treat the probabilities as proportions. Sometimes, however, we only wish to test the hypothesis that the set of data came from a population having an un-specified one of a whole family of distributions, differing in one or more parameters. In that event, we estimate the parameters from the set of sample data and then test the hypothesis that the sample data came from a population having the specified *type* of distribution with the *particular* parameters estimated from the sample. We use essentially the same chi-square test we would have used if H_0 had specified the population distribution completely, except that degrees of freedom are reduced by the number of parameters estimated from the sample.

For example, a subject was given 2,520 trials at a task. In 2,259 trials he committed no errors, in each of 249 trials he committed 1 error, in each of 11 trials he committed 2 errors, and in 1 trial he committed 3 errors. The relative sample frequencies suggest that the subject's true error probabilities may follow essentially a Poisson probability law. But which one? There are as many Poisson probability laws as there are values of the parameter m, where $m = E(x) = \sum_{x=0}^{\infty} xP(x)$. We can estimate m from our sample data, as follows, by using our relative sample frequencies as estimates of $P(x)$ (and using the "hat" sign $\frown$ to denote an estimate).

$$\hat{m} = \sum_x x\widehat{P(x)} = 0\left(\frac{2{,}259}{2{,}520}\right) + 1\left(\frac{249}{2{,}520}\right) + 2\left(\frac{11}{2{,}520}\right) + 3\left(\frac{1}{2{,}520}\right)$$

$$= \frac{0 + 249 + 22 + 3}{2{,}520} = \frac{274}{2{,}520} = .1087$$

Consulting tables of probabilities for the Poisson distribution, we find that for $m = .1087$, $P(x = 0) = .897000$, $P(x = 1) = .097504$, $P(x = 2) = .005299$, $P(x = 3) = .000192$, and $P(x > 3) = .000005$. Multiplying each of these probabilities by 2,520, we obtain the expected frequencies 2260.44, 245.71008, 13.35348, .48384, and .0126, respectively, for 0, 1, 2, 3, and more than 3 errors in a sample of size $n = 2{,}520$. Since we have estimated one population parameter from the sample, the number of degrees of freedom for chi-square will be $(K - 1) - 1 = K - 2$ where K is the number of categories of numbers-of-errors; so the test statistic

$$\sum_{i=1}^{K} \frac{(o_i - e_i)^2}{e_i} \tag{15-52}$$

will be approximately distributed as chi-square with $K - 2$ degrees of freedom.

Now we can test at nominal $\alpha = .05$ the H_0 that our subject's error probabilities have a Poisson distribution (at least over the portion where probabilities are relatively large) by first making all error frequencies ≥ 2 into a single category in order to avoid very small expected frequencies and their consequent bad effect upon the chi-square approximation, then calculating

$$\sum_{i=1}^{3} \frac{(o_i - e_i)^2}{e_i} = \frac{(2259 - 2260.44)^2}{2260.44} + \frac{(249 - 245.71008)^2}{245.71008} + \frac{(12 - 13.84992)^2}{13.84992}$$

$$= \frac{(-1.44)^2}{2260.44} + \frac{3.28992^2}{245.71008} + \frac{(-1.84992)^2}{13.84992}$$

$$= \frac{2.0736}{2260.44} + \frac{10.8235736}{245.71008} + \frac{3.422204}{13.84992}$$

$$= .000917 + .044050 + .247092 = .292059$$

and, finally since this value is less than 3.84, the critical value of chi-square with $K - 2 = 3 - 2 = 1$ degree of freedom for nominal $\alpha = .05$, we accept the H_0. The probability level of the test is approximately

$$P(\chi^2_{3-2} \geq .292)$$

which is about .59. Of course the H_0 we have accepted applies only to the categories that were actually used, i.e., our H_0 is that the subject's error probabilities for 0, 1, and ≥ 2 errors are the same as the Poisson probabilities for x's of 0, 1, and ≥ 2, respectively.

Table 15.5

Summary of Requirements for Validity of Normal-Approximation Statistics Used in Tests or Estimates

	Sampling requirements			Sample-size requirements (Partly sampling, partly population, requirements)		Population requirements	
	Sampling is random	Within-sample observations are independent of each other	Samples are independent of each other	Samples are large enough to have caused distributions of sample means (or frequencies, in the case of Chi-square) to have become at least Quasi-normal	Samples are large enough to have caused the variance of $\sigma^2_{\bar{X}}$ to have become negligible relative to the variance of $\bar{X}$	All sampled populations are at least Quasi-normally distributed	All sampled populations have the same variance
One-sample Z	X			X			
Difference-score Z	X			X			
Two-sample Z	X		X	X			
One-sample Z	X	X		X	X		
Difference-score Z	X	X		X	X		
Two-sample Z	X	X	X	X	X		
One-sample t	X	X				X	
Difference-score t	X	X				X	
Two-sample t	X	X	X			X	X
Chi-square	X	X		X			

Note: In the case of Z and t statistics, the second requirement is necessitated by the particular expression used to estimate the population variance, and the third requirement will be met if the first two are.

15-9   Problems

1. If ϕ is a variable having a standardized normal distribution, what are
    (a) $P(\phi > 1.00)$, (b) the value of C for which $P(\phi < C) = .07$, (c) the
    value of K for which $P(|\phi| \geq K) = .40$, (d) $P(.22 \leq |\phi| \leq 1.04)$, (e) the
    value of K for which $P(.10 < |\phi| \leq K) = .260$?

2. If X is a quasi-normally distributed variable with a mean of 5 and a
    variance of 4, (a) approximately what proportion of the X distribution
    lies below 3, (b) what is the approximate probability that the mean $\bar{X}$
    of a random sample of 16 X-observations will exceed 6?

3. A random sample of $n = 25$ observations is drawn from a quasi-normal
    population with mean μ and variance $\sigma^2 = 100$. The mean $\bar{X}$ of the
    sample is 6. Calculate a symmetric 95% confidence interval for μ.

4. A random sample of 4 observations is drawn from a quasi-normally dis-
    tributed population whose variance is 25. The sample observations are:
    9, 16, 12, 23. Using this information, construct a symmetric confidence
    interval for μ corresponding to the 95% level of confidence.

5. A random sample of 4 observations is to be drawn from an essentially
    normally distributed population whose variance is 25. A Z test is then
    to be applied at $\alpha = .05$ to test the $H_0: \mu \geq 17$ against the $H_a: \mu < 17$.
    What is the power of the test if actually $\mu = 18$?

6. A sample of 4 observations is to be drawn from an essentially normally
    distributed population whose variance is 36 and whose mean μ is un-
    known.

    (a) For what values of $\bar{X}$ would one reject, at the .05 level of signifi-
        cance, the hypothesis that $\mu \geq 5$ against the alternative hypothesis
        that $\mu < 5$?
    (b) What would be the power of the above test to reject the false null
        hypothesis $H_0: \mu \geq 5$ if actually $\mu = -5.815$?
    (c) Give the critical region for a test of the $H_0: \mu = 9$ against the H_a:
        $\mu \neq 9$ at the .01 level of significance.
    (d) What would be the power of the above test to reject the false null
        hypothesis that $\mu = 9$ if actually $\mu = 10$?
    (e) The following four observations are randomly drawn from the
        population, 2, -2, -2, -2. Give 99% confidence limits for μ.

7. The ball bearings produced by a certain manufacturing process are
    intended to have a diameter of .5 inches. Actually the diameters of the
    ball bearings produced have a variance of .00000025 which remains
    essentially constant, but the mean of the distribution tends to drift
    (very slowly) rather than remain constant at .5. The shape of the distri-

bution is quasi-normal. A random sample of four ball bearings is drawn from the infinite process, and their diameters are .5009, .5007, .4999, and .5001. Using this information:

(a) Construct a 95% confidence interval for the mean μ of the population of ball bearing diameters from which the sample was drawn.

(b) Construct a 99% confidence interval for μ.

(c) Test at the .05 level of significance the null hypothesis that $\mu = .5000$ against the alternative hypothesis that $\mu \neq .5000$. Give the obtained value of Z_0, its probability level, the appropriate decision, and the critical value.

(d) Test at the .01 level of significance the null hypothesis that $\mu \leq .5000$ against the alternative hypothesis that $\mu > .5000$. Give the obtained value of Z_0, its probability level, the statistical decision, and critical value.

(e) Calculate the power of the test in the above problem if actually μ is: (i) .5010, (ii) .5005, (iii) .5001, and (iv) .4995.

(f) Calculate the smallest sample size necessary for a test of H_0: $\mu \leq .5000$ against H_a: $\mu > .5000$ at $\alpha = .01$ to have a power of at least .90 if actually $\mu = .5002$.

8. An essentially normally distributed population has a mean of μ and a variance of 64. A one-sample Z test is to be applied to a random sample of n observations from this population. Calculate and plot the power function of the one-sample Z test for each of the following conditions: (a) H_0: $\mu = 100$, H_a: $\mu \neq 100$, $\alpha = .05$, $n = 4$; (b) H_0: $\mu \geq 100$, H_a: $\mu < 100$, $\alpha = .05$, $n = 4$; (c) H_0: $\mu \leq 100$, H_a: $\mu > 100$, $\alpha = .05$, $n = 4$; (d) H_0: $\mu \geq 100$, H_a: $\mu < 100$, $\alpha = .01$, $n = 4$; (e) H_0: $\mu \leq 100$, H_a: $\mu > 100$, $\alpha = .05$, $n = 16$. Which test is more powerful, (b) or (d)? Why? Which test is more powerful, (c) or (e)? Why? Arrange tests (a), (b), and (c) in order of increasing power when $\mu < 100$. Arrange tests (a), (b), and c in order of increasing power when $\mu > 100$. Upon what factors does power depend?

9. Machine A bores a circular hole in a piece of metal. The diameter of the circle has an essentially constant variance of .00000004 and a mean that can vary somewhat with the state of adjustment of the machine. Machine B makes a cylindrical plug that is supposed to fit into the hole bored by Machine A. The plug's diameter has a virtually constant variance of .00000081 and a mean that is virtually constant over the short run but may vary in the long run. The distributions of diameters produced by both machines are quasi-normal. A random sample of five holes drilled by Machine A have diameters of .37521, .37487, .37492, .37524, .37511, and a random sample of three plugs made by Machine B have diameters of .37301, .37224, .37216. Using this information:

(a) Construct a 95% confidence interval for $\mu_A - \mu_B$, the difference between population means.

(b) Construct a 99% confidence interval for $\mu_A - \mu_B$.

(c) Test at the .05 level of significance the null hypothesis that

$$\mu_A - \mu_B = .001$$

against the alternative hypothesis that $\mu_A - \mu_B \neq .001$. Give the obtained value of Z_0, its probability level, the critical value, and the decision.

(d) Test at $\alpha = .01$ the null hypothesis that $\mu_A - \mu_B \geq .001$ against the alternative hypothesis that $\mu_A - \mu_B < .001$. Give the obtained value of Z_0, its probability level, the critical value, and the statistical decision.

(e) Calculate the power of the test in part (d) if actually $\mu_A - \mu_B$ is: (i) 0, (ii) .0005, (iii) .001, (iv) .0015.

(f) Calculate the smallest common sample size $n = n_A = n_B$ necessary for a test of $H_0: \mu_A - \mu_B = .001$ against $H_a: \mu_A - \mu_B \neq .001$ conducted at $\alpha = .05$ to have a power of at least .99 if actually

$$\mu_A - \mu_B = .003$$

10. Two populations X and Y are known to be essentially normally distributed with $\sigma_X^2 = 63$ and $\sigma_Y^2 = 36$. An experimenter decides to test the $H_0: \mu_X \leq \mu_Y$ against the $H_a: \mu_X > \mu_Y$ by means of a two-sample Z test based upon random samples of sizes $n_X = 9$ and $n_Y = 4$ and conducted at the .05 level of significance. If actually μ_X lies 5 points above μ_Y, what is the power of the test?

11. A random sample of $n = 100$ observations has been drawn from a population with mean μ, and calculations upon the sample observations reveal that $\bar{X} = 81.4$ and $\hat{\sigma}^2 = 1024$. Using this information and the one-sample $\hat{Z}$ statistic:

(a) Construct a symmetric confidence interval for μ corresponding to the .90 level of confidence.

(b) Test at the .01 level of significance the $H_0: \mu \leq 75$ against the $H_a: \mu > 75$. Give the obtained value of $\hat{Z}_0$, its approximate probability level, the critical value, and the decision.

(c) Give the power of the test if actually $\mu = 77$.

12. Table 12-6 (in Section 12-3-4) gave data used by a physiologist to test the $H_0: \mu_X \leq \mu_Y$ against the $H_a: \mu_X > \mu_Y$. Suppose that the difference-score population is quasi-normally distributed.

(a) Use the difference-score $\hat{Z}$ statistic to test the physiologist's hypotheses at the .05 level of significance. Give the obtained value of $\hat{Z}_0$, its approximate probability level, the critical value, and the decision.

(b) Use the difference-score $\hat{Z}$ statistic to establish a symmetric 90% confidence interval for $\mu_X - \mu_Y$.

13. Independent random samples of $n_X = 100$ and $n_Y = 200$ observations have been drawn, respectively, from an X population with mean μ_X and a Y population with mean μ_Y. Calculations upon the sample observations produce the following values for sample statistics: $\bar{X} = 81.4$, $\bar{Y} = 98.6$, $\hat{\sigma}_X^2 = 1{,}024$, $\hat{\sigma}_Y^2 = 7{,}752$. Using the two-sample $\hat{Z}$ statistic:

   (a) Construct a symmetric 99% confidence interval for $\mu_X - \mu_Y$.

   (b) Test $H_0: \mu_X - \mu_Y \geq -4$ against $H_a: \mu_X - \mu_Y < -4$ at $\alpha = .001$. Give the obtained value of $\hat{Z}_0$, its approximate probability level, the critical value, and the decision.

   (c) Calculate the approximate power of the test if actually

   $$\mu_X - \mu_Y = -18$$

14. If t_v is a variable having a t with v degrees of freedom distribution, what are (a) the value of C for which $P(t_{24} < C) = .05$, (b) the value of K for which $P(t_3 > K) = .99$, (c) the value of K for which

   $$P(-K \leq |t_8| \leq K) = .99$$

   (d) the value of v for which $P(|t_v| > 1.96) = .05$, (e) $P(t_{10} < 4.144)$?

15. Work parts a, b, c, and d of Problem 7, assuming that the population variance is *un*known so that you must use the one-sample t statistic rather than the one-sample Z statistic.

16. Work Problem 11, using the one-sample t statistic instead of the one-sample $\hat{Z}$ statistic.

17. A random sample of 5 graduate students in the Psychology Dept. at Humanities U. made scores on the verbal and quantitative tests of the G.R.E. as shown below. Test at the .05 level of significance the H_0 that in the potential population to which they belong the mean verbal score is less than or equal to the mean quantitative score against the H_a that it is greater. Give the obtained value of the test statistic, the critical value, the statistical decision, and the probability level of the test.

Student	Verbal	Quantitative
Carl	498	510
Alfred	741	534
Sigmund	418	423
Karen	614	377
Harry	557	584

18. Work Problem 12 using the difference-score t statistic instead of the difference-score $\hat{Z}$ statistic.

19. The Zap gasoline company believes that Zap gas gets at least 1 more mile per gallon than Brand X gas, and to prove it they produce the data below. Assuming that all mileage distributions concerned are normally distributed with equal variances and that sampling was random, use these data to test Zap's hypothesis at the .01 level. What is H_0? What is H_a? What is the appropriate test? What is the obtained value of the test statistic, the critical value, the probability level, the statistical decision?

	Mileage per Gallon	
	Zap Gas	Brand X Gas
Car A	18.7	17.2
Car B	16.1	14.7
Car C	12.9	11.3

20. The speed of a cat in pursuit of a road runner has been clocked on two randomly selected occasions to be 18.5 and 19.5 miles per hour. The speed of the road runner, fleeing the cat, has been measured on three other random occasions to be 19.0, 22.0, and 22.0 miles per hour. Both speeds are normally distributed and they have the same variance. Using the two-sample t statistic:
    (a) Calculate a 99% confidence interval for $\mu_C - \mu_R$.
    (b) Test the H_0: $\mu_C \geq \mu_R$ against the H_a: $\mu_C < \mu_R$ at the .05 level of significance. Give the obtained value of t_0, its probability level, the critical value, and the decision.
    (c) If cat-speed and road runner-speed had been measured on the *same* occasions, would they be independent? Why, or why not?

21. In the year 1637, of 9,160 babies born in the city of London, 4,703 were boys. What is the probability that there would be as many or more boy-babies in a total of 9,160 if in each birth $P(\text{boy}) = .5$? (Use the normal approximation to the binomial.)

22. If χ_ν^2 is a random variable having a chi-square with ν degrees of freedom distribution, what are (a) the value of C for which $P(\chi_6^2 \leq C) = .99$, (b) the value of K for which $P(\chi_{10}^2 > K) = .05$, (c) $P(\chi_1^2 > 10.828)$, (d) the value of ν for which $P(|\chi_\nu| > 1.9600) = .05$?

23. A gambler tosses a single die 360 times and gets 59 ones, 52 twos, 46 threes, 69 fours, 73 fives, and 61 sixes. Test at the .05 level of significance the H_0 that the die is equally likely to come up each of the six possible ways. Give the obtained value of the test statistic, the critical value, the probability level of the test, and the statistical decision.

24. A highway safety engineer hypothesizes that the distribution of fatal one-car accidents is as follows: 40% at night, male driver; 30% at night,

female driver; 20% in daytime, male driver; 10% in daytime, female driver. To test his hypothesis, he applies a chi-square test, at the .05 level, to the following random sample of fatal accidents. What are (a) the obtained value of chi-square, (b) the critical value, (c) the number of degrees of freedom, (d) the probability level, (e) the statistical decision?

	Men	Women
Day	12	8
Night	21	9

25. Normally, 98% of all pregnancies result in a single child, 1.98% result in twins, and .02% result in three or more children. Of 100,000 women who took a certain drug before pregnancy, 97,850 gave birth to a single child, 2,120 gave birth to twins, and 30 gave birth to three or more children. Test at the .001 level of significance the H_0 that the drug had no influence upon the relative likelihood of "singles," "twins," and "more." What is the obtained value of the test statistic, the critical value, the decision?

26. Table 5-1 in Section 5-4-4 gives the outcomes of 3,600 tosses of a pair of dice. Test whether or not the dice are "fair" by applying the appropriate chi-square test (a) to the 36 observed frequencies for the 36 possible outcomes, (b) to the 6 marginal observed frequencies representing the 6 possible outcomes for the red die, (c) to the 6 marginal observed frequencies representing the 6 possible outcomes for the green die, (d) to the two observed frequencies representing the total number of times a 7 came up (in whatever manner) and the total number of times it did not come up. In each case, state the number of degrees of freedom, the obtained value of the test statistic, its probability level, the critical value for a test at the .05 level of significance, and the statistical decision.

27. In a random sample of 100 people, each person was diagnosed as either hypertense or not hypertense and was categorized as belonging to one and only one of the three social classes, lower, middle, and upper. Results are given in the table below. Test at the .05 level of significance the H_0 that hypertension is independent of social class. Give the obtained value of the test statistic, the critical value, the probability level, and the decision.

		Lower	CLASS *Middle*	*Upper*
CONDITION	Hypertense	10	15	5
	Not Hypertense	40	15	15

28. A business magazine reports that business declined after only 6 of the past 18 presidential elections. Apply a chi-square test to these data, at an α-level of .05, to test the H_0 that after an election

$$P(\text{decline}) = P(\text{nondecline}) = .5$$

What are (a) the obtained value of the test statistic, (b) the number of degrees of freedom, (c) the decision?

29. During the past 5 years, the Fine Arts Dept. of Territorial University admitted 250 graduate students, all of whom took both the verbal and quantitative tests of the G.R.E. Exactly 200 of the students scored above 500 on the verbal test, exactly 150 scored above 500 on the quantitative, and exactly 127 scored above 500 on both tests. What is the probability level for a test of the H_0 that scoring above 500 on the verbal test is independent of scoring above 500 on the quantitative test? What is the obtained value of the test statistic? What would the critical value of the test be for $\alpha = .01$? What would the statistical decision be for that test?

30. Given the data in Table 5-1 (Section 5-4-4), using a chi-square test at the .05 level of significance, test whether or not the outcomes on the red die are independent of those on the green die.
    (a) How many degrees of freedom are there?
    (b) What is the obtained value of chi-square?
    (c) What is the critical value?
    (d) What is the probability level of the test?
    (e) What is the statistical decision?

31. In a random sample of 500 automobile tires taken from a manufacturer's output, 120 had no defects, 200 had one defect, 110 had two, 50 had three, 15 had four, and 5 had five. Test at $\alpha = .05$ the H_0 that the sampled population has a Poisson distribution. What is the obtained value of the test statistic, the critical value, the probability level, the statistical decision?

32. A psychologist hypothesizes that for a certain task errors follow a known and completely specified distribution quite similar to the Poisson distribution with a mean of $m = 1$. On the basis of his hypothesis he calculates that for a random sample of 20 trials there should be 7 errorless trials, 7 with one error, 4 with two, 1 with three, and 1 with more than three. He then draws the sample and finds that actually there were 11 errorless trials, 2 with one error, 2 with two, 3 with three, and 2 with more than three. Using these data, test the hypothesis at $\alpha = .01$. What is the obtained value of the test statistic, the critical value, the probability level, the statistical decision?

33. The personnel manager of a large corporation has accumulated a record of 2,000 cases of single-day sick leave taken by the plant's 800 employees. His records are broken down into the day of the week on which the sick leave was taken and the sex of the employee involved. The personnel manager suspects a tendency for sick leave to be taken on days adjacent to week ends and suspects that men are more guilty of this than women.

   (a) Would it be appropriate for him to use chi-square to test the H_0 that sick leave has the same probability of being taken on each of the 5 days of the work week? Why?

   (b) Would it be appropriate for him to use chi-square to test whether sex of employee is independent of day of week in determining the taking of sick leave? Why?

appendix a

an algebraic proof of the central limit theorem[1]

Introduction and Abstract

Modern proofs of the Central Limit Theorem are based on the moment generating function and are both concise and elegant. However, while they show what happens when[2] N actually "equals" infinity, by their very conciseness they yield virtually no insight into the behavior of the distribution of $\bar{X}$ at increasing finite values of N. And in particular (with rare exceptions—see Discussion) they neither mention nor throw any light upon an important qualification for practical applications, namely, that at any given N the fit to normality tends to worsen rapidly at increasingly remote tail values, or, conversely, that larger and larger N values are required to produce a "good fit" to the approached normal distribution as the tail value at which fit is measured becomes more and more remote. The present algebraic proof does supply these missing insights and has the further advantage of requiring no calculus. It is therefore of pedagogical and explicative value.

[1]Reproduced from New Mexico State University Statistical Laboratory Technical Report No. 6, "A Simple Proof of the Central Limit Theorem That Elucidates an Important Qualification," James V. Bradley, September 1968, with the permission of the New Mexico State University.

[2]In this Appendix sample size will generally be denoted by N rather than n.

Proof

Notation and Definitions

Let

$$\mu_r(\theta) = E\{[\theta - E\{\theta\}]^r\} = r\text{th central moment of } \theta$$

$$\alpha_r(\theta) = E\left\{\left[\frac{\theta - E\{\theta\}}{\text{Stand. Dev. }(\theta)}\right]^r\right\} = r\text{th standardized central moment of } \theta$$

and note that

when $r = 0$, $\quad \alpha_r(\theta) = E\left\{\left[\frac{\theta - E\{\theta\}}{\sqrt{\text{Var }(\theta)}}\right]^0\right\} = E\{1\} = 1$

when $r = 1$, $\quad \alpha_r(\theta) = E\left\{\frac{\theta - E\{\theta\}}{\sqrt{\text{Var }(\theta)}}\right\} = \frac{E\{\theta\} - E\{\theta\}}{\sqrt{\text{Var }(\theta)}} = 0$

when $r = 2$, $\quad \alpha_r(\theta) = E\left\{\left[\frac{\theta - E\{\theta\}}{\sqrt{\text{Var }(\theta)}}\right]^2\right\} = \frac{E[\theta - E\{\theta\}]^2\}}{\text{Var }(\theta)} = \frac{\text{Var }(\theta)}{\text{Var }(\theta)} = 1$

Now let

$$\bar{X} = \frac{\sum_{i=1}^{N} X_i}{N}$$

be the mean of N observations upon a random variable X with mean μ and variance σ^2. Also let $Z_i = (X_i - \mu)/\sigma$ and

$$\bar{Z} = \frac{\sum_{i=1}^{N} Z_i}{N} = \frac{\sum_{i=1}^{N} \dfrac{X_i - \mu}{\sigma}}{N} = \frac{\sum_{i=1}^{N} X_i - N\mu}{N\sigma} = \frac{N\bar{X} - N\mu}{N\sigma} = \frac{\bar{X} - \mu}{\sigma}$$

Furthermore, note that

$$E\{Z_i^r\} = E\left\{\left[\frac{X_i - \mu}{\sigma}\right]^r\right\} = \alpha_r(X_i) = \alpha_r(X)$$

and that since the X_i's, and therefore the Z_i's, are independent

$$E\{Z_1^{r_1} Z_2^{r_2} \cdots Z_N^{r_N}\} = E\{Z_1^{r_1}\}E\{Z_2^{r_2}\} \cdots E\{Z_N^{r_N}\} = \alpha_{r_1}(X)\alpha_{r_2}(X) \cdots \alpha_{r_N}(X)$$

Derivation

$$\alpha_r(\bar{X}) = E\left\{\left[\frac{\bar{X} - \mu}{\sigma_{\bar{X}}}\right]^r\right\} = E\left\{\left[\frac{\bar{X} - \mu}{\sigma/\sqrt{N}}\right]^r\right\} = N^{r/2}E\left\{\left[\frac{\bar{X} - \mu}{\sigma}\right]^r\right\}$$

$$= N^{r/2}E\{\bar{Z}^r\} = N^{r/2}E\left\{\left[\frac{Z_1 + Z_2 + \cdots + Z_N}{N}\right]^r\right\}$$

$$= \frac{1}{N^{r/2}}E\{[Z_1 + Z_2 + \cdots + Z_N]^r\}$$

The bracketed expression is simply a multinomial expansion whose general term is

$$\frac{r!}{r_1! r_2! \cdots r_N!} Z_1^{r_1} Z_2^{r_2} \cdots Z_N^{r_N}$$

where the r_i's are integers such that $0 \le r_i \le r$ and $\sum_{i=1}^{N} r_i = r$. Therefore,

$$\alpha_r(\bar{X}) = \frac{1}{N^{r/2}} E\left\{ \sum \frac{r!}{r_1! r_2! \cdots r_N!} Z_1^{r_1} Z_2^{r_2} \cdots Z_N^{r_N} \right\}$$

$$= \sum \frac{r!}{N^{r/2} r_1! r_2! \cdots r_N!} E\{Z_1^{r_1}\} E\{Z_2^{r_2}\} \cdots E\{Z_N^{r_N}\}$$

where the summation is taken both over every distinguishably different set of N r_i's whose sum is r, and for a given such set, over every distinguishable assignment of the N r_i's to the N Z_i's as exponents. The number of such distinguishable assignments is

$$\frac{N!}{(N - n)! n_1! n_2! \cdots n_k! \cdots n_r!} = \frac{N!}{(N - n)! \prod_{k=1}^{r} n_k!}$$

where n is the number of nonzero r_i's (or $N - n$ is the number of r_i's equal to zero) and n_k is the number of r_i's whose value is the positive integer k. Therefore, we may multiply the summed term by the above value and regard the summation as *now* extending only over the distinguishably different sets of r_i whose sum is r. Doing this and substituting $\alpha_{r_i}(X)$ for $E\{Z_i^{r_i}\}$, we obtain

$$\alpha_r(\bar{X}) = \sum \frac{N!}{(N - n)! \prod_{k=1}^{r} n_k! N^{r/2} \prod_{i=1}^{N} r_i!} \alpha_{r_1}(X) \alpha_{r_2}(X) \cdots \alpha_{r_N}(X)$$

$$= \sum \frac{N(N-1) \cdots (N-n+1)}{N^{r/2}} \frac{r!}{\prod_{k=1}^{r} n_k! \prod_{i=1}^{N} r_i!} \alpha_{r_1}(X) \alpha_{r_2}(X) \cdots \alpha_{r_N}(X)$$

Now if any of the r_i's equals 1, the corresponding $\alpha_{r_i}(X)$ is an $\alpha_1(X)$ which, as we have seen, equals zero. And this causes the entire expression following the summation sign to become zero. So we may regard the summation as extending only over distinguishably different sets of r_i whose sum is r *and* which contain no r_i's that are 1's. But if the number n of nonzero r_i's exceeds half the sum of the r_i's, i.e., if $n > r/2$, then at least one of the r_i's must be a 1. So we may ignore cases where $n > r/2$. Furthermore, when $n = r/2$, the only way that the $n = r/2$ positive nonzero integers can sum to r without any of the integers being 1's is if all of the n integers are 2's. Since n and r are both integers, n can equal $r/2$ only when r is an even integer, i.e., this case can occur only when $\alpha_r(\bar{X})$ is an even numbered moment. And when r *is* even, this case contributes a single term to the summation: a term in which (a) each of the $n = r/2$ elements in the product $N(N - 1) \cdots (N - n + 1)$ can be matched with an N from the product $N^{r/2}$, (b) the only nonzero r_i's are 2's, (c) the only nonzero n_k is n_2 which equals $n = r/2$, (d) all $\alpha_{r_i}(X)$'s are either $\alpha_2(X)$'s or $\alpha_0(X)$'s both of which equal, and can be replaced by, 1's. This term therefore reduces to

$$T = \left(\frac{N}{N}\right)\left(\frac{N-1}{N}\right)\cdots\left(\frac{N-\frac{r}{2}+1}{N}\right)\frac{r!}{\left(\frac{r}{2}\right)!(2!)^{r/2}}$$

$$= \left(\frac{N}{N}\right)\left(\frac{N-1}{N}\right)\cdots\left(\frac{N-\frac{r}{2}+1}{N}\right)\frac{r(r-1)(r-2)(r-3)(r-4)\cdots(3)(2)(1)}{2\left(\frac{r}{2}\right)\ 2\left(\frac{r}{2}-1\right)\ 2\left(\frac{r}{2}-2\right)\cdots 2}$$

$$= \left(\frac{N}{N}\right)\left(\frac{N-1}{N}\right)\cdots\left(\frac{N-\frac{r}{2}+1}{N}\right)\frac{r(r-1)(r-2)(r-3)(r-4)\cdots(3)(2)(1)}{r\ (r-2)\ (r-4)\ (2)}$$

$$= \left(\frac{N}{N}\right)\left(\frac{N-1}{N}\right)\cdots\left(\frac{N-\frac{r}{2}+1}{N}\right)1\cdot3\cdot5\cdots(r-3)(r-1)$$

The formula for $\alpha_r(\bar{X})$ can therefore be written

$$\alpha_r(\bar{X}) = \left\{\sum_{n<r/2}\left(\frac{N}{N}\right)\left(\frac{N-1}{N}\right)\cdots\left(\frac{N-n+1}{N}\right)\frac{1}{N^{(r/2)-n}}\frac{r!}{\prod_{k=1}^{r}n_k!\ \prod_{i=1}^{N}r_i!}\right.$$

$$\left.\times\ \alpha_{r_1}(X)\alpha_{r_2}(X)\cdots\alpha_{r_N}(X)\right\}\ \underset{\text{(only if }r\text{ is even)}}{+}$$

$$\left[\left(\frac{N}{N}\right)\left(\frac{N-1}{N}\right)\cdots\left(\frac{N-\frac{r}{2}+1}{N}\right)1\cdot3\cdot5\cdots(r-3)(r-1)\right]$$

where the $n < r/2$ below the summation sign means only that n must be $< r/2$, not that the summation takes place over all such values.[3]

Now let N become infinite while everything else remains finite. In that event, each of the fractions of the type $(N-j)/N$ becomes 1, as does their product, and the fraction $1/N^{(r/2)-n}$ becomes zero causing the entire summation to become zero, leaving

$$\alpha_r(\bar{X}) = \begin{cases} 0 & \text{if } r \text{ is odd} \\ 1\cdot3\cdot5\cdots(r-3)(r-1) & \text{if } r \text{ is even} \end{cases}$$

But these are identically the standardized central moments of a normal distribution [4].[4] And since the normal "distribution is uniquely determined

---

[3] Actually, the summation takes place over every distinguishably different combination of r_i's that contains no r_i of 1 and whose sum is r. There may be several such sets of r_i's yielding the same value of n. And not only must n, the number of nonzero r_i's, be less than $r/2$, it must also be less than or equal to N, the total number of r_i's of all types.

[4] Numbers in brackets refer to the References at the end of Appendix A.

by its moments" [4], the demonstration of normal moments for the distribution of $\bar{X}$ when N is infinite tentatively concludes the proof of the Central Limit Theorem.

Interpretation and Discussion

In the preceding section the Central Limit Theorem was "proved" by letting "N become infinite *while everything else remains finite.*" The qualification is important and will now be examined further.

Suppose that all the $\alpha_r(X)$ are > 0. (This will generally be the case if the X distribution is positively skewed; and if it is negatively skewed, we may deal with its mirror image.) Then every nonzero term of which $\alpha_r(\bar{X})$ is the summation will be positive. Consequently, in order for

$$\sum_{n<r/2}\left(\frac{N}{N}\right)\left(\frac{N-1}{N}\right)\cdots\left(\frac{N-n+1}{N}\right)\frac{1}{N^{(r/2)-n}}\frac{r!}{\prod_{k=1}^{r}n_k!\prod_{i=1}^{N}r_i!}$$

$$\times\ \alpha_{r_1}(X)\alpha_{r_2}(X)\cdots\alpha_{r_N}(X)$$

to become essentially zero, as required in the derivation, when N becomes infinite, *every one* of the summed terms must become infinitesimal as $N \longrightarrow \infty$. But several things happen as r increases, all of which make it more difficult, i.e., necessitate a larger N value, for the summation to become essentially zero: (a) The *number* of nonzero terms (all positive) contributing to the summation tends to increase rapidly.[5] (b) Both the number of moments and the highest order of moment $r_i\,max \le r$ in the product $\alpha_{r_1}(X)\alpha_{r_2}(X)\cdots\alpha_{r_N}(X)$, for nonzero terms, tends to increase, and this is true irrespective of whether one is referring to averages or to maxima; and since the value of $\alpha_r(X)$ tends to increase over that of $\alpha_{r-2}(X)$ as r increases, the product itself tends to increase with increasing r.[6] (c) Both the average and the maximum value of

$$\frac{r!}{\prod_{k=1}^{r}n_k!\prod_{i=1}^{N}r_i!}$$

for nonzero terms increases rapidly. Some of these effects are partially offset by the fact that both the average and the minimum values of the product

$$\left(\frac{N}{N}\right)\left(\frac{N-1}{N}\right)\cdots\left(\frac{N-n+1}{N}\right)$$

[5]For example, if the sampled population is asymmetric, and $N \ge r$, as r goes from 3 to 12 the number of nonzero terms (including T) goes through the sequence 1, 2, 2, 4, 4, 7, 8, 12, 14, 20.

[6]In a population studied by the writer [1], as r increased from 3 to 10 $\alpha_r(X)$ went through the sequence 3; 14; 71; 426; 2,910; 21,720; 172,200; 1,425,000.

tend to diminish as r increases and by the fact that the average value and the maximum value $(r/2) - 1$ of the exponent $(r/2) - n$ of $1/N$ increase as r increases, but these latter effects fall far short of being compensatory.

To be more specific, replacing each r_i equal to the integer k by a k, a single nonzero term contributing to the summation can be written as

$$t = \left(\frac{N}{N}\right)\left(\frac{N-1}{N}\right) \cdots \left(\frac{N-n+1}{N}\right) \frac{r!}{N^{(r/2)-n}} \prod_{k=2}^{r} \frac{1}{n_k!}\left(\frac{\alpha_k(X)}{k!}\right)^{n_k}$$

where $\sum k n_k = r$. At one extreme, when n takes its minimum value of 1, there is a single nonzero r_i, equal to r, and

$$t = \frac{\alpha_r(X)}{N^{(r/2)-1}}$$

so for this term in the summation to be less than 1, N must exceed the $(r/2) - 1$th root of $\alpha_r(X)$,[7] and for the term to become essentially zero, it must considerably exceed it, at least when r is small-to-moderate. As r increases, the denominator increases rapidly, but the numerator tends to do likewise, and if $\alpha_r(X)$ increased faster with r than did $N^{(r/2)-1}$, the term would, of course, increase rather than diminish. At the opposite extreme, n takes its maximum value, for nonzero terms and for $N \geq r$, when all but one of the nonzero r_i's are 2's and one is a 3, so that $r = 2(n-1) + 3 = 2n + 1$ and $n = (r-1)/2$, causing $1/N^{(r/2)-n}$ to assume its largest value $1/\sqrt{N}$. In this case the term becomes

$$t = \left(\frac{N}{N}\right)\left(\frac{N-1}{N}\right) \cdots \left(\frac{N-\dfrac{r-3}{2}}{N}\right) \frac{\alpha_3(X)}{6\sqrt{N}} r(r-1)[(r-2)(r-4)(r-6)\cdots]$$

$$\geq \left(\frac{N}{N}\right)\left(\frac{N-1}{N}\right) \cdots \left(\frac{N-\dfrac{r-3}{2}}{N}\right) \frac{\alpha_3(X)}{6}\sqrt{\frac{r![r(r-1)]}{N}}$$

or, replacing each of the $n = (r-1)/2$ fractions of the type $(N-j)/N$ by $1/2$ (which can only yield a smaller product) and replacing $r!$ by its Sterling approximation, we have

$$t > \sqrt{\frac{.139[\alpha_3(X)]^2 r(r-1)\sqrt{r}\left(\dfrac{r}{2e}\right)^r}{N}} \cong \sqrt{\frac{r^{5/2}\left(\dfrac{r}{2e}\right)^r}{N}}$$

So this term in the summation will be greater than 1 so long as the numerator under the first radical exceeds N, and it is clear that in order to reduce t to essentially zero, $\sqrt{N}$ must become very large relative to

$$\sqrt{r^{5/2}\left(\frac{r}{2e}\right)^r}$$

[7] For the writer's population mentioned in the preceding footnote, $\alpha_{10}(X) = 1,425,000$, the $(r/2) - 1 = $ 4th root of which is about 35. So in this case $t = (35/N)^4$ which is > 1 as long as $N < 35$ and becomes essentially zero at a much higher value of N.

which, of course, increases rapidly with increasing r. Finally, even the single term

$$T = \left(\frac{N}{N}\right)\left(\frac{N-1}{N}\right) \cdots \left(\frac{N - \frac{r}{2} + 1}{N}\right) 1 \cdot 3 \cdot 5 \cdots (r-3)(r-1)$$

which lies outside the summation and appears in the equation for $\alpha_r(\bar{X})$ only when r is even, does not become essentially $1 \cdot 3 \cdot 5 \cdots (r-3)(r-1)$, as required in the derivation, until the product

$$\left(\frac{N}{N}\right)\left(\frac{N-1}{N}\right) \cdots \left(\frac{N - \frac{r}{2} + 1}{N}\right)$$

becomes essentially 1, which it surely cannot do if N is not large relative to the $r/2$ in the last fraction.

These examples illustrate the fact that, for every individual nonzero term contributing to $\alpha_r(\bar{X})$, N, or some power of it, must become infinite not merely relative to unity but relative to some function of r (such as $r^{5/2}(r/2e)^r$), and/or $\alpha_k(X)$'s where $k \leq r$. These restrictions cause little trouble when r is small, e.g., 3. But the "proof" of the Central Limit Theorem requires that they be met *at each and every one of the infinite number of values of r from 3 to ∞*. Clearly, some difficulty arises when we demand, in effect, that N become infinite relative to $r^{5/2}(r/2e)^r$ when r is infinite, and the difficulty is not merely "practical" but is intrinsic and fundamental to the theoretical validity of the Central Limit Theorem: r can be "equated" with any infinity to which N is "equated" and simultaneously. Furthermore, as $r \rightarrow \infty$ so also does the number of nonzero terms (all of which may be positive) contributing to the summation; so not only must each term $\rightarrow 0$ in order for the Central Limit Theorem to hold, but the sum of an *infinite number* of such infinitesimal terms must also be infinitesimal.

The conclusion then is clear: In order to make $\alpha_r(\bar{X})$ an essentially normal moment the ratio N/r must be large when r is small, enormous when r is moderate, "infinite" when r is large, and "superinfinite" when r is infinite. Thus, for any given finite N we can only expect the $\alpha_r(\bar{X})$ to be essentially normal for r's $\ll N$. And as a practical matter the finite values of N necessary to make $|\alpha_r(\bar{X}) - \alpha_r(Y)| \leq \epsilon$, where Y is a normally distributed variate and ϵ is a sufficiently small constant, increase rapidly with increasing values of r.

From this information about the normality of the moments of the distribution of $\bar{X}$, we can draw important practical implications about the normality of its tails. In general, the larger the value of r the greater is the relative contribution of the tail values of the distribution of θ to the magnitude of $\alpha_r(\theta)$. Indeed as r increases, $\alpha_r(\theta)$ owes the overwhelming proportion of its magnitude to θ's comprising more and more remote tail regions in the distribution of θ. Thus, as r increases, $\alpha_r(X)$ "reflects the influence" of increasingly remote tails of the X distribution and $\alpha_r(\bar{X})$ "represents" increasingly remote tails of the $\bar{X}$ distribution. Therefore, the increasingly large N's required to make an effectively "normal moment" out of $\alpha_r(\bar{X})$ as r increases implies that increasingly large N's are required to produce effective normality at increasingly remote tail regions in the distribution of $\bar{X}$. And since $\alpha_r(\bar{X})$ is a function only of $\alpha_k(X)$'s of order $k \leq r$, there is the further implication that the nonnormality at the tails of $\bar{X}$'s distribution is largely attributable to the nonnormality at correspondingly remote tail regions in the distribution of X.

The preceding proof of the Central Limit Theorem, then, both implies and explains an important qualification of the theorem—that it "holds" only over an abscissa range that is a function of r, the practical consequence of which is that increasingly large N values are required to produce essential normality at increasingly remote distances from μ in the distribution of $\bar{X}$. This qualification is particularly well explicated, at a simple mathematical level, by the proof given, but it is in no way an artifact of that particular proof. Chernoff [2], designating sample size by n, has called attention to the fact that although the Central Limit Theorem says that

$$P\left(\frac{\bar{X}_n - \mu}{\sigma/\sqrt{n}} \leq a\right) \longrightarrow \int_{-\infty}^{a} \frac{e^{-t^2/2}}{\sqrt{2\pi}} \, dt \qquad \text{as } n \longrightarrow \infty$$

a new situation arises if "a is not fixed but is replaced by a_n, where $a_n \to -\infty$ as $n \to \infty$" in which case Cramer [3] in an obscure publication "has essentially shown that as long as a_n does not approach $-\infty$ too rapidly, the two sides are roughly equivalent. However, this result fails to hold when a_n is of the order of magnitude of $\sqrt{n}$." Or, in other words, the standardized cumulative probability distribution of $\bar{X}$ is not well fitted by the standardized cumulative normal probability distribution at tail values $\geq \sqrt{n}$. Thus, Cramer's little known result, arrived at by much more sophisticated mathematical methods, has the same implications as those of the present paper.

References

1. BRADLEY, J. V., Studies in Research Methodology: VI The Central Limit Effect for a Variety of Populations and the Robustness of Z, t, and F; VII The Central Limit Effect for Two Dozen Populations and Its Correlation with Population Moments, *Dissertation Abstracts* (B), **28** (1968), 4815–4816.

2. CHERNOFF, H., Large-Sample Theory: Parametric Case, *Annals of Mathematical Statistics*, **27** (1956), 1–22.

3. CRAMER, H., Sur un Nouveau Théoreme-Limite de la Théorie des Probabilités, *Actualités Scientifiques et Industrielles*, No. 736. Paris: Hermann & Cie., 1938.

4. KENDALL, M. G., and A. STUART, *The Advanced Theory of Statistics*, Vol. I. New York: Hafner, 1958, 109–111.

appendix b

tables

Table B-1

Table of Random Digits

99275	48612	21216	65329	47006	81802	25488	71983	53254	49130
93487	83437	17776	85952	25025	36775	48054	51516	34250	80153
77638	96966	33408	52567	97394	44738	82972	74641	10629	27328
13771	86099	85457	58571	84468	13977	68318	67241	81211	50595
82422	08792	07864	18509	17314	57017	41471	01334	83346	52276
79394	17748	90396	44781	82857	73286	98281	05360	64793	89261
15053	36157	18243	03108	81691	95429	61175	63251	51012	30760
30831	20830	21955	49948	55957	13305	47123	90904	83261	63990
61512	40712	24620	52733	44361	69509	14503	27507	11978	53166
52270	08623	15054	40687	92854	34969	25843	05886	23848	37815
69103	56560	84932	19661	06307	56052	78502	97906	53173	20703
73582	54875	08117	98969	71713	04025	91705	35226	26130	20635
91009	80511	72733	51864	93842	41062	34805	54917	58877	06764
57300	92584	28176	93034	51982	67445	81231	35754	08862	16353
92344	93509	93644	67316	55786	46115	81226	81580	56533	28894
92749	83442	68882	04917	60100	60970	89024	26833	22520	14141
42831	12508	45484	77529	12684	74732	97088	83012	10059	29112
07745	53194	02923	61074	63854	76031	53621	82550	48493	04735
45498	56291	87859	53926	31865	91287	32266	51667	11757	85115
09182	74356	98937	05305	44177	41774	56282	83432	18409	59954
66926	27548	13360	74005	81863	47185	45649	98578	33918	26819
29434	65828	22535	71027	81963	89151	85892	46253	32153	62668
42571	68007	13720	72268	36932	83209	50630	86254	99242	84871
44191	51743	00755	11342	96156	70667	92793	64207	00908	20585
74446	78425	06622	18035	10454	00769	82604	23280	31513	80496
01361	80115	72935	17163	22449	53094	44751	89764	63157	10031
12978	51982	49118	56362	57711	86799	99400	99659	94439	72369
61774	37757	07413	39789	32552	79649	50218	18721	43988	44445
38043	01415	45193	84171	52383	92519	94553	32365	46859	74074
15829	42557	52816	20654	41230	92652	20289	18515	59762	77168
65919	70072	05256	51571	82292	43086	03342	00323	34974	27538
42819	32053	60657	82161	35440	03414	06554	36616	95862	14257
23139	91838	52934	54293	84798	35913	01191	24708	34713	37519
19405	64597	82153	40667	49899	84924	49530	56359	42477	08924
07288	45243	60807	81236	70539	14403	92228	85675	29187	93505
69116	89344	12029	01539	93894	65818	59005	07675	24657	68510
63194	11499	17269	88825	33188	65639	63508	66633	68749	38774
45783	92486	36117	48960	28351	14145	10509	87554	58285	72322
07396	88455	49617	31643	59540	77294	84426	31160	54194	05018
89360	23385	55210	93043	02738	29451	16953	50118	05254	65817
73831	35096	11635	69913	59826	49421	30885	99775	94291	39334
52665	80524	66871	10050	20685	15957	17210	61591	80857	13700
65846	39590	21711	76563	70545	01898	13150	61557	28181	68003
53627	91900	09405	76537	98930	77889	57807	75958	91341	58920
70669	44663	26954	84605	51598	23696	28725	09031	66099	05956
62677	12248	24848	02140	01131	56569	13528	24145	45199	09776
83555	56534	67758	71368	26859	95717	28288	18551	30987	20330
83697	77759	52073	04864	55059	94511	15250	26670	78996	64549
44903	84173	97008	84837	53936	01667	37797	55720	19801	69681
49014	65018	85266	95013	10543	74324	00370	44692	12436	62278

Table B-2

Squares of Integers From 1 to 999

(Column heading gives last digit of integer; row heading gives preceding digits)

n	0	1	2	3	4	5	6	7	8	9
0	0	1	4	9	16	25	36	49	64	81
1	100	121	144	169	196	225	256	289	324	361
2	400	441	484	529	576	625	676	729	784	841
3	900	961	1024	1089	1156	1225	1296	1369	1444	1521
4	1600	1681	1764	1849	1936	2025	2116	2209	2304	2401
5	2500	2601	2704	2809	2916	3025	3136	3249	3364	3481
6	3600	3721	3844	3969	4096	4225	4356	4489	4624	4761
7	4900	5041	5184	5329	5476	5625	5776	5929	6084	6241
8	6400	6561	6724	6889	7056	7225	7396	7569	7744	7921
9	8100	8281	8464	8649	8836	9025	9216	9409	9604	9801
10	10000	10201	10404	10609	10816	11025	11236	11449	11664	11881
11	12100	12321	12544	12769	12996	13225	13456	13689	13924	14161
12	14400	14641	14884	15129	15376	15625	15876	16129	16384	16641
13	16900	17161	17424	17689	17956	18225	18496	18769	19044	19321
14	19600	19881	20164	20449	20736	21025	21316	21609	21904	22201
15	22500	22801	23104	23409	23716	24025	24336	24649	24964	25281
16	25600	25921	26244	26569	26896	27225	27556	27889	28224	28561
17	28900	29241	29584	29929	30276	30625	30976	31329	31684	32041
18	32400	32761	33124	33489	33856	34225	34596	34969	35344	35721
19	36100	36481	36864	37249	37636	38025	38416	38809	39204	39601
20	40000	40401	40804	41209	41616	42025	42436	42849	43264	43681
21	44100	44521	44944	45369	45796	46225	46656	47089	47524	47961
22	48400	48841	49284	49729	50176	50625	51076	51529	51984	52441
23	52900	53361	53824	54289	54756	55225	55696	56169	56644	57121
24	57600	58081	58564	59049	59536	60025	60516	61009	61504	62001
25	62500	63001	63504	64009	64516	65025	65536	66049	66564	67081
26	67600	68121	68644	69169	69696	70225	70756	71289	71824	72361
27	72900	73441	73984	74529	75076	75625	76176	76729	77284	77841
28	78400	78961	79524	80089	80656	81225	81796	82369	82944	83521
29	84100	84681	85264	85849	86436	87025	87616	88209	88804	89401
30	90000	90601	91204	91809	92416	93025	93636	94249	94864	95481
31	96100	96721	97344	97969	98596	99225	99856	100489	101124	101761
32	102400	103041	103684	104329	104976	105625	106276	106929	107584	108241
33	108900	109561	110224	110889	111556	112225	112896	113569	114244	114921
34	115600	116281	116964	117649	118336	119025	119716	120409	121104	121801
35	122500	123201	123904	124609	125316	126025	126736	127449	128164	128881
36	129600	130321	131044	131769	132496	133225	133956	134689	135424	136161
37	136900	137641	138384	139129	139876	140625	141376	142129	142884	143641
38	144400	145161	145924	146689	147456	148225	148996	149769	150544	151321
39	152100	152881	153664	154449	155236	156025	156816	157609	158404	159201

Body of table is reproduced from Table 50, "Squares of Integers," in *Biometrika Tables for Statisticians,* Vol. I, eds., E. S. Pearson and H. O. Hartley, Cambridge University Press, 1954, with the permission of the Trustees of *Biometrika.*

Table B-2 cont.

n	0	1	2	3	4	5	6	7	8	9
40	160000	160801	161604	162409	163216	164025	164836	165649	166464	167281
41	168100	168921	169744	170569	171396	172225	173056	173889	174724	175561
42	176400	177241	178084	178929	179776	180625	181476	182329	183184	184041
43	184900	185761	186624	187489	188356	189225	190096	190969	191844	192721
44	193600	194481	195364	196249	197136	198025	198916	199809	200704	201601
45	202500	203401	204304	205209	206116	207025	207936	208849	209764	210681
46	211600	212521	213444	214369	215296	216225	217156	218089	219024	219961
47	220900	221841	222784	223729	224676	225625	226576	227529	228484	229441
48	230400	231361	232324	233289	234256	235225	236196	237169	238144	239121
49	240100	241081	242064	243049	244036	245025	246016	247009	248004	249001
50	250000	251001	252004	253009	254016	255025	256036	257049	258064	259081
51	260100	261121	262144	263169	264196	265225	266256	267289	268324	269361
52	270400	271441	272484	273529	274576	275625	276676	277729	278784	279841
53	280900	281961	283024	284089	285156	286225	287296	288369	289444	290521
54	291600	292681	293764	294849	295936	297025	298116	299209	300304	301401
55	302500	303601	304704	305809	306916	308025	309136	310249	311364	312481
56	313600	314721	315844	316969	318096	319225	320356	321489	322624	323761
57	324900	326041	327184	328329	329476	330625	331776	332929	334084	335241
58	336400	337561	338724	339889	341056	342225	343396	344569	345744	346921
59	348100	349281	350464	351649	352836	354025	355216	356409	357604	358801
60	360000	361201	362404	363609	364816	366025	367236	368449	369664	370881
61	372100	373321	374544	375769	376996	378225	379456	380689	381924	383161
62	384400	385641	386884	388129	389376	390625	391876	393129	394384	395641
63	396900	398161	399424	400689	401956	403225	404496	405769	407044	408321
64	409600	410881	412164	413449	414736	416025	417316	418609	419904	421201
65	422500	423801	425104	426409	427716	429025	430336	431649	432964	434281
66	435600	436921	438244	439569	440896	442225	443556	444889	446224	447561
67	448900	450241	451584	452929	454276	455625	456976	458329	459684	461041
68	462400	463761	465124	466489	467856	469225	470596	471969	473344	474721
69	476100	477481	478864	480249	481636	483025	484416	485809	487204	488601
70	490000	491401	492804	494209	495616	497025	498436	499849	501264	502681
71	504100	505521	506944	508369	509796	511225	512656	514089	515524	516961
72	518400	519841	521284	522729	524176	525625	527076	528529	529984	531441
73	532900	534361	535824	537289	538756	540225	541696	543169	544644	546121
74	547600	549081	550564	552049	553536	555025	556516	558009	559504	561001
75	562500	564001	565504	567009	568516	570025	571536	573049	574564	576081
76	577600	579121	580644	582169	583696	585225	586756	588289	589824	591361
77	592900	594441	595984	597529	599076	600625	602176	603729	605284	606841
78	608400	609961	611524	613089	614656	616225	617796	619369	620944	622521
79	624100	625681	627264	628849	630436	632025	633616	635209	636804	638401
80	640000	641601	643204	644809	646416	648025	649636	651249	652864	654481
81	656100	657721	659344	660969	662596	664225	665856	667489	669124	670761
82	672400	674041	675684	677329	678976	680625	682276	683929	685584	687241
83	688900	690561	692224	693889	695556	697225	698896	700569	702244	703921
84	705600	707281	708964	710649	712336	714025	715716	717409	719104	720801

n	0	1	2	3	4	5	6	7	8	9
85	722500	724201	725904	727609	729316	731025	732736	734449	736164	737881
86	739600	741321	743044	744769	746496	748225	749956	751689	753424	755161
87	756900	758641	760384	762129	763876	765625	767376	769129	770884	772641
88	774400	776161	777924	779689	781456	783225	784996	786769	788544	790321
89	792100	793881	795664	797449	799236	801025	802816	804609	806404	808201
90	810000	811801	813604	815409	817216	819025	820836	822649	824464	826281
91	828100	829921	831744	833569	835396	837225	839056	840889	842724	844561
92	846400	848241	850084	851929	853776	855625	857476	859329	861184	863041
93	864900	866761	868624	870489	872356	874225	876096	877969	879844	881721
94	883600	885481	887364	889249	891136	893025	894916	896809	898704	900601
95	902500	904401	906304	908209	910116	912025	913936	915849	917764	919681
96	921600	923521	925444	927369	929296	931225	933156	935089	937024	938961
97	940900	942841	944784	946729	948676	950625	952576	954529	956484	958441
98	960400	962361	964324	966289	968256	970225	972196	974169	976144	978121
99	980100	982081	984064	986049	988036	990025	992016	994009	996004	998001

Table B-3 (a)
Cumulative Binomial Probabilities

$$p(r) = \binom{n}{r}\, p^r (1-p)^{n-r}$$

Cell Entries Give $p(r)$

n	r	.001	.01	.05	.1	.15	.2	.25	.3	.35	.4	.45	.5
2	0	.99800	.98010	.90250	.81000	.72250	.64000	.56250	.49000	.42250	.36000	.30250	.25000
2	1	.00200	.01980	.09500	.18000	.25500	.32000	.37500	.42000	.45500	.48000	.49500	.50000
2	2	.00000	.00010	.00250	.01000	.02250	.04000	.06250	.09000	.12250	.16000	.20250	.25000
3	0	.99700	.97030	.85738	.72900	.61413	.51200	.42188	.34300	.27463	.21600	.16638	.12500
3	1	.00299	.02940	.13538	.24300	.32513	.38400	.42188	.44100	.44363	.43200	.40838	.37500
3	2	.00000	.00030	.00713	.02700	.05738	.09600	.14063	.18900	.23888	.28800	.33413	.37500
3	3	.00000	.00000	.00013	.00100	.00338	.00800	.01563	.02700	.04288	.06400	.09113	.12500
4	0	.99601	.96060	.81451	.65610	.52201	.40960	.31641	.24010	.17851	.12960	.09151	.06250
4	1	.00399	.03881	.17148	.29160	.36848	.40960	.42188	.41160	.38448	.34560	.29948	.25000
4	2	.00001	.00059	.01354	.04860	.09754	.15360	.21094	.26460	.31054	.34560	.36754	.37500
4	3	.00000	.00000	.00048	.00360	.01148	.02560	.04688	.07560	.11148	.15360	.20048	.25000
4	4	.00000	.00000	.00000	.00010	.00051	.00160	.00391	.00810	.01501	.02560	.04101	.06250
5	0	.99501	.95099	.77378	.59049	.44371	.32768	.23730	.16807	.11603	.07776	.05033	.03125
5	1	.00498	.04803	.20363	.32805	.39150	.40960	.39551	.36015	.31239	.25920	.20589	.15625
5	2	.00001	.00097	.02143	.07290	.13818	.20480	.26367	.30870	.33642	.34560	.33691	.31250
5	3	.00000	.00001	.00113	.00810	.02438	.05120	.08789	.13230	.18115	.23040	.27565	.31250
5	4	.00000	.00000	.00003	.00045	.00215	.00640	.01465	.02835	.04877	.07680	.11277	.15625
5	5	.00000	.00000	.00000	.00001	.00008	.00032	.00098	.00243	.00525	.01024	.01845	.03125
6	0	.99401	.94148	.73509	.53144	.37715	.26214	.17798	.11765	.07542	.04666	.02768	.01563
6	1	.00597	.05706	.23213	.35429	.39933	.39322	.35596	.30253	.24366	.18662	.13589	.09375
6	2	.00001	.00144	.03054	.09842	.17618	.24576	.29663	.32414	.32801	.31104	.27795	.23438
6	3	.00000	.00002	.00214	.01458	.04145	.08192	.13184	.18522	.23549	.27648	.30322	.31250
6	4	.00000	.00000	.00008	.00122	.00549	.01536	.03296	.05954	.09510	.13824	.18607	.23438
6	5	.00000	.00000	.00000	.00005	.00039	.00154	.00439	.01021	.02048	.03686	.06089	.09375
6	6	.00000	.00000	.00000	.00000	.00001	.00006	.00024	.00073	.00184	.00410	.00830	.01563
7	0	.99302	.93207	.69834	.47830	.32058	.20972	.13348	.08235	.04902	.02799	.01522	.00781
7	1	.00696	.06590	.25728	.37201	.39601	.36700	.31146	.24706	.18478	.13064	.08719	.05469
7	2	.00002	.00200	.04062	.12400	.20965	.27525	.31146	.31765	.29848	.26127	.21402	.16406
7	3	.00000	.00003	.00375	.02296	.06164	.11469	.17303	.22689	.26787	.29030	.29185	.27344
7	4	.00000	.00000	.00020	.00255	.01088	.02867	.05768	.09724	.14424	.19354	.23878	.27344
7	5	.00000	.00000	.00001	.00017	.00115	.00430	.01154	.02501	.04660	.07741	.11722	.16406
7	6	.00000	.00000	.00000	.00001	.00007	.00036	.00128	.00357	.00836	.01720	.03197	.05469
7	7	.00000	.00000	.00000	.00000	.00000	.00001	.00006	.00022	.00064	.00164	.00374	.00781
8	0	.99203	.92274	.66342	.43047	.27249	.16777	.10011	.05765	.03186	.01680	.00837	.00391
8	1	.00794	.07457	.27933	.38264	.38469	.33554	.26697	.19765	.13726	.08958	.05481	.03125
8	2	.00003	.00264	.05146	.14880	.23760	.29360	.31146	.29648	.25869	.20902	.15695	.10938
8	3	.00000	.00005	.00542	.03307	.08386	.14680	.20764	.25412	.27859	.27869	.25683	.21875
8	4	.00000	.00000	.00036	.00459	.01850	.04588	.08652	.13614	.18751	.23224	.26266	.27344
8	5	.00000	.00000	.00002	.00041	.00261	.00918	.02307	.04668	.08077	.12386	.17192	.21875
8	6	.00000	.00000	.00000	.00002	.00023	.00115	.00385	.01000	.02175	.04129	.07033	.10938
8	7	.00000	.00000	.00000	.00000	.00001	.00008	.00037	.00122	.00335	.00786	.01644	.03125
8	8	.00000	.00000	.00000	.00000	.00000	.00000	.00002	.00007	.00023	.00066	.00168	.00391
9	0	.99104	.91352	.63025	.38742	.23162	.13422	.07508	.04035	.02071	.01008	.00461	.00195
9	1	.00893	.08305	.29854	.38742	.36786	.30199	.22525	.15565	.10037	.06047	.03391	.01758
9	2	.00004	.00336	.06285	.17219	.25967	.30199	.30034	.26683	.21619	.16124	.11099	.07031
9	3	.00000	.00008	.00772	.04464	.10692	.17616	.23360	.26683	.27162	.25082	.21188	.16406
9	4	.00000	.00000	.00061	.00744	.02830	.06606	.11680	.17153	.21939	.25082	.26004	.24609
9	5	.00000	.00000	.00003	.00083	.00499	.01652	.03893	.07351	.11813	.16722	.21276	.24609
9	6	.00000	.00000	.00000	.00006	.00059	.00275	.00865	.02100	.04241	.07432	.11605	.16406
9	7	.00000	.00000	.00000	.00000	.00004	.00029	.00124	.00386	.00979	.02123	.04069	.07031
9	8	.00000	.00000	.00000	.00000	.00000	.00002	.00010	.00041	.00132	.00354	.00832	.01758
9	9	.00000	.00000	.00000	.00000	.00000	.00000	.00000	.00002	.00008	.00026	.00076	.00195

Cumulative binomial probabilities — $P(X \le r)$ for $n = 9$ to 13

.5	.45	.4	.35	.3	.25	.2 (ρ)	.15	.1	.05	.01	.001	r	n
1.00000	1.00000	1.00000	1.00000	1.00000	1.00000	1.00000	1.00000	1.00000	1.00000	1.00000	1.00000	9	9
0.00098	0.00253	0.00605	0.01346	0.02825	0.05631	0.10737	0.19687	0.34868	0.59874	0.90438	0.99004	0	10
0.01074	0.02326	0.04636	0.08595	0.14931	0.24403	0.37581	0.54430	0.73610	0.91386	0.99573	0.99996	1	10
0.05469	0.09956	0.16729	0.26161	0.38278	0.52559	0.67780	0.82020	0.92981	0.98850	0.99989	1.00000	2	10
0.17188	0.26604	0.38228	0.51383	0.64973	0.77588	0.87913	0.95003	0.98720	0.99897	1.00000	1.00000	3	10
0.37695	0.50440	0.63310	0.75150	0.84973	0.92187	0.96721	0.99013	0.99837	0.99994	1.00000	1.00000	4	10
0.62305	0.73844	0.83376	0.90507	0.95265	0.98027	0.99363	0.99862	0.99985	1.00000	1.00000	1.00000	5	10
0.82813	0.89801	0.94524	0.97398	0.98941	0.99649	0.99914	0.99987	0.99999	1.00000	1.00000	1.00000	6	10
0.94531	0.97261	0.98771	0.99518	0.99841	0.99958	0.99992	0.99999	1.00000	1.00000	1.00000	1.00000	7	10
0.98926	0.99550	0.99832	0.99946	0.99986	0.99997	0.99999	1.00000	1.00000	1.00000	1.00000	1.00000	8	10
0.99902	0.99966	0.99990	0.99997	0.99999	1.00000	1.00000	1.00000	1.00000	1.00000	1.00000	1.00000	9	10
1.00000	1.00000	1.00000	1.00000	1.00000	1.00000	1.00000	1.00000	1.00000	1.00000	1.00000	1.00000	10	10
0.00049	0.00139	0.00363	0.00875	0.01977	0.04224	0.08590	0.16734	0.31381	0.56880	0.89534	0.98905	0	11
0.00586	0.01393	0.03023	0.06058	0.11299	0.19710	0.32212	0.49219	0.69736	0.89811	0.99482	0.99995	1	11
0.03271	0.06522	0.11892	0.20013	0.31274	0.45520	0.61740	0.77881	0.91044	0.98476	0.99984	1.00000	2	11
0.11328	0.19112	0.29628	0.42555	0.56956	0.71330	0.83886	0.93056	0.98147	0.99845	1.00000	1.00000	3	11
0.27441	0.39714	0.53277	0.66831	0.78970	0.88537	0.94959	0.98411	0.99725	0.99989	1.00000	1.00000	4	11
0.50000	0.63312	0.75350	0.85132	0.92178	0.96567	0.98835	0.99734	0.99970	0.99999	1.00000	1.00000	5	11
0.72559	0.82620	0.90065	0.94986	0.97838	0.99244	0.99803	0.99968	0.99998	1.00000	1.00000	1.00000	6	11
0.88672	0.93904	0.97072	0.98776	0.99571	0.99882	0.99977	0.99997	1.00000	1.00000	1.00000	1.00000	7	11
0.96729	0.98520	0.99408	0.99796	0.99942	0.99988	0.99998	1.00000	1.00000	1.00000	1.00000	1.00000	8	11
0.99414	0.99779	0.99927	0.99979	0.99995	1.00000	1.00000	1.00000	1.00000	1.00000	1.00000	1.00000	9	11
0.99951	0.99985	0.99996	0.99999	1.00000	1.00000	1.00000	1.00000	1.00000	1.00000	1.00000	1.00000	10	11
1.00000	1.00000	1.00000	1.00000	1.00000	1.00000	1.00000	1.00000	1.00000	1.00000	1.00000	1.00000	11	11
0.00024	0.00077	0.00218	0.00569	0.01384	0.03168	0.06872	0.14224	0.28243	0.54036	0.88638	0.98807	0	12
0.00317	0.00829	0.01959	0.04244	0.08503	0.15838	0.27488	0.44346	0.65900	0.88164	0.99383	0.99993	1	12
0.01929	0.04214	0.08344	0.15129	0.25282	0.39068	0.55835	0.73582	0.88913	0.97776	0.99979	1.00000	2	12
0.07300	0.13447	0.22534	0.34666	0.49252	0.64878	0.79457	0.90779	0.97436	0.99776	1.00000	1.00000	3	12
0.19385	0.30443	0.43818	0.58337	0.72366	0.84236	0.92744	0.97608	0.99567	0.99982	1.00000	1.00000	4	12
0.38721	0.52693	0.66521	0.78730	0.88215	0.94560	0.98059	0.99536	0.99946	0.99999	1.00000	1.00000	5	12
0.61279	0.73931	0.84179	0.91538	0.96140	0.98575	0.99610	0.99933	0.99995	1.00000	1.00000	1.00000	6	12
0.80615	0.88826	0.94269	0.97452	0.99051	0.99722	0.99942	0.99993	1.00000	1.00000	1.00000	1.00000	7	12
0.92700	0.96443	0.98473	0.99442	0.99831	0.99961	0.99994	0.99999	1.00000	1.00000	1.00000	1.00000	8	12
0.98071	0.99212	0.99719	0.99918	0.99979	0.99996	1.00000	1.00000	1.00000	1.00000	1.00000	1.00000	9	12
0.99683	0.99892	0.99968	0.99995	0.99998	1.00000	1.00000	1.00000	1.00000	1.00000	1.00000	1.00000	10	12
0.99976	0.99993	0.99999	1.00000	1.00000	1.00000	1.00000	1.00000	1.00000	1.00000	1.00000	1.00000	11	12
1.00000	1.00000	1.00000	1.00000	1.00000	1.00000	1.00000	1.00000	1.00000	1.00000	1.00000	1.00000	12	12
0.00012	0.00042	0.00131	0.00370	0.00969	0.02376	0.05498	0.12091	0.25419	0.51334	0.87752	0.98708	0	13
0.00171	0.00490	0.01263	0.02958	0.06367	0.12671	0.23365	0.39828	0.62134	0.86458	0.99275	0.99992	1	13
0.01123	0.02691	0.05791	0.11319	0.20248	0.33260	0.50165	0.69196	0.86612	0.97549	0.99973	1.00000	2	13
0.04614	0.09292	0.16858	0.27827	0.42061	0.58425	0.74732	0.88200	0.96584	0.99671	0.99999	1.00000	3	13
0.13342	0.22795	0.35305	0.50050	0.65431	0.79396	0.90087	0.96584	0.99354	0.99971	1.00000	1.00000	4	13
0.29053	0.42681	0.57440	0.71589	0.83460	0.91979	0.96965	0.99243	0.99908	0.99998	1.00000	1.00000	5	13
0.50000	0.64374	0.77116	0.87053	0.93762	0.97571	0.99300	0.99873	0.99990	1.00000	1.00000	1.00000	6	13
0.70947	0.82123	0.90234	0.95380	0.98178	0.99435	0.99875	0.99983	0.99999	1.00000	1.00000	1.00000	7	13
0.86658	0.93015	0.96792	0.98743	0.99597	0.99901	0.99983	0.99998	1.00000	1.00000	1.00000	1.00000	8	13
0.95386	0.97966	0.99222	0.99751	0.99935	0.99987	0.99998	1.00000	1.00000	1.00000	1.00000	1.00000	9	13
0.98877	0.99586	0.99869	0.99972	0.99993	0.99999	1.00000	1.00000	1.00000	1.00000	1.00000	1.00000	10	13
0.99829	0.99947	0.99987	0.99997	1.00000	1.00000	1.00000	1.00000	1.00000	1.00000	1.00000	1.00000	11	13
0.99988	0.99997	0.99999	1.00000	1.00000	1.00000	1.00000	1.00000	1.00000	1.00000	1.00000	1.00000	12	13
1.00000	1.00000	1.00000	1.00000	1.00000	1.00000	1.00000	1.00000	1.00000	1.00000	1.00000	1.00000	13	13

n	r	.001	.01	.05	.1	.15	.2 (ρ)	.25	.3	.35	.4	.45	.5
14	0	0.98609	0.86875	0.48767	0.22877	0.10277	0.04398	0.01782	0.00678	0.00240	0.00078	0.00023	0.00006
14	1	0.99991	0.99160	0.84701	0.58463	0.35667	0.19791	0.10097	0.04748	0.02052	0.00810	0.00289	0.00092
14	2	1.00000	0.99966	0.96995	0.84164	0.64791	0.44805	0.28113	0.16084	0.08393	0.03979	0.01701	0.00647
14	3	1.00000	0.99999	0.99583	0.95587	0.85349	0.69819	0.52134	0.35517	0.22050	0.12431	0.06322	0.02869
14	4	1.00000	1.00000	0.99957	0.99077	0.95326	0.87016	0.74153	0.58420	0.42272	0.27926	0.16719	0.08978
14	5	1.00000	1.00000	0.99997	0.99853	0.98847	0.95615	0.88833	0.78052	0.64051	0.48585	0.33732	0.21198
14	6	1.00000	1.00000	1.00000	0.99982	0.99779	0.98839	0.96173	0.90653	0.81641	0.69245	0.54612	0.39526
14	7	1.00000	1.00000	1.00000	0.99998	0.99967	0.99760	0.98969	0.96853	0.92466	0.84986	0.74136	0.60474
14	8	1.00000	1.00000	1.00000	1.00000	0.99996	0.99962	0.99785	0.99171	0.97566	0.94168	0.88114	0.78802
14	9	1.00000	1.00000	1.00000	1.00000	1.00000	0.99995	0.99966	0.99833	0.99396	0.98249	0.95738	0.91022
14	10	1.00000	1.00000	1.00000	1.00000	1.00000	1.00000	0.99996	0.99975	0.99886	0.99609	0.98857	0.97131
14	11	1.00000	1.00000	1.00000	1.00000	1.00000	1.00000	1.00000	0.99997	0.99986	0.99939	0.99785	0.99353
14	12	1.00000	1.00000	1.00000	1.00000	1.00000	1.00000	1.00000	1.00000	0.99999	0.99994	0.99975	0.99908
14	13	1.00000	1.00000	1.00000	1.00000	1.00000	1.00000	1.00000	1.00000	1.00000	1.00000	0.99999	0.99994
14	14	1.00000	1.00000	1.00000	1.00000	1.00000	1.00000	1.00000	1.00000	1.00000	1.00000	1.00000	1.00000
15	0	0.98510	0.86006	0.46329	0.20589	0.08735	0.03518	0.01336	0.00475	0.00156	0.00047	0.00013	0.00003
15	1	0.99990	0.99037	0.82905	0.54904	0.31859	0.16713	0.08018	0.03527	0.01418	0.00517	0.00169	0.00049
15	2	1.00000	0.99958	0.96380	0.81594	0.60423	0.39802	0.23609	0.12683	0.06173	0.02711	0.01065	0.00369
15	3	1.00000	0.99999	0.99453	0.94444	0.82266	0.64816	0.46129	0.29687	0.17270	0.09050	0.04242	0.01758
15	4	1.00000	1.00000	0.99939	0.98728	0.93829	0.83577	0.68649	0.51549	0.35194	0.21728	0.12040	0.05923
15	5	1.00000	1.00000	0.99995	0.99775	0.98319	0.93895	0.85163	0.72162	0.56428	0.40322	0.26076	0.15088
15	6	1.00000	1.00000	1.00000	0.99969	0.99639	0.98194	0.94338	0.86886	0.75484	0.60981	0.45216	0.30362
15	7	1.00000	1.00000	1.00000	0.99997	0.99939	0.99576	0.98270	0.94999	0.88677	0.78690	0.65350	0.50000
15	8	1.00000	1.00000	1.00000	1.00000	0.99992	0.99922	0.99581	0.98476	0.95781	0.90495	0.81824	0.69638
15	9	1.00000	1.00000	1.00000	1.00000	0.99999	0.99989	0.99921	0.99635	0.98756	0.96617	0.92307	0.84912
15	10	1.00000	1.00000	1.00000	1.00000	1.00000	0.99999	0.99988	0.99933	0.99717	0.99065	0.97453	0.94077
15	11	1.00000	1.00000	1.00000	1.00000	1.00000	1.00000	0.99999	0.99991	0.99952	0.99807	0.99367	0.98242
15	12	1.00000	1.00000	1.00000	1.00000	1.00000	1.00000	1.00000	0.99999	0.99994	0.99972	0.99889	0.99631
15	13	1.00000	1.00000	1.00000	1.00000	1.00000	1.00000	1.00000	1.00000	0.99999	0.99997	0.99988	0.99951
15	14	1.00000	1.00000	1.00000	1.00000	1.00000	1.00000	1.00000	1.00000	1.00000	1.00000	0.99999	0.99997
15	15	1.00000	1.00000	1.00000	1.00000	1.00000	1.00000	1.00000	1.00000	1.00000	1.00000	1.00000	1.00000
16	0	0.98412	0.85146	0.44013	0.18530	0.07425	0.02815	0.01002	0.00332	0.00102	0.00028	0.00007	0.00002
16	1	0.99988	0.98907	0.81076	0.51473	0.28390	0.14074	0.06348	0.02611	0.00976	0.00329	0.00099	0.00026
16	2	1.00000	0.99950	0.95707	0.78925	0.56138	0.35184	0.19711	0.09936	0.04509	0.01834	0.00662	0.00209
16	3	1.00000	0.99998	0.99300	0.93159	0.78989	0.59813	0.40499	0.24586	0.13386	0.06515	0.02813	0.01064
16	4	1.00000	1.00000	0.99916	0.98300	0.92095	0.79825	0.63019	0.44990	0.28921	0.16657	0.08531	0.03841
16	5	1.00000	1.00000	0.99994	0.99670	0.97646	0.91831	0.81035	0.65978	0.48996	0.32884	0.19760	0.10506
16	6	1.00000	1.00000	0.99999	0.99950	0.99441	0.97334	0.92044	0.82469	0.68815	0.52717	0.36603	0.22725
16	7	1.00000	1.00000	1.00000	0.99994	0.99894	0.99300	0.97287	0.92565	0.84059	0.71606	0.56290	0.40181
16	8	1.00000	1.00000	1.00000	0.99999	0.99984	0.99853	0.99253	0.97433	0.93294	0.85773	0.74411	0.59819
16	9	1.00000	1.00000	1.00000	1.00000	0.99998	0.99975	0.99836	0.99287	0.97714	0.94168	0.87590	0.77275
16	10	1.00000	1.00000	1.00000	1.00000	1.00000	0.99997	0.99971	0.99843	0.99380	0.98086	0.95138	0.89494
16	11	1.00000	1.00000	1.00000	1.00000	1.00000	1.00000	0.99996	0.99973	0.99870	0.99510	0.98506	0.96159
16	12	1.00000	1.00000	1.00000	1.00000	1.00000	1.00000	1.00000	0.99996	0.99978	0.99906	0.99654	0.98936
16	13	1.00000	1.00000	1.00000	1.00000	1.00000	1.00000	1.00000	1.00000	0.99997	0.99987	0.99944	0.99791
16	14	1.00000	1.00000	1.00000	1.00000	1.00000	1.00000	1.00000	1.00000	1.00000	0.99999	0.99994	0.99974
16	15	1.00000	1.00000	1.00000	1.00000	1.00000	1.00000	1.00000	1.00000	1.00000	1.00000	1.00000	0.99998
16	16	1.00000	1.00000	1.00000	1.00000	1.00000	1.00000	1.00000	1.00000	1.00000	1.00000	1.00000	1.00000
17	0	0.98314	0.84294	0.41812	0.16677	0.06311	0.02252	0.00752	0.00233	0.00066	0.00017	0.00004	0.00001
17	1	0.99987	0.98769	0.79223	0.48179	0.25245	0.11822	0.05011	0.01928	0.00670	0.00209	0.00057	0.00014
17	2	1.00000	0.99939	0.94975	0.76180	0.51976	0.30962	0.16370	0.07739	0.03273	0.01232	0.00409	0.00117

Table of binomial probabilities $P(r)$ for $n = 17, 18, 19$ as a function of ρ.

.5	.45	.4	.35	.3	ρ .25	.2	.15	.1	.05	.01	.001	r	n
.00519	.01436	.03410	.07006	.12452	.18932	.23925	.23586	.15556	.04145	.00059	.00000	3	17
.01816	.04113	.07958	.13205	.18678	.22087	.20935	.14568	.06050	.00764	.00002	.00000	4	17
.04721	.08749	.13793	.18486	.20813	.19142	.13608	.06684	.01748	.00104	.00000	.00000	5	17
.09442	.14317	.18391	.19908	.17840	.12761	.06804	.02359	.00388	.00011	.00000	.00000	6	17
.14838	.18408	.19267	.16846	.12014	.06684	.02673	.00654	.00068	.00001	.00000	.00000	7	17
.18547	.18828	.16056	.11338	.06436	.02785	.00835	.00144	.00009	.00000	.00000	.00000	8	17
.18547	.15403	.10704	.06105	.02758	.00928	.00209	.00025	.00001	.00000	.00000	.00000	9	17
.14838	.10082	.05709	.02630	.00946	.00248	.00042	.00004	.00000	.00000	.00000	.00000	10	17
.09442	.05249	.02422	.00901	.00258	.00053	.00007	.00000	.00000	.00000	.00000	.00000	11	17
.04721	.02147	.00807	.00243	.00055	.00009	.00001	.00000	.00000	.00000	.00000	.00000	12	17
.01816	.00676	.00207	.00050	.00009	.00001	.00000	.00000	.00000	.00000	.00000	.00000	13	17
.00519	.00158	.00039	.00008	.00001	.00000	.00000	.00000	.00000	.00000	.00000	.00000	14	17
.00104	.00026	.00005	.00001	.00000	.00000	.00000	.00000	.00000	.00000	.00000	.00000	15	17
.00013	.00003	.00000	.00000	.00000	.00000	.00000	.00000	.00000	.00000	.00000	.00000	16	17
.00001	.00000	.00000	.00000	.00000	.00000	.00000	.00000	.00000	.00000	.00000	.00000	17	17
.00000	.00002	.00010	.00043	.00163	.00564	.01801	.05365	.15009	.39721	.83451	.98215	0	18
.00007	.00031	.00122	.00416	.01256	.03383	.08106	.17041	.30019	.37631	.15173	.01770	1	18
.00058	.00217	.00691	.01903	.04576	.09584	.17226	.25561	.28351	.16835	.01303	.00015	2	18
.00311	.00948	.02455	.05465	.10460	.17038	.22968	.24057	.16801	.04726	.00070	.00000	3	18
.01167	.02908	.06139	.11035	.16810	.21298	.21533	.15920	.07000	.00933	.00003	.00000	4	18
.03268	.06663	.11459	.16638	.20173	.19878	.15073	.07866	.02178	.00137	.00000	.00000	5	18
.07082	.11811	.16552	.19411	.18732	.14356	.08165	.03008	.00524	.00016	.00000	.00000	6	18
.12140	.16567	.18916	.17918	.13762	.08204	.03499	.00910	.00100	.00001	.00000	.00000	7	18
.16692	.18637	.17340	.13266	.08110	.03760	.01203	.00221	.00015	.00000	.00000	.00000	8	18
.18547	.16943	.12844	.07937	.03862	.01393	.00334	.00043	.00002	.00000	.00000	.00000	9	18
.16692	.12476	.07707	.03846	.01490	.00418	.00075	.00007	.00000	.00000	.00000	.00000	10	18
.12140	.07424	.03737	.01506	.00464	.00101	.00014	.00001	.00000	.00000	.00000	.00000	11	18
.07082	.03543	.01453	.00473	.00116	.00020	.00002	.00000	.00000	.00000	.00000	.00000	12	18
.03268	.01338	.00447	.00118	.00023	.00003	.00000	.00000	.00000	.00000	.00000	.00000	13	18
.01167	.00391	.00106	.00023	.00004	.00000	.00000	.00000	.00000	.00000	.00000	.00000	14	18
.00311	.00085	.00019	.00003	.00000	.00000	.00000	.00000	.00000	.00000	.00000	.00000	15	18
.00058	.00013	.00002	.00000	.00000	.00000	.00000	.00000	.00000	.00000	.00000	.00000	16	18
.00007	.00001	.00000	.00000	.00000	.00000	.00000	.00000	.00000	.00000	.00000	.00000	17	18
.00000	.00001	.00006	.00028	.00114	.00423	.01441	.04560	.13509	.37735	.82617	.98117	0	19
.00004	.00018	.00077	.00285	.00928	.02678	.06845	.15289	.28518	.37735	.15856	.01866	1	19
.00033	.00134	.00463	.01382	.03580	.08034	.15402	.24283	.28518	.17875	.01441	.00017	2	19
.00185	.00619	.01750	.04218	.08695	.15175	.21820	.24283	.17956	.05331	.00083	.00000	3	19
.00739	.02026	.04665	.09086	.14905	.20233	.21820	.17141	.07980	.01122	.00003	.00000	4	19
.02218	.04973	.09331	.14677	.19164	.20233	.16365	.09075	.02660	.00177	.00000	.00000	5	19
.05175	.09494	.14515	.18440	.19164	.15737	.09546	.03737	.00690	.00022	.00000	.00000	6	19
.09611	.14427	.17771	.18440	.15253	.09742	.04432	.01225	.00142	.00002	.00000	.00000	7	19
.14416	.17705	.17771	.14894	.09805	.04871	.01662	.00324	.00024	.00000	.00000	.00000	8	19
.17620	.17705	.14643	.09802	.05136	.01984	.00508	.00070	.00003	.00000	.00000	.00000	9	19
.17620	.14643	.09762	.05278	.02201	.00661	.00127	.00012	.00000	.00000	.00000	.00000	10	19
.14416	.09697	.05325	.02325	.00772	.00180	.00026	.00002	.00000	.00000	.00000	.00000	11	19
.09611	.05290	.02366	.00835	.00221	.00040	.00004	.00000	.00000	.00000	.00000	.00000	12	19
.05175	.02330	.00850	.00242	.00051	.00007	.00001	.00000	.00000	.00000	.00000	.00000	13	19
.02218	.00817	.00243	.00056	.00009	.00001	.00000	.00000	.00000	.00000	.00000	.00000	14	19
.00739	.00223	.00054	.00010	.00001	.00000	.00000	.00000	.00000	.00000	.00000	.00000	15	19
.00185	.00046	.00009	.00001	.00000	.00000	.00000	.00000	.00000	.00000	.00000	.00000	16	19

Table entries give the binomial probability for given n, r, and ρ.

.5	.45	.4	.35	.3	ρ=.25	.2	.15	.1	.05	.01	.001	n	r
.00033	.00007	.00001	.00000	.00000	.00000	.00000	.00000	.00000	.00000	.00000	.00000	19	17
.00004	.00001	.00000	.00000	.00000	.00000	.00000	.00000	.00000	.00000	.00000	.00000	19	18
.00000	.00000	.00000	.00000	.00000	.00000	.00000	.00000	.00000	.00000	.00000	.00000	19	19
.00000	.00001	.00004	.00018	.00080	.00317	.01153	.03876	.12158	.35849	.81791	.98019	20	0
.00002	.00010	.00049	.00195	.00684	.02114	.05765	.13680	.27017	.37735	.16523	.01962	20	1
.00018	.00082	.00309	.00998	.02785	.06695	.13691	.22934	.28518	.18868	.01586	.00019	20	2
.00109	.00401	.01235	.03226	.07160	.13390	.20536	.24283	.19012	.05958	.00096	.00000	20	3
.00462	.01393	.03499	.07382	.13042	.18969	.21820	.18212	.08978	.01333	.00004	.00000	20	4
.01479	.03647	.07465	.12720	.17886	.20233	.17456	.10285	.03192	.00224	.00000	.00000	20	5
.03696	.07460	.12441	.17123	.19164	.16861	.10910	.04537	.00887	.00030	.00000	.00000	20	6
.07393	.12207	.16588	.18440	.16426	.11241	.05455	.01601	.00197	.00003	.00000	.00000	20	7
.12013	.16230	.17971	.16135	.11440	.06089	.02216	.00459	.00036	.00000	.00000	.00000	20	8
.16018	.17705	.15974	.11584	.06537	.02706	.00739	.00108	.00005	.00000	.00000	.00000	20	9
.17620	.15935	.11714	.06861	.03082	.00992	.00203	.00021	.00001	.00000	.00000	.00000	20	10
.16018	.11852	.07099	.03359	.01201	.00301	.00046	.00003	.00000	.00000	.00000	.00000	20	11
.12013	.07273	.03550	.01356	.00386	.00075	.00009	.00000	.00000	.00000	.00000	.00000	20	12
.07393	.03662	.01456	.00449	.00102	.00015	.00001	.00000	.00000	.00000	.00000	.00000	20	13
.03696	.01498	.00485	.00121	.00022	.00003	.00000	.00000	.00000	.00000	.00000	.00000	20	14
.01479	.00490	.00129	.00026	.00004	.00000	.00000	.00000	.00000	.00000	.00000	.00000	20	15
.00462	.00125	.00027	.00004	.00001	.00000	.00000	.00000	.00000	.00000	.00000	.00000	20	16
.00109	.00024	.00004	.00001	.00000	.00000	.00000	.00000	.00000	.00000	.00000	.00000	20	17
.00018	.00003	.00000	.00000	.00000	.00000	.00000	.00000	.00000	.00000	.00000	.00000	20	18
.00002	.00000	.00000	.00000	.00000	.00000	.00000	.00000	.00000	.00000	.00000	.00000	20	19
.00000	.00000	.00000	.00000	.00000	.00000	.00000	.00000	.00000	.00000	.00000	.00000	20	20
.00000	.00000	.00002	.00012	.00056	.00238	.00922	.03295	.10942	.34056	.80973	.97913	21	0
.00001	.00006	.00031	.00133	.00503	.01665	.04842	.12209	.25531	.37641	.17176	.02058	21	1
.00010	.00050	.00205	.00717	.02154	.05550	.12106	.21546	.28368	.19811	.01735	.00021	21	2
.00063	.00257	.00864	.02446	.05848	.11716	.19167	.24081	.19963	.06604	.00111	.00000	21	3
.00285	.00946	.02593	.05927	.11278	.17574	.21563	.19123	.09981	.01564	.00005	.00000	21	4
.00970	.02633	.05878	.10852	.16433	.19917	.18329	.11474	.03771	.00280	.00000	.00000	21	5
.02588	.05744	.10451	.15582	.18781	.17704	.12219	.05399	.01117	.00039	.00000	.00000	21	6
.05545	.10071	.14929	.17979	.17248	.12646	.06546	.02042	.00266	.00004	.00000	.00000	21	7
.09703	.14420	.17418	.16942	.12936	.07377	.02864	.00631	.00052	.00000	.00000	.00000	21	8
.14016	.17042	.16773	.13177	.08008	.03552	.01034	.00161	.00008	.00000	.00000	.00000	21	9
.16819	.16732	.13418	.08514	.04118	.01421	.00310	.00034	.00001	.00000	.00000	.00000	21	10
.16819	.13690	.08945	.04585	.01765	.00474	.00078	.00006	.00000	.00000	.00000	.00000	21	11
.14016	.09334	.04970	.02057	.00630	.00132	.00016	.00001	.00000	.00000	.00000	.00000	21	12
.09703	.05287	.02294	.00767	.00187	.00030	.00003	.00000	.00000	.00000	.00000	.00000	21	13
.05545	.02472	.00874	.00236	.00046	.00006	.00000	.00000	.00000	.00000	.00000	.00000	21	14
.02588	.00944	.00272	.00059	.00009	.00001	.00000	.00000	.00000	.00000	.00000	.00000	21	15
.00970	.00290	.00068	.00012	.00001	.00000	.00000	.00000	.00000	.00000	.00000	.00000	21	16
.00285	.00070	.00013	.00002	.00000	.00000	.00000	.00000	.00000	.00000	.00000	.00000	21	17
.00063	.00013	.00002	.00000	.00000	.00000	.00000	.00000	.00000	.00000	.00000	.00000	21	18
.00010	.00002	.00000	.00000	.00000	.00000	.00000	.00000	.00000	.00000	.00000	.00000	21	19
.00001	.00000	.00000	.00000	.00000	.00000	.00000	.00000	.00000	.00000	.00000	.00000	21	20
.00000	.00000	.00000	.00000	.00000	.00000	.00000	.00000	.00000	.00000	.00000	.00000	21	21
.00000	.00000	.00001	.00008	.00039	.00178	.00738	.02800	.09848	.32353	.80163	.97823	22	0
.00001	.00003	.00019	.00091	.00369	.01308	.04058	.10872	.24073	.37462	.17814	.02154	22	1
.00006	.00030	.00135	.00513	.01659	.04578	.10653	.20145	.28085	.20703	.01889	.00023	22	2
.00037	.00164	.00601	.01841	.04740	.10174	.17755	.23700	.20803	.07264	.00127	.00000	22	3
.00174	.00636	.01902	.04709	.09649	.16109	.21084	.19866	.10980	.01816	.00006	.00000	22	4

Table of binomial probabilities $P(X = r)$ for $n = 22, 23, 24$ (column headings give ρ, the probability value).

n	r	.001	.01	.05	.1	.15	.2	.25	.3	.35	.4	.45	.5
22	5	.00000	.00000	.00344	.04392	.12621	.18976	.19331	.14886	.09128	.04564	.01874	.00628
22	6	.00000	.00000	.00051	.01383	.06311	.13441	.18257	.18076	.13926	.08622	.04344	.01779
22	7	.00000	.00000	.00006	.00351	.02545	.07681	.13910	.17707	.17140	.13138	.08124	.04066
22	8	.00000	.00000	.00001	.00073	.00842	.03600	.08694	.14229	.17305	.16422	.12463	.07624
22	9	.00000	.00000	.00000	.00013	.00231	.01400	.04508	.09486	.14495	.17031	.15862	.11859
22	10	.00000	.00000	.00000	.00002	.00053	.00455	.01953	.05285	.10146	.14760	.16871	.15417
22	11	.00000	.00000	.00000	.00000	.00010	.00124	.00710	.02471	.05960	.10734	.15059	.16819
22	12	.00000	.00000	.00000	.00000	.00002	.00028	.00217	.00971	.02942	.06560	.11294	.15417
22	13	.00000	.00000	.00000	.00000	.00000	.00005	.00056	.00320	.01219	.03364	.07108	.11859
22	14	.00000	.00000	.00000	.00000	.00000	.00001	.00012	.00088	.00422	.01442	.03739	.07624
22	15	.00000	.00000	.00000	.00000	.00000	.00000	.00002	.00020	.00121	.00513	.01631	.04066
22	16	.00000	.00000	.00000	.00000	.00000	.00000	.00000	.00004	.00029	.00150	.00584	.01779
22	17	.00000	.00000	.00000	.00000	.00000	.00000	.00000	.00001	.00005	.00035	.00169	.00628
22	18	.00000	.00000	.00000	.00000	.00000	.00000	.00000	.00000	.00001	.00007	.00038	.00174
22	19	.00000	.00000	.00000	.00000	.00000	.00000	.00000	.00000	.00000	.00001	.00007	.00037
22	20	.00000	.00000	.00000	.00000	.00000	.00000	.00000	.00000	.00000	.00000	.00001	.00006
22	21	.00000	.00000	.00000	.00000	.00000	.00000	.00000	.00000	.00000	.00000	.00000	.00001
22	22	.00000	.00000	.00000	.00000	.00000	.00000	.00000	.00000	.00000	.00000	.00000	.00000
23	0	.97725	.79361	.30736	.08863	.02380	.00590	.00134	.00027	.00005	.00001	.00000	.00000
23	1	.02250	.18438	.37206	.22650	.09661	.03394	.01026	.00270	.00062	.00012	.00002	.00000
23	2	.00025	.02049	.21541	.27683	.18757	.09334	.03761	.01272	.00365	.00089	.00018	.00003
23	3	.00000	.00145	.07936	.21531	.23164	.16335	.08775	.03815	.01376	.00414	.00104	.00021
23	4	.00000	.00007	.02088	.11962	.20442	.20418	.14626	.08176	.03705	.01381	.00424	.00106
23	5	.00000	.00000	.00418	.05050	.13708	.19397	.18526	.13315	.07581	.03499	.01317	.00401
23	6	.00000	.00000	.00066	.01684	.07257	.14548	.18526	.17121	.12247	.06999	.03232	.01203
23	7	.00000	.00000	.00008	.00454	.03110	.08833	.14997	.17818	.16015	.11331	.06423	.02922
23	8	.00000	.00000	.00001	.00101	.01098	.04417	.09998	.15272	.17249	.15109	.10510	.05845
23	9	.00000	.00000	.00000	.00019	.00323	.01840	.05554	.10912	.15481	.16787	.14333	.09742
23	10	.00000	.00000	.00000	.00003	.00080	.00644	.02592	.06546	.11670	.15668	.16419	.13638
23	11	.00000	.00000	.00000	.00000	.00017	.00190	.01021	.03315	.07427	.12345	.15876	.16118
23	12	.00000	.00000	.00000	.00000	.00003	.00048	.00340	.01421	.03999	.08230	.12990	.16118
23	13	.00000	.00000	.00000	.00000	.00000	.00010	.00096	.00515	.01822	.04642	.08992	.13638
23	14	.00000	.00000	.00000	.00000	.00000	.00002	.00023	.00158	.00701	.02211	.05255	.09742
23	15	.00000	.00000	.00000	.00000	.00000	.00000	.00005	.00041	.00226	.00884	.02580	.05845
23	16	.00000	.00000	.00000	.00000	.00000	.00000	.00001	.00009	.00061	.00295	.01055	.02922
23	17	.00000	.00000	.00000	.00000	.00000	.00000	.00000	.00002	.00014	.00081	.00356	.01203
23	18	.00000	.00000	.00000	.00000	.00000	.00000	.00000	.00000	.00002	.00018	.00097	.00401
23	19	.00000	.00000	.00000	.00000	.00000	.00000	.00000	.00000	.00000	.00003	.00021	.00106
23	20	.00000	.00000	.00000	.00000	.00000	.00000	.00000	.00000	.00000	.00000	.00003	.00021
23	21	.00000	.00000	.00000	.00000	.00000	.00000	.00000	.00000	.00000	.00000	.00000	.00003
23	22	.00000	.00000	.00000	.00000	.00000	.00000	.00000	.00000	.00000	.00000	.00000	.00000
24	0	.97627	.78568	.29199	.07977	.02023	.00472	.00100	.00019	.00003	.00000	.00000	.00000
24	1	.02345	.19047	.36883	.21271	.08569	.02833	.00803	.00197	.00042	.00008	.00001	.00000
24	2	.00027	.02213	.22324	.27180	.17390	.08146	.03077	.00971	.00259	.00058	.00011	.00002
24	3	.00000	.00164	.08616	.22146	.22505	.14934	.07522	.03052	.01022	.00284	.00065	.00012
24	4	.00000	.00008	.02381	.12919	.20851	.19602	.13163	.06868	.02890	.00995	.00280	.00063
24	5	.00000	.00000	.00501	.05742	.14718	.19602	.17551	.11773	.06225	.02652	.00915	.00253
24	6	.00000	.00000	.00084	.02020	.08225	.15518	.18526	.15978	.10614	.05599	.02370	.00802
24	7	.00000	.00000	.00011	.00577	.03732	.09976	.15879	.17608	.14696	.09599	.04987	.02063
24	8	.00000	.00000	.00001	.00136	.01400	.05300	.11248	.16036	.16816	.13598	.08671	.04384

n	r	.001	.01	.05	.1	.15	.2 ρ	.25	.3	.35	.4	.45	.5
24	9	1.00000	1.00000	1.00000	0.99995	0.99852	0.98738	0.94534	0.84722	0.68665	0.48908	0.29913	0.15373
24	10	1.00000	1.00000	1.00000	0.99999	0.99968	0.99621	0.97866	0.92576	0.81667	0.65024	0.45391	0.27063
24	11	1.00000	1.00000	1.00000	1.00000	0.99994	0.99902	0.99280	0.96861	0.90577	0.78698	0.61510	0.41941
24	12	1.00000	1.00000	1.00000	1.00000	0.99999	0.99978	0.99791	0.98850	0.95775	0.88573	0.75797	0.58059
24	13	1.00000	1.00000	1.00000	1.00000	1.00000	0.99996	0.99948	0.99637	0.98358	0.94651	0.86587	0.72937
24	14	1.00000	1.00000	1.00000	1.00000	1.00000	0.99999	0.99988	0.99902	0.99451	0.97834	0.93523	0.84627
24	15	1.00000	1.00000	1.00000	1.00000	1.00000	1.00000	0.99998	0.99977	0.99844	0.99249	0.97307	0.92421
24	16	1.00000	1.00000	1.00000	1.00000	1.00000	1.00000	1.00000	0.99995	0.99962	0.99780	0.99047	0.96804
24	17	1.00000	1.00000	1.00000	1.00000	1.00000	1.00000	1.00000	0.99999	0.99992	0.99946	0.99718	0.98867
24	18	1.00000	1.00000	1.00000	1.00000	1.00000	1.00000	1.00000	1.00000	0.99999	0.99989	0.99931	0.99669
24	19	1.00000	1.00000	1.00000	1.00000	1.00000	1.00000	1.00000	1.00000	1.00000	0.99998	0.99986	0.99923
24	20	1.00000	1.00000	1.00000	1.00000	1.00000	1.00000	1.00000	1.00000	1.00000	1.00000	0.99998	0.99986
24	21	1.00000	1.00000	1.00000	1.00000	1.00000	1.00000	1.00000	1.00000	1.00000	1.00000	1.00000	0.99998
24	22	1.00000	1.00000	1.00000	1.00000	1.00000	1.00000	1.00000	1.00000	1.00000	1.00000	1.00000	1.00000
24	23	1.00000	1.00000	1.00000	1.00000	1.00000	1.00000	1.00000	1.00000	1.00000	1.00000	1.00000	1.00000
24	24	1.00000	1.00000	1.00000	1.00000	1.00000	1.00000	1.00000	1.00000	1.00000	1.00000	1.00000	1.00000
25	0	0.97530	0.77782	0.27739	0.07179	0.01720	0.00378	0.00075	0.00013	0.00002	0.00000	0.00000	0.00000
25	1	0.99970	0.97424	0.64238	0.27121	0.09307	0.02739	0.00702	0.00157	0.00030	0.00005	0.00007	0.00001
25	2	1.00000	0.99805	0.87289	0.53739	0.25374	0.09823	0.03211	0.00896	0.00213	0.00043	0.00048	0.00008
25	3	1.00000	0.99989	0.96591	0.76359	0.47112	0.23399	0.09621	0.03324	0.00968	0.00237	0.00231	0.00046
25	4	1.00000	1.00000	0.99284	0.90201	0.68211	0.42067	0.21374	0.09047	0.03205	0.00947	0.00860	0.00204
25	5	1.00000	1.00000	0.99879	0.96660	0.83848	0.61669	0.37828	0.19349	0.08262	0.02936	0.02575	0.00732
25	6	1.00000	1.00000	0.99983	0.99052	0.93047	0.78004	0.56110	0.34065	0.17340	0.07357	0.06385	0.02164
25	7	1.00000	1.00000	0.99998	0.99770	0.97453	0.89088	0.72651	0.51185	0.30608	0.15355	0.13398	0.05388
25	8	1.00000	1.00000	1.00000	0.99954	0.99203	0.95323	0.85056	0.67693	0.46682	0.27353	0.24237	0.11476
25	9	1.00000	1.00000	1.00000	0.99992	0.99786	0.98267	0.92867	0.81056	0.63031	0.42462	0.38426	0.21218
25	10	1.00000	1.00000	1.00000	0.99999	0.99951	0.99445	0.97033	0.90220	0.77116	0.58577	0.54257	0.34502
25	11	1.00000	1.00000	1.00000	1.00000	0.99990	0.99841	0.98927	0.95575	0.87458	0.73228	0.69368	0.50000
25	12	1.00000	1.00000	1.00000	1.00000	0.99998	0.99960	0.99663	0.98253	0.93956	0.84623	0.81731	0.65498
25	13	1.00000	1.00000	1.00000	1.00000	1.00000	0.99991	0.99908	0.99401	0.97454	0.92220	0.90402	0.78782
25	14	1.00000	1.00000	1.00000	1.00000	1.00000	0.99998	0.99979	0.99822	0.99069	0.96561	0.95604	0.88524
25	15	1.00000	1.00000	1.00000	1.00000	1.00000	1.00000	0.99996	0.99955	0.99706	0.98683	0.98264	0.94612
25	16	1.00000	1.00000	1.00000	1.00000	1.00000	1.00000	0.99999	0.99990	0.99921	0.99567	0.99417	0.97836
25	17	1.00000	1.00000	1.00000	1.00000	1.00000	1.00000	1.00000	0.99998	0.99982	0.99879	0.99836	0.99268
25	18	1.00000	1.00000	1.00000	1.00000	1.00000	1.00000	1.00000	1.00000	0.99996	0.99971	0.99962	0.99796
25	19	1.00000	1.00000	1.00000	1.00000	1.00000	1.00000	1.00000	1.00000	0.99999	0.99995	0.99993	0.99954
25	20	1.00000	1.00000	1.00000	1.00000	1.00000	1.00000	1.00000	1.00000	1.00000	0.99999	0.99999	0.99992
25	21	1.00000	1.00000	1.00000	1.00000	1.00000	1.00000	1.00000	1.00000	1.00000	1.00000	1.00000	0.99999
25	22	1.00000	1.00000	1.00000	1.00000	1.00000	1.00000	1.00000	1.00000	1.00000	1.00000	1.00000	1.00000
25	23	1.00000	1.00000	1.00000	1.00000	1.00000	1.00000	1.00000	1.00000	1.00000	1.00000	1.00000	1.00000
25	24	1.00000	1.00000	1.00000	1.00000	1.00000	1.00000	1.00000	1.00000	1.00000	1.00000	1.00000	1.00000
25	25	1.00000	1.00000	1.00000	1.00000	1.00000	1.00000	1.00000	1.00000	1.00000	1.00000	1.00000	1.00000

Table B-3 (b)
Point Binomial Probabilities

Cell Entries Give $\sum_{j=0}^{r} \binom{n}{i} \rho^j (1-\rho)^{n-i}$

n	r	.001	.01	.05	.1	.15	.2	.25	.3	.35	.4	.45	.5
2	0	0.99800	0.98010	0.90250	0.81000	0.72250	0.64000	0.56250	0.49000	0.42250	0.36000	0.30250	0.25000
2	1	1.00000	0.99990	0.99750	0.99000	0.97750	0.96000	0.93750	0.91000	0.87750	0.84000	0.79750	0.75000
3	0	0.99700	0.97030	0.85738	0.72900	0.61413	0.51200	0.42188	0.34300	0.27463	0.21600	0.16638	0.12500
3	1	1.00000	0.99970	0.99275	0.97200	0.93925	0.89600	0.84375	0.78400	0.71825	0.64800	0.57475	0.50000
3	2	1.00000	1.00000	0.99988	0.99900	0.99663	0.99200	0.98438	0.97300	0.95713	0.93600	0.90888	0.87500
4	0	0.99601	0.96060	0.81451	0.65610	0.52201	0.40960	0.31641	0.24010	0.17851	0.12960	0.09151	0.06250
4	1	0.99999	0.99941	0.98598	0.94770	0.89048	0.81920	0.73828	0.65170	0.56298	0.47520	0.39098	0.31250
4	2	1.00000	1.00000	0.99952	0.99630	0.98802	0.97280	0.94922	0.91630	0.87352	0.82080	0.75852	0.68750
4	3	1.00000	1.00000	0.99999	0.99990	0.99949	0.99840	0.99609	0.99190	0.98499	0.97440	0.95899	0.93750
5	0	0.99501	0.95099	0.77378	0.59049	0.44371	0.32768	0.23730	0.16807	0.11603	0.07776	0.05033	0.03125
5	1	0.99999	0.99902	0.97741	0.91854	0.83521	0.73728	0.63281	0.52822	0.42842	0.33696	0.25622	0.18750
5	2	1.00000	0.99999	0.99884	0.99144	0.97339	0.94208	0.89648	0.83692	0.76483	0.68256	0.59313	0.50000
5	3	1.00000	1.00000	0.99997	0.99954	0.99777	0.99328	0.98438	0.96922	0.94598	0.91296	0.86878	0.81250
5	4	1.00000	1.00000	1.00000	0.99999	0.99992	0.99968	0.99902	0.99757	0.99475	0.98976	0.98155	0.96875
6	0	0.99401	0.94148	0.73509	0.53144	0.37715	0.26214	0.17798	0.11765	0.07542	0.04666	0.02768	0.01563
6	1	0.99999	0.99854	0.96723	0.88574	0.77648	0.65536	0.53394	0.42018	0.31908	0.23328	0.16357	0.10938
6	2	1.00000	0.99998	0.99777	0.98415	0.95266	0.90112	0.83057	0.74431	0.64709	0.54432	0.44152	0.34375
6	3	1.00000	1.00000	0.99991	0.99873	0.99411	0.98304	0.96240	0.92953	0.88258	0.82080	0.74474	0.65625
6	4	1.00000	1.00000	1.00000	0.99995	0.99960	0.99840	0.99536	0.98907	0.97768	0.95904	0.93080	0.89063
6	5	1.00000	1.00000	1.00000	1.00000	0.99999	0.99994	0.99976	0.99927	0.99816	0.99590	0.99170	0.98438
7	0	0.99302	0.93207	0.69834	0.47830	0.32058	0.20972	0.13348	0.08235	0.04902	0.02799	0.01522	0.00781
7	1	0.99998	0.99797	0.95562	0.85031	0.71658	0.57672	0.44495	0.32942	0.23380	0.15863	0.10242	0.06250
7	2	1.00000	0.99997	0.99624	0.97431	0.92623	0.85197	0.75641	0.64707	0.53228	0.41990	0.31644	0.22656
7	3	1.00000	1.00000	0.99981	0.99727	0.98790	0.96666	0.92944	0.87396	0.80015	0.71021	0.60829	0.50000
7	4	1.00000	1.00000	0.99999	0.99982	0.99878	0.99533	0.98712	0.97120	0.94439	0.90374	0.84707	0.77344
7	5	1.00000	1.00000	1.00000	0.99999	0.99993	0.99963	0.99866	0.99621	0.99099	0.98116	0.96429	0.93750
7	6	1.00000	1.00000	1.00000	1.00000	1.00000	0.99999	0.99994	0.99978	0.99936	0.99836	0.99626	0.99219
8	0	0.99203	0.92274	0.66342	0.43047	0.27249	0.16777	0.10011	0.05765	0.03186	0.01680	0.00837	0.00391
8	1	0.99997	0.99731	0.94276	0.81310	0.65718	0.50332	0.36708	0.25530	0.16913	0.10638	0.06318	0.03516
8	2	1.00000	0.99995	0.99421	0.96191	0.89479	0.79692	0.67854	0.55177	0.42781	0.31539	0.22013	0.14453
8	3	1.00000	1.00000	0.99963	0.99498	0.97865	0.94372	0.88618	0.80590	0.70640	0.59409	0.47696	0.36328
8	4	1.00000	1.00000	0.99998	0.99957	0.99715	0.98959	0.97270	0.94203	0.89391	0.82633	0.73962	0.63672
8	5	1.00000	1.00000	1.00000	0.99998	0.99976	0.99877	0.99577	0.98871	0.97468	0.95019	0.91154	0.85547
8	6	1.00000	1.00000	1.00000	1.00000	0.99999	0.99992	0.99962	0.99871	0.99643	0.99148	0.98188	0.96484
8	7	1.00000	1.00000	1.00000	1.00000	1.00000	1.00000	0.99998	0.99993	0.99977	0.99934	0.99832	0.99609
9	0	0.99104	0.91352	0.63025	0.38742	0.23162	0.13422	0.07508	0.04035	0.02071	0.01008	0.00461	0.00195
9	1	0.99996	0.99656	0.92879	0.77484	0.59948	0.43621	0.30034	0.19600	0.12109	0.07054	0.03852	0.01953
9	2	1.00000	0.99992	0.99164	0.94703	0.85915	0.73820	0.60068	0.46283	0.33727	0.23179	0.14950	0.08984
9	3	1.00000	1.00000	0.99936	0.99167	0.96607	0.91436	0.83427	0.72966	0.60889	0.48261	0.36138	0.25391
9	4	1.00000	1.00000	0.99997	0.99911	0.99437	0.98042	0.95107	0.90119	0.82828	0.73343	0.62142	0.50000
9	5	1.00000	1.00000	1.00000	0.99994	0.99937	0.99693	0.99001	0.97470	0.94641	0.90065	0.83418	0.74609
9	6	1.00000	1.00000	1.00000	1.00000	0.99995	0.99969	0.99866	0.99571	0.98882	0.97497	0.95023	0.91016
9	7	1.00000	1.00000	1.00000	1.00000	1.00000	0.99998	0.99989	0.99957	0.99860	0.99620	0.99092	0.98047
9	8	1.00000	1.00000	1.00000	1.00000	1.00000	1.00000	1.00000	0.99998	0.99992	0.99974	0.99924	0.99805

Binomial probability table — individual terms $b(r; n, \rho) = \binom{n}{r}\rho^r(1-\rho)^{n-r}$

r	n	ρ = .5	.45	.4	.35	.3	.25	.2	.15	.1	.05	.01	.001
9	9	.00195	.00076	.00026	.00008	.00002	.00000	.00000	.00000	.00000	.00000	.00000	.00000
0	10	.00098	.00253	.00605	.01346	.02825	.05631	.10737	.19687	.34868	.59874	.90438	.99004
1	10	.00977	.02072	.04031	.07249	.12106	.18771	.26844	.34743	.38742	.31512	.09135	.00991
2	10	.04395	.07630	.12093	.17565	.23347	.28157	.30199	.27590	.19371	.07463	.00415	.00004
3	10	.11719	.16648	.21499	.25222	.26683	.25028	.20133	.12983	.05740	.01048	.00011	.00000
4	10	.20508	.23837	.25082	.23767	.20012	.14600	.08808	.04009	.01116	.00096	.00000	.00000
5	10	.24609	.23403	.20066	.15357	.10292	.05840	.02642	.00849	.00149	.00006	.00000	.00000
6	10	.20508	.15957	.11148	.06891	.03676	.01622	.00551	.00125	.00014	.00000	.00000	.00000
7	10	.11719	.07460	.04247	.02120	.00900	.00309	.00079	.00013	.00001	.00000	.00000	.00000
8	10	.04395	.02289	.01062	.00428	.00145	.00039	.00007	.00001	.00000	.00000	.00000	.00000
9	10	.00977	.00416	.00157	.00051	.00014	.00003	.00000	.00000	.00000	.00000	.00000	.00000
10	10	.00098	.00034	.00010	.00003	.00001	.00000	.00000	.00000	.00000	.00000	.00000	.00000
0	11	.00049	.00139	.00363	.00875	.01977	.04224	.08590	.16734	.31381	.56880	.89534	.98905
1	11	.00537	.01254	.02661	.05182	.09322	.15486	.23622	.32484	.38355	.32931	.09948	.01089
2	11	.02686	.05129	.08868	.13955	.19975	.25810	.29528	.28663	.21308	.08666	.00502	.00005
3	11	.08057	.12590	.17737	.22542	.25682	.25810	.22146	.15174	.07103	.01368	.00015	.00000
4	11	.16113	.20601	.23649	.24273	.22013	.17207	.11073	.05356	.01578	.00144	.00000	.00000
5	11	.22559	.23598	.22072	.18300	.13208	.08030	.03876	.01323	.00246	.00011	.00000	.00000
6	11	.22559	.19307	.14715	.09855	.05661	.02677	.00969	.00234	.00027	.00001	.00000	.00000
7	11	.16113	.11284	.07007	.03790	.01733	.00637	.00173	.00029	.00002	.00000	.00000	.00000
8	11	.08057	.04616	.02336	.01020	.00371	.00106	.00022	.00003	.00000	.00000	.00000	.00000
9	11	.02686	.01259	.00519	.00183	.00053	.00012	.00002	.00000	.00000	.00000	.00000	.00000
10	11	.00537	.00206	.00069	.00020	.00005	.00001	.00000	.00000	.00000	.00000	.00000	.00000
11	11	.00049	.00015	.00004	.00001	.00000	.00000	.00000	.00000	.00000	.00000	.00000	.00000
0	12	.00024	.00077	.00218	.00569	.01384	.03168	.06872	.14224	.28243	.54036	.88638	.98807
1	12	.00293	.00752	.01741	.03675	.07118	.12671	.20616	.30122	.37657	.34128	.10744	.01187
2	12	.01611	.03385	.06385	.10885	.16779	.23229	.28347	.29236	.23013	.09879	.00597	.00007
3	12	.05371	.09233	.14189	.19537	.23970	.25810	.23622	.17198	.08523	.01733	.00020	.00000
4	12	.12085	.16997	.21284	.23670	.23114	.19358	.13288	.06828	.02131	.00205	.00000	.00000
5	12	.19336	.22251	.22703	.20392	.15850	.10324	.05315	.01928	.00379	.00017	.00000	.00000
6	12	.22559	.21239	.17658	.12810	.07925	.04015	.01550	.00397	.00049	.00001	.00000	.00000
7	12	.19336	.14895	.10090	.05912	.02911	.01147	.00332	.00060	.00005	.00000	.00000	.00000
8	12	.12085	.07617	.04204	.01990	.00780	.00239	.00052	.00007	.00000	.00000	.00000	.00000
9	12	.05371	.02770	.01246	.00476	.00149	.00035	.00006	.00001	.00000	.00000	.00000	.00000
10	12	.01611	.00680	.00249	.00077	.00019	.00004	.00000	.00000	.00000	.00000	.00000	.00000
11	12	.00293	.00101	.00030	.00008	.00001	.00000	.00000	.00000	.00000	.00000	.00000	.00000
12	12	.00024	.00007	.00002	.00000	.00000	.00000	.00000	.00000	.00000	.00000	.00000	.00000
0	13	.00012	.00042	.00131	.00370	.00969	.02376	.05498	.12091	.25419	.51334	.87752	.98708
1	13	.00159	.00448	.01132	.02588	.05398	.10295	.17867	.27737	.36716	.35123	.11523	.01284
2	13	.00952	.02200	.04528	.08361	.13881	.20590	.26801	.29369	.24477	.11092	.00698	.00008
3	13	.03491	.06601	.11068	.16508	.21813	.25165	.24567	.19003	.09972	.02140	.00026	.00000
4	13	.08728	.13503	.18446	.22223	.23371	.20971	.15355	.08384	.02770	.00282	.00001	.00000
5	13	.15710	.19888	.22135	.21539	.18029	.12583	.06910	.02663	.00554	.00027	.00000	.00000
6	13	.20947	.21694	.19676	.15464	.10302	.05592	.02303	.00627	.00082	.00002	.00000	.00000
7	13	.20947	.17749	.13117	.08327	.04415	.01864	.00576	.00111	.00009	.00000	.00000	.00000
8	13	.15710	.10892	.06559	.03363	.01419	.00466	.00108	.00015	.00001	.00000	.00000	.00000
9	13	.08728	.04951	.02429	.01006	.00338	.00086	.00015	.00001	.00000	.00000	.00000	.00000
10	13	.03491	.01620	.00648	.00217	.00058	.00012	.00002	.00000	.00000	.00000	.00000	.00000
11	13	.00952	.00362	.00118	.00032	.00007	.00001	.00000	.00000	.00000	.00000	.00000	.00000
12	13	.00159	.00049	.00013	.00003	.00000	.00000	.00000	.00000	.00000	.00000	.00000	.00000
13	13	.00012	.00003	.00001	.00000	.00000	.00000	.00000	.00000	.00000	.00000	.00000	.00000

ρ

.5	.45	.4	.35	.3	.25	.2	.15	.1	.05	.01	.001	r	n
.00006	.00023	.00078	.00240	.00678	.01782	.04398	.10277	.22877	.48767	.86875	.98609	0	14
.00085	.00265	.00731	.01812	.04069	.08315	.15393	.25390	.35586	.35934	.12285	.01382	1	14
.00555	.01412	.03169	.06341	.11336	.18016	.25014	.29124	.25701	.12293	.00807	.00009	2	14
.02222	.04621	.08452	.13657	.19433	.24021	.25014	.20558	.11423	.02588	.00033	.00000	3	14
.06110	.10397	.15495	.20223	.22903	.22019	.17197	.09977	.03490	.00375	.00001	.00000	4	14
.12219	.17013	.20660	.21778	.19631	.14680	.08598	.03521	.00776	.00039	.00000	.00000	5	14
.18329	.20880	.20660	.17590	.12620	.07340	.03224	.00932	.00129	.00003	.00000	.00000	6	14
.20947	.19524	.15741	.10825	.06181	.02796	.00921	.00188	.00016	.00000	.00000	.00000	7	14
.18329	.13978	.09182	.05100	.02318	.00816	.00202	.00029	.00002	.00000	.00000	.00000	8	14
.12219	.07624	.04081	.01831	.00662	.00181	.00034	.00003	.00000	.00000	.00000	.00000	9	14
.06110	.03119	.01360	.00493	.00142	.00030	.00004	.00000	.00000	.00000	.00000	.00000	10	14
.02222	.00928	.00330	.00097	.00022	.00004	.00000	.00000	.00000	.00000	.00000	.00000	11	14
.00555	.00190	.00055	.00013	.00002	.00000	.00000	.00000	.00000	.00000	.00000	.00000	12	14
.00085	.00024	.00006	.00001	.00000	.00000	.00000	.00000	.00000	.00000	.00000	.00000	13	14
.00006	.00001	.00000	.00000	.00000	.00000	.00000	.00000	.00000	.00000	.00000	.00000	14	14
.00003	.00013	.00047	.00156	.00475	.01336	.03518	.08735	.20589	.46329	.86006	.98510	0	15
.00046	.00156	.00470	.01262	.03052	.06682	.13194	.23123	.34315	.36576	.13031	.01479	1	15
.00320	.00896	.02194	.04756	.09156	.15591	.23090	.28564	.26690	.13475	.00921	.00010	2	15
.01389	.03177	.06339	.11096	.17004	.22520	.25014	.21843	.12851	.03073	.00040	.00000	3	15
.04166	.07798	.12678	.17925	.21862	.22520	.18760	.11564	.04284	.00485	.00001	.00000	4	15
.09164	.14036	.18594	.21234	.20613	.16515	.10318	.04490	.01047	.00056	.00000	.00000	5	15
.15274	.19140	.20660	.19056	.14724	.09175	.04299	.01320	.00194	.00005	.00000	.00000	6	15
.19638	.20134	.17708	.13193	.08113	.03932	.01382	.00300	.00028	.00000	.00000	.00000	7	15
.19638	.16474	.11806	.07104	.03477	.01311	.00345	.00053	.00003	.00000	.00000	.00000	8	15
.15274	.10483	.06121	.02975	.01159	.00340	.00067	.00007	.00000	.00000	.00000	.00000	9	15
.09164	.05146	.02449	.00961	.00298	.00068	.00010	.00001	.00000	.00000	.00000	.00000	10	15
.04166	.01914	.00742	.00235	.00058	.00010	.00001	.00000	.00000	.00000	.00000	.00000	11	15
.01389	.00522	.00165	.00042	.00008	.00001	.00000	.00000	.00000	.00000	.00000	.00000	12	15
.00320	.00099	.00025	.00005	.00001	.00000	.00000	.00000	.00000	.00000	.00000	.00000	13	15
.00046	.00012	.00002	.00000	.00000	.00000	.00000	.00000	.00000	.00000	.00000	.00000	14	15
.00003	.00001	.00000	.00000	.00000	.00000	.00000	.00000	.00000	.00000	.00000	.00000	15	15
.00002	.00007	.00028	.00102	.00332	.01002	.02815	.07425	.18530	.44013	.85146	.98412	0	16
.00024	.00092	.00301	.00875	.02279	.05345	.11259	.20965	.32943	.37063	.13761	.01576	1	16
.00183	.00563	.01505	.03533	.07325	.13363	.21111	.27748	.27452	.14630	.01042	.00012	2	16
.00854	.02151	.04681	.08877	.14650	.20788	.24629	.22851	.14234	.03593	.00049	.00000	3	16
.02777	.05718	.10142	.15535	.20405	.22520	.20011	.13106	.05140	.00615	.00002	.00000	4	16
.06665	.11229	.16227	.20076	.20988	.18016	.12007	.05551	.01371	.00078	.00000	.00000	5	16
.12219	.16843	.19833	.19818	.16490	.11010	.05503	.01796	.00279	.00007	.00000	.00000	6	16
.17456	.19687	.18889	.15245	.10096	.05243	.01965	.00453	.00044	.00000	.00000	.00000	7	16
.19638	.18121	.14167	.09235	.04868	.01966	.00553	.00090	.00006	.00000	.00000	.00000	8	16
.17456	.13179	.08395	.04420	.01854	.00583	.00123	.00014	.00001	.00000	.00000	.00000	9	16
.12219	.07548	.03918	.01666	.00556	.00136	.00021	.00002	.00000	.00000	.00000	.00000	10	16
.06665	.03368	.01425	.00489	.00130	.00025	.00003	.00000	.00000	.00000	.00000	.00000	11	16
.02777	.01148	.00396	.00110	.00023	.00003	.00000	.00000	.00000	.00000	.00000	.00000	12	16
.00854	.00289	.00081	.00018	.00003	.00000	.00000	.00000	.00000	.00000	.00000	.00000	13	16
.00183	.00051	.00012	.00002	.00000	.00000	.00000	.00000	.00000	.00000	.00000	.00000	14	16
.00024	.00006	.00001	.00000	.00000	.00000	.00000	.00000	.00000	.00000	.00000	.00000	15	16
.00002	.00000	.00000	.00000	.00000	.00000	.00000	.00000	.00000	.00000	.00000	.00000	16	16
.00001	.00004	.00017	.00066	.00233	.00752	.02252	.06311	.16677	.41812	.84294	.98314	0	17
.00006	.00054	.00192	.00604	.01695	.04260	.09570	.18934	.31501	.37411	.14475	.01673	1	17
.00104	.00351	.01023	.02602	.05811	.11359	.19140	.26730	.28001	.15752	.01170	.00013	2	17

ρ													
.5	.45	.4	.35	.3	.25	.2	.15	.1	.05	.01	.001	r	n
0.00636	0.01845	0.04642	0.10279	0.20191	0.35302	0.54888	0.75561	0.91736	0.99120	0.99998	1.00000	3	17
0.02452	0.05958	0.12600	0.23484	0.38869	0.57389	0.75822	0.90129	0.97786	0.99884	1.00000	1.00000	4	17
0.07173	0.14707	0.26393	0.41970	0.59682	0.76531	0.89430	0.96813	0.99533	0.99988	1.00000	1.00000	5	17
0.16615	0.29024	0.44784	0.61878	0.77522	0.89292	0.96234	0.99172	0.99922	0.99999	1.00000	1.00000	6	17
0.31453	0.47431	0.64051	0.78724	0.89536	0.95976	0.98907	0.99826	0.99989	1.00000	1.00000	1.00000	7	17
0.50000	0.66256	0.80106	0.90062	0.95972	0.98762	0.99742	0.99966	0.99999	1.00000	1.00000	1.00000	8	17
0.68547	0.81659	0.90810	0.96167	0.98731	0.99690	0.99951	0.99996	1.00000	1.00000	1.00000	1.00000	9	17
0.83385	0.91741	0.96519	0.98797	0.99676	0.99937	0.99992	1.00000	1.00000	1.00000	1.00000	1.00000	10	17
0.92827	0.96990	0.98941	0.99699	0.99934	0.99990	0.99999	1.00000	1.00000	1.00000	1.00000	1.00000	11	17
0.97548	0.99138	0.99748	0.99941	0.99990	0.99999	1.00000	1.00000	1.00000	1.00000	1.00000	1.00000	12	17
0.99364	0.99771	0.99955	0.99991	0.99999	1.00000	1.00000	1.00000	1.00000	1.00000	1.00000	1.00000	13	17
0.99883	0.99955	0.99994	0.99999	1.00000	1.00000	1.00000	1.00000	1.00000	1.00000	1.00000	1.00000	14	17
0.99986	0.99997	1.00000	1.00000	1.00000	1.00000	1.00000	1.00000	1.00000	1.00000	1.00000	1.00000	15	17
0.99999	1.00000	1.00000	1.00000	1.00000	1.00000	1.00000	1.00000	1.00000	1.00000	1.00000	1.00000	16	17
1.00000	1.00000	1.00000	1.00000	1.00000	1.00000	1.00000	1.00000	1.00000	1.00000	1.00000	1.00000	17	17
0.00000	0.00002	0.00010	0.00043	0.00163	0.00564	0.01801	0.05365	0.15009	0.39721	0.83451	0.98215	0	18
0.00007	0.00033	0.00132	0.00459	0.01419	0.03946	0.09908	0.22405	0.45328	0.77352	0.98624	0.99985	1	18
0.00066	0.00251	0.00823	0.02362	0.05995	0.13531	0.27134	0.47966	0.73380	0.94187	0.99927	1.00000	2	18
0.00377	0.01198	0.03278	0.07827	0.16455	0.30559	0.50103	0.72024	0.90180	0.98913	0.99997	1.00000	3	18
0.01544	0.04107	0.09417	0.18862	0.33265	0.51867	0.71635	0.87944	0.97181	0.99846	1.00000	1.00000	4	18
0.04813	0.10770	0.20876	0.35500	0.53438	0.71745	0.86708	0.95810	0.99358	0.99983	1.00000	1.00000	5	18
0.11894	0.22581	0.37428	0.54318	0.72170	0.86102	0.94873	0.98818	0.99883	0.99998	1.00000	1.00000	6	18
0.24034	0.39148	0.56344	0.72828	0.85932	0.94305	0.98372	0.99728	0.99983	1.00000	1.00000	1.00000	7	18
0.40726	0.57785	0.73684	0.86094	0.94041	0.98065	0.99575	0.99949	0.99998	1.00000	1.00000	1.00000	8	18
0.59274	0.74728	0.86529	0.94031	0.97903	0.99458	0.99909	0.99992	1.00000	1.00000	1.00000	1.00000	9	18
0.75966	0.87204	0.94235	0.97877	0.99393	0.99876	0.99984	0.99999	1.00000	1.00000	1.00000	1.00000	10	18
0.88106	0.94628	0.97972	0.99383	0.99857	0.99977	0.99998	1.00000	1.00000	1.00000	1.00000	1.00000	11	18
0.95187	0.98171	0.99425	0.99856	0.99973	0.99996	1.00000	1.00000	1.00000	1.00000	1.00000	1.00000	12	18
0.98456	0.99509	0.99872	0.99974	0.99996	0.99999	1.00000	1.00000	1.00000	1.00000	1.00000	1.00000	13	18
0.99623	0.99899	0.99979	0.99996	0.99999	1.00000	1.00000	1.00000	1.00000	1.00000	1.00000	1.00000	14	18
0.99934	0.99986	0.99997	0.99999	1.00000	1.00000	1.00000	1.00000	1.00000	1.00000	1.00000	1.00000	15	18
0.99993	0.99999	1.00000	1.00000	1.00000	1.00000	1.00000	1.00000	1.00000	1.00000	1.00000	1.00000	16	18
0.99999	1.00000	1.00000	1.00000	1.00000	1.00000	1.00000	1.00000	1.00000	1.00000	1.00000	1.00000	17	18
1.00000	1.00000	1.00000	1.00000	1.00000	1.00000	1.00000	1.00000	1.00000	1.00000	1.00000	1.00000	18	18
0.00000	0.00001	0.00006	0.00028	0.00114	0.00423	0.01441	0.04560	0.13509	0.37735	0.82617	0.98117	0	19
0.00004	0.00019	0.00083	0.00313	0.01042	0.03101	0.08287	0.19849	0.42026	0.75471	0.98473	0.99983	1	19
0.00036	0.00153	0.00546	0.01696	0.04622	0.11134	0.23689	0.44132	0.70544	0.93345	0.99917	1.00000	2	19
0.00221	0.00772	0.02296	0.05914	0.13317	0.26309	0.45509	0.68415	0.88503	0.98676	0.99997	1.00000	3	19
0.00961	0.02798	0.06961	0.15000	0.28222	0.46542	0.67329	0.85556	0.96481	0.99799	1.00000	1.00000	4	19
0.03178	0.07771	0.16292	0.29676	0.47386	0.66776	0.83394	0.94630	0.99141	0.99976	1.00000	1.00000	5	19
0.08353	0.17266	0.30807	0.48117	0.66550	0.82512	0.93240	0.98367	0.99830	0.99998	1.00000	1.00000	6	19
0.17964	0.31693	0.48778	0.66557	0.81803	0.92254	0.97672	0.99592	0.99973	1.00000	1.00000	1.00000	7	19
0.32380	0.49398	0.66748	0.81451	0.91608	0.97125	0.99334	0.99916	0.99996	1.00000	1.00000	1.00000	8	19
0.50000	0.67104	0.81451	0.91253	0.96745	0.99110	0.99842	0.99986	1.00000	1.00000	1.00000	1.00000	9	19
0.67620	0.81590	0.91153	0.96531	0.98946	0.99771	0.99969	0.99998	1.00000	1.00000	1.00000	1.00000	10	19
0.82036	0.91287	0.96531	0.98716	0.99718	0.99946	0.99995	1.00000	1.00000	1.00000	1.00000	1.00000	11	19
0.91647	0.96577	0.98856	0.99591	0.99938	0.99989	0.99999	1.00000	1.00000	1.00000	1.00000	1.00000	12	19
0.96822	0.98907	0.99691	0.99889	0.99987	0.99998	1.00000	1.00000	1.00000	1.00000	1.00000	1.00000	13	19
0.99039	0.99724	0.99933	0.99976	0.99998	1.00000	1.00000	1.00000	1.00000	1.00000	1.00000	1.00000	14	19
0.99779	0.99947	0.99989	0.99996	1.00000	1.00000	1.00000	1.00000	1.00000	1.00000	1.00000	1.00000	15	19
0.99964	0.99993	0.99999	0.99999	1.00000	1.00000	1.00000	1.00000	1.00000	1.00000	1.00000	1.00000	16	19

Cumulative binomial distribution — the column group heading is ρ (probability of success).

n	r	ρ = .001	.01	.05	.1	.15	.2	.25	.3	.35	.4	.45	.5
19	17	1.00000	1.00000	1.00000	1.00000	1.00000	1.00000	1.00000	1.00000	1.00000	1.00000	0.99999	0.99996
19	18	1.00000	1.00000	1.00000	1.00000	1.00000	1.00000	1.00000	1.00000	1.00000	1.00000	1.00000	1.00000
19	19	1.00000	1.00000	1.00000	1.00000	1.00000	1.00000	1.00000	1.00000	1.00000	1.00000	1.00000	1.00000
20	0	0.98019	0.81791	0.35849	0.12158	0.03876	0.01153	0.00317	0.00080	0.00018	0.00004	0.00001	0.00000
20	1	0.99981	0.98314	0.73584	0.39175	0.17556	0.06918	0.02431	0.00764	0.00213	0.00052	0.00011	0.00002
20	2	1.00000	0.99900	0.92452	0.67693	0.40490	0.20608	0.09126	0.03548	0.01212	0.00361	0.00093	0.00020
20	3	1.00000	0.99996	0.98410	0.86705	0.64773	0.41145	0.22516	0.10709	0.04438	0.01596	0.00493	0.00129
20	4	1.00000	1.00000	0.99743	0.95683	0.82985	0.62965	0.41484	0.23751	0.11820	0.05095	0.01886	0.00591
20	5	1.00000	1.00000	0.99967	0.98875	0.93269	0.80421	0.61717	0.41637	0.24540	0.12560	0.05533	0.02069
20	6	1.00000	1.00000	0.99997	0.99761	0.97806	0.91331	0.78578	0.60801	0.41663	0.25001	0.12993	0.05766
20	7	1.00000	1.00000	1.00000	0.99958	0.99408	0.96786	0.89819	0.77227	0.60103	0.41589	0.25201	0.13159
20	8	1.00000	1.00000	1.00000	0.99994	0.99867	0.99002	0.95907	0.88667	0.76238	0.59560	0.41431	0.25172
20	9	1.00000	1.00000	1.00000	0.99999	0.99975	0.99741	0.98614	0.95204	0.87822	0.75534	0.59136	0.41190
20	10	1.00000	1.00000	1.00000	1.00000	0.99996	0.99944	0.99606	0.98286	0.94683	0.87248	0.75071	0.58810
20	11	1.00000	1.00000	1.00000	1.00000	1.00000	0.99990	0.99906	0.99486	0.98042	0.94347	0.86924	0.74828
20	12	1.00000	1.00000	1.00000	1.00000	1.00000	0.99999	0.99982	0.99872	0.99398	0.97897	0.94197	0.86841
20	13	1.00000	1.00000	1.00000	1.00000	1.00000	1.00000	0.99997	0.99969	0.99848	0.99353	0.97859	0.94234
20	14	1.00000	1.00000	1.00000	1.00000	1.00000	1.00000	1.00000	0.99994	0.99969	0.99839	0.99357	0.97931
20	15	1.00000	1.00000	1.00000	1.00000	1.00000	1.00000	1.00000	0.99999	0.99995	0.99968	0.99847	0.99409
20	16	1.00000	1.00000	1.00000	1.00000	1.00000	1.00000	1.00000	1.00000	0.99999	0.99995	0.99972	0.99871
20	17	1.00000	1.00000	1.00000	1.00000	1.00000	1.00000	1.00000	1.00000	1.00000	0.99999	0.99996	0.99980
20	18	1.00000	1.00000	1.00000	1.00000	1.00000	1.00000	1.00000	1.00000	1.00000	1.00000	1.00000	0.99998
20	19	1.00000	1.00000	1.00000	1.00000	1.00000	1.00000	1.00000	1.00000	1.00000	1.00000	1.00000	1.00000
20	20	1.00000	1.00000	1.00000	1.00000	1.00000	1.00000	1.00000	1.00000	1.00000	1.00000	1.00000	1.00000
21	0	0.97921	0.80973	0.34056	0.10942	0.03295	0.00922	0.00238	0.00056	0.00012	0.00002	0.00000	0.00000
21	1	0.99979	0.98149	0.71697	0.36473	0.15504	0.05765	0.01903	0.00559	0.00145	0.00033	0.00006	0.00001
21	2	1.00000	0.99884	0.91508	0.64841	0.37050	0.17870	0.07452	0.02713	0.00862	0.00236	0.00056	0.00011
21	3	1.00000	0.99995	0.98112	0.84803	0.61130	0.37038	0.19168	0.08561	0.03309	0.01102	0.00313	0.00074
21	4	1.00000	1.00000	0.99676	0.94785	0.80255	0.58601	0.36743	0.19838	0.09236	0.03696	0.01259	0.00360
21	5	1.00000	1.00000	0.99956	0.98555	0.91729	0.76931	0.56666	0.36271	0.20088	0.09574	0.03892	0.01330
21	6	1.00000	1.00000	0.99995	0.99673	0.97129	0.89150	0.74371	0.55052	0.35670	0.20025	0.09636	0.03918
21	7	1.00000	1.00000	1.00000	0.99939	0.99171	0.95696	0.87016	0.72299	0.53649	0.34954	0.19707	0.09462
21	8	1.00000	1.00000	1.00000	0.99990	0.99801	0.98560	0.94394	0.85235	0.70590	0.52372	0.34127	0.19166
21	9	1.00000	1.00000	1.00000	0.99999	0.99962	0.99594	0.97945	0.93243	0.83767	0.69144	0.51169	0.33181
21	10	1.00000	1.00000	1.00000	1.00000	0.99996	0.99904	0.99366	0.97361	0.92282	0.82562	0.67900	0.50000
21	11	1.00000	1.00000	1.00000	1.00000	0.99999	0.99982	0.99840	0.99126	0.96867	0.91508	0.81508	0.66819
21	12	1.00000	1.00000	1.00000	1.00000	1.00000	0.99998	0.99970	0.99756	0.98924	0.96477	0.90924	0.80834
21	13	1.00000	1.00000	1.00000	1.00000	1.00000	1.00000	0.99996	0.99943	0.99691	0.98771	0.96211	0.90538
21	14	1.00000	1.00000	1.00000	1.00000	1.00000	1.00000	0.99999	0.99989	0.99927	0.99645	0.98683	0.96082
21	15	1.00000	1.00000	1.00000	1.00000	1.00000	1.00000	1.00000	0.99998	0.99986	0.99917	0.99626	0.98670
21	16	1.00000	1.00000	1.00000	1.00000	1.00000	1.00000	1.00000	1.00000	0.99998	0.99984	0.99916	0.99640
21	17	1.00000	1.00000	1.00000	1.00000	1.00000	1.00000	1.00000	1.00000	1.00000	0.99998	0.99986	0.99926
21	18	1.00000	1.00000	1.00000	1.00000	1.00000	1.00000	1.00000	1.00000	1.00000	1.00000	0.99998	0.99989
21	19	1.00000	1.00000	1.00000	1.00000	1.00000	1.00000	1.00000	1.00000	1.00000	1.00000	1.00000	0.99999
21	20	1.00000	1.00000	1.00000	1.00000	1.00000	1.00000	1.00000	1.00000	1.00000	1.00000	1.00000	1.00000
21	21	1.00000	1.00000	1.00000	1.00000	1.00000	1.00000	1.00000	1.00000	1.00000	1.00000	1.00000	1.00000
22	0	0.97823	0.80163	0.32353	0.09848	0.02800	0.00738	0.00178	0.00039	0.00008	0.00001	0.00000	0.00000
22	1	0.99977	0.97977	0.69815	0.33920	0.13672	0.04796	0.01487	0.00408	0.00098	0.00021	0.00004	0.00001
22	2	1.00000	0.99866	0.90518	0.62004	0.33818	0.15449	0.06065	0.02067	0.00611	0.00156	0.00034	0.00006
22	3	1.00000	0.99994	0.97782	0.82807	0.57518	0.33204	0.16239	0.06806	0.02452	0.00756	0.00197	0.00043
22	4	1.00000	1.00000	0.99598	0.93787	0.77384	0.54288	0.32349	0.16455	0.07161	0.02658	0.00834	0.00217

Table of ρ

.5	.45	.4	.35	.3	.25	.2	.15	.1	.05	.01	.001	r	n
0.00845	0.02707	0.07223	0.16290	0.31341	0.51680	0.73264	0.90005	0.98178	0.99942	1.00000	1.00000	5	22
0.02624	0.07052	0.15844	0.30216	0.49418	0.69937	0.86705	0.96316	0.99561	0.99993	1.00000	1.00000	6	22
0.06690	0.15175	0.28982	0.47356	0.67125	0.83847	0.94386	0.98861	0.99912	0.99999	1.00000	1.00000	7	22
0.14314	0.27638	0.45405	0.64661	0.81354	0.92541	0.97986	0.99704	0.99984	1.00000	1.00000	1.00000	8	22
0.26173	0.43500	0.62435	0.79156	0.90840	0.97049	0.99386	0.99935	0.99998	1.00000	1.00000	1.00000	9	22
0.41591	0.60371	0.77195	0.89302	0.96126	0.99003	0.99841	0.99988	1.00000	1.00000	1.00000	1.00000	10	22
0.58409	0.75430	0.87929	0.95262	0.98596	0.99713	0.99965	0.99998	1.00000	1.00000	1.00000	1.00000	11	22
0.73827	0.86723	0.94489	0.98204	0.99567	0.99930	0.99994	1.00000	1.00000	1.00000	1.00000	1.00000	12	22
0.85686	0.93832	0.97853	0.99422	0.99887	0.99986	0.99999	1.00000	1.00000	1.00000	1.00000	1.00000	13	22
0.93310	0.97570	0.99295	0.99844	0.99975	0.99998	1.00000	1.00000	1.00000	1.00000	1.00000	1.00000	14	22
0.97376	0.99202	0.99808	0.99965	0.99996	1.00000	1.00000	1.00000	1.00000	1.00000	1.00000	1.00000	15	22
0.99155	0.99786	0.99957	0.99999	1.00000	1.00000	1.00000	1.00000	1.00000	1.00000	1.00000	1.00000	16	22
0.99783	0.99954	0.99992	1.00000	1.00000	1.00000	1.00000	1.00000	1.00000	1.00000	1.00000	1.00000	17	22
0.99957	0.99993	0.99999	1.00000	1.00000	1.00000	1.00000	1.00000	1.00000	1.00000	1.00000	1.00000	18	22
0.99994	0.99999	1.00000	1.00000	1.00000	1.00000	1.00000	1.00000	1.00000	1.00000	1.00000	1.00000	19	22
0.99999	1.00000	1.00000	1.00000	1.00000	1.00000	1.00000	1.00000	1.00000	1.00000	1.00000	1.00000	20	22
1.00000	1.00000	1.00000	1.00000	1.00000	1.00000	1.00000	1.00000	1.00000	1.00000	1.00000	1.00000	21	22
1.00000	1.00000	1.00000	1.00000	1.00000	1.00000	1.00000	1.00000	1.00000	1.00000	1.00000	1.00000	22	22
0.00000	0.00000	0.00001	0.00005	0.00027	0.00134	0.00590	0.02380	0.08863	0.30736	0.79361	0.97361	1	23
0.00003	0.00002	0.00013	0.00067	0.00297	0.01159	0.03984	0.12042	0.31513	0.67942	0.97799	0.99975	2	23
0.00024	0.00020	0.00102	0.00432	0.01569	0.04920	0.13319	0.30796	0.59196	0.89483	0.99992	1.00000	3	23
0.00130	0.00124	0.00516	0.01808	0.05384	0.13696	0.29653	0.53963	0.80727	0.97419	1.00000	1.00000	4	23
0.00531	0.00547	0.01897	0.05513	0.13560	0.28321	0.50171	0.74404	0.92689	0.99527	1.00000	1.00000	5	23
0.01734	0.01864	0.05397	0.13095	0.26875	0.46847	0.69469	0.88112	0.97739	0.99925	1.00000	1.00000	6	23
0.04657	0.05097	0.12396	0.25342	0.43995	0.65373	0.84017	0.95369	0.99423	0.99991	1.00000	1.00000	7	23
0.10502	0.11520	0.23342	0.41357	0.61813	0.80370	0.92849	0.98480	0.99878	0.99999	1.00000	1.00000	8	23
0.20244	0.22030	0.38836	0.58604	0.77086	0.90368	0.97266	0.99577	0.99978	1.00000	1.00000	1.00000	9	23
0.33882	0.36362	0.55623	0.74082	0.87999	0.95922	0.99106	0.99888	0.99997	1.00000	1.00000	1.00000	10	23
0.50000	0.52779	0.71291	0.85751	0.94540	0.98514	0.99750	0.99980	1.00000	1.00000	1.00000	1.00000	11	23
0.66118	0.68653	0.83636	0.93176	0.97855	0.99535	0.99940	0.99997	1.00000	1.00000	1.00000	1.00000	12	23
0.79756	0.81641	0.91865	0.97174	0.99276	0.99876	0.99988	1.00000	1.00000	1.00000	1.00000	1.00000	13	23
0.89498	0.90633	0.96508	0.98996	0.99791	0.99972	0.99998	1.00000	1.00000	1.00000	1.00000	1.00000	14	23
0.95343	0.95888	0.98718	0.99696	0.99990	0.99995	1.00000	1.00000	1.00000	1.00000	1.00000	1.00000	15	23
0.98266	0.98467	0.99603	0.99923	0.99998	0.99999	1.00000	1.00000	1.00000	1.00000	1.00000	1.00000	16	23
0.99469	0.99523	0.99897	0.99984	1.00000	1.00000	1.00000	1.00000	1.00000	1.00000	1.00000	1.00000	17	23
0.99870	0.99878	0.99978	0.99997	1.00000	1.00000	1.00000	1.00000	1.00000	1.00000	1.00000	1.00000	18	23
0.99976	0.99978	0.99996	1.00000	1.00000	1.00000	1.00000	1.00000	1.00000	1.00000	1.00000	1.00000	19	23
0.99997	0.99996	1.00000	1.00000	1.00000	1.00000	1.00000	1.00000	1.00000	1.00000	1.00000	1.00000	20	23
1.00000	1.00000	1.00000	1.00000	1.00000	1.00000	1.00000	1.00000	1.00000	1.00000	1.00000	1.00000	21	23
1.00000	1.00000	1.00000	1.00000	1.00000	1.00000	1.00000	1.00000	1.00000	1.00000	1.00000	1.00000	22	23
1.00000	1.00000	1.00000	1.00000	1.00000	1.00000	1.00000	1.00000	1.00000	1.00000	1.00000	1.00000	23	23
0.00000	0.00000	0.00000	0.00003	0.00019	0.00100	0.00472	0.02023	0.07977	0.29199	0.78568	0.97627	0	24
0.00000	0.00001	0.00008	0.00045	0.00216	0.00910	0.03306	0.10592	0.29248	0.66082	0.97827	0.99973	1	24
0.00002	0.00012	0.00066	0.00304	0.01187	0.03980	0.11452	0.27983	0.56427	0.88406	0.99827	1.00000	2	24
0.00014	0.00077	0.00350	0.01326	0.04240	0.11502	0.26386	0.54627	0.78574	0.97022	0.99991	1.00000	3	24
0.00077	0.00357	0.01345	0.04216	0.11108	0.24665	0.45988	0.71338	0.91493	0.99403	1.00000	1.00000	4	24
0.00331	0.01272	0.03997	0.10441	0.22881	0.42216	0.65589	0.86281	0.97234	0.99904	1.00000	1.00000	5	24
0.01133	0.03642	0.09596	0.21055	0.38859	0.60741	0.81107	0.94281	0.99254	0.99987	1.00000	1.00000	6	24
0.03196	0.08629	0.19195	0.35752	0.54467	0.76620	0.91383	0.98013	0.99832	0.99999	1.00000	1.00000	7	24
0.07579	0.17300	0.32792	0.52568	0.72504	0.87868	0.96383	0.99413	0.99968	1.00000	1.00000	1.00000	8	24

Table of individual binomial probabilities $b(r; n, \rho)$. The parameter ρ heads the columns.

ρ=.5	.45	.4	.35	.3	.25	.2	.15	.1	.05	.01	.001	r	n
.07793	.12612	.16116	.16097	.12218	.06665	.02355	.00439	.00027	.00000	.00000	.00000	9	24
.11690	.15479	.16116	.13002	.07855	.03333	.00883	.00116	.00004	.00000	.00000	.00000	10	24
.14878	.16118	.13674	.08910	.04284	.01414	.00281	.00026	.00001	.00000	.00000	.00000	11	24
.16118	.14287	.09876	.05198	.01989	.00511	.00076	.00005	.00000	.00000	.00000	.00000	12	24
.14878	.10790	.06077	.02583	.00787	.00157	.00018	.00001	.00000	.00000	.00000	.00000	13	24
.11690	.06936	.03183	.01093	.00265	.00041	.00003	.00000	.00000	.00000	.00000	.00000	14	24
.07793	.03784	.01415	.00392	.00076	.00009	.00001	.00000	.00000	.00000	.00000	.00000	15	24
.04384	.01741	.00531	.00119	.00018	.00002	.00000	.00000	.00000	.00000	.00000	.00000	16	24
.02063	.00670	.00166	.00030	.00004	.00000	.00000	.00000	.00000	.00000	.00000	.00000	17	24
.00802	.00213	.00043	.00006	.00001	.00000	.00000	.00000	.00000	.00000	.00000	.00000	18	24
.00253	.00055	.00009	.00001	.00000	.00000	.00000	.00000	.00000	.00000	.00000	.00000	19	24
.00063	.00011	.00002	.00000	.00000	.00000	.00000	.00000	.00000	.00000	.00000	.00000	20	24
.00012	.00002	.00000	.00000	.00000	.00000	.00000	.00000	.00000	.00000	.00000	.00000	21	24
.00002	.00000	.00000	.00000	.00000	.00000	.00000	.00000	.00000	.00000	.00000	.00000	22	24
.00000	.00000	.00000	.00000	.00000	.00000	.00000	.00000	.00000	.00000	.00000	.00000	23	24
.00000	.00000	.00000	.00000	.00000	.00000	.00000	.00000	.00000	.00000	.00000	.00000	24	24
.00000	.00000	.00000	.00002	.00013	.00075	.00378	.01720	.07179	.27739	.77782	.97530	0	25
.00000	.00001	.00005	.00028	.00144	.00627	.02361	.07587	.19942	.36499	.19642	.02441	1	25
.00001	.00006	.00038	.00183	.00739	.02508	.07084	.16067	.26589	.23052	.02381	.00029	2	25
.00007	.00041	.00194	.00755	.02428	.06411	.13577	.21738	.22650	.09302	.00184	.00000	3	25
.00038	.00183	.00710	.02236	.05723	.11753	.18668	.21099	.13842	.02693	.00010	.00000	4	25
.00158	.00629	.01989	.05058	.10302	.16454	.19602	.15638	.06459	.00595	.00000	.00000	5	25
.00528	.01715	.04420	.09078	.14717	.18282	.16335	.09199	.02392	.00104	.00000	.00000	6	25
.01433	.03810	.07999	.13268	.17119	.16541	.11084	.04406	.00722	.00015	.00000	.00000	7	25
.03223	.07013	.11998	.16074	.16508	.12406	.06235	.01749	.00180	.00002	.00000	.00000	8	25
.06089	.10839	.15109	.16349	.13364	.07811	.02944	.00583	.00038	.00000	.00000	.00000	9	25
.09742	.14189	.16116	.14085	.09164	.04166	.01178	.00165	.00007	.00000	.00000	.00000	10	25
.13284	.15831	.14651	.10342	.05355	.01894	.00401	.00040	.00001	.00000	.00000	.00000	11	25
.15498	.15111	.11395	.06497	.02678	.00736	.00117	.00008	.00000	.00000	.00000	.00000	12	25
.15498	.12364	.07597	.03498	.01148	.00245	.00029	.00001	.00000	.00000	.00000	.00000	13	25
.13284	.08671	.04341	.01615	.00422	.00070	.00006	.00000	.00000	.00000	.00000	.00000	14	25
.09742	.05202	.02122	.00638	.00132	.00017	.00001	.00000	.00000	.00000	.00000	.00000	15	25
.06089	.02660	.00884	.00215	.00035	.00004	.00000	.00000	.00000	.00000	.00000	.00000	16	25
.03223	.01152	.00312	.00061	.00008	.00001	.00000	.00000	.00000	.00000	.00000	.00000	17	25
.01433	.00419	.00092	.00015	.00002	.00000	.00000	.00000	.00000	.00000	.00000	.00000	18	25
.00528	.00126	.00023	.00003	.00000	.00000	.00000	.00000	.00000	.00000	.00000	.00000	19	25
.00158	.00031	.00005	.00000	.00000	.00000	.00000	.00000	.00000	.00000	.00000	.00000	20	25
.00038	.00006	.00001	.00000	.00000	.00000	.00000	.00000	.00000	.00000	.00000	.00000	21	25
.00007	.00001	.00000	.00000	.00000	.00000	.00000	.00000	.00000	.00000	.00000	.00000	22	25
.00001	.00000	.00000	.00000	.00000	.00000	.00000	.00000	.00000	.00000	.00000	.00000	23	25
.00000	.00000	.00000	.00000	.00000	.00000	.00000	.00000	.00000	.00000	.00000	.00000	24	25
.00000	.00000	.00000	.00000	.00000	.00000	.00000	.00000	.00000	.00000	.00000	.00000	25	25

Table B-4

Table of Point and Cumulative Poisson Probabilities*

$$P(x) = \frac{m^x}{x!e^m} \qquad C(x) = \sum_{i=0}^{x} \frac{m^i}{i!e^m} \qquad D(x) = \sum_{i=x}^{\infty} \frac{m^i}{i!e^m}$$

x	P(x)	C(x)	D(x)
		m = 0.0001	
0	.99990001	.99990001	1.00000000
1	.00009999	.99999999	.00009999
		m = 0.0005	
0	.99950013	.99950013	1.00000000
1	.00049975	.99999987	.00049987
2	.00000012	.99999999	.00000012
		m = 0.001	
0	.99900050	.99900050	1.00000000
1	.00099900	.99999950	.00099950
2	.00000050	.99999999	.00000050
		m = 0.005	
0	.99501248	.99501248	1.00000000
1	.00497506	.99998754	.00498752
2	.00001244	.99999998	.00001246
3	.00000002	.99999999	.00000002
		m = 0.01	
0	.99004983	.99004983	1.00000000
1	.00990050	.99995033	.00995017
2	.00004950	.99999983	.00004967
3	.00000016	.99999999	.00000017
		m = 0.02	
0	.98019867	.98019867	1.00000000
1	.01960397	.99980264	.01980133
2	.00019604	.99999868	.00019735
3	.00000131	.99999999	.00000131
		m = 0.03	
0	.97044554	.97044554	1.00000000
1	.02911337	.99955890	.02955447
2	.00043670	.99999560	.00044110
3	.00000437	.99999996	.00000440
4	.00000003	.99999999	.00000003
		m = 0.04	
0	.96078944	.96078944	1.00000000
1	.03843158	.99922102	.03921056
2	.00076863	.99998964	.00077898
3	.00001025	.99999990	.00001035
4	.00000010	.99999999	.00000010

x	P(x)	C(x)	D(x)
		m = 0.05	
0	.95122942	.95122942	1.00000000
1	.04756147	.99879089	.04877058
2	.00118904	.99997993	.00120910
3	.00001982	.99999975	.00002007
4	.00000025	.99999999	.00000025
		m = 0.06	
0	.94176454	.94176454	1.00000000
1	.05650587	.99827041	.05823547
2	.00169518	.99996558	.00172959
3	.00003390	.99999948	.00003442
4	.00000051	.99999999	.00000051
		m = 0.07	
0	.93239383	.93239383	1.00000000
1	.06526757	.99766139	.06760618
2	.00228436	.99994574	.00233861
3	.00005330	.99999905	.00005425
4	.00000093	.99999999	.00000095
		m = 0.08	
0	.92311635	.92311635	1.00000000
1	.07384931	.99696565	.07688365
2	.00295397	.99991962	.00303435
3	.00007877	.99999840	.00008037
4	.00000158	.99999997	.00000160
5	.00000003	.99999999	.00000002
		m = 0.09	
0	.91393118	.91393118	1.00000000
1	.08225381	.99618499	.08606881
2	.00370142	.99988641	.00381501
3	.00011104	.99999745	.00011359
4	.00000250	.99999995	.00000254
5	.00000004	.99999999	.00000005
		m = 0.10	
0	.90483742	.90483742	1.00000000
1	.09048374	.99532116	.09516258
2	.00452419	.99984535	.00467884
3	.00015081	.99999615	.00015465
4	.00000377	.99999992	.00000385
5	.00000008	.99999999	.00000008
		m = 0.20	
0	.81873076	.81873076	1.00000000
1	.16374615	.98247690	.18126925
2	.01637461	.99885152	.01752310
3	.00109164	.99994316	.00114848
4	.00005458	.99999774	.00005684
5	.00000218	.99999992	.00000226
6	.00000007	.99999999	.00000007

Table entries are reproduced from the far more extensive *Tables of the Individual and Cumulative Terms of Poisson Distribution*, authored by the Defense Systems Department of the General Electric Company and published by the Van Nostrand Reinhold Company, 1962, with the permission of author and publisher.

x	P(x)	C(x)	D(x)	x	P(x)	C(x)	D(x)
		m = .30				m = .80	
0	.74081822	.74081822	1.00000000	0	.44932897	.44932897	1.00000000
1	.22224547	.96306369	.25918178	1	.35946317	.80879214	.55067103
2	.03333682	.99640051	.03693631	2	.14378527	.95257740	.19120786
3	.00333368	.99973419	.00359949	3	.03834274	.99092014	.04742259
4	.00025003	.99998421	.00026581	4	.00766855	.99858869	.00907985
5	.00001500	.99999922	.00001578	5	.00122697	.99981566	.00141131
6	.00000075	.99999996	.00000078	6	.00016360	.99997925	.00018434
7	.00000003	.99999999	.00000003	7	.00001870	.99999795	.00002074
				8	.00000187	.99999982	.00000205
		m = .40		9	.00000017	.99999999	.00000018
0	.67032005	.67032005	1.00000000			m = .90	
1	.26812802	.93844806	.32967995				
2	.05362560	.99207336	.06155194	0	.40656966	.40656966	1.00000000
3	.00715008	.99922375	.00792633	1	.36591269	.77248235	.59343034
4	.00071501	.99993876	.00077625	2	.16466071	.93714307	.22751765
5	.00005720	.99999595	.00006124	3	.04939821	.98654128	.06285693
6	.00000381	.99999977	.00000404	4	.01111460	.99765588	.01345872
7	.00000022	.99999999	.00000023	5	.00200063	.99965651	.00234412
				6	.00030009	.99995660	.00034349
		m = .50		7	.00003858	.99999518	.00004340
				8	.00000434	.99999952	.00000482
0	.60653067	.60653067	1.00000000	9	.00000043	.99999996	.00000048
1	.30326533	.90979599	.39346934	10	.00000004	.99999999	.00000004
2	.07581633	.98561233	.09020401				
3	.01263606	.99824838	.01438768			m = 1.00	
4	.00157951	.99982788	.00175162				
5	.00015795	.99998583	.00017212	0	.36787944	.36787944	1.00000000
6	.00001316	.99999899	.00001416	1	.36787944	.73575888	.63212056
7	.00000094	.99999993	.00000100	2	.18393972	.91969860	.26424112
8	.00000006	.99999999	.00000006	3	.06131324	.98101184	.08030140
				4	.01532831	.99634015	.01898816
		m = .60		5	.00306566	.99940581	.00365985
				6	.00051094	.99991675	.00059418
0	.54881164	.54881164	1.00000000	7	.00007299	.99998975	.00008324
1	.32928698	.87809862	.45118836	8	.00000912	.99999887	.00001025
2	.09878609	.97688472	.12190138	9	.00000101	.99999989	.00000113
3	.01975722	.99664193	.02311528	10	.00000010	.99999999	.00000011
4	.00296358	.99960551	.00335806				
5	.00035563	.99996115	.00039448			m = 1.10	
6	.00003556	.99999671	.00003885				
7	.00000305	.99999975	.00000329	0	.33287108	.33287108	1.00000000
8	.00000023	.99999999	.00000024	1	.36615819	.69902927	.66712892
				2	.20138700	.90041628	.30097073
		m = .70		3	.07384190	.97425818	.09958372
				4	.02030652	.99456470	.02574182
0	.49658531	.49658531	1.00000000	5	.00446743	.99903214	.00543529
1	.34760971	.84419502	.50341470	6	.00081903	.99985117	.00096786
2	.12166340	.96585841	.15580498	7	.00012870	.99997988	.00014883
3	.02838813	.99424654	.03414158	8	.00001770	.99999757	.00002012
4	.00496792	.99921446	.00575345	9	.00000216	.99999973	.00000243
5	.00069551	.99990997	.00078553	10	.00000024	.99999997	.00000026
6	.00008114	.99999112	.00009002	11	.00000002	.99999999	.00000003
7	.00000811	.99999923	.00000888				
8	.00000071	.99999994	.00000076				
9	.00000006	.99999999	.00000005				

x	P(x)	C(x)	D(x)	x	P(x)	C(x)	D(x)
	m	=	1.20		m	=	1.60
0	.30119421	.30119421	1.00000000	0	.20189652	.20189652	1.00000000
1	.36143306	.66262727	.69880579	1	.32303443	.52493095	.79810348
2	.21685983	.87948710	.33737273	2	.25842754	.78335849	.47506905
3	.08674393	.96623104	.12051289	3	.13782802	.92118651	.21664151
4	.02602318	.99225422	.03376896	4	.05513121	.97631773	.07881348
5	.00624556	.99849978	.00774578	5	.01764199	.99395971	.02368227
6	.00124911	.99974889	.00150021	6	.00470453	.99866424	.00604029
7	.00021413	.99996303	.00025110	7	.00107532	.99973956	.00133576
8	.00003212	.99999515	.00003697	8	.00021506	.99995463	.00026043
9	.00000428	.99999943	.00000485	9	.00003823	.99999286	.00004537
10	.00000051	.99999995	.00000056	10	.00000612	.99999898	.00000714
11	.00000006	.99999999	.00000005	11	.00000089	.99999987	.00000102
				12	.00000012	.99999999	.00000013
	m	=	1.30				
					m	=	1.70
0	.27253179	.27253179	1.00000000				
1	.35429133	.62682312	.72746821	0	.18268352	.18268352	1.00000000
2	.23028936	.85711249	.37317688	1	.31056199	.49324552	.81731648
3	.09979206	.95690455	.14288751	2	.26397769	.75722320	.50675449
4	.03243242	.98933697	.04309545	3	.14958736	.90681057	.24277679
5	.00843243	.99776939	.01066303	4	.06357463	.97038519	.09318943
6	.00182703	.99959642	.00223060	5	.02161537	.99200056	.02961480
7	.00033930	.99993572	.00040358	6	.00612436	.99812492	.00799943
8	.00005514	.99999087	.00006427	7	.00148734	.99961226	.00187507
9	.00000796	.99999883	.00000913	8	.00031606	.99992833	.00038773
10	.00000104	.99999987	.00000117	9	.00005970	.99998803	.00007167
11	.00000012	.99999999	.00000013	10	.00001015	.99999817	.00001197
				11	.00000157	.99999975	.00000182
	m	=	1.40	12	.00000022	.99999997	.00000025
				13	.00000003	.99999999	.00000003
0	.24659697	.24659697	1.00000000				
1	.34523575	.59183271	.75340304		m	=	1.80
2	.24166503	.83349774	.40816728				
3	.11277701	.94627476	.16650226	0	.16529889	.16529889	1.00000000
4	.03947195	.98574671	.05372524	1	.29753800	.46283689	.83470111
5	.01105215	.99679886	.01425329	2	.26778420	.73062108	.53716312
6	.00257883	.99937769	.00320114	3	.16067052	.89129160	.26937892
7	.00051577	.99989346	.00062231	4	.07230173	.96359334	.10870839
8	.00009026	.99998371	.00010654	5	.02602862	.98962396	.03640666
9	.00001404	.99999776	.00001628	6	.00780859	.99743055	.01037803
10	.00000197	.99999972	.00000224	7	.00200792	.99943847	.00256945
11	.00000025	.99999997	.00000027	8	.00045178	.99989025	.00056152
12	.00000003	.99999999	.00000002	9	.00009036	.99998061	.00010974
				10	.00001626	.99999688	.00001939
	m	=	1.50	11	.00000266	.99999954	.00000312
				12	.00000040	.99999993	.00000046
0	.22313016	.22313016	1.00000000	13	.00000006	.99999999	.00000006
1	.33469524	.55782539	.77686984				
2	.25102143	.80884682	.44217460		m	=	1.90
3	.12551071	.93435754	.19115317				
4	.04706652	.98142406	.06564245	0	.14956862	.14956862	1.00000000
5	.01411996	.99554402	.01857594	1	.28418038	.43374900	.85043138
6	.00352999	.99907400	.00445598	2	.26997136	.70372035	.56625100
7	.00075643	.99983043	.00092599	3	.17098186	.87470222	.29627964
8	.00014183	.99997226	.00016957	4	.08121638	.95591860	.12529778
9	.00002364	.99999590	.00002774	5	.03086223	.98678083	.04408140
10	.00000355	.99999944	.00000410	6	.00977304	.99655387	.01321917
11	.00000048	.99999993	.00000055	7	.00265268	.99920655	.00344613
12	.00000006	.99999999	.00000007	8	.00063001	.99983656	.00079345
				9	.00013300	.99996956	.00016344
				10	.00002527	.99999483	.00003043
				11	.00000436	.99999920	.00000516
				12	.00000069	.99999989	.00000080
				13	.00000010	.99999999	.00000011

Table B-4 cont.

x	P(x)	C(x)	D(x)	x	P(x)	C(x)	D(x)
	m	=	2.00		m	=	2.30
0	.13533528	.13533528	1.00000000	0	.10025884	.10025884	1.00000000
1	.27067056	.40600585	.86466472	1	.23059534	.33085419	.89974116
2	.27067056	.67667641	.59399416	2	.26518464	.59603883	.66914582
3	.18044704	.85712346	.32332359	3	.20330823	.79934707	.40396117
4	.09022352	.94734698	.14287654	4	.11690223	.91624929	.20065293
5	.03608941	.98343639	.05265302	5	.05377503	.97002432	.08375070
6	.01202980	.99546619	.01656361	6	.02061376	.99063808	.02997568
7	.00343709	.99890327	.00453381	7	.00677309	.99741117	.00936192
8	.00085927	.99976255	.00109672	8	.00194726	.99935844	.00258882
9	.00019095	.99995350	.00023745	9	.00049763	.99985607	.00064156
10	.00003819	.99999169	.00004650	10	.00011446	.99997053	.00014393
11	.00000694	.99999863	.00000831	11	.00002393	.99999446	.00002947
12	.00000116	.99999979	.00000136	12	.00000459	.99999905	.00000554
13	.00000018	.99999997	.00000021	13	.00000081	.99999986	.00000095
14	.00000003	.99999999	.00000003	14	.00000013	.99999999	.00000014
	m	=	2.10		m	=	2.40
0	.12245643	.12245643	1.00000000	0	.09071796	.09071796	1.00000000
1	.25715850	.37961493	.87754358	1	.21772309	.30844105	.90928205
2	.27001642	.64963135	.62038507	2	.26126771	.56970876	.69155896
3	.18901150	.83864284	.35036865	3	.20901417	.77872293	.43029124
4	.09923103	.93787389	.16135715	4	.12540850	.90413143	.22127707
5	.04167703	.97955091	.06212611	5	.06019608	.96432751	.09586857
6	.01458696	.99413788	.02044908	6	.02407843	.98840594	.03567249
7	.00437609	.99851397	.00586212	7	.00825546	.99666140	.01159406
8	.00114872	.99966269	.00148603	8	.00247664	.99913804	.00333859
9	.00026804	.99993072	.00033731	9	.00066044	.99979848	.00086195
10	.00005629	.99998701	.00006927	10	.00015850	.99995698	.00020152
11	.00001075	.99999776	.00001298	11	.00003458	.99999157	.00004301
12	.00000188	.99999964	.00000224	12	.00000692	.99999848	.00000843
13	.00000030	.99999994	.00000036	13	.00000128	.99999976	.00000151
14	.00000005	.99999999	.00000005	14	.00000022	.99999998	.00000024
				15	.00000004	.99999999	.00000002
	m	=	2.20		m	=	2.50
0	.11080316	.11080316	1.00000000	0	.08208500	.08208500	1.00000000
1	.24376695	.35457011	.88919684	1	.20521250	.28729749	.91791500
2	.26814365	.62271375	.64542989	2	.25651562	.54381312	.71270251
3	.19663867	.81935243	.37728624	3	.21376302	.75757613	.45618688
4	.10815127	.92750370	.18064757	4	.13360189	.89117801	.24242387
5	.04758656	.97509026	.07249630	5	.06680094	.95797896	.10882198
6	.01744840	.99253866	.02490974	6	.02783373	.98581269	.04202104
7	.00548378	.99802244	.00746133	7	.00994062	.99575330	.01418731
8	.00150804	.99953049	.00197755	8	.00310644	.99885974	.00424670
9	.00036863	.99989912	.00046951	9	.00086290	.99972264	.00114025
10	.00008110	.99998022	.00010088	10	.00021573	.99993837	.00027735
11	.00001622	.99999644	.00001978	11	.00004903	.99998740	.00006163
12	.00000297	.99999941	.00000356	12	.00001021	.99999761	.00001260
13	.00000050	.99999992	.00000058	13	.00000196	.99999958	.00000238
14	.00000008	.99999999	.00000008	14	.00000035	.99999993	.00000042
				15	.00000006	.99999999	.00000007

x	P(x)	C(x)	D(x)	x	P(x)	C(x)	D(x)

	m	=	2.60		m	=	2.90
0	.07427358	.07427358	1.00000000	0	.05502322	.05502322	1.00000000
1	.19311130	.26738488	.92572643	1	.15956734	.21459056	.94497678
2	.25104469	.51842958	.73261512	2	.23137265	.44596321	.78540944
3	.21757207	.73600165	.48157042	3	.22366022	.66962343	.55403680
4	.14142185	.87742349	.26399835	4	.16215366	.83177710	.33037657
5	.07353936	.95096285	.12257651	5	.09404912	.92582621	.16822290
6	.03186706	.98282991	.04903715	6	.04545708	.97128329	.07417378
7	.01183633	.99466624	.01717009	7	.01883222	.99011551	.02871670
8	.00384681	.99851305	.00533376	8	.00682668	.99694219	.00988449
9	.00111130	.99962435	.00148695	9	.00219971	.99914189	.00305781
10	.00028894	.99991329	.00037565	10	.00063792	.99977981	.00085810
11	.00006829	.99998158	.00008671	11	.00016818	.99994799	.00022018
12	.00001480	.99999638	.00001841	12	.00004064	.99998863	.00005201
13	.00000296	.99999934	.00000362	13	.00000907	.99999770	.00001136
14	.00000055	.99999989	.00000066	14	.00000188	.99999958	.00000230
15	.00000010	.99999999	.00000011	15	.00000036	.99999994	.00000042
				16	.00000007	.99999999	.00000006

	m	=	2.70		m	=	3.00
0	.06720551	.06720551	1.00000000	0	.04978707	.04978707	1.00000000
1	.18145488	.24866040	.93279449	1	.14936120	.19914827	.95021293
2	.24496409	.49362449	.75133961	2	.22404181	.42319008	.80085173
3	.22046768	.71409217	.50637551	3	.22404181	.64723188	.57680992
4	.14881569	.86290786	.28590782	4	.16803136	.81526324	.35276811
5	.08036047	.94326834	.13709213	5	.10081881	.91608205	.18473676
6	.03616221	.97943055	.05673166	6	.05040941	.96649146	.08391794
7	.01394828	.99337883	.02056945	7	.02160403	.98809549	.03350854
8	.00470755	.99808637	.00662117	8	.00810151	.99619701	.01190450
9	.00141226	.99949864	.00191362	9	.00270050	.99889751	.00380299
10	.00038131	.99987995	.00050136	10	.00081015	.99970766	.00110249
11	.00009359	.99997354	.00012005	11	.00022095	.99992861	.00029234
12	.00002106	.99999460	.00002645	12	.00005524	.99998385	.00007139
13	.00000437	.99999897	.00000540	13	.00001275	.99999660	.00001615
14	.00000084	.99999982	.00000102	14	.00000273	.99999933	.00000340
15	.00000015	.99999997	.00000018	15	.00000055	.99999987	.00000067
16	.00000003	.99999999	.00000003	16	.00000010	.99999998	.00000012
				17	.00000002	.99999999	.00000002

	m	=	2.80		m	=	3.10
0	.06081006	.06081006	1.00000000				
1	.17026818	.23107824	.93918994	0	.04504920	.04504920	1.00000000
2	.23837545	.46945369	.76892176	1	.13965253	.18470173	.95495080
3	.22248375	.69193744	.53054632	2	.21646142	.40116315	.81529827
4	.15573863	.84767607	.30806256	3	.22367680	.62483995	.59883685
5	.08721363	.93488970	.15232393	4	.17334952	.79818946	.37516005
6	.04069969	.97558939	.06511030	5	.10747670	.90566617	.20181053
7	.01627988	.99186927	.02441060	6	.05552963	.96119580	.09433383
8	.00569796	.99756723	.00813072	7	.02459169	.98578749	.03880420
9	.00177270	.99933993	.00243277	8	.00952928	.99531677	.01421251
10	.00049636	.99983628	.00066007	9	.00328231	.99859907	.00468323
11	.00012634	.99996263	.00016371	10	.00101752	.99961659	.00140092
12	.00002948	.99999211	.00003737	11	.00028675	.99990335	.00038340
13	.00000635	.99999846	.00000789	12	.00007408	.99997742	.00009665
14	.00000127	.99999972	.00000154	13	.00001766	.99999509	.00002257
15	.00000024	.99999996	.00000027	14	.00000391	.99999900	.00000490
16	.00000004	.99999999	.00000003	15	.00000081	.99999981	.00000099
				16	.00000016	.99999997	.00000019
				17	.00000003	.99999999	.00000003

Table B-4 cont.

x	P(x)	C(x)	D(x)	x	P(x)	C(x)	D(x)
	m	=	3.20		m	=	3.50
0	.04076220	.04076220	1.00000000	0	.03019738	.03019738	1.00000000
1	.13043905	.17120126	.95923780	1	.10569084	.13588822	.96980262
2	.20870249	.37990374	.82879875	2	.18495897	.32084720	.86411178
3	.22261599	.60251973	.62009626	3	.21578547	.53663266	.67915280
4	.17809279	.78061251	.39748027	4	.18881228	.72544495	.46336734
5	.11397938	.89459191	.21938748	5	.13216860	.85761355	.27455505
6	.06078900	.95538091	.10540809	6	.07709835	.93471190	.14238645
7	.02778926	.98317017	.04461909	7	.03854917	.97326107	.06528810
8	.01111570	.99428587	.01682983	8	.01686526	.99012634	.02673892
9	.00395225	.99823812	.00571413	9	.00655871	.99668505	.00987366
10	.00126472	.99950284	.00176188	10	.00229555	.99898060	.00331494
11	.00036792	.99987076	.00049716	11	.00073040	.99971101	.00101939
12	.00009811	.99996887	.00012924	12	.00021303	.99992404	.00028899
13	.00002415	.99999302	.00003113	13	.00005736	.99998140	.00007596
14	.00000552	.99999854	.00000698	14	.00001434	.99999573	.00001860
15	.00000118	.99999972	.00000146	15	.00000335	.99999908	.00000426
16	.00000024	.99999996	.00000028	16	.00000073	.99999981	.00000092
17	.00000004	.99999999	.00000004	17	.00000015	.99999996	.00000019
				18	.00000003	.99999999	.00000004
	m	=	3.30		m	=	3.60
0	.03688317	.03688317	1.00000000				
1	.12171445	.15859762	.96311683	0	.02732372	.02732372	1.00000000
2	.20082885	.35942647	.84140238	1	.09836540	.12568912	.97267628
3	.22091173	.58033820	.64057353	2	.17705772	.30274684	.87431088
4	.18225218	.76259039	.41966179	3	.21246927	.51521611	.69725316
5	.12028644	.88287683	.23740961	4	.19122234	.70643845	.48478389
6	.06615754	.94903437	.11712317	5	.13768008	.84411854	.29356155
7	.03118856	.98022293	.05096563	6	.08260805	.92672659	.15588146
8	.01286528	.99308820	.01977707	7	.04248414	.96921073	.07327341
9	.00471727	.99780547	.00691179	8	.01911786	.98832859	.03078927
10	.00155670	.99936217	.00219453	9	.00764715	.99597573	.01167141
11	.00046701	.99982918	.00063783	10	.00275297	.99872871	.00402426
12	.00012843	.99995761	.00017082	11	.00090097	.99962968	.00127129
13	.00003260	.99999021	.00004239	12	.00027029	.99989997	.00037032
14	.00000768	.99999789	.00000979	13	.00007485	.99997482	.00010002
15	.00000169	.99999958	.00000210	14	.00001925	.99999407	.00002517
16	.00000035	.99999993	.00000041	15	.00000462	.99999869	.00000593
17	.00000007	.99999999	.00000006	16	.00000104	.99999972	.00000131
				17	.00000022	.99999995	.00000027
	m	=	3.40	18	.00000004	.99999999	.00000005
0	.03337327	.03337327	1.00000000		m	=	3.70
1	.11346912	.14684239	.96662673				
2	.19289750	.33973989	.85315761	0	.02472353	.02472353	1.00000000
3	.21861717	.55835707	.66026011	1	.09147705	.11620057	.97527648
4	.18582460	.74418166	.44164293	2	.16923254	.28543311	.88379943
5	.12636072	.87054238	.25581834	3	.20872013	.49415325	.71456689
6	.07160441	.94214680	.12945761	4	.19306612	.68721937	.50584675
7	.03477929	.97692608	.05785320	5	.14286893	.83008830	.31278063
8	.01478120	.99170728	.02307391	6	.08810251	.91819080	.16991170
9	.00558401	.99729129	.00829272	7	.04656847	.96475927	.08180919
10	.00189856	.99918985	.00270871	8	.02153792	.98629719	.03524073
11	.00058683	.99977668	.00081015	9	.00885448	.99515166	.01370281
12	.00016627	.99994294	.00022332	10	.00327616	.99842782	.00484833
13	.00004349	.99998643	.00005705.	11	.00110198	.99952980	.00157218
14	.00001056	.99999699	.00001356	12	.00033978	.99986958	00047020
15	.00000239	.99999939	.00000300	13	.00009671	.99996629	.00013042
16	.00000051	.99999990	.00000061	14	.00002556	.99999184	.00003371
17	.00000010	.99999999	.00000010	15	.00000630	.99999814	.00000816
				16	.00000146	.99999961	.00000185
				17	.00000032	.99999992	.00000039
				18	.00000007	.99999999	.00000008

Table B-4 cont.

x	P(x)	C(x)	D(x)	x	P(x)	C(x)	D(x)
	m	=	3.80		m	=	4.10
0	.02237077	.02237077	1.00000000	0	.01657268	.01657268	1.00000000
1	.08500893	.10737971	.97762923	1	.06794797	.08452065	.98342733
2	.16151698	.26889668	.89262030	2	.13929334	.22381399	.91547935
3	.20458817	.47348485	.73110332	3	.19036757	.41418155	.77618601
4	.19435876	.66784361	.52651515	4	.19512676	.60930831	.58581845
5	.14771266	.81555627	.33215639	5	.16000394	.76931225	.39069169
6	.09355135	.90910762	.18444373	6	.10933603	.87864827	.23068775
7	.05078502	.95989264	.09089238	7	.06403967	.94268794	.12135172
8	.02412288	.98401552	.04010736	8	.03282033	.97550828	.05731205
9	.01018522	.99420074	.01598448	9	.01495148	.99045976	.02449172
10	.00387038	.99807112	.00579926	10	.00613011	.99658987	.00954023
11	.00133704	.99940816	.00192888	11	.00228486	.99887473	.00341012
12	.00042340	.99983156	.00059184	12	.00078066	.99965539	.00112527
13	.00012376	.99995532	.00016844	13	.00024621	.99990160	.00034461
14	.00003359	.99998891	.00004468	14	.00007210	.99997370	.00009840
15	.00000851	.99999742	.00001108	15	.00001971	.99999341	.00002629
16	.00000202	.99999944	.00000257	16	.00000505	.99999847	.00000659
17	.00000045	.99999990	.00000055	17	.00000122	.99999968	.00000153
18	.00000010	.99999999	.00000010	18	.00000028	.99999996	.00000032
				19	.00000006	.99999999	.00000004
	m	=	3.90				
					m	=	4.20
0	.02024191	.02024191	1.00000000				
1	.07894346	.09918537	.97975809	0	.01499558	.01499558	1.00000000
2	.15393974	.25312511	.90081464	1	.06298142	.07797700	.98500443
3	.20012166	.45324677	.74687490	2	.13226099	.21023799	.92202301
4	.19511862	.64836539	.54675324	3	.18516538	.39540337	.78976201
5	.15219252	.80055791	.35163461	4	.19442365	.58982702	.60459663
6	.09892514	.89948305	.19944208	5	.16331587	.75314289	.41017297
7	.05511544	.95459849	.10051694	6	.11432111	.86746401	.24685710
8	.02686877	.98146726	.04540151	7	.06859266	.93605667	.13253599
9	.01164314	.99311040	.01853273	8	.03601115	.97206782	.06394333
10	.00454082	.99765122	.00688960	9	.01680520	.98887302	.02793218
11	.00160993	.99926115	.00234877	10	.00705819	.99593121	.01112698
12	.00052323	.99978438	.00073885	11	.00269494	.99862615	.00406879
13	.00015697	.99994134	.00021562	12	.00094323	.99956938	.00137385
14	.00004373	.99998507	.00005865	13	.00030474	.99987411	.00043062
15	.00001137	.99999644	.00001492	14	.00009142	.99996553	.00012588
16	.00000277	.99999921	.00000356	15	.00002560	.99999113	.00003446
17	.00000064	.99999985	.00000078	16	.00000672	.99999785	.00000886
18	.00000014	.99999999	.00000015	17	.00000166	.99999952	.00000214
				18	.00000039	.99999990	.00000048
	m	=	4.00	19	.00000009	.99999999	.00000010
0	.01831564	.01831564	1.00000000		m	=	4.30
1	.07326256	.09157819	.98168436				
2	.14652511	.23810330	.90842181	0	.01356856	.01356856	1.00000000
3	.19536681	.43347012	.76189670	1	.05834481	.07191337	.98643144
4	.19536681	.62883693	.56652988	2	.12544133	.19735470	.92808664
5	.15629345	.78513038	.37116307	3	.17979924	.37715394	.80264530
6	.10419563	.88932602	.21486961	4	.19328419	.57043813	.62284606
7	.05954036	.94886638	.11067398	5	.16622440	.73666254	.42956187
8	.02977018	.97863656	.05113362	6	.11912749	.85579002	.26333746
9	.01323119	.99186775	.02136343	7	.07317831	.92896833	.14420998
10	.00529248	.99716023	.00813224	8	.03933334	.96830168	.07103166
11	.00192454	.99908477	.00283977	9	.01879260	.98709428	.03169832
12	.00064151	.99972628	.00091523	10	.00808082	.99517509	.01290572
13	.00019739	.99992367	.00027372	11	.00315886	.99833396	.00482490
14	.00005640	.99998006	.00007633	12	.00113193	.99946588	.00166604
15	.00001504	.99999510	.00001993	13	.00037441	.99984029	.00053411
16	.00000376	.99999886	.00000489	14	.00011500	.99995529	.00015970
17	.00000088	.99999975	.00000113	15	.00003297	.99998826	.00004471
18	.00000020	.99999995	.00000025	16	.00000886	.99999712	.00001174
19	.00000004	.99999999	.00000005	17	.00000224	.99999935	.00000288
				18	.00000054	.99999989	.00000064
				19	.00000012	.99999999	.00000011

Table B-4 cont.

x	P(x)	C(x)	D(x)	x	P(x)	C(x)	D(x)
	m	=	4.40		m	=	4.60
0	.01227734	.01227734	1.00000000	0	.01005184	.01005184	1.00000000
1	.05402030	.06629764	.98772267	1	.04623845	.05629028	.98994817
2	.11884465	.18514229	.93370236	2	.10634843	.16263871	.94370972
3	.17430549	.35944778	.81485771	3	.16306758	.32570629	.83736130
4	.19173604	.55118382	.64055222	4	.18752772	.51323401	.67429371
5	.16872771	.71991153	.44881618	5	.17252551	.68575952	.48676598
6	.12373366	.84364519	.28008847	6	.13226955	.81802908	.31424048
7	.07777544	.92142063	.15635481	7	.08691999	.90494907	.18197092
8	.04277649	.96419713	.07857936	8	.04997900	.95492806	.09505093
9	.02091295	.98511008	.03580287	9	.02554482	.98047289	.04507193
10	.00920170	.99431178	.01488992	10	.01175062	.99222350	.01952711
11	.00368068	.99799246	.00568822	11	.00491389	.99713740	.00777649
12	.00134958	.99934204	.00200754	12	.00188366	.99902106	.00286260
13	.00045678	.99979882	.00065796	13	.00066653	.99968758	.00097894
14	.00014356	.99994238	.00020117	14	.00021900	.99990658	.00031241
15	.00004211	.99998450	.00005761	15	.00006716	.99997374	.00009341
16	.00001158	.99999607	.00001550	16	.00001931	.99999306	.00002625
17	.00000300	.99999907	.00000392	17	.00000522	.99999828	.00000694
18	.00000073	.99999981	.00000093	18	.00000134	.99999961	.00000172
19	.00000017	.99999997	.00000019	19	.00000032	.99999994	.00000038
20	.00000004	.99999999	.00000002	20	.00000007	.99999999	.00000006
	m	=	4.50		m	=	4.70
0	.01110900	.01110900	1.00000000	0	.00909528	.00909528	1.00000000
1	.04999048	.06109948	.98889101	1	.04274780	.05184308	.99090473
2	.11247859	.17357807	.93890052	2	.10045734	.15230042	.94815692
3	.16871788	.34229595	.82642193	3	.15738316	.30968358	.84769959
4	.18980762	.53210358	.65770405	4	.18492521	.49460879	.69031642
5	.17082686	.70293043	.46789642	5	.17382970	.66843849	.50539121
6	.12812014	.83105057	.29706957	6	.13616660	.80460509	.33156151
7	.08236295	.91341352	.16894942	7	.09142614	.89603123	.19539491
8	.04632916	.95974268	.08658647	8	.05371286	.94974410	.10396876
9	.02316458	.98290727	.04025731	9	.02805005	.97779414	.05025590
10	.01042406	.99333132	.01709273	10	.01318352	.99097767	.02220585
11	.00426439	.99759571	.00666867	11	.00563296	.99661063	.00902233
12	.00159915	.99919486	.00240428	12	.00220624	.99881687	.00338937
13	.00055355	.99974841	.00080514	13	.00079764	.99961451	.00118313
14	.00017793	.99992634	.00025159	14	.00026778	.99988229	.00038549
15	.00005338	.99997971	.00007366	15	.00008390	.99996620	.00011771
16	.00001501	.99999472	.00002028	16	.00002465	.99999084	.00003380
17	.00000397	.99999870	.00000527	17	.00000681	.99999765	.00000915
18	.00000099	.99999969	.00000130	18	.00000178	.99999943	.00000234
19	.00000024	.99999993	.00000030	19	.00000044	.99999987	.00000056
20	.00000005	.99999999	.00000007	20	.00000010	.99999998	.00000012
				21	.00000002	.99999999	.00000002

Table B-4 cont.

x	P(x)	C(x)	D(x)	x	P(x)	C(x)	D(x)
	m	=	4.80		m	=	5.00
0	.00822975	.00822975	1.00000000	0	.00673795	.00673795	1.00000000
1	.03950279.	.04773253	.99177025	1	.03368973	.04042768	.99326206
2	.09480669	.14253923	.95226747	2	.08422434	.12465202	.95957232
3	.15169070	.29422993	.85746077	3	.14037389	.26502591	.87534799
4	.18202884	.47625877	.70577008	4	.17546737	.44049328	.73497409
5	.17474769	.65100647	.52374123	5	.17546737	.61596065	.55950672
6	.13979815	.79080462	.34899353	6	.14622281	.76218346	.38403935
7	.09586159	.88666621	.20919538	7	.10444486	.86662832	.23781654
8	.05751695	.94418316	.11333379	8	.06527804	.93190636	.13337168
9	.03067571	.97485887	.05581683	9	.03626558	.96817194	.06809364
10	.01472434	.98958322	.02514112	10	.01813279	.98630473	.03182806
11	.00642517	.99600838	.01041678	11	.00824218	.99454691	.01369527
12	.00257007	.99857844	.00399162	12	.00343424	.99798115	.00545309
13	.00094895	.99952739	.00142155	13	.00132086	.99930201	.00201885
14	.00032535	.99985275	.00047260	14	.00047174	.99977374	.00069799
15	.00010411	.99995686	.00014725	15	.00015725	.99993099	.00022625
16	.00003123	.99998809	.00004314	16	.00004914	.99998013	.00006901
17	.00000882	.99999692	.00001190	17	.00001445	.99999458	.00001987
18	.00000235	.99999926	.00000308	18	.00000401	.99999859	.00000542
19	.00000059	.99999986	.00000073	19	.00000106	.99999965	.00000140
20	.00000014	.99999999	.00000014	20	.00000026	.99999992	.00000035
				21	.00000006	.99999998	.00000008
	m	=	4.90	22	.00000001	.99999999	.00000002
0	.00744658	.00744658	1.00000000				
1	.03648826	.04393484	.99255342				
2	.08939623	.13333107	.95606516				
3	.14601384	.27934492	.86666893				
4	.17886696	.45821188	.72065508				
5	.17528962	.63350150	.54178812				
6	.14315319	.77665469	.36649850				
7	.10020723	.87686192	.22334531				
8	.06137693	.93823885	.12313808				
9	.03341633	.97165518	.06176114				
10	.01637400	.98802918	.02834482				
11	.00729387	.99532305	.01197081				
12	.00297833	.99830139	.00467694				
13	.00112260	.99942399	.00169861				
14	.00039291	.99981690	.00057601				
15	.00012835	.99994525	.00018310				
16	.00003931	.99998455	.00005475				
17	.00001133	.99999589	.00001544				
18	.00000308	..99999897	.00000411				
19	.00000080	.99999977	.00000102				
20	.00000019	.99999996	.00000023				
21	.00000005	.99999999	.00000003				

Lower Critical Values for Two-Tail Binomial Tests for Median, Median Difference, or Trend (Where H_0 Implies That $\rho = .5$)

Largest value of r' for which $P(r \leqslant r' \mid \rho = .5) + P(r \geqslant n - r' \mid \rho = .5) \leqslant \alpha$ or for which
$$P(r \leqslant r' \mid \rho = .5) \leqslant \alpha/2$$

n	Two-tailed probability, α						n	Two-tailed probability, α					
	.001	.01	.02	.05	.10	.50		.001	.01	.02	.05	.10	.50
1	–	–	–	–	–	–	51	13	15	16	18	19	22
2	–	–	–	–	–	0	52	13	16	17	18	19	23
3	–	–	–	–	–	0	53	14	16	17	18	20	23
4	–	–	–	–	–	0	54	14	17	18	19	20	24
5	–	–	–	–	0	1	55	14	17	18	19	20	24
6	–	–	–	0	0	1	56	15	17	18	20	21	24
7	–	–	0	0	0	2	57	15	18	19	20	21	25
8	–	0	0	0	1	2	58	16	18	19	21	22	25
9	–	0	0	1	1	2	59	16	19	20	21	22	26
10	–	0	0	1	1	3	60	16	19	20	21	23	26
11	0	0	1	1	2	3	61	17	20	20	22	23	27
12	0	1	1	2	2	4	62	17	20	21	22	24	27
13	0	1	1	2	3	4	63	18	20	21	23	24	28
14	0	1	2	2	3	5	64	18	21	22	23	24	28
15	1	2	2	3	3	5	65	18	21	22	24	25	29
16	1	2	2	3	4	6	66	19	22	23	24	25	29
17	1	2	3	4	4	6	67	19	22	23	25	26	30
18	1	3	3	4	5	7	68	20	22	23	25	26	30
19	2	3	4	4	5	7	69	20	23	24	25	27	31
20	2	3	4	5	5	7	70	20	23	24	26	27	31
21	2	4	4	5	6	8	71	21	24	25	26	28	32
22	3	4	5	5	6	8	72	21	24	25	27	28	32
23	3	4	5	6	7	9	73	22	25	26	27	28	33
24	3	5	5	6	7	9	74	22	25	26	28	29	33
25	4	5	6	7	7	10	75	22	25	26	28	29	34
26	4	6	6	7	8	10	76	23	26	27	28	30	34
27	4	6	7	7	8	11	77	23	26	27	29	30	35
28	5	6	7	8	9	11	78	24	27	28	29	31	35
29	5	7	7	8	9	12	79	24	27	28	30	31	36
30	5	7	8	9	10	12	80	24	28	29	30	32	36
31	6	7	8	9	10	13	81	25	28	29	31	32	36
32	6	8	8	9	10	13	82	25	28	30	31	33	37
33	6	8	9	10	11	14	83	26	29	30	32	33	37
34	7	9	9	10	11	14	84	26	29	30	32	33	38
35	7	9	10	11	12	15	85	26	30	31	32	34	38
36	7	9	10	11	12	15	86	27	30	31	33	34	39
37	8	10	10	12	13	15	87	27	31	32	33	35	39
38	8	10	11	12	13	16	88	28	31	32	34	35	40
39	8	11	11	12	13	16	89	28	31	33	34	36	40
40	9	11	12	13	14	17	90	29	32	33	35	36	41
41	9	11	12	13	14	17	91	29	32	33	35	37	41
42	10	12	13	14	15	18	92	29	33	34	36	37	42
43	10	12	13	14	15	18	93	30	33	34	36	38	42
44	10	13	13	15	16	19	94	30	34	35	37	38	43
45	11	13	14	15	16	19	95	31	34	35	37	38	43
46	11	13	14	15	16	20	96	31	34	36	37	39	44
47	11	14	15	16	17	20	97	31	35	36	38	39	44
48	12	14	15	16	17	21	98	32	35	37	38	40	45
49	12	15	15	17	18	21	99	32	36	37	39	40	45
50	13	15	16	17	18	22	100	33	36	37	39	41	46

Body of table is reproduced, with changes only in notation, from Table 1 in W.J. MacKinnon's "Table for Both the Sign Test and Distribution-Free Confidence Intervals of the Median for Sample Sizes to 1,000, "*Journal of the American Statistical Association,* Vol. 59 (1964), pp. 935-956, with permission of author and editor.

Table B-5 cont.

n	.001	.01	.02	.05	.10	.50	*n*	.001	.01	.02	.05	.10	.50
	Two-tailed probability, α							Two-tailed probability, α					
101	33	37	38	40	41	46	151	54	59	60	62	64	70
102	34	37	38	40	42	47	152	55	59	61	63	65	71
103	34	37	39	41	42	47	153	55	60	61	63	65	71
104	34	38	39	41	43	48	154	56	60	62	64	66	72
105	35	38	40	41	43	48	155	56	61	62	64	66	72
106	35	39	40	42	44	49	156	57	61	63	65	67	73
107	36	39	41	42	44	49	157	57	61	63	65	67	73
108	36	40	41	43	44	49	158	57	62	63	66	68	74
109	36	40	41	43	45	50	159	58	62	64	66	68	74
110	37	41	42	44	45	50	160	58	63	64	67	69	75
111	37	41	42	44	46	51	161	59	63	65	67	69	75
112	38	41	43	45	46	51	162	59	64	65	68	70	76
113	38	42	43	45	47	52	163	60	64	66	68	70	76
114	39	42	44	46	47	52	164	60	65	66	68	70	77
115	39	43	44	46	48	53	165	60	65	67	69	71	77
116	39	43	45	46	48	53	166	61	65	67	69	71	78
117	40	44	45	47	49	54	167	61	66	68	70	72	78
118	40	44	45	47	49	54	168	62	66	68	70	72	79
119	41	45	46	48	50	55	169	62	67	68	71	73	79
120	41	45	46	48	50	55	170	63	67	69	71	73	80
121	42	45	47	49	50	56	171	63	68	69	72	74	80
122	42	46	47	49	51	56	172	64	68	70	72	74	81
123	42	46	48	50	51	57	173	64	69	70	73	75	81
124	43	47	48	50	52	57	174	64	69	71	73	75	82
125	43	47	49	51	52	58	175	65	70	71	74	76	82
126	44	48	49	51	53	58	176	65	70	72	74	76	83
127	44	48	49	51	53	59	177	66	70	72	74	77	83
128	45	48	50	52	54	59	178	66	71	73	75	77	83
129	45	49	50	52	54	60	179	67	71	73	75	78	84
130	45	49	51	53	55	60	180	67	72	73	76	78	84
131	46	50	51	53	55	61	181	67	72	74	76	78	85
132	46	50	52	54	56	61	182	68	73	74	77	79	85
133	47	51	52	54	56	62	183	68	73	75	77	79	86
134	47	51	53	55	56	62	184	69	74	75	78	80	86
135	48	52	53	55	57	63	185	69	74	76	78	80	87
136	48	52	53	56	57	63	186	70	74	76	79	81	87
137	48	52	54	56	58	64	187	70	75	77	79	81	88
138	49	53	54	57	58	64	188	71	75	77	80	82	88
139	49	53	55	57	59	65	189	71	76	78	80	82	89
140	50	54	55	57	59	65	190	71	76	78	81	83	89
141	50	54	56	58	60	65	191	72	77	78	81	83	90
142	51	55	56	58	60	66	192	72	77	79	81	84	90
143	51	55	57	59	61	66	193	73	78	79	82	84	91
144	51	56	57	59	61	67	194	73	78	80	82	85	91
145	52	56	58	60	62	67	195	74	79	80	83	85	92
146	52	56	58	60	62	68	196	74	79	81	83	85	92
147	53	57	58	61	63	68	197	75	79	81	84	86	93
148	53	57	59	61	63	69	198	75	80	82	84	86	93
149	54	58	59	62	63	69	199	75	80	82	85	87	94
150	54	58	60	62	64	70	200	76	81	83	85	87	94

Table B-5 cont

n	Two-tailed probability, α						n	Two-tailed probability, α					
	.001	.01	.02	.05	.10	.50		.001	.01	.02	.05	.10	.50
201	76	81	83	86	88	95	251	99	104	106	109	111	119
202	77	82	83	86	88	95	252	99	105	107	109	112	120
203	77	82	84	87	89	96	253	99	105	107	110	112	120
204	78	83	84	87	89	96	254	100	106	107	110	113	121
205	78	83	85	87	90	97	255	100	106	108	111	113	121
206	78	84	85	88	90	97	256	101	106	108	111	114	122
207	79	84	86	88	91	98	257	101	107	109	112	114	122
208	79	84	86	89	91	98	258	102	107	109	112	115	123
209	80	85	87	89	92	99	259	102	108	110	113	115	123
210	80	85	87	90	92	99	260	103	108	110	113	116	124
211	81	86	88	90	93	100	261	103	109	111	114	116	124
212	81	86	88	91	93	100	262	103	109	111	114	117	125
213	82	87	89	91	94	101	263	104	110	112	115	117	125
214	82	87	89	92	94	101	264	104	110	112	115	118	126
215	82	88	89	92	94	102	265	105	111	113	116	118	126
216	83	88	90	93	95	102	266	105	111	113	116	119	126
217	83	89	90	93	95	103	267	106	111	114	117	119	127
218	84	89	91	94	96	103	268	106	112	114	117	120	127
219	84	89	91	94	96	104	269	107	112	114	117	120	128
220	85	90	92	94	97	104	270	107	113	115	118	120	128
221	85	90	92	95	97	104	271	107	113	115	118	121	129
222	86	91	93	95	98	105	272	108	114	116	119	121	129
223	86	91	93	96	98	105	273	108	114	116	119	122	130
224	86	92	94	96	99	106	274	109	115	117	120	122	130
225	87	92	94	97	99	106	275	109	115	117	120	123	131
226	87	93	95	97	100	107	276	110	116	118	121	123	131
227	88	93	95	98	100	107	277	110	116	118	121	124	132
228	88	94	95	98	101	108	278	111	117	119	122	124	132
229	89	94	96	99	101	108	279	111	117	119	122	125	133
230	89	95	96	99	102	109	280	112	117	120	123	125	133
231	90	95	97	100	102	109	281	112	118	120	123	126	134
232	90	95	97	100	102	110	282	112	118	120	124	126	134
233	90	96	98	101	103	110	283	113	119	121	124	127	135
234	91	96	98	101	103	111	284	113	119	121	124	127	135
235	91	97	99	101	104	111	285	114	120	122	125	128	136
236	92	97	99	102	104	112	286	114	120	122	125	128	136
237	92	98	100	102	105	112	287	115	121	123	126	129	137
238	93	98	100	103	105	113	288	115	121	123	126	129	137
239	93	99	101	103	106	113	289	116	122	124	127	130	138
240	94	99	101	104	106	114	290	116	122	124	127	130	138
241	94	100	101	104	107	114	291	117	123	125	128	130	139
242	94	100	102	105	107	115	292	117	123	125	128	131	139
243	95	100	102	105	108	115	293	117	123	126	129	131	140
244	95	101	103	106	108	116	294	118	124	126	129	132	140
245	96	101	103	106	109	116	295	118	124	127	130	132	141
246	96	102	104	107	109	117	296	119	125	127	130	133	141
247	97	102	104	107	110	117	297	119	125	127	131	133	142
248	97	103	105	108	110	118	298	120	126	128	131	134	142
249	98	103	105	108	111	118	299	120	126	128	132	134	143
250	98	104	106	109	111	119	300	121	127	129	132	135	143

Table B-5 cont.

n	Two-tailed probability, α						n	Two-tailed probability, α					
	.001	.01	.02	.05	.10	.50		.001	.01	.02	.05	.10	.50
301	121	127	129	133	135	144	351	144	150	153	156	159	168
302	121	128	130	133	136	144	352	144	151	153	157	160	169
303	122	128	130	133	136	145	353	145	151	154	157	160	169
304	122	129	131	134	137	145	354	145	152	154	158	161	170
305	123	129	131	134	137	146	355	146	152	155	158	161	170
306	123	130	132	135	138	146	356	146	153	155	159	161	171
307	124	130	132	135	138	147	357	146	153	156	159	162	171
308	124	130	133	136	139	147	358	147	154	156	159	162	172
309	125	131	133	136	139	148	359	147	154	156	160	163	172
310	125	131	134	137	140	148	360	148	155	157	160	163	173
311	126	132	134	137	140	149	361	148	155	157	161	164	173
312	126	132	134	138	140	149	362	149	156	158	161	164	174
313	126	133	135	138	141	150	363	149	156	158	162	165	174
314	127	133	135	139	141	150	364	150	156	159	162	165	175
315	127	134	136	139	142	151	365	150	157	159	163	166	175
316	128	134	136	140	142	151	366	151	157	160	163	166	176
317	128	135	137	140	143	151	367	151	158	160	164	167	176
318	129	135	137	141	143	152	368	152	158	161	164	167	177
319	129	136	138	141	144	152	369	152	159	161	165	168	177
320	130	136	138	141	144	153	370	152	159	162	165	168	178
321	130	136	139	142	145	153	371	153	160	162	166	169	178
322	131	137	139	142	145	154	372	153	160	163	166	169	178
323	131	137	140	143	146	154	373	154	161	163	167	170	179
324	131	138	140	143	146	155	374	154	161	164	167	170	179
325	132	138	141	144	147	155	375	155	162	164	168	171	180
326	132	139	141	144	147	156	376	155	162	164	168	171	180
327	133	139	141	145	148	156	377	156	163	165	168	172	181
328	133	140	142	145	148	157	378	156	163	165	169	172	181
329	134	140	142	146	149	157	379	157	163	166	169	172	182
330	134	141	143	146	149	158	380	157	164	166	170	173	182
331	135	141	143	147	150	158	381	157	164	167	170	173	183
332	135	142	144	147	150	159	382	158	165	167	171	174	183
333	136	142	144	148	150	159	383	158	165	168	171	174	184
334	136	142	145	148	151	160	384	159	166	168	172	175	184
335	136	143	145	149	151	160	385	159	166	169	172	175	185
336	137	143	146	149	152	161	386	160	167	169	173	176	185
337	137	144	146	150	152	161	387	160	167	170	173	176	186
338	138	144	147	150	153	162	388	161	168	170	174	177	186
339	138	145	147	150	153	162	389	161	168	171	174	177	187
340	139	145	148	151	154	163	390	162	169	171	175	178	187
341	139	146	148	151	154	163	391	162	169	172	175	178	188
342	140	146	149	152	155	164	392	162	170	172	176	179	188
343	140	147	149	152	155	164	393	163	170	172	176	179	189
344	141	147	149	153	156	165	394	163	170	173	177	180	189
345	141	148	150	153	156	165	395	164	171	173	177	180	190
346	141	148	150	154	157	166	396	164	171	174	178	181	190
347	142	149	151	154	157	166	397	165	172	174	178	181	191
348	142	149	151	155	158	167	398	165	172	175	178	182	191
349	143	149	152	155	158	167	399	166	173	175	179	182	192
350	143	150	152	156	159	168	400	166	173	176	179	183	192

n	Two-tailed probability, α						n	Two-tailed probability, α					
	.001	.01	.02	.05	.10	.50		.001	.01	.02	.05	.10	.50
401	167	174	176	180	183	193	451	190	197	200	204	207	217
402	167	174	177	180	184	193	452	190	198	200	204	208	218
403	168	175	177	181	184	194	453	191	198	201	205	208	218
404	168	175	178	181	184	194	454	191	199	201	205	208	219
405	168	176	178	182	185	195	455	191	199	202	206	209	219
406	169	176	179	182	185	195	456	192	200	202	206	209	220
407	169	177	179	183	186	196	457	192	200	203	207	210	220
408	170	177	180	183	186	196	458	193	200	203	207	210	221
409	170	177	180	184	187	197	459	193	201	204	208	211	221
410	171	178	180	184	187	197	460	194	201	204	208	211	222
411	171	178	181	185	188	198	461	194	202	205	208	212	222
412	172	179	181	185	188	198	462	195	202	205	209	212	223
413	172	179	182	186	189	199	463	195	203	205	209	213	223
414	173	180	182	186	189	199	464	196	203	206	210	213	224
415	173	180	183	187	190	200	465	196	204	206	210	214	224
416	174	181	183	187	190	200	466	197	204	207	211	214	225
417	174	181	184	187	191	201	467	197	205	207	211	215	225
418	174	182	184	188	191	201	468	197	205	208	212	215	226
419	175	182	185	188	192	202	469	198	206	208	212	216	226
420	175	183	185	189	192	202	470	198	206	209	213	216	227
421	176	183	186	189	193	203	471	199	207	209	213	217	227
422	176	184	186	190	193	203	472	199	207	210	214	217	228
423	177	184	187	190	194	204	473	200	208	210	214	218	228
424	177	185	187	191	194	204	474	200	208	211	215	218	229
425	178	185	188	191	195	205	475	201	208	211	215	219	229
426	178	185	188	192	195	205	476	201	209	212	216	219	230
427	179	186	188	192	196	206	477	202	209	212	216	220	230
428	179	186	189	193	196	206	478	202	210	213	217	220	231
429	179	187	189	193	196	207	479	203	210	213	217	221	231
430	180	187	190	194	197	207	480	203	211	214	218	221	232
431	180	188	190	194	197	207	481	203	211	214	218	221	232
432	181	188	191	195	198	208	482	204	212	214	218	222	233
433	181	189	191	195	198	208	483	204	212	215	219	222	233
434	182	189	192	196	199	209	484	205	213	215	219	223	234
435	182	190	192	196	199	209	485	205	213	216	220	223	234
436	183	190	193	197	200	210	486	206	214	216	220	224	235
437	183	191	193	197	200	210	487	206	214	217	221	224	235
438	184	191	194	198	201	211	488	207	215	217	221	225	236
439	184	192	194	198	201	211	489	207	215	218	222	225	236
440	185	192	195	198	202	212	490	208	216	218	222	226	237
441	185	192	195	199	202	212	491	208	216	219	223	226	237
442	185	193	196	199	203	213	492	209	216	219	223	227	238
443	186	193	196	200	203	213	493	209	217	220	224	227	238
444	186	194	197	200	204	214	494	209	217	220	224	228	239
445	187	194	197	201	204	214	495	210	218	221	225	228	239
446	187	195	197	201	205	215	496	210	218	221	225	229	239
447	188	195	198	202	205	215	497	211	219	222	226	229	240
448	188	196	198	202	206	216	498	211	219	222	226	230	240
449	189	196	199	203	206	216	499	212	220	223	227	230	241
450	189	197	199	203	207	217	500	212	220	223	227	231	241

n	Two-tailed probability, α						n	Two-tailed probability, α					
	.001	.01	.02	.05	.10	.50		.001	.01	.02	.05	.10	.50
501	213	221	223	228	231	242	551	236	244	247	252	255	267
502	213	221	224	228	232	242	552	236	245	248	252	256	267
503	214	222	224	229	232	243	553	237	245	248	252	256	268
504	214	222	225	229	233	243	554	237	246	249	253	257	268
505	215	223	225	229	233	244	555	238	246	249	253	257	269
506	215	223	226	230	234	244	556	238	247	250	254	258	269
507	216	224	226	230	234	245	557	239	247	250	254	258	270
508	216	224	227	231	234	245	558	239	248	251	255	259	270
509	216	224	227	231	235	246	559	240	248	251	255	259	271
510	217	225	228	232	235	246	560	240	249	251	256	260	271
511	217	225	228	232	236	247	561	241	249	252	256	260	272
512	218	226	229	233	236	247	562	241	249	252	257	261	272
513	218	226	229	233	237	248	563	242	250	253	257	261	272
514	219	227	230	234	237	248	564	242	250	253	258	261	273
515	219	227	230	234	238	249	565	242	251	254	258	262	273
516	220	228	231	235	238	249	566	243	251	254	259	262	274
517	220	228	231	235	239	250	567	243	252	255	259	263	274
518	221	229	232	236	239	250	568	244	252	255	260	263	275
519	221	229	232	236	240	251	569	244	253	256	260	264	275
520	222	230	232	237	240	251	570	245	253	256	261	264	276
521	222	230	233	237	241	252	571	245	254	257	261	265	276
522	222	231	233	238	241	252	572	246	254	257	262	265	277
523	223	231	234	238	242	253	573	246	255	258	262	266	277
524	223	232	234	239	242	253	574	247	255	258	263	266	278
525	224	232	235	239	243	254	575	247	256	259	263	267	278
526	224	232	235	240	243	254	576	248	256	259	263	267	279
527	225	233	236	240	244	255	577	248	257	260	264	268	279
528	225	233	236	240	244	255	578	249	257	260	264	268	280
529	226	234	237	241	245	256	579	249	258	261	265	269	280
530	226	234	237	241	245	256	580	249	258	261	265	269	281
531	227	235	238	242	246	257	581	250	258	261	266	270	281
532	227	235	238	242	246	257	582	250	259	262	266	270	282
533	228	236	239	243	247	258	583	251	259	262	267	271	282
534	228	236	239	243	247	258	584	251	260	263	267	271	283
535	229	237	240	244	247	259	585	252	260	263	268	272	283
536	229	237	240	244	248	259	586	252	261	264	268	272	284
537	229	238	241	245	248	260	587	253	261	264	269	273	284
538	230	238	241	245	249	260	588	253	262	265	269	273	285
539	230	239	242	246	249	261	589	254	262	265	270	274	285
540	231	239	242	246	250	261	590	254	263	266	270	274	286
541	231	240	242	247	250	262	591	255	263	266	271	275	286
542	232	240	243	247	251	262	592	255	264	267	271	275	287
543	232	241	243	248	251	263	593	255	264	267	272	275	287
544	233	241	244	248	252	263	594	256	265	268	272	276	288
545	233	241	244	249	252	264	595	256	265	268	273	276	288
546	234	242	245	249	253	264	596	257	266	269	273	277	289
547	234	242	245	250	253	265	597	257	266	269	274	277	289
548	235	243	246	250	254	265	598	258	267	270	274	278	290
549	235	243	246	251	254	266	599	258	267	270	275	278	290
550	235	244	247	251	255	266	600	259	267	271	275	279	291

n	Two-tailed probability, α						n	Two-tailed probability, α					
	.001	.01	.02	.05	.10	.50		.001	.01	.02	.05	.10	.50
601	259	268	271	275	279	291	651	283	292	295	300	304	316
602	260	268	271	276	280	292	652	283	292	295	300	304	316
603	260	269	272	276	280	292	653	284	293	296	300	304	317
604	261	269	272	277	281	293	654	284	293	296	301	305	317
605	261	270	273	277	281	293	655	284	294	297	301	305	318
606	262	270	273	278	282	294	656	285	294	297	302	306	318
607	262	271	274	278	282	294	657	285	295	298	302	306	319
608	262	271	274	279	283	295	658	286	295	298	303	307	319
609	263	272	275	279	283	295	659	286	295	299	303	307	320
610	263	272	275	280	284	296	660	287	296	299	304	308	320
611	264	273	276	280	284	296	661	287	296	300	304	308	321
612	264	273	276	281	285	297	662	288	297	300	305	309	321
613	265	274	277	281	285	297	663	288	297	301	305	309	322
614	265	274	277	282	286	298	664	289	298	301	306	310	322
615	266	275	278	282	286	298	665	289	298	302	306	310	323
616	266	275	278	283	287	299	666	290	299	302	307	311	323
617	267	276	279	283	287	299	667	290	299	302	307	311	324
618	267	276	279	284	288	300	668	291	300	303	308	312	324
619	268	276	280	284	288	300	669	291	300	303	308	312	325
620	268	277	280	285	289	301	670	291	301	304	309	313	325
621	269	277	281	285	289	301	671	292	301	304	309	313	326
622	269	278	281	286	289	302	672	292	302	305	310	314	326
623	269	278	281	286	290	302	673	293	302	305	310	314	327
624	270	279	282	287	290	303	674	293	303	306	311	315	327
625	270	279	282	287	291	303	675	294	303	306	311	315	328
626	271	280	283	287	291	304	676	294	304	307	312	316	328
627	271	280	283	288	292	304	677	295	304	307	312	316	329
628	272	281	284	288	292	305	678	295	304	308	312	317	329
629	272	281	284	289	293	305	679	296	305	308	313	317	330
630	273	282	285	289	293	306	680	296	305	309	313	318	330
631	273	282	285	290	294	306	681	297	306	309	314	318	331
632	274	283	286	290	294	307	682	297	306	310	314	319	331
633	274	283	286	291	295	307	683	298	307	310	315	319	332
634	275	284	287	291	295	308	684	298	307	311	315	319	332
635	275	284	287	292	296	308	685	298	308	311	316	320	333
636	276	285	288	292	296	308	686	299	308	312	316	320	333
637	276	285	288	293	297	309	687	299	309	312	317	321	334
638	276	285	289	293	297	309	688	300	309	313	317	321	334
639	277	286	289	294	298	310	689	300	310	313	318	322	335
640	277	286	290	294	298	310	690	301	310	313	318	322	335
641	278	287	290	295	299	311	691	301	311	314	319	323	336
642	278	287	291	295	299	311	692	302	311	314	319	323	336
643	279	288	291	296	300	312	693	302	312	315	320	324	337
644	279	288	291	296	300	312	694	303	312	315	320	324	337
645	280	289	292	297	301	313	695	303	313	316	321	325	338
646	280	289	292	297	301	313	696	304	313	316	321	325	338
647	281	290	293	298	302	314	697	304	314	317	322	326	339
648	281	290	293	298	302	314	698	305	314	317	322	326	339
649	282	291	294	299	303	315	699	305	314	318	323	327	340
650	282	291	294	299	303	315	700	306	315	318	323	327	340

n	Two-tailed probability, α						n	Two-tailed probability, α					
	.001	.01	.02	.05	.10	.50		.001	.01	.02	.05	.10	.50
701	306	315	319	324	328	341	751	329	339	343	348	352	365
702	306	316	319	324	328	341	752	330	340	343	348	352	366
703	307	316	320	325	329	342	753	330	340	344	349	353	366
704	307	317	320	325	329	342	754	331	341	344	349	353	367
705	308	317	321	325	330	343	755	331	341	345	350	354	367
706	308	318	321	326	330	343	756	332	342	345	350	354	368
707	309	318	322	326	331	344	757	332	342	346	351	355	368
708	309	319	322	327	331	344	758	333	343	346	351	355	369
709	310	319	323	327	332	345	759	333	343	346	352	356	369
710	310	320	323	328	332	345	760	334	344	347	352	356	370
711	311	320	324	328	333	346	761	334	344	347	352	357	370
712	311	321	324	329	333	346	762	335	344	348	353	357	371
713	312	321	324	329	334	346	763	335	345	348	353	358	371
714	312	322	325	330	334	347	764	336	345	349	354	358	372
715	313	322	325	330	335	347	765	336	346	349	354	359	372
716	313	323	326	331	335	348	766	337	346	350	355	359	373
717	313	323	326	331	335	348	767	337	347	350	355	360	373
718	314	324	327	332	336	349	768	337	347	351	356	360	374
719	314	324	327	332	336	349	769	338	348	351	356	361	374
720	315	324	328	333	337	350	770	338	348	352	357	361	375
721	315	325	328	333	337	350	771	339	349	352	357	362	375
722	316	325	329	334	338	351	772	339	349	353	358	362	376
723	316	326	329	334	338	351	773	340	350	353	358	363	376
724	317	326	330	335	339	352	774	340	350	354	359	363	377
725	317	327	330	335	339	352	775	341	351	354	359	364	377
726	318	327	331	336	340	353	776	341	351	355	360	364	378
727	318	328	331	336	340	353	777	342	352	355	360	365	378
728	319	328	332	337	341	354	778	342	352	356	361	365	379
729	319	329	332	337	341	354	779	343	353	356	361	366	379
730	320	329	333	338	342	355	780	343	353	357	362	366	380
731	320	330	333	338	342	355	781	344	354	357	362	367	380
732	321	330	334	338	343	356	782	344	354	357	363	367	381
733	321	331	334	339	343	356	783	345	354	358	363	367	381
734	321	331	335	339	344	357	784	345	355	358	364	368	382
735	322	332	335	340	344	357	785	345	355	359	364	368	382
736	322	332	335	340	345	358	786	346	356	359	365	369	383
737	323	333	336	341	345	358	787	346	356	360	365	369	383
738	323	333	336	341	346	359	788	347	357	360	365	370	384
739	324	334	337	342	346	359	789	347	357	361	366	370	384
740	324	334	337	342	347	360	790	348	358	361	366	371	385
741	325	334	338	343	347	360	791	348	358	362	367	371	385
742	325	335	338	343	348	361	792	349	359	362	367	372	386
743	326	335	339	344	348	361	793	349	359	363	368	372	386
744	326	336	339	344	349	362	794	350	360	363	368	373	386
745	327	336	340	345	349	362	795	350	360	364	369	373	387
746	327	337	340	345	350	363	796	351	361	364	369	374	387
747	328	337	341	346	350	363	797	351	361	365	370	374	388
748	328	338	341	346	351	364	798	352	362	365	370	375	388
749	329	338	342	347	351	364	799	352	362	366	371	375	389
750	329	339	342	347	351	365	800	353	363	366	371	376	389

Table B-5 cont.

n	Two-tailed probability, α						n	Two-tailed probability, α					
	.001	.01	.02	.05	.10	.50		.001	.01	.02	.05	.10	.50
801	353	363	367	372	376	390	851	377	387	391	396	401	415
802	353	364	367	372	377	390	852	377	387	391	396	401	415
803	354	364	368	373	377	391	853	377	388	392	397	401	416
804	354	365	368	373	378	391	854	378	388	392	397	402	416
805	355	365	369	374	378	392	855	378	389	393	398	402	417
806	355	365	369	374	379	392	856	379	389	393	398	403	417
807	356	366	369	375	379	393	857	379	390	393	399	403	418
808	356	366	370	375	380	393	858	380	390	394	399	404	418
809	357	367	370	376	380	394	859	380	391	394	400	404	419
810	357	367	371	376	381	394	860	381	391	395	400	405	419
811	358	368	371	377	381	395	861	381	392	395	401	405	420
812	358	368	372	377	382	395	862	382	392	396	401	406	420
813	359	369	372	378	382	396	863	382	393	396	402	406	421
814	359	369	373	378	383	396	864	383	393	397	402	407	421
815	360	370	373	379	383	397	865	383	394	397	403	407	422
816	360	370	374	379	384	397	866	384	394	398	403	408	422
817	361	371	374	379	384	398	867	384	395	398	404	408	423
818	361	371	375	380	384	398	868	385	395	399	404	409	423
819	361	372	375	380	385	399	869	385	396	399	405	409	424
820	362	372	376	381	385	399	870	386	396	400	405	410	424
821	362	373	376	381	386	400	871	386	397	400	406	410	425
822	363	373	377	382	386	400	872	386	397	401	406	411	425
823	363	374	377	382	387	401	873	387	397	401	407	411	426
824	364	374	378	383	387	401	874	387	398	402	407	412	426
825	364	375	378	383	388	402	875	388	398	402	408	412	427
826	365	375	379	384	388	402	876	388	399	403	408	413	427
827	365	375	379	384	389	403	877	389	399	403	408	413	428
828	366	376	380	385	389	403	878	389	400	404	409	414	428
829	366	376	380	385	390	404	879	390	400	404	409	414	429
830	367	377	381	386	390	404	880	390	401	405	410	415	429
831	367	377	381	386	391	405	881	391	401	405	410	415	429
832	368	378	381	387	391	405	882	391	402	405	411	416	430
833	368	378	382	387	392	406	883	392	402	406	411	416	430
834	369	379	382	388	392	406	884	392	403	406	412	417	431
835	369	379	383	388	393	407	885	393	403	407	412	417	431
836	369	380	383	389	393	407	886	393	404	407	413	418	432
837	370	380	384	389	394	408	887	394	404	408	413	418	432
838	370	381	384	390	394	408	888	394	405	408	414	418	433
839	371	381	385	390	395	409	889	394	405	409	414	419	433
840	371	382	385	391	395	409	890	395	406	409	415	419	434
841	372	382	386	391	396	410	891	395	406	410	415	420	434
842	372	383	386	392	396	410	892	396	407	410	416	420	435
843	373	383	387	392	397	411	893	396	407	411	416	421	435
844	373	384	387	393	397	411	894	397	408	411	417	421	436
845	374	384	388	393	398	412	895	397	408	412	417	422	436
846	374	385	388	394	398	412	896	398	408	412	418	422	437
847	375	385	389	394	399	413	897	398	409	413	418	423	437
848	375	386	389	394	399	413	898	399	409	413	419	423	438
849	376	386	390	395	400	414	899	399	410	414	419	424	438
850	376	386	390	395	400	414	900	400	410	414	420	424	439

n	Two-tailed probability, α						n	Two-tailed probability, α					
	.001	.01	.02	.05	.10	.50		.001	.01	.02	.05	.10	.50
901	400	411	415	420	425	439	951	424	435	439	444	449	464
902	401	411	415	421	425	440	952	424	435	439	445	450	465
903	401	412	416	421	426	440	953	425	436	440	445	450	465
904	402	412	416	422	426	441	954	425	436	440	446	451	466
905	402	413	417	422	427	441	955	426	437	441	446	451	466
906	403	413	417	423	427	442	956	426	437	441	447	452	467
907	403	414	417	423	428	442	957	427	438	442	447	452	467
908	403	414	418	423	428	443	958	427	438	442	448	453	468
909	404	415	418	424	429	443	959	428	439	442	448	453	468
910	404	415	419	424	429	444	960	428	439	443	449	454	469
911	405	416	419	425	430	444	961	429	440	443	449	454	469
912	405	416	420	425	430	445	962	429	440	444	450	454	470
913	406	417	420	426	431	445	963	429	441	444	450	455	470
914	406	417	421	426	431	446	964	430	441	445	451	455	471
915	407	418	421	427	432	446	965	430	442	445	451	456	471
916	407	418	422	427	432	447	966	431	442	446	452	456	472
917	408	419	422	428	433	447	967	431	442	446	452	457	472
918	408	419	423	428	433	448	968	432	443	447	453	457	473
919	409	419	423	429	434	448	969	432	443	447	453	458	473
920	409	420	424	429	434	449	970	433	444	448	453	458	473
921	410	420	424	430	435	449	971	433	444	448	454	459	474
922	410	421	425	430	435	450	972	434	445	449	454	459	474
923	411	421	425	431	436	450	973	434	445	449	455	460	475
924	411	422	426	431	436	451	974	435	446	450	455	460	475
925	412	422	426	432	436	451	975	435	446	450	456	461	476
926	412	423	427	432	437	452	976	436	447	451	456	461	476
927	412	423	427	433	437	452	977	436	447	451	457	462	477
928	413	424	428	433	438	453	978	437	448	452	457	462	477
929	413	424	428	434	438	453	979	437	448	452	458	463	478
930	414	425	429	434	439	454	980	438	449	453	458	463	478
931	414	425	429	435	439	454	981	438	449	453	459	464	479
932	415	426	430	435	440	455	982	438	450	454	459	464	479
933	415	426	430	436	440	455	983	439	450	454	460	465	480
934	416	427	430	436	441	456	984	439	451	455	460	465	480
935	416	427	431	437	441	456	985	440	451	455	461	466	481
936	417	428	431	437	442	457	986	440	452	455	461	466	481
937	417	428	432	438	442	457	987	441	452	456	462	467	482
938	418	429	432	438	443	458	988	441	453	456	462	467	482
939	418	429	433	438	443	458	989	442	453	457	463	468	483
940	419	430	433	439	444	459	990	442	453	457	463	468	483
941	419	430	434	439	444	459	991	443	454	458	464	469	484
942	420	430	434	440	445	460	992	443	454	458	464	469	484
943	420	431	435	440	445	460	993	444	455	459	465	470	485
944	420	431	435	441	446	461	994	444	455	459	465	470	485
945	421	432	436	441	446	461	995	445	456	460	466	471	486
946	421	432	436	442	447	462	996	445	456	460	466	471	486
947	422	433	437	442	447	462	997	446	457	461	467	472	487
948	422	433	437	443	448	463	998	446	457	461	467	472	487
949	423	434	438	443	448	463	999	447	458	462	468	473	488
950	423	434	438	444	449	464	1000	447	458	462	468	473	488

Table B-6

Critical Lower-Tail Values of b for Fisher's Hypergeometric Test for Dependence, Followed by Corresponding *Exact* Significance Level

Largest value of b, given *fixed* values of A, B and a, for the ratio a/A to be significantly larger than b/B in the following 2 × 2 table

a	$A - a$	A	$A \geqslant B$
b	$B - b$	B	$a/A \geqslant b/B$
$a + b$	$(A + B) - (a + b)$	$A + B$	

Thus largest value of b' for which

$$P(b \leqslant b' \mid A, B, a) = P\left\{ \left(\frac{a}{A} - \frac{b}{B} \geqslant \frac{a}{A} - \frac{b'}{B} \right) \right\} \leqslant \alpha$$

Followed by $P(b \leqslant b' \mid A, B, a + b')$

Note that the probabilities are conditional upon fixed marginal frequencies A, B, $a + b'$, and $(A + B) - (a + b')$.

In Table 12-1 of Chapter 12 the marginal frequencies are fixed. Therefore, when r takes small (or large) values so does $(N - R) - (n - r)$ in the diagonally opposite cell, whereas $R - r$ and $n - r$, in the adjacent cells, do the opposite. Therefore, for lower-tail values of r, b (in the above table) will correspond to r or to $(N - R) - (n - r)$, whereas for *upper*-tail values of r, b will correspond to $n - r$ or $(N - R) - (n - r)$, and a *lower*-tail test of $n - r$ will be equivalent to an *upper*-tail test of r.

Body of table is reproduced from Table 38, "Significance tests in a 2 x 2 Contingency Table" in Biometrika Tables for Statistcians, Vol. I, eds., E. S. Pearson and H. O. Hartley, Cambridge University Press, 1954, with the permission of the Trustees of Biometrika and of D. J. Finney, the author of the table.

Table B-6 cont.

A	B		a	0.05	0.025	0.01	0.005
A = 3	B = 3	3	3	0 .050			
A = 4	B = 4	4	4	0 .014	0 .014		
		3	4	0 .029			
A = 5	B = 5	5	5	1 .024	1 .024	0 .004	0 .004
			4	0 .024	0 .024		
		4	5	1 .048	0 .008	0 .008	
			4	0 .040			
		3	5	0 .018	0 .018		
		2	5	0 .048			
A = 6	B = 6	6	6	2 .030	1 .008	1 .008	0 .001
			5	1 .040	0 .008	0 .008	
			4	0 .030			
		5	6	1 .015+	0 .015+	0 .002	0 .002
			5	0 .013	0 .013		
			4	0 .045+			
		4	6	1 .033	0 .005-	0 .005-	0 .005-
			5	0 .024	0 .024		
		3	6	0 .012	0 .012		
			5	0 .048			
		2	6	0 .036			
A = 7	B = 7	7	7	3 .035-	2 .010+	1 .002	1 .002
			6	1 .015-	1 .015-	0 .002	0 .002
			5	0 .010+	0 .010+		
			4	0 .035-			
		6	7	2 .021	2 .021	1 .005-	1 .005-
			6	1 .025+	0 .004	0 .004	0 .004
			5	0 .016	0 .016		
			4	0 .049			
		5	7	2 .045+	1 .010+	0 .001	0 .001
			6	1 .045+	0 .008	0 .008	
			5	0 .027			
		4	7	1 .024	1 .024	0 .003	0 .003
			6	0 .015+	0 .015+		
			5	0 .045+			
		3	7	0 .008	0 .008	0 .008	
			6	0 .033			
		2	7	0 .028			

A	B		a	0.05	0.025	0.01	0.005
A = 8	B = 8	8	8	4 .038	3 .013	2 .003	2 .003
			7	2 .020	2 .020	1 .005+	0 .001
			6	1 .020	1 .020	0 .003	0 .003
			5	0 .013	0 .013		
			4	0 .038			
		7	8	3 .026	2 .007	2 .007	1 .001
			7	2 .035-	1 .009	1 .009	0 .001
			6	1 .032	0 .006	0 .006	
			5	0 .019	0 .019		
		6	8	2 .015-	2 .015-	1 .003	1 .003
			7	1 .016	1 .016	0 .002	0 .002
			6	0 .009	0 .009	0 .009	
			5	0 .028			
		5	8	2 .035-	1 .007	1 .007	0 .001
			7	1 .032	0 .005-	0 .005-	0 .005-
			6	0 .016	0 .016		
			5	0 .044			
		4	8	1 .018	1 .018	0 .002	0 .002
			7	0 .010+	0 .010+		
			6	0 .030			
		3	8	0 .006	0 .006	0 .006	
			7	0 .024	0 .024		
		2	8	0 .022	0 .022		
A = 9	B = 9	9	9	5 .041	4 .015-	3 .005-	3 .005-
			8	3 .025-	3 .025-	2 .008	1 .002
			7	2 .028	1 .008	1 .008	0 .001
			6	1 .025-	1 .025-	0 .005-	0 .005-
			5	0 .015-	0 .015-		
			4	0 .041			
		8	9	4 .029	3 .009	3 .009	2 .002
			8	3 .043	2 .013	1 .003	1 .003
			7	2 .044	1 .012	0 .002	0 .002
			6	1 .036	0 .007	0 .007	
			5	0 .020	0 .020		
		7	9	3 .019	3 .019	2 .005-	2 .005-
			8	2 .024	2 .024	1 .006	0 .001
			7	1 .020	1 .020	0 .003	0 .003
			6	0 .010+	0 .010+		
			5	0 .029			
		6	9	3 .044	2 .011	1 .002	1 .002
			8	2 .047	1 .011	0 .001	0 .001
			7	1 .035-	0 .006	0 .006	
			6	0 .017	0 .017		
			5	0 .042			

Left panel:

A	B	a	0.05	0.025	0.01	0.005
A = 9	B = 5	9	2 .027	1 .005-	1 .005-	1 .005-
		8	1 .023	1 .023	0 .003	0 .003
		7	0 .010+	0 .010+		
		6	0 .028			
	4	9	1 .014	1 .014	0 .001	0 .001
		8	0 .007	0 .007	0 .007	
		7	0 .021	0 .021		
		6	0 .049			
	3	9	1 .045+	1 .005-	0 .005-	0 .005-
		8	0 .018	0 .018		
		7	0 .045+			
	2	9	0 .018	0 .018		
A = 10	B = 10	10	6 .043	5 .016	4 .005+	3 .002
		9	4 .029	3 .010-	3 .010-	2 .003
		8	3 .035-	2 .012	1 .003	1 .003
		7	2 .035-	1 .010-	1 .010-	0 .002
		6	1 .029	0 .005+	0 .005+	
		5	0 .016	0 .016		
		4	0 .043			
	9	10	5 .033	4 .011	3 .003	3 .003
		9	4 .050-	3 .017	2 .005-	2 .005-
		8	2 .019	2 .019	1 .004	1 .004
		7	1 .015-	1 .015-	0 .002	0 .002
		6	1 .040	0 .008	0 .008	
		5	0 .022	0 .022		
	8	10	4 .023	4 .023	3 .007	2 .002
		9	3 .032	2 .009	2 .009	1 .002
		8	2 .031	1 .008	1 .008	0 .001
		7	1 .023	1 .023	0 .004	0 .004
		6	0 .011	0 .011		
		5	0 .029			
	7	10	3 .015-	3 .015-	2 .003	2 .003
		9	2 .018	2 .018	1 .004	1 .004
		8	1 .013	1 .013	0 .002	0 .002
		7	1 .036	0 .006	0 .006	
		6	0 .017	0 .017		
		5	0 .041			
	6	10	3 .036	2 .008	2 .008	1 .001
		9	2 .036	1 .008	1 .008	0 .001
		8	1 .024	1 .024	0 .003	0 .003
		7	0 .010+	0 .010+		
		6	0 .026			
	5	10	2 .022	2 .022	1 .004	1 .004
		9	1 .017	1 .017	0 .002	0 .002
		8	1 .047	0 .007	0 .007	
		7	0 .019	0 .019		
		6	0 .042			

Right panel:

A	B	a	0.05	0.025	0.01	0.005
A = 10	B = 4	10	1 .011	1 .011	0 .001	0 .001
		9	1 .041	0 .005-	0 .005-	0 .005-
		8	0 .015-	0 .015-		
		7	0 .035			
	3	10	1 .038	0 .003	0 .003	0 .003
		9	0 .014	0 .014		
		8	0 .035-			
	2	10	0 .015+	0 .015+		
		9	0 .045+			
A = 11	B = 11	11	7 .045+	6 .018	5 .006	4 .002
		10	5 .032	4 .012	3 .004	3 .004
		9	4 .040	3 .015-	2 .004	2 .004
		8	3 .043	2 .015-	1 .004	1 .004
		7	2 .040	1 .012	0 .002	0 .002
		6	1 .032	0 .006	0 .006	
		5	0 .018	0 .018		
		4	0 .045+			
	10	11	6 .035+	5 .012	4 .004	4 .004
		10	4 .021	4 .021	3 .007	2 .002
		9	3 .024	3 .024	2 .007	1 .002
		8	2 .023	2 .023	1 .006	0 .001
		7	1 .017	1 .017	0 .003	0 .003
		6	1 .043	0 .009	0 .009	
		5	0 .023	0 .023		
	9	11	5 .026	4 .008	4 .008	3 .002
		10	4 .038	3 .012	2 .003	2 .003
		9	3 .040	2 .012	1 .003	1 .003
		8	2 .035-	1 .009	1 .009	0 .001
		7	1 .025-	1 .025-	0 .004	0 .004
		6	0 .012	0 .012		
		5	0 .030			
	8	11	4 .018	4 .018	3 .005-	3 .005-
		10	3 .024	3 .024	2 .006	1 .001
		9	2 .022	2 .022	1 .005-	1 .005-
		8	1 .015-	1 .015-	0 .002	0 .002
		7	1 .037	0 .007	0 .007	
		6	0 .017	0 .017		
		5	0 .040			
	7	11	4 .043	3 .011	2 .002	2 .002
		10	3 .047	2 .013	1 .002	1 .002
		9	2 .039	1 .009	1 .009	0 .001
		8	1 .025-	1 .025-	0 .004	0 .004
		7	0 .010+	0 .010+		
		6	0 .025-	0 .025-		
	6	11	3 .029	2 .006	2 .006	1 .001
		10	2 .028	1 .005+	1 .005+	0 .001
		9	1 .018	1 .018	0 .002	0 .002

A	B	a	0.05	0.025	0.01	0.005
A = 11	B = 6	8	1 .043	0 .007	0 .007	—
		7	0 .017	0 .017		
		6	0 .037			
	5	11	2 .018	2 .018	1 .003	1 .003
		10	1 .013	1 .013	0 .001	0 .001
		9	1 .036	0 .005-	0 .005-	0 .005-
		8	0 .013	0 .013		
		7	0 .029			
	4	11	1 .009	1 .009	1 .009	0 .001
		10	1 .033	0 .004	0 .004	0 .004
		9	0 .011	0 .011		
		8	0 .026			
	3	11	1 .033	0 .003	0 .003	0 .003
		10	0 .011	0 .011		
		9	0 .027			
	2	11	0 .013	0 .013		
		10	0 .038			
A = 12	B = 12	12	8 .047	7 .019	6 .007	5 .002
		11	6 .034	5 .014	4 .005-	4 .005-
		10	5 .045-	4 .018	3 .006	2 .002
		9	4 .050-	3 .020	2 .006	1 .001
		8	3 .050-	2 .018	1 .005-	1 .005-
		7	2 .045-	1 .014	0 .002	0 .002
		6	1 .034	0 .007	0 .007	
		5	0 .019	0 .019		
		4	0 .047			
	11	12	7 .037	6 .014	5 .005-	5 .005-
		11	5 .024	5 .024	4 .008	3 .002
		10	4 .029	3 .010+	2 .003	2 .003
		9	3 .030	2 .009	2 .009	1 .002
		8	2 .026	1 .007	1 .007	0 .001
		7	1 .019	1 .019	0 .003	0 .003
		6	1 .045-	0 .009	0 .009	
		5	0 .024	0 .024		
	10	12	6 .029	5 .010-	5 .010-	4 .003
		11	5 .043	4 .015+	3 .005-	3 .005-
		10	4 .048	3 .017	2 .005-	2 .005-
		9	3 .046	2 .015-	1 .004	1 .004
		8	2 .038	1 .010+	0 .002	0 .002
		7	1 .026	0 .005-	0 .005-	0 .005-
		6	0 .012	0 .012		
		5	0 .030			
	9	12	5 .021	5 .021	4 .006	3 .002
		11	4 .029	3 .009	3 .009	2 .002
		10	3 .029	2 .008	2 .008	1 .002
		9	2 .024	2 .024	1 .006	0 .001
		8	1 .016	1 .016	0 .002	0 .002

A	B	a	0.05	0.025	0.01	0.005
A = 12	B = 9	7	1 .037	0 .007	0 .007	
		6	0 .017	0 .017		
		5	0 .039			
	8	12	5 .049	4 .014	3 .004	3 .004
		11	3 .018	3 .018	2 .004	2 .004
		10	2 .015+	2 .015+	1 .003	1 .003
		9	2 .040	1 .010-	1 .010-	0 .001
		8	1 .025-	1 .025-	0 .004	0 .004
		7	0 .010+	0 .010+		
		6	0 .024	0 .024		
	7	12	4 .036	3 .009	3 .009	2 .002
		11	3 .038	2 .010-	2 .010-	1 .002
		10	1 .017	1 .017	0 .002	0 .002
		9	1 .017	1 .017	0 .002	0 .002
		8	1 .040	0 .007	0 .007	
		7	0 .016	0 .016		
		6	0 .034			
	6	12	3 .025-	3 .025-	2 .005-	2 .005-
		11	2 .022	2 .022	1 .004	1 .004
		10	1 .013	1 .013	0 .002	0 .002
		9	1 .032	0 .005-	0 .005-	0 .005-
		8	0 .011	0 .011		
		7	0 .025-	0 .025-		
		6	0 .050-			
	5	12	2 .015-	2 .015-	1 .002	1 .002
		11	1 .010-	1 .010-	1 .010-	0 .001
		10	1 .028	0 .003	0 .003	0 .003
		9	0 .009	0 .009	0 .009	
			0 .020	0 .020		
			0 .041			
	4		2 .050	1 .007	1 .007	0 .001
		11	1 .027	0 .003	0 .003	0 .003
		10	0 .008	0 .008	0 .008	
		9	0 .019	0 .019		
		8	0 .038			
	3	12	1 .029	0 .002	0 .002	0 .002
		11	0 .009	0 .009	0 .009	
		10	0 .022	0 .022		
		9	0 .044			
	2	12	0 .011	0 .011		
		11	0 .033			
A = 13	B = 13	13	9 .048	8 .020	7 .007	6 .003
		12	7 .037	6 .015+	5 .006	4 .002
		11	6 .048	5 .021	4 .008	3 .002
		10	4 .024	4 .024	3 .008	2 .002
		9	3 .024	3 .024	2 .008	1 .002
		8	2 .021	2 .021	1 .006	0 .001

Table B-6 cont.

A = 13	B = 13	a	0.05	0.025	0.01	0.005
		7	2 .048	1 .015+	0 .003	0 .003
		6	1 .037	0 .007	0 .007	
		5	0 .020	0 .020		
		4	0 .048			
	12	13	8 .039	7 .015-	6 .005+	5 .002
		12	6 .027	5 .010-	5 .010-	4 .003
		11	5 .033	4 .013	3 .004	3 .004
		10	4 .036	3 .013	2 .004	2 .004
		9	3 .034	2 .011	1 .003	1 .003
		8	2 .029	1 .008	1 .008	0 .001
		7	1 .020	1 .020	0 .004	0 .004
		6	1 .046	0 .010-	0 .010-	
		5	0 .024	0 .024		
	11	13	7 .031	6 .011	5 .003	5 .003
		12	6 .048	5 .018	4 .006	3 .002
		11	4 .021	4 .021	3 .007	2 .002
		10	3 .021	3 .021	2 .006	1 .001
		9	3 .050-	2 .017	1 .004	1 .004
		8	2 .040	1 .011	0 .002	0 .002
		7	1 .027	0 .005-	0 .005-	0 .005-
		6	0 .013	0 .013		
		5	0 .030			
	10	13	6 .024	6 .024	5 .007	4 .002
		12	5 .035-	4 .012	3 .003	3 .003
		11	4 .037	3 .012	2 .003	2 .003
		10	3 .033	2 .010+	1 .002	1 .002
		9	2 .026	1 .006	1 .006	0 .001
		8	1 .017	1 .017	0 .003	0 .003
		7	1 .038	0 .007	0 .007	
		6	0 .017	0 .017		
		5	0 .038			
	9	13	5 .017	5 .017	4 .005-	4 .005-
		12	4 .023	4 .023	3 .007	2 .001
		11	3 .022	3 .022	2 .004	1 .001
		10	2 .017	2 .017	1 .004	1 .004
		9	2 .040	1 .010+	0 .001	0 .001
		8	1 .025-	1 .025-	0 .004	0 .004
		7	0 .010+	0 .010+		
		6	0 .023	0 .023		
		5	0 .049			
	8	13	5 .042	4 .012	3 .003	3 .003
		12	4 .047	3 .014	2 .003	2 .003
		11	3 .041	2 .011	1 .002	1 .002
		10	2 .029	1 .007	1 .007	0 .001
		9	1 .017	1 .017	0 .002	0 .002
		8	1 .037	0 .006	0 .006	
		7	0 .015-	0 .015-		
		6	0 .032			
	7	13	4 .031	3 .007	3 .007	2 .001
		12	3 .031	2 .007	2 .007	1 .001

A = 13	B = 7	a	0.05	0.025	0.01	0.005
		11	2 .022	2 .022	1 .004	1 .004
		10	1 .012	1 .012	0 .002	0 .002
		9	1 .029	0 .004	0 .004	0 .004
		8	0 .010+	0 .010+		
		7	0 .022	0 .022		
		6	0 .044			
	6	13	3 .021	3 .021	2 .004	2 .004
		12	2 .017	2 .017	1 .003	1 .003
		11	2 .046	1 .010-	1 .010-	0 .001
		10	1 .024	1 .024	0 .003	0 .003
		9	1 .050-	0 .008	0 .008	
		8	0 .017	0 .017		
		7	0 .034			
	5	13	2 .012	2 .012	1 .002	1 .002
		12	2 .044	1 .008	1 .008	0 .001
		11	1 .022	1 .022	0 .002	0 .002
		10	1 .047	0 .007	0 .007	
		9	0 .015-	0 .015-		
		8	0 .029			
	4	13	2 .044	1 .006	1 .006	1 .000
		12	1 .022	1 .022	0 .002	0 .002
		11	0 .006	0 .006	0 .006	
		10	0 .015-	0 .015-		
		9	0 .029			
	3	13	1 .025	1 .025	0 .002	0 .002
		12	0 .007	0 .007	0 .007	
		11	0 .018	0 .018		
		10	0 .036			
	2	13	0 .010-	0 .010-	0 .010-	
		12	0 .029			
A = 14	B = 14	14	10 .049	9 .020	8 .008	7 .003
		13	8 .038	7 .016	6 .006	5 .002
		12	6 .023	6 .023	5 .009	4 .003
		11	5 .027	4 .011	3 .004	3 .004
		10	4 .028	3 .011	2 .003	2 .003
		9	3 .027	2 .009	2 .009	1 .002
		8	2 .023	2 .023	1 .006	0 .001
		7	1 .016	1 .016	0 .003	0 .003
		6	1 .038	0 .008	0 .008	
		5	0 .020	0 .020		
		4	0 .049			
	13	14	9 .041	8 .016	7 .006	6 .002
		13	7 .029	6 .011	5 .004	5 .004
		12	6 .037	5 .015+	4 .005+	3 .002
		11	5 .041	4 .017	3 .006	2 .001
		10	4 .041	3 .016	2 .005-	2 .005-
		9	3 .038	2 .013	1 .003	1 .003
		8	2 .031	1 .009	1 .009	0 .001

A = 14	B = 13	a	0.05	0.025	0.01	0.005
		7	1 .021	1 .021	0 .004	0 .004
		6	1 .048	0 .010+		
		5	0 .025-	0 .025-		
	12	14	8 .033	7 .012	6 .004	6 .004
		13	6 .021	6 .021	5 .007	4 .002
		12	5 .025+	4 .009	4 .009	3 .003
		11	4 .026	3 .009	3 .009	2 .002
		10	3 .024	3 .024	2 .007	1 .002
		9	2 .019	2 .019	1 .005-	1 .005-
		8	2 .042	1 .012	0 .002	0 .002
		7	1 .028	0 .005+	0 .005+	
		6	0 .013	0 .013		
		5	0 .030			
	11	14	7 .026	6 .009	6 .009	5 .003
		13	6 .039	5 .014	4 .004	4 .004
		12	5 .043	4 .016	3 .005-	3 .005-
		11	4 .042	3 .015-	2 .004	2 .004
		10	3 .036	2 .011	1 .003	1 .003
		9	2 .027	1 .007	1 .007	0 .001
		8	1 .017	1 .017	0 .003	0 .003
		7	1 .038	0 .007	0 .007	
		6	0 .017	0 .017		
		5	0 .038			
	10	14	6 .020	6 .020	5 .006	4 .002
		13	5 .028	4 .009	4 .009	3 .002
		12	4 .028	3 .009	3 .009	2 .002
		11	3 .024	3 .024	2 .007	1 .001
		10	2 .018	2 .018	1 .004	1 .004
		9	2 .040	1 .011	0 .002	0 .002
		8	1 .024	1 .024	0 .004	0 .004
		7	0 .010-	0 .010-	0 .010-	
		6	0 .022	0 .022		
		5	0 .047			
	9	14	6 .047	5 .014	4 .004	4 .004
		13	4 .018	4 .018	3 .005-	3 .005-
		12	3 .017	3 .017	2 .004	2 .004
		11	3 .042	2 .012	1 .002	1 .002
		10	2 .029	1 .007	1 .007	0 .001
		9	1 .017	1 .017	0 .002	0 .002
		8	1 .036	0 .006	0 .006	
		7	0 .014	0 .014		
		6	0 .030			
	8	14	5 .036	4 .010-	4 .010-	3 .002
		13	4 .039	3 .011	2 .002	2 .002
		12	3 .032	2 .008	2 .008	1 .001
		11	2 .022	2 .022	1 .005-	1 .005-
		10	2 .048	1 .012	0 .002	0 .002
		9	1 .026	0 .004	0 .004	0 .004
		8	0 .009	0 .009	0 .009	
		7	0 .020	0 .020		
		6	0 .040			

A = 14	B = 7	a	0.05	0.025	0.005	0.01
		14	4 .026	3 .006	3 .006	2 .001
		13	3 .025	2 .006	2 .006	1 .001
		12	2 .017	2 .017	1 .003	1 .003
		11	2 .041	1 .009	1 .009	0 .001
		10	1 .021	1 .021	0 .003	0 .003
		9	1 .043	0 .007	0 .007	
		8	0 .015-	0 .015-		
		7	0 .030			
	6	14	3 .018	3 .018	2 .003	2 .003
		13	2 .014	2 .014	1 .002	1 .002
		12	2 .037	1 .007	1 .007	0 .001
		11	1 .018	1 .018	0 .002	0 .002
		10	1 .038	0 .005+	0 .005+	
		9	0 .012	0 .012		
		8	0 .024	0 .024		
		7	0 .044			
	5	14	2 .010+	2 .010+	1 .001	1 .001
		13	2 .037	1 .006	1 .006	0 .001
		12	1 .017	1 .017	0 .002	0 .002
		11	1 .038	0 .005-	0 .005-	0 .005-
		10	0 .011	0 .011		
		9	0 .022	0 .022		
		8	0 .040			
	4	14	2 .039	1 .005-	1 .005-	1 .005-
		13	1 .019	1 .019	0 .022	0 .002
		12	1 .044	0 .005-	0 .005-	0 .005-
		11	0 .011	0 .011		
		10	0 .023	0 .023		
		9	0 .041			
	3	14	1 .022	1 .022	0 .011	0 .001
		13	0 .006	0 .006	0 .006	
		12	0 .015-	0 .015-		
		11	0 .029			
	2	14	0 .008	0 .008	0 .008	
		13	0 .025	0 .025		
		12	0 .050			
A = 15	B = 15	15	11 .050-	10 .021	9 .008	8 .003
		14	9 .040	8 .018	7 .007	6 .003
		13	7 .025+	6 .010+	5 .004	5 .004
		12	6 .030	5 .013	4 .005-	4 .005-
		11	5 .033	4 .013	3 .005-	3 .005-
		10	4 .033	3 .013	2 .004	2 .004
		9	3 .030	2 .010+	1 .003	1 .003
		8	2 .025+	1 .007	1 .007	0 .004
		7	1 .018	1 .018	0 .003	0 .003
		6	1 .040	0 .008	0 .008	
		5	0 .021	0 .021		
		4	0 .050-			

Table B-6 cont.

A = 15, B = 14

	a	0.05	0.025	0.01	
15	15	10 .042	9 .017	8 .006	7 .002
	14	8 .031	7 .013	6 .005-	6 .005-
	13	7 .041	6 .017	5 .007	4 .002
	12	6 .046	5 .020	4 .007	3 .002
	11	5 .048	4 .020	3 .007	2 .002
	10	4 .046	3 .018	2 .006	1 .001
	9	3 .041	2 .014	1 .004	1 .004
	8	2 .033	1 .009	1 .009	0 .001
	7	1 .022	1 .022	0 .004	0 .004
	6	1 .049	0 .011		
	5	0 .025+			
13	15	9 .035-	8 .013	7 .005-	7 .005-
	14	7 .023	7 .023	6 .009	5 .003
	13	6 .029	5 .011	4 .004	4 .004
	12	5 .031	4 .012	3 .004	3 .004
	11	4 .030	3 .011	2 .003	2 .003
	10	3 .026	2 .008	2 .008	1 .002
	9	2 .020	2 .020	1 .005+	0 .001
	8	2 .043	1 .013	0 .002	0 .002
	7	1 .029	0 .005+	0 .005+	
	6	0 .013	0 .013		
	5	0 .031			
12	15	8 .028	7 .010-	7 .010-	6 .003
	14	7 .043	6 .016	5 .006	4 .002
	13	6 .049	5 .019	4 .007	3 .002
	12	5 .049	4 .019	3 .006	2 .002
	11	4 .045+	3 .017	2 .005-	2 .005-
	10	3 .038	2 .012	1 .003	1 .003
	9	2 .028	1 .007	1 .007	0 .001
	8	1 .018	1 .018	0 .003	0 .003
	7	1 .038	0 .007	0 .007	
	6	0 .017	0 .017		
	5	0 .037			
11	15	7 .022	7 .022	6 .007	5 .002
	14	6 .032	5 .011	4 .003	4 .003
	13	5 .034	4 .012	3 .003	3 .003
	12	4 .032	3 .010+	2 .003	2 .003
	11	3 .026	2 .008	2 .008	1 .002
	10	2 .019	2 .019	1 .004	1 .004
	9	2 .040	1 .011	0 .002	0 .002
	8	1 .024	1 .024	0 .004	0 .004
	7	1 .049	0 .010-	0 .010-	
	6	0 .022	0 .022		
	5	0 .046			
10	15	6 .017	6 .017	5 .005-	5 .005-
	14	5 .023	5 .023	4 .007	3 .002
	13	4 .022	4 .022	3 .007	2 .001
	12	3 .018	3 .018	2 .005-	2 .005-
	11	3 .042	2 .013	1 .003	1 .003
	10	2 .029	1 .007	1 .007	0 .001
	9	1 .016	1 .016	0 .002	0 .002
	8	1 .034	0 .006	0 .006	
	7	0 .013	0 .013		
	6	0 .028			
9	15	6 .042	5 .012	4 .003	4 .003
	14	5 .047	4 .015-	3 .004	3 .004

A = 15, B = 9

	a	0.05	0.025	0.01	
13	13	4 .042	3 .013	2 .003	2 .003
	12	3 .032	2 .009	2 .009	1 .002
	11	2 .021	2 .021	1 .005-	1 .005-
	10	2 .045-	1 .011	0 .002	0 .002
	9	1 .024	1 .024	0 .004	0 .004
	8	1 .048	0 .009	0 .009	
	7	0 .019	0 .019		
	6	0 .037			
8	15	5 .032	4 .008	4 .008	3 .002
	14	4 .033	3 .009	3 .009	2 .002
	13	3 .026	2 .006	2 .006	1 .001
	12	2 .017	2 .017	1 .003	1 .003
	11	2 .037	1 .008	1 .008	0 .001
	10	1 .019	1 .019	0 .003	0 .003
	9	1 .038	0 .006	0 .006	
	8	0 .013	0 .013		
	7	0 .026			
	6	0 .050-			
7	15	4 .023	4 .023	3 .005-	3 .005-
	14	3 .021	3 .021	2 .004	2 .004
	13	2 .014	2 .014	1 .002	1 .002
	12	2 .032	1 .007	1 .007	0 .001
	11	1 .015+	1 .015+	0 .002	0 .002
	10	1 .032	0 .005-	0 .005-	0 .005-
	9	0 .010+	0 .010+		
	8	0 .020	0 .020		
	7	0 .038			
6	15	3 .015+	3 .015+	2 .003	2 .003
	14	2 .011	2 .011	1 .002	1 .002
	13	2 .031	1 .006	1 .006	0 .001
	12	1 .014	1 .014	0 .002	0 .002
	11	1 .029	0 .004	0 .004	0 .004
	10	0 .009	0 .009	0 .009	
	9	0 .017	0 .017		
	8	0 .032			
5	15	2 .009	2 .009	2 .009	1 .001
	14	2 .032	1 .005-	1 .005-	1 .005-
	13	1 .014	1 .014	0 .001	0 .001
	12	1 .031	0 .004	0 .004	0 .004
	11	0 .008	0 .008	0 .008	
	10	0 .016	0 .016		
	9	0 .030			
4	15	2 .035+	1 .004	1 .004	1 .004
	14	1 .016	1 .016	0 .001	0 .001
	13	1 .037	0 .004	0 .004	0 .004
	12	0 .009	0 .009	0 .009	
	11	0 .018	0 .018		
	10	0 .033			
3	15	1 .020	1 .020	0 .001	0 .001
	14	0 .005-	0 .005-	0 .005-	0 .005-
	13	0 .012	0 .012		
	12	0 .025-	0 .025-		
	11	0 .043			
2	15	0 .007	0 .007	0 .007	
	14	0 .022	0 .022		
	13	0 .044			

Table B-7

Critical Lower-Tail Values of D (Indicating Positive Correlation) for Hotelling and Pabst's Test for Correlation

Largest value of D' for which $P(D \leqslant D') \leqslant \alpha$

Significance level, α

n	.001	.005	.010	.025	.050	.100
4	—	—	—	—	0	0
5	—	—	0	0	2	4
6	—	0	2	4	6	12
7	0	4	6	12	16	24
8	4	10	14	24	32	42
9	10	20	26	36	48	62
10	20	34	42	58	72	90
11	32	52	64	84	102	126
12	50	76	92	118	142	170
13	74	108	128	160	188	224
14	104	146	170	210	244	288
15	140	192	222	268	310	362
16	184	248	282	338	388	448
17	236	312	354	418	478	548
18	298	388	436	510	580	662
19	370	474	530	616	694	788
20	452	572	636	736	824	932
21	544	684	756	868	970	1090
22	650	808	890	1018	1132	1268
23	770	948	1040	1182	1310	1462
24	902	1102	1206	1364	1508	1676
25	1048	1272	1388	1564	1724	1910
26	1210	1460	1588	1784	1958	2166
27	1388	1664	1806	2022	2214	2442
28	1584	1888	2044	2282	2492	2742
29	1798	2132	2304	2562	2794	3066
30	2030	2396	2582	2866	3118	3414

Critical *upper*-tail values of D for significance level α are obtained by subtracting cell entry from $n(n^2 - 1)/3$. That is, the smallest value of D' for which $P(D \geqslant D') \leqslant \alpha$ is $\lceil n(n^2 - 1)/3 \rceil$ − cell entry. (Example: Critical value of D for upper-tail test at level α = .01 when n = 6 is $n(n^2 - 1)/3$ minus cell entry corresponding to n = 6 and α = .01. So it is $\lceil 6(35)/3 \rceil - 2 = 68$.)

The table is exact for $n \leqslant 10$ and is based upon an approximation at higher n-values. Table is adapted from Table 2 in G. J. Glasser and R. F. Winter's "Critical Values of the Coefficient of Rank Correlation for Testing the Hypothesis of Independence," *Biometrika*, Vol. 48 (1961), pp. 444-448, with permission of the authors and the Trustees of *Biometrika*.

Critical and Quasi-Critical Lower-Tail Values of W_+
(and Their Corresponding Cumulative Probabilities)
for Wilcoxon's Signed-Rank Test

W_+' followed by $P(W_+ \leqslant W_+')$

n	$\alpha = .05$		$\alpha = .025$		$\alpha = .01$		$\alpha = .005$	
5	0	.0313						
	1	.0625						
6	2	.0469	0	.0156				
	3	.0781	1	.0313				
7	3	.0391	2	.0234	0	.0078		
	4	.0547	3	.0391	1	.0156		
8	5	.0391	3	.0195	1	.0078	0	.0039
	6	.0547	4	.0273	2	.0117	1	.0078
9	8	.0488	5	.0195	3	.0098	1	.0039
	9	.0645	6	.0273	4	.0137	2	.0059
10	10	.0420	8	.0244	5	.0098	3	.0049
	11	.0527	9	.0322	6	.0137	4	.0068
11	13	.0415	10	.0210	7	.0093	5	.0049
	14	.0508	11	.0269	8	.0122	6	.0068
12	17	.0461	13	.0212	9	.0081	7	.0046
	18	.0549	14	.0261	10	.0105	8	.0061
13	21	.0471	17	.0239	12	.0085	9	.0040
	22	.0549	18	.0287	13	.0107	10	.0052
14	25	.0453	21	.0247	15	.0083	12	.0043
	26	.0520	22	.0290	16	.0101	13	.0054
15	30	.0473	25	.0240	19	.0090	15	.0042
	31	.0535	26	.0277	20	.0108	16	.0051
16	35	.0467	29	.0222	23	.0091	19	.0046
	36	.0523	30	.0253	24	.0107	20	.0055
17	41	.0492	34	.0224	27	.0087	23	.0047
	42	.0544	35	.0253	28	.0101	24	.0055
18	47	.0494	40	.0241	32	.0091	27	.0045
	48	.0542	41	.0269	33	.0104	28	.0052
19	53	.0478	46	.0247	37	.0090	32	.0047
	54	.0521	47	.0273	38	.0102	33	.0054
20	60	.0487	52	.0242	43	.0096	37	.0047
	61	.0527	53	.0266	44	.0107	38	.0053
21	67	.0479	58	.0230	49	.0097	42	.0045
	68	.0516	59	.0251	50	.0108	43	.0051
22	75	.0492	65	.0231	55	.0095	48	.0046
	76	.0527	66	.0250	56	.0104	49	.0052
23	83	.0490	73	.0242	62	.0098	54	.0046
	84	.0523	74	.0261	63	.0107	55	.0051
24	91	.0475	81	.0245	69	.0097	61	.0048
	92	.0505	82	.0263	70	.0106	62	.0053
25	100	.0479	89	.0241	76	.0094	68	.0048
	101	.0507	90	.0258	77	.0101	69	.0053

Critical *upper*-tail values of W_+ for significance level α are obtained by subtracting first cell entry from $n(n + 1)/2$. (Example: Critical value of W_+ for upper-tail test at level $\alpha = .05$ when $n = 9$ is $n(n + 1)/2$ minus cell entry corresponding to $n = 9$ and $\alpha = .05$. So it is $[9(10)/2] - 8 = 37$
Body of table is reproduced, with changes only in notation, from Table II in F. Wilcoxon, S. K. Katti, and Roberta A. Wilcox, *Crtical Values and Probability Levels for the Wilcoxon Rank-Sum Test and the Wilcoxon Signed-Rank Test,* American Cyanamid Company (Lederle Laboratories Division, Pearl River, N.Y.) and the Florida State University (Department of Statistics, Tallahassee, Fla.), August 1963, with permission of authors and publishers.

n	$\alpha = .05$		$\alpha = .025$		$\alpha = .01$		$\alpha = .005$	
26	110	.0497	98	.0247	84	.0095	75	.0047
	111	.0524	99	.0263	85	.0102	76	.0051
27	119	.0477	107	.0246	92	.0093	83	.0048
	120	.0502	108	.0260	93	.0100	84	.0052
28	130	.0496	116	.0239	101	.0096	91	.0048
	131	.0521	117	.0252	102	.0102	92	.0051
29	140	.0482	126	.0240	110	.0095	100	.0049
	141	.0504	127	.0253	111	.0101	101	.0053
30	151	.0481	137	.0249	120	.0098	109	.0050
	152	.0502	138	.0261	121	.0104	110	.0053
31	163	.0491	147	.0239	130	.0099	118	.0049
	164	.0512	148	.0251	131	.0105	119	.0052
32	175	.0492	159	.0249	140	.0097	128	.0050
	176	.0512	160	.0260	141	.0103	129	.0053
33	187	.0485	170	.0242	151	.0099	138	.0049
	188	.0503	171	.0253	152	.0104	139	.0052
34	200	.0488	182	.0242	162	.0098	148	.0048
	201	.0506	183	.0252	163	.0103	149	.0051
35	213	.0484	195	.0247	173	.0096	159	.0048
	214	.0501	196	.0257	174	.0100	160	.0051
36	227	.0489	208	.0248	185	.0096	171	.0050
	228	.0505	209	.0258	186	.0100	172	.0052
37	241	.0487	221	.0245	198	.0099	182	.0048
	242	.0503	222	.0254	199	.0103	183	.0050
38	256	.0493	235	.0247	211	.0099	194	.0048
	257	.0509	236	.0256	212	.0104	195	.0050
39	271	.0493	249	.0246	224	.0099	207	.0049
	272	.0507	250	.0254	225	.0103	208	.0051
40	286	.0486	264	.0249	238	.0100	220	.0049
	287	.0500	265	.0257	239	.0104	221	.0051
41	302	.0488	279	.0248	252	.0100	233	.0048
	303	.0501	280	.0256	253	.0103	234	.0050
42	319	.0496	294	.0245	266	.0098	247	.0049
	320	.0509	295	.0252	267	.0102	248	.0051
43	336	.0498	310	.0245	281	.0098	261	.0048
	337	.0511	311	.0252	282	.0102	262	.0050
44	353	.0495	327	.0250	296	.0097	276	.0049
	354	.0507	328	.0257	297	.0101	277	.0051
45	371	.0498	343	.0244	312	.0098	291	.0049
	372	.0510	344	.0251	313	.0101	292	.0051
46	389	.0497	361	.0249	328	.0098	307	.0050
	390	.0508	362	.0256	329	.0101	308	.0052
47	407	.0490	378	.0245	345	.0099	322	.0048
	408	.0501	379	.0251	346	.0102	323	.0050
48	426	.0490	396	.0244	362	.0099	339	.0050
	427	.0500	397	.0251	363	.0102	340	.0051
49	446	.0495	415	.0247	379	.0098	355	.0049
	447	.0505	416	.0253	380	.0100	356	.0050
50	466	.0495	434	.0247	397	.0098	373	.0050
	467	.0506	435	.0253	398	.0101	374	.0051

Table B-9

Critical Lower-Tail Values of W_n for Wilcoxon's Rank-Sum Test

Largest value of W_n' for which $P(W_n \leqslant W_n') \leqslant \alpha$ where α is boldface column heading

m	n=1 0.001	0.005	0.010	0.025	0.05	0.10	$2\bar{W}$	n=2 0.001	0.005	0.010	0.025	0.05	0.10	$2\bar{W}$	m
2							4						−	10	2
3							5						3	12	3
4							6					−	3	14	4
5							7					3	4	16	5
6							8					3	4	18	6
7							9				−	3	4	20	7
8						−	10				3	4	5	22	8
9						1	11				3	4	5	24	9
10						1	12				3	4	6	26	10
11						1	13				3	4	6	28	11
12						1	14			−	4	5	7	30	12
13						1	15			3	4	5	7	32	13
14						1	16			3	4	6	8	34	14
15						1	17			3	4	6	8	36	15
16						1	18			3	4	6	8	38	16
17						1	19			3	5	6	9	40	17
18					−	1	20		−	3	5	7	9	42	18
19					1	2	21		3	4	5	7	10	44	19
20					1	2	22		3	4	5	7	10	46	20
21					1	2	23		3	4	6	8	11	48	21
22					1	2	24		3	4	6	8	11	50	22
23					1	2	25		3	4	6	8	12	52	23
24					1	2	26		3	4	6	9	12	54	24
25	−	−	−	−	1	2	27	−	3	4	6	9	12	56	25

m	n=3 0.001	0.005	0.010	0.025	0.05	0.10	$2\bar{W}$	n=4 0.001	0.005	0.010	0.025	0.05	0.10	$2\bar{W}$	m
3					6	7	21								
4				−	6	7	24			−	10	11	13	36	4
5				6	7	8	27		−	10	11	12	14	40	5
6			−	7	8	9	30		10	11	12	13	15	44	6
7			6	7	8	10	33		10	11	13	14	16	48	7
8		−	6	8	9	11	36		11	12	14	15	17	52	8
9		6	7	8	10	11	39	−	11	13	14	16	19	56	9
10		6	7	9	10	12	42	10	12	13	15	17	20	60	10
11		6	7	9	11	13	45	10	12	14	16	18	21	64	11
12		7	8	10	11	14	48	10	13	15	17	19	22	68	12
13		7	8	10	12	15	51	11	13	15	18	20	23	72	13
14		7	8	11	13	16	54	11	14	16	19	21	25	76	14
15		8	9	11	13	16	57	11	15	17	20	22	26	80	15
16	−	8	9	12	14	17	60	12	15	17	21	24	27	84	16
17	6	8	10	12	15	18	63	12	16	18	21	25	28	88	17
18	6	8	10	13	15	19	66	13	16	19	22	26	30	92	18
19	6	9	10	13	16	20	69	13	17	19	23	27	31	96	19
20	6	9	11	14	17	21	72	13	18	20	24	28	32	100	20
21	7	9	11	14	17	21	75	14	18	21	25	29	33	104	21
22	7	10	12	15	18	22	78	14	19	21	26	30	35	108	22
23	7	10	12	15	19	23	81	14	19	22	27	31	36	112	23
24	7	10	12	16	19	24	84	15	20	23	27	32	38	116	24
25	7	11	13	16	20	25	87	15	20	23	28	33	38	120	25

Critical *upper*-tail values of W_n for significance level α are obtained by subtracting cell centry from $2\bar{W}$, which is given in the table. That is, the *smallest* value of W_n' for which $P(W_n \geqslant W_n') \leqslant \alpha$ is $2\bar{W}$ -cell entry. (Example: Critical value of W_n for upper-tail test at level $\alpha = .001$ when $n = 3$ and $m = 22$ is obtained by subtracting cell entry corresponding to $n = 3$, $m = 22$ and $\alpha = .001$ from value of $2\bar{W}$ listed in the same row. So it is $78 - 7 = 71$.)

Body of table is reproduced, with changes only in notation, from Table 1 in L. R. Verdooren's "Extended Tables of Critical Values for Wilcoxon's Test Statistic," *Biometrika*, Vol. 50 (1963), pp. 177-186, with permission of the author and the Trustees of *Biometrika*.

Table B-9 cont.

m	n = 5							n = 6							m
	0.001	0.005	0.010	0.025	0.05	0.10	$2\bar{W}$	0.001	0.005	0.010	0.025	0.05	0.10	$2\bar{W}$	
5		15	16	17	19	20	55								
6		16	17	18	20	22	60	−	23	24	26	28	30	78	6
7	−	16	18	20	21	23	65	21	24	25	27	29	32	84	7
8	15	17	19	21	23	25	70	22	25	27	29	31	34	90	8
9	16	18	20	22	24	27	75	23	26	28	31	33	36	96	9
10	16	19	21	23	26	28	80	24	27	29	32	35	38	102	10
11	17	20	22	24	27	30	85	25	28	30	34	37	40	108	11
12	17	21	23	26	28	32	90	25	30	32	35	38	42	114	12
13	18	22	24	27	30	33	95	26	31	33	37	40	44	120	13
14	18	22	25	28	31	35	100	27	32	34	38	42	46	126	14
15	19	23	26	29	33	37	105	28	33	36	40	44	48	132	15
16	20	24	27	30	34	38	110	29	34	37	42	46	50	138	16
17	20	25	28	32	35	40	115	30	36	39	43	47	52	144	17
18	21	26	29	33	37	42	120	31	37	40	45	49	55	150	18
19	22	27	30	34	38	43	125	32	38	41	46	51	57	156	19
20	22	28	31	35	40	45	130	33	39	43	48	53	59	162	20
21	23	29	32	37	41	47	135	33	40	44	50	55	61	168	21
22	23	29	33	38	43	48	140	34	42	45	51	57	63	174	22
23	24	30	34	39	44	50	145	35	43	47	53	58	65	180	23
24	25	31	35	40	45	51	150	36	44	48	54	60	67	186	24
25	25	32	36	42	47	53	155	37	45	50	56	62	69	192	25

m	n = 7							n = 8							m
	0.001	0.005	0.010	0.025	0.05	0.10	$2\bar{W}$	0.001	0.005	0.010	0.025	0.05	0.10	$2\bar{W}$	
7	29	32	34	36	39	41	105								
8	30	34	35	38	41	44	112	40	43	45	49	51	55	136	8
9	31	35	37	40	43	46	119	41	45	47	51	54	58	144	9
10	33	37	39	42	45	49	126	42	47	49	53	56	60	152	10
11	34	38	40	44	47	51	133	44	49	51	55	59	63	160	11
12	35	40	42	46	49	54	140	45	51	53	58	62	66	168	12
13	36	41	44	48	52	56	147	47	53	56	60	64	69	176	13
14	37	43	45	50	54	59	154	48	54	58	62	67	72	184	14
15	38	44	47	52	56	61	161	50	56	60	65	69	75	192	15
16	39	46	49	54	58	64	168	51	58	62	67	72	78	200	16
17	41	47	51	56	61	66	175	53	60	64	70	75	81	208	17
18	42	49	52	58	63	69	182	54	62	66	72	77	84	216	18
19	43	50	54	60	65	71	189	56	64	68	74	80	87	224	19
20	44	52	56	62	67	74	196	57	66	70	77	83	90	232	20
21	46	53	58	64	69	76	203	59	68	72	79	85	92	240	21
22	47	55	59	66	72	79	210	60	70	74	81	88	95	248	22
23	48	57	61	68	74	81	217	62	71	76	84	90	98	256	23
24	49	58	63	70	76	84	224	64	73	78	86	93	101	264	24
25	50	60	64	72	78	86	231	65	75	81	89	96	104	272	25

m	n = 9							n = 10							m
	0.001	0.005	0.010	0.025	0.05	0.10	$2\bar{W}$	0.001	0.005	0.010	0.025	0.05	0.10	$2\bar{W}$	
9	52	56	59	62	66	70	171								
10	53	58	61	65	69	73	180	65	71	74	78	82	87	210	10
11	55	61	63	68	72	76	189	67	73	77	81	86	91	220	11
12	57	63	66	71	75	80	198	69	76	79	84	89	94	230	12
13	59	65	68	73	78	83	207	72	79	82	88	92	98	240	13
14	60	67	71	76	81	86	216	74	81	85	91	96	102	250	14
15	62	69	73	79	84	90	225	76	84	88	94	99	106	260	15
16	64	72	76	82	87	93	234	78	86	91	97	103	109	270	16
17	66	74	78	84	90	97	243	80	89	93	100	106	113	280	17
18	68	76	81	87	93	100	252	82	92	96	103	110	117	290	18
19	70	78	83	90	96	103	261	84	94	99	107	113	121	300	19
20	71	81	85	93	99	107	270	87	97	102	110	117	125	310	20
21	73	83	88	95	102	110	279	89	99	105	113	120	128	320	21
22	75	85	90	98	105	113	288	91	102	108	116	123	132	330	22
23	77	88	93	101	108	117	297	93	105	110	119	127	136	340	23
24	79	90	95	104	111	120	306	95	107	113	122	130	140	350	24
25	81	92	98	107	114	123	315	98	110	116	126	134	144	360	25

Table B-9 cont.

m	n = 11 0.001	0.005	0.010	0.025	0.05	0.10	2W̄	n = 12 0.001	0.005	0.010	0.025	0.05	0.10	2W̄	m
11	81	87	91	96	100	106	253								
12	83	90	94	99	104	110	264	98	105	109	115	120	127	300	12
13	86	93	97	103	108	114	275	101	109	113	119	125	131	312	13
14	88	96	100	106	112	118	286	103	112	116	123	129	136	324	14
15	90	99	103	110	116	123	297	106	115	120	127	133	141	336	15
16	93	102	107	113	120	127	308	109	119	124	131	138	145	348	16
17	95	105	110	117	123	131	319	112	122	127	135	142	150	360	17
18	98	108	113	121	127	135	330	115	125	131	139	146	155	372	18
19	100	111	116	124	131	139	341	118	129	134	143	150	159	384	19
20	103	114	119	128	135	144	352	120	132	138	147	155	164	396	20
21	106	117	123	131	139	148	363	123	136	142	151	159	169	408	21
22	108	120	126	135	143	152	374	126	139	145	155	163	173	420	22
23	111	123	129	139	147	156	385	129	142	149	159	168	178	432	23
24	113	126	132	142	151	161	396	132	146	153	163	172	183	444	24
25	116	129	136	146	155	165	407	135	149	156	167	176	187	456	25

m	n = 13 0.001	0.005	0.010	0.025	0.05	0.10	2W̄	n = 14 0.001	0.005	0.010	0.025	0.05	0.10	2W̄	m
13	117	125	130	136	142	149	351								
14	120	129	134	141	147	154	364	137	147	152	160	166	174	406	14
15	123	133	138	145	152	159	377	141	151	156	164	171	179	420	15
16	126	136	142	150	156	165	390	144	155	161	169	176	185	434	16
17	129	140	146	154	161	170	403	148	159	165	174	182	190	448	17
18	133	144	150	158	166	175	416	151	163	170	179	187	196	462	18
19	136	148	154	163	171	180	429	155	168	174	183	192	202	476	19
20	139	151	158	167	175	185	442	159	172	178	188	197	207	490	20
21	142	155	162	171	180	190	455	162	176	183	193	202	213	504	21
22	145	159	166	176	185	195	468	166	180	187	198	207	218	518	22
23	149	163	170	180	189	200	481	169	184	192	203	212	224	532	23
24	152	166	174	185	194	205	494	173	188	196	207	218	229	546	24
25	155	170	178	189	199	211	507	177	192	200	212	223	235	560	25

m	n = 15 0.001	0.005	0.010	0.025	0.05	0.10	2W̄	n = 16 0.001	0.005	0.010	0.025	0.05	0.10	2W̄	m
15	160	171	176	184	192	200	465								
16	163	175	181	190	197	206	480	184	196	202	211	219	229	528	16
17	167	180	186	195	203	212	495	188	201	207	217	225	235	544	17
18	171	184	190	200	208	218	510	192	206	212	222	231	242	560	18
19	175	189	195	205	214	224	525	196	210	218	228	237	248	576	19
20	179	193	200	210	220	230	540	201	215	223	234	243	255	592	20
21	183	198	205	216	225	236	555	205	220	228	239	249	261	608	21
22	187	202	210	221	231	242	570	209	225	233	245	255	267	624	22
23	191	207	214	226	236	248	585	214	230	238	251	261	274	640	23
24	195	211	219	231	242	254	600	218	235	244	256	267	280	656	24
25	199	216	224	237	248	260	615	222	240	249	262	273	287	672	25

m	n = 17 0.001	0.005	0.010	0.025	0.05	0.10	2W̄	n = 18 0.001	0.005	0.010	0.025	0.05	0.10	2W̄	m
17	210	223	230	240	249	259	595								
18	214	228	235	246	255	266	612	237	252	259	270	280	291	666	18
19	219	234	241	252	262	273	629	242	258	265	277	287	299	684	19
20	223	239	246	258	268	280	646	247	263	271	283	294	306	702	20
21	228	244	252	264	274	287	663	252	269	277	290	301	313	720	21
22	233	249	258	270	281	294	680	257	275	283	296	307	321	738	22
23	238	255	263	276	287	300	697	262	280	289	303	314	328	756	23
24	242	260	269	282	294	307	714	267	286	295	309	321	335	774	24
25	247	265	275	288	300	314	731	273	292	301	316	328	343	792	25

Table B-9 cont.

m	$n = 19$							$n = 20$							m
	0.001	0.005	0.010	0.025	0.05	0.10	$2\bar{W}$	0.001	0.005	0.010	0.025	0.05	0.10	$2\bar{W}$	
19	267	283	291	303	313	325	741								
20	272	289	297	309	320	333	760	298	315	324	337	348	361	820	20
21	277	295	303	316	328	341	779	304	322	331	344	356	370	840	21
22	283	301	310	323	335	349	798	309	328	337	351	364	378	860	22
23	288	307	316	330	342	357	817	315	335	344	359	371	386	880	23
24	294	313	323	337	350	364	836	321	341	351	366	379	394	900	24
25	299	319	329	344	357	372	855	327	348	358	373	387	403	920	25

m	$n = 21$							$n = 22$							m
	0.001	0.005	0.010	0.025	0.05	0.10	$2\bar{W}$	0.001	0.005	0.010	0.025	0.05	0.10	$2\bar{W}$	
21	331	349	359	373	385	399	903								
22	337	356	366	381	393	408	924	365	386	396	411	424	439	990	22
23	343	363	373	388	401	417	945	372	393	403	419	432	448	1012	23
24	349	370	381	396	410	425	966	379	400	411	427	441	457	1034	24
25	356	377	388	404	418	434	987	385	408	419	435	450	467	1056	25

m	$n = 23$							$n = 24$							m
	0.001	0.005	0.010	0.025	0.05	0.10	$2\bar{W}$	0.001	0.005	0.010	0.025	0.05	0.10	$2\bar{W}$	
23	402	424	434	451	465	481	1081								
24	409	431	443	459	474	491	1104	440	464	475	492	507	525	1176	24
25	416	439	451	468	483	500	1127	448	472	484	501	517	535	1200	25

m	$n = 25$						
	0.001	0.005	0.010	0.025	0.05	0.10	$2\bar{W}$
25	480	505	517	536	552	570	1275

Table B-10

"Critical Values" of S for Friedman Test,
Based Upon an Approximation

Approximation to smallest value of S' for which $P(S \geqslant S') \leqslant \alpha$

R	C					Additional values for $C = 3$	
	3	4	5	6	7	R	S'
Values at .05 level of significance							
3			64.4	103.9	157.3	9	54.0
4		49.5	88.4	143.3	217.0	12	71.9
5		62.6	112.3	182.4	276.2	14	83.8
6		75.7	136.1	221.4	335.2	16	95.8
8	48.1	101.7	183.7	299.0	453.1	18	107.7
10	60.0	127.8	231.2	376.7	571.0		
15	89.8	192.9	349.8	570.5	864.9		
20	119.7	258.0	468.5	764.4	1158.7		
Values at .01 level of significance							
3			75.6	122.8	185.6	9	75.9
4		61.4	109.3	176.2	265.0	12	103.5
5		80.5	142.8	229.4	343.8	14	121.9
6		99.5	176.1	282.4	422.6	16	140.2
8	66.8	137.4	242.7	388.3	579.9	18	158.6
10	85.1	175.3	309.1	494.0	737.0		
15	131.0	269.8	475.2	758.2	1129.5		
20	177.0	364.2	641.2	1022.2	1521.9		

Body of table is reproduced, with changes only in notation, from Table III in M. Friedman's "A Comparison of Alternative Tests of Significance for the Problem of m Rankings," *Annals of Mathematical Statistics*, Vol. 11 (1940), pp. 86-92, with permission of author and editor.

Cumulative Probabilities for a Standardized Normal Variable ϕ

Q = left row heading + upper column heading = $P(\phi \geqslant$ cell entry$) = P(\phi \leqslant -$ cell entry$)$
P = right row heading + lower column heading = $P(\phi \leqslant$ cell entry$) = P(\phi \geqslant -$ cell entry$)$

Q	.000	.001	.002	.003	.004	.005	.006	.007	.008	.009	.010	
.00	∞	3.0902	2.8782	2.7478	2.6521	2.5758	2.5121	2.4573	2.4089	2.3656	2.3263	.99
.01	2.3263	2.2904	2.2571	2.2262	2.1973	2.1701	2.1444	2.1201	2.0969	2.0749	2.0537	.98
.02	2.0537	2.0335	2.0141	1.9954	1.9774	1.9600	1.9431	1.9268	1.9110	1.8957	1.8808	.97
.03	1.8808	1.8663	1.8522	1.8384	1.8250	1.8119	1.7991	1.7866	1.7744	1.7624	1.7507	.96
.04	1.7507	1.7392	1.7279	1.7169	1.7060	1.6954	1.6849	1.6747	1.6646	1.6546	1.6449	.95
.05	1.6449	1.6352	1.6258	1.6164	1.6072	1.5982	1.5893	1.5805	1.5718	1.5632	1.5548	.94
.06	1.5548	1.5464	1.5382	1.5301	1.5220	1.5141	1.5063	1.4985	1.4909	1.4833	1.4758	.93
.07	1.4758	1.4684	1.4611	1.4538	1.4466	1.4395	1.4325	1.4255	1.4187	1.4118	1.4651	.92
.08	1.4051	1.3984	1.3917	1.3852	1.3787	1.3722	1.3658	1.3595	1.3532	1.3469	1.3408	.91
.09	1.3408	1.3346	1.3285	1.3225	1.3165	1.3106	1.3047	1.2988	1.2930	1.2873	1.2816	.90
.10	1.2816	1.2759	1.2702	1.2646	1.2591	1.2536	1.2481	1.2426	1.2372	1.2319	1.2265	.89
.11	1.2265	1.2212	1.2160	1.2107	1.2055	1.2004	1.1952	1.1901	1.1850	1.1800	1.1750	.88
.12	1.1750	1.1700	1.1650	1.1601	1.1552	1.1503	1.1455	1.1407	1.1359	1.1311	1.1264	.87
.13	1.1264	1.1217	1.1170	1.1123	1.1077	1.1031	1.0985	1.0939	1.0893	1.0848	1.0803	.86
.14	1.0803	1.0758	1.0714	1.0669	1.0625	1.0581	1.0537	1.0494	1.0450	1.0407	1.0364	.85
.15	1.0364	1.0322	1.0279	1.0237	1.0194	1.0152	1.0110	1.0069	1.0027	0.9986	0.9945	.84
.16	0.9945	0.9904	0.9863	0.9822	0.9782	0.9741	0.9701	0.9661	0.9621	0.9581	0.9542	.83
.17	0.9542	0.9502	0.9463	0.9424	0.9385	0.9346	0.9307	0.9269	0.9230	0.9192	0.9154	.82
.18	0.9154	0.9116	0.9078	0.9040	0.9002	0.8965	0.8927	0.8890	0.8853	0.8816	0.8779	.81
.19	0.8779	0.8742	0.8705	0.8669	0.8633	0.8596	0.8560	0.8524	0.8488	0.8452	0.8416	.80
.20	0.8416	0.8381	0.8345	0.8310	0.8274	0.8239	0.8204	0.8169	0.8134	0.8099	0.8064	.79
.21	0.8064	0.8030	0.7995	0.7961	0.7926	0.7892	0.7858	0.7824	0.7790	0.7756	0.7722	.78
.22	0.7722	0.7688	0.7655	0.7621	0.7588	0.7554	0.7521	0.7488	0.7454	0.7421	0.7388	.77
.23	0.7388	0.7356	0.7323	0.7290	0.7257	0.7225	0.7192	0.7160	0.7128	0.7095	0.7063	.76
.24	0.7063	0.7031	0.6999	0.6967	0.6935	0.6903	0.6871	0.6840	0.6808	0.6776	0.6745	.75
.25	0.6745	0.6713	0.6682	0.6651	0.6620	0.6588	0.6557	0.6526	0.6495	0.6464	0.6433	.74
.26	0.6433	0.6403	0.6372	0.6341	0.6311	0.6280	0.6250	0.6219	0.6189	0.6158	0.6128	.73
.27	0.6128	0.6098	0.6068	0.6038	0.6008	0.5978	0.5948	0.5918	0.5888	0.5858	0.5828	.72
.28	0.5828	0.5799	0.5769	0.5740	0.5710	0.5681	0.5651	0.5622	0.5592	0.5563	0.5534	.71
.29	0.5534	0.5505	0.5476	0.5446	0.5417	0.5388	0.5359	0.5330	0.5302	0.5273	0.5244	.70
.30	0.5244	0.5215	0.5187	0.5158	0.5129	0.5101	0.5072	0.5044	0.5015	0.4987	0.4959	.69
.31	0.4959	0.4930	0.4902	0.4874	0.4845	0.4817	0.4789	0.4761	0.4733	0.4705	0.4677	.68
.32	0.4677	0.4649	0.4621	0.4593	0.4565	0.4538	0.4510	0.4482	0.4454	0.4427	0.4399	.67
.33	0.4399	0.4372	0.4344	0.4316	0.4289	0.4261	0.4234	0.4207	0.4179	0.4152	0.4125	.66
.34	0.4125	0.4097	0.4070	0.4043	0.4016	0.3989	0.3961	0.3934	0.3907	0.3880	0.3853	.65
.35	0.3853	0.3826	0.3799	0.3772	0.3745	0.3719	0.3692	0.3665	0.3638	0.3611	0.3585	.64
.36	0.3585	0.3558	0.3531	0.3505	0.3478	0.3451	0.3425	0.3398	0.3372	0.3345	0.3319	.63
.37	0.3319	0.3292	0.3266	0.3239	0.3213	0.3186	0.3160	0.3134	0.3107	0.3081	0.3055	.62
.38	0.3055	0.3029	0.3002	0.2976	0.2950	0.2924	0.2898	0.2871	0.2845	0.2819	0.2793	.61
.39	0.2793	0.2767	0.2741	0.2715	0.2689	0.2663	0.2637	0.2611	0.2585	0.2559	0.2533	.60
.40	0.2533	0.2508	0.2482	0.2456	0.2430	0.2404	0.2378	0.2353	0.2327	0.2301	0.2275	.59
.41	0.2275	0.2250	0.2224	0.2198	0.2173	0.2147	0.2121	0.2096	0.2070	0.2045	0.2019	.58
.42	0.2019	0.1993	0.1968	0.1942	0.1917	0.1891	0.1866	0.1840	0.1815	0.1789	0.1764	.57
.43	0.1764	0.1738	0.1713	0.1687	0.1662	0.1637	0.1611	0.1586	0.1560	0.1535	0.1510	.56
.44	0.1510	0.1484	0.1459	0.1434	0.1408	0.1383	0.1358	0.1332	0.1307	0.1282	0.1257	.55
.45	0.1257	0.1231	0.1206	0.1181	0.1156	0.1130	0.1105	0.1080	0.1055	0.1030	0.1004	.54
.46	0.1004	0.0979	0.0954	0.0929	0.0904	0.0878	0.0853	0.0828	0.0803	0.0778	0.0753	.53
.47	0.0753	0.0728	0.0702	0.0677	0.0652	0.0627	0.0602	0.0577	0.0552	0.0527	0.0502	.52
.48	0.0502	0.0476	0.0451	0.0426	0.0401	0.0376	0.0351	0.0326	0.0301	0.0276	0.0251	.51
.49	0.0251	0.0226	0.0201	0.0175	0.0150	0.0125	0.0100	0.0075	0.0050	0.0025	0.0000	.50
	.010	.009	.008	.007	.006	.005	.004	.003	.002	.001	.000	P

Extreme values:

Q	10^{-4}	10^{-5}	10^{-6}	10^{-7}	10^{-8}	10^{-9}
Cell entry	3.7190	4.2649	4.7534	5.1993	5.6120	5.9978

For an extension in the range Q = 0.0000 (0.0001) 0.0200, see Table B-11b.

Body of table is reproduced from Table 4, "The Normal probability function," in *Biometrika Tables for Statisticians,* Vol. I, eds., E. S. Pearson and H. O. Hartley, Cambridge University Press, 1954, with the permission of the Trustees of *Biometrika.*

Table B-11b

Cumulative Probabilities for a Standardized Normal Variable φ
(Extended to Extreme Probabilities)

Q = left row heading + upper column heading = $P(\phi \geq$ cell entry$) = P(\phi \leq -$ cell entry$)$
P = right row heading + lower column heading = $P(\phi \leq$ cell entry$) = P(\phi \geq -$ cell entry$)$

Q	0.0000	0.0001	0.0002	0.0003	0.0004	0.0005	0.0006	0.0007	0.0008	0.0009	0.0010	P
0.000	∞	3.7190	3.5401	3.4316	3.3528	3.2905	3.2389	3.1947	3.1559	3.1214	3.0902	0.999
.001	3.0902	3.0618	3.0357	3.0115	2.9889	2.9677	2.9478	2.9290	2.9112	2.8943	2.8782	.998
.002	2.8782	2.8627	2.8480	2.8338	2.8202	2.8070	2.7944	2.7822	2.7703	2.7589	2.7478	.997
.003	2.7478	2.7370	2.7266	2.7164	2.7065	2.6968	2.6874	2.6783	2.6693	2.6606	2.6521	.996
.004	2.6521	2.6437	2.6356	2.6276	2.6197	2.6121	2.6045	2.5972	2.5899	2.5828	2.5758	.995
.005	2.5758	2.5690	2.5622	2.5556	2.5491	2.5427	2.5364	2.5302	2.5241	2.5181	2.5121	0.994
.006	2.5121	2.5063	2.5006	2.4949	2.4893	2.4838	2.4783	2.4730	2.4677	2.4624	2.4573	.993
.007	2.4573	2.4522	2.4471	2.4422	2.4372	2.4324	2.4276	2.4228	2.4181	2.4135	2.4089	.992
.008	2.4089	2.4044	2.3999	2.3954	2.3911	2.3867	2.3824	2.3781	2.3739	2.3698	2.3656	.991
.009	2.3656	2.3615	2.3575	2.3535	2.3495	2.3455	2.3416	2.3378	2.3339	2.3301	2.3263	.990
.010	2.3263	2.3226	2.3189	2.3152	2.3116	2.3080	2.3044	2.3009	2.2973	2.2938	2.2904	.989
.011	2.2904	2.2869	2.2835	2.2801	2.2768	2.2734	2.2701	2.2668	2.2636	2.2603	2.2571	.988
.012	2.2571	2.2539	2.2508	2.2476	2.2445	2.2414	2.2383	2.2353	2.2322	2.2292	2.2262	.987
.013	2.2262	2.2232	2.2203	2.2173	2.2144	2.2115	2.2086	2.2058	2.2029	2.2001	2.1973	.986
.014	2.1973	2.1945	2.1917	2.1890	2.1862	2.1835	2.1808	2.1781	2.1754	2.1727	2.1701	.985
.015	2.1701	2.1675	2.1648	2.1622	2.1596	2.1571	2.1545	2.1520	2.1494	2.1469	2.1444	0.984
.016	2.1444	2.1419	2.1394	2.1370	2.1345	2.1321	2.1297	2.1272	2.1248	2.1224	2.1201	.983
.017	2.1201	2.1177	2.1154	2.1130	2.1107	2.1084	2.1061	2.1038	2.1015	2.0992	2.0969	.982
.018	2.0969	2.0947	2.0924	2.0902	2.0880	2.0858	2.0836	2.0814	2.0792	2.0770	2.0749	.981
.019	2.0749	2.0727	2.0706	2.0684	2.0663	2.0642	2.0621	2.0600	2.0579	2.0558	2.0537	.980
	0.0010	0.0009	0.0008	0.0007	0.0006	0.0005	0.0004	0.0003	0.0002	0.0001	0.0000	P

Table B-12

Values of t With ν Degrees of Freedom Having Certain Cumulative Probabilities

$Q = P(t \geq \text{cell entry}) = P(t \leq -\text{cell entry})$
$2Q = P(t \geq \text{cell entry}) + P(t \leq -\text{cell entry})$

ν	$Q = 0.4$ $2Q = 0.8$	0.25 0.5	0.1 0.2	0.05 0.1	0.025 0.05	0.01 0.02	0.005 0.01	0.0025 0.005	0.001 0.002	0.0005 0.001
1	0.325	1.000	3.078	6.314	12.706	31.821	63.657	127.32	318.31	636.62
2	.289	0.816	1.886	2.920	4.303	6.965	9.925	14.089	22.326	31.598
3	.277	.765	1.638	2.353	3.182	4.541	5.841	7.453	10.213	12.924
4	.271	.741	1.533	2.132	2.776	3.747	4.604	5.598	7.173	8.610
5	0.267	0.727	1.476	2.015	2.571	3.365	4.032	4.773	5.893	6.869
6	.265	.718	1.440	1.943	2.447	3.143	3.707	4.317	5.208	5.959
7	.263	.711	1.415	1.895	2.365	2.998	3.499	4.029	4.785	5.408
8	.262	.706	1.397	1.860	2.306	2.896	3.355	3.833	4.501	5.041
9	.261	.703	1.383	1.833	2.262	2.821	3.250	3.690	4.297	4.781
10	0.260	0.700	1.372	1.812	2.228	2.764	3.169	3.581	4.144	4.587
11	.260	.697	1.363	1.796	2.201	2.718	3.106	3.497	4.025	4.437
12	.259	.695	1.356	1.782	2.179	2.681	3.055	3.428	3.930	4.318
13	.259	.694	1.350	1.771	2.160	2.650	3.012	3.372	3.852	4.221
14	.258	.692	1.345	1.761	2.145	2.624	2.977	3.326	3.787	4.140
15	0.258	0.691	1.341	1.753	2.131	2.602	2.947	3.286	3.733	4.073
16	.258	.690	1.337	1.746	2.120	2.583	2.921	3.252	3.686	4.015
17	.257	.689	1.333	1.740	2.110	2.567	2.898	3.222	3.646	3.965
18	.257	.688	1.330	1.734	2.101	2.552	2.878	3.197	3.610	3.922
19	.257	.688	1.328	1.729	2.093	2.539	2.861	3.174	3.579	3.883
20	0.257	0.687	1.325	1.725	2.086	2.528	2.845	3.153	3.552	3.850
21	.257	.686	1.323	1.721	2.080	2.518	2.831	3.135	3.527	3.819
22	.256	.686	1.321	1.717	2.074	2.508	2.819	3.119	3.505	3.792
23	.256	.685	1.319	1.714	2.069	2.500	2.807	3.104	3.485	3.767
24	.256	.685	1.318	1.711	2.064	2.492	2.797	3.091	3.467	3.745
25	0.256	0.684	1.316	1.708	2.060	2.485	2.787	3.078	3.450	3.725
26	.256	.684	1.315	1.706	2.056	2.479	2.779	3.067	3.435	3.707
27	.256	.684	1.314	1.703	2.052	2.473	2.771	3.057	3.421	3.690
28	.256	.683	1.313	1.701	2.048	2.467	2.763	3.047	3.408	3.674
29	.256	.683	1.311	1.699	2.045	2.462	2.756	3.038	3.396	3.659
30	0.256	0.683	1.310	1.697	2.042	2.457	2.750	3.030	3.385	3.646
40	.255	.681	1.303	1.684	2.021	2.423	2.704	2.971	3.307	3.551
60	.254	.679	1.296	1.671	2.000	2.390	2.660	2.915	3.232	3.460
120	.254	.677	1.289	1.658	1.980	2.358	2.617	2.860	3.160	3.373
∞	.253	.674	1.282	1.645	1.960	2.326	2.576	2.807	3.090	3.291

Q is the upper-tail area of the distribution for ν degrees of freedom, appropriate for use in a single-tail test. For a two-tail test, $2Q$ must be used.

Body of table is reproduced from Table 12, "Percentage points of the t-distribution," in *Biometrika Tables for Statisticians*, Vol. I, eds., E. S. Pearson and H. O. Hartley, Cambridge University Press, 1954, with the permission of the Trustees of *Biometrika*.

Values of Chi-Square With V Degrees of Freedom Having Certain
Cumulative Probabilities

$Q = P(\chi_\nu^2 \geqslant \text{cell entry})$

ν \ Q	0.250	0.100	0.050	0.025	0.010	0.005	0.001
1	1.32330	2.70554	3.84146	5.02389	6.63490	7.87944	10.828
2	2.77259	4.60517	5.99146	7.37776	9.21034	10.5966	13.816
3	4.10834	6.25139	7.81473	9.34840	11.3449	12.8382	16.266
4	5.38527	7.77944	9.48773	11.1433	13.2767	14.8603	18.467
5	6.62568	9.23636	11.0705	12.8325	15.0863	16.7496	20.515
6	7.84080	10.6446	12.5916	14.4494	16.8119	18.5476	22.458
7	9.03715	12.0170	14.0671	16.0128	18.4753	20.2777	24.322
8	10.2189	13.3616	15.5073	17.5345	20.0902	21.9550	26.125
9	11.3888	14.6837	16.9190	19.0228	21.6660	23.5894	27.877
10	12.5489	15.9872	18.3070	20.4832	23.2093	25.1882	29.588
11	13.7007	17.2750	19.6751	21.9200	24.7250	26.7568	31.264
12	14.8454	18.5493	21.0261	23.3367	26.2170	28.2995	32.909
13	15.9839	19.8119	22.3620	24.7356	27.6882	29.8195	34.528
14	17.1169	21.0641	23.6848	26.1189	29.1412	31.3194	36.123
15	18.2451	22.3071	24.9958	27.4884	30.5779	32.8013	37.697
16	19.3689	23.5418	26.2962	28.8454	31.9999	34.2672	39.252
17	20.4887	24.7690	27.5871	30.1910	33.4087	35.7185	40.790
18	21.6049	25.9894	28.8693	31.5264	34.8053	37.1565	42.312
19	22.7178	27.2036	30.1435	32.8523	36.1909	38.5823	43.820
20	23.8277	28.4120	31.4104	34.1696	37.5662	39.9968	45.315
21	24.9348	29.6151	32.6706	35.4789	38.9322	41.4011	46.797
22	26.0393	30.8133	33.9244	36.7807	40.2894	42.7957	48.268
23	27.1413	32.0069	35.1725	38.0756	41.6384	44.1813	49.728
24	28.2412	33.1962	36.4150	39.3641	42.9798	45.5585	51.179
25	29.3389	34.3816	37.6525	40.6465	44.3141	46.9279	52.618
26	30.4346	35.5632	38.8851	41.9232	45.6417	48.2899	54.052
27	31.5284	36.7412	40.1133	43.1945	46.9629	49.6449	55.476
28	32.6205	37.9159	41.3371	44.4608	48.2782	50.9934	56.892
29	33.7109	39.0875	42.5570	45.7223	49.5879	52.3356	58.301
30	34.7997	40.2560	43.7730	46.9792	50.8922	53.6720	59.703
40	45.6160	51.8051	55.7585	59.3417	63.6907	66.7660	73.402
50	56.3336	63.1671	67.5048	71.4202	76.1539	79.4900	86.661
60	66.9815	74.3970	79.0819	83.2977	88.3794	91.9517	99.607
70	77.5767	85.5270	90.5312	95.0232	100.425	104.215	112.317
80	88.1303	96.5782	101.879	106.629	112.329	116.321	124.839
90	98.6499	107.565	113.145	118.136	124.116	128.299	137.208
100	109.141	118.498	124.342	129.561	135.807	140.169	149.449
X	+ 0.6745	+ 1.2816	+ 1.6449	+ 1.9600	+ 2.3263	+ 2.5758	+ 3.0902

For $\nu > 100$ take

$$\chi^2 = \nu \left\{ 1 - \frac{2}{9\nu} + X\sqrt{\frac{2}{9\nu}} \right\}^3 \quad \text{or} \quad \chi^2 = \frac{1}{2}\left\{ X + \sqrt{(2\nu - 1)} \right\}^2 ,$$

according to the degree of accuracy required. X is the standardized normal deviate corresponding to $P = 1 - Q$, and is shown in the bottom line of the table.

appendix c

answers to odd-numbered problems

Chapter 2

1. (a) 3, (b) 4, (c) 4.625, (d) 2, (e) 9, (f) 9, (g) 7, (h) 3
    (i) 1.5, (j) 6.141, (k) 2.478
3. (a) frequency distribution, (b) frequency distribution,
    (c) histogram, (d) histogram,
    (e) frequency polygon (or histogram)

Chapter 3

1. (a) 11!, (b) 1152, (c) 34,650
3. 17,325
5. (a) 252, (b) 252, (c) 252, (d) 126
7. (a) 10,000, (b) 16
9. 31,104
11. 60,480
13. 288
15. (a) 16,276, (b) 18,278
17. 65
19. 3360
21. (a) 29,760, (b) 32,768, (c) 2,976

Chapter 4

1. No. Farmers seem to prefer red or white.
3. (a) Yes, (b) Yes, (c) Yes, (d) No
5. d(least) c b a(most)

7. (a) Random sample requires that every unit be equally likely to be chosen; representative sample does not. (b) No (c) Extremely atypical cases (d) No (e) The typicalness of its contents or of some statistic calculated from its contents.

Chapter 5

1. (a) A priori (symmetry), (b) frequency, (c) a priori (symmetry), (d) frequency, (e) subjective, (f) subjective
3. (a) No, (b) No, (c) Doubtful, (d) No, (e) No, (f) Yes

Chapter 6

1. (a) .360, (b) .750, (c) .050, (d) .230, (e) .580, (f) 0, (g) .600, (h) .375, (i) 0
3. .615
5. (i) .778, (ii) 1.00
7. .0433
9. (a) .856, (b) No
11. 7
13. $\frac{1}{3}$
15. .987
17. .140
19. .847
21. (a) $1 - \dfrac{12!/(12 - n)!}{12^n}$ (b) 5
23. .760
25. (a) $.999^{1000}$, (b) $.001(.999)^{999}$, (c) $\binom{1000}{1}.001(.999)^{999}$
27. .020
29. .0286
31. .335

Chapter 7

1. .5
3. .536
5. (a) .531, (b) .469, (c) .114, (d) .000001
7. .159
9. .114
11. (a) 7, (b) .0531, (c) .0387
13. (a) .0588, (b) .2499, (c) .8319, (d) .1681
15. (a) .139, (b) .593, (c) .912, (d) .984
17. (a) .00103, (b) .249, (c) .718
19. .393
21. (a) .0988, (b) .296
23. .00995
25. .0952
27. (a) .138, (b) .198
29. .918

Chapter 8

1. $-.6$ 3. 1
5. 6.45   7. 5
9. (a) a_2,   (b) b_3,   (c) a_1, a_4, and a_5 are each dominated by a_2 and a_3
   (d) b_1, b_2, b_5, and b_6 are each dominated by b_3 and b_4

   (e)

50	100
90	3

   (f) $P(a_2) = 87/137, P(a_3) = 50/137,$
   $E(\text{payoff}) = 64.6$

   (g) $P(b_3) = 97/137, P(b_4) = 40/137, E(\text{payoff}) = 64.6$
   (h) Player A, because if both players play maximin-minimax A gets
   only 50
11. (a) 8, (b) B, (c) $P(b_1) = .5, P(b_2) = .5, E(\text{payoff}) = 5$
13. (a) -2, (b) A, (c) $P(a_1) = 9/14, P(a_2) = 5/14, E(\text{payoff}) = .571$
15. (a) 100 dollars, (b) 1/1024, (c) $.5^{100}$

Chapter 9

1. a_3 because its the action corresponding to the largest utility under
   θ_2 — company pays half.
3. a_3 — Stock 2. Method would be reasonable if merchant were going
   out of business, unreasonable if merchant expects to play the game
   often and can afford to take occasional losses.
5. 0, 28, 35, 27, 10, a_3
7. (a) .105, .291, .604, (b) 400, $-666, a_1$
9. .890
11. .754, .246, a_1
13. S_5 has largest minimum expected utility. Its expected utilities under
   θ_1 and θ_2 are 6.3 and 3.9.
15.

	WOULD HAVE WON					Expected
	1st Prize	2nd Prize	3rd Prize	4th Prize	No Prize	Utility
Enter	99,999	9,999	99	4	-1	$-.9175$
Don't	0	0	0	0	0	0

(a) not enter, (b) not enter, (c) not enter

17.

	CONDITION OF PATIENT			Expected	Appropriate
	A	B	C	Utility	Action
Operate	10	2	12	9.88	
Don't	0	5	30	16.1	X

(a) don't operate, (b) don't operate, (c) operate

19.

	NUMBER OF SPOILED CANS IN ORIGINAL DOZEN				Revised Expected Utility	Appropriate Action
	0	*1*	*2*	*12*		
a_1	—	100	50	—	84.4	X
a_2	—	70	46	—	62.5	
a_3	—	50	40	—	46.9	

Revised
State   0   .688   .312   0
Probabilities

21. 5, 1.695 dollars

Chapter 10

1. (a) .2, .7, .1, (b) -5.4, -5.9, -11.6, (c) a_1, (d) θ_2
3. (a) 2, (b) .385
5. Needs minor adjustment, .0828, .5030, .4142
7. Needs overhaul, .010, .458, .532
9. .493
11. .99999109
13. (a) .2, (b) Route manager's hypothesis, .0000612, .1761608
    (c) Route manager's hypothesis, .000347
15. (a) .0914, .0867, (b) popular, (c) unpopular, (d) .949
17. 0, 0, 1
19. (a)

CHIEF	WITCH DOCTOR	GRAND VIZIER	PROBABILITY
A	B	C	.117
A	C	B	.039
B	A	C	.058
B	C	A	.006
C	A	B	.010
C	B	A	.003

(b) A, B, C

Chapter 11

1. (a) .000432, (b) .00502, (c) .99498, (d) .000432, (e) .0000007
3. (a) 0, 1, (b) .99964, (c) .00036, (d) .00351
5. (a) .111, (b) .262, (c) .738, (d) .407
7. (a) .461, (b) .461
9. (a) H_o: B was killer, H_a: A was killer, (b) $r = 3$, (c) H_o
   (d) .578, (e) .729

11. (a) number of trout caught, (b) hypergeometric, (c) H_o: fish caught in Pond B, H_a: fish caught in Pond A, (d) θ_o: $R = 5$, $N = 8$; θ_a: $R = 3$, $N = 10$, (e) $r = 0$, (f) .0179, (g) .536 (h) .708, (i) .292, (j) H_o
13. (a) .0168, (b) .000655, (c) .900, (d) .100, (e) .999345
15. (a) 3 to 9, (b) .0530, (c) .0000803, (d) .601, (e) .399, (f) .0000001
17. (a) 5 to 12, (b) .00433, (c) .0000001, (d) .842, (e) .158, (f) .0000034
19. (a) 0, 1, (b) .0464, (c) .0000042, (d) .264, (e) .0464, (f) .167
21. (a) 0, (b) .00605, (c) .0000001, (d) .651, (e) .00605, (f) .167
23. (a) 6, 7, 8, (b) .0498, (c) .145
25. (a) $p \geq .02$, (b) $p < .02$, (c) 0, 1, (d) .0404, (e) .265, (f) claim is false
27. (a) $r = 0$, (b) .0198, (c) .0198, (d) mild form, (e) .895
29. (a) 3, (b) .677
31. (a) $r = 0$, (b) .0422, (c) accept, .455, (d) .314

Chapter 12

1. (a) 2, (b) 2, (c) reject H_o
3. .104   5. .349
7. (a) 8, (b) 0 and 8, (c) reject H_o
9. (a) 0, 1; .109, (b) .344, (c) No
11. (a) Yes, (b) .0207
13. (a) maximum value is 3, (b) .0352, (c) ≤ 3 and ≥ 12, (d) reject H_o
15. (a) 12, (b) .99631, (c) 3, (d) accept H_o
17. (a) No, (b) .0699
19. (a) .0406, (b) reject H_o
21. (a) 22, (b) ≤ 20 and ≥ 220, (c) accept H_o
23. .00833
25. (a) 108, (b) 90, (c) reject H_o
27. (a) .0357, (b) .00794, (c) .00833, (d) .000666, (e) .000194
29. (a) 154, (b) 184 and 281, (c) reject H_o
31. (a) 28, (b) 55, (c) accept H_o
33. (a) 14, (b) 14, (c) reject H_o, (d) .194
35. (a) 2, (b) .600, (c) there is none, (d) accept H_o
37. (a) 74, 109.3, accept H_o, (b) 2, there is none, accept H_o, (c) 136, 109.3, reject H_o
39. .961, .031, .928   41. .998, .978, .891, .366
43. (a) 6, (b) 9   45. .961

47. 74,707   49. .103
51. (a) .00046, (b) .0156   53. .711

Chapter 13

1. (a) -11.5, (b) 2, (c) 2.3, (d) 10, (e) 6.3, (f) 11.6, (g) .5,
   (h) 4.3, (i) 8.2, (j) 4.6, (k) 14.6, (l) $2.17 + 1.36\sqrt{-1}$,
   (m) 134, (n) 9.65, (o) 3.57
3. (a) -1, (b) 10.4, (c) 1.145, (d) 844.8
5. 1, 0, 3, 1.15, 2.33   7. $-.333$, .389
9. 3.5, 16.16
11. $\mu_X - \mu_Y$, $\text{Var}(X) + \text{Var}(Y) - 2\,\text{Cov}(XY)$
13. (a) 0, (b) 5, (c) 1, (d) 5, (e) 1, (f) 6, (g) 6, (h) 4,
    (i) 21, (j) 4, (k) 25, (l) 16, (m) 25, (n) 37
15. .08
17. (a) 0, (b) 0, (c) 0, (d) .5, (e) .25, (f) .6875
19. 95, 3.60
21. 23.5 days, 10.512 days
23. 5 pill approach is preferable because the standard deviation of the
    total is .4472 which is smaller than that for the single 5 grain pill.

Chapter 14

1. (a) .75, .433, (b) $P(\bar{X} = 0) = .0625$, $P(\bar{X} = .5) = .3750$,
   $P(\bar{X} = 1) = .5625$, (c) .75, .306
3. (a) .444, (b) .741   14-5. 100
7. .25   14-9. 25,000
11. (a) .0278, (b) .00694, (c) .000025, (d) .00000025
13. 3,200,000
15. (a) no larger than 250, (b) no larger than 250,000
17. (a) .333, (b) .333, (c) .100, (d) .067, (e) 0, (f) 0,
    (g) 1.000, (h) 1.000, (i) .667, (j) .500, (k) .200, (l) 0
19. -6.32
21. 2.28, 13.59, 68.27, 13.59, 2.28
23. .0228
25. (a) .875, (b) .711, (c) .875, (d) .930
27. (a) .5, .5, .5, .5, .3, .3, (b) 1.581, 1.118, .913, .791, .707, .5
    (c) .822, .582, .475, .411, .368, .260
29. (a) .01, .01, .01, .01, .01, .01, (b) .315, .222, .182, .157, .141, .099,
    (c) .164, .116, .0944, .0818, .0732, .0518

Chapter 15

1. (a) .159,   (b) −1.4758,   (c) .8416,   (d) .528,   (e) .44
3. $2.08 < \mu < 9.92$
5. .0204
7. (a) $.49991 < \mu < .50089$,   (b) $.499756 < \mu < .501044$,   (c) 1.6,
    .110, accept H_o, ±1.96,   (d) 1.6, .055, accept H_o, 2.3263,
    (e) .953, .372, .027, <.0001,   (f) 7
9. (a) $.00157 < \mu_A - \mu_B < .00363$,   (b) $.00124 < \mu_A - \mu_B < .00396$
    (c) 3.034, .0024, ±1.96, reject H_o,   (d) 3.034, .9988, −2.3263,
    accept H_o,   (e) .3336, .0838, .0100, .0005,   (f) 4
11. (a) $76.14 < \mu < 86.66$,   (b) 2, .023, 2.3263, accept H_o,   (c) .044
13. (a) $-35.23 < \mu_X - \mu_Y < .83$,   (b) −1.886, .030, −3.0902,
    accept H_o,   (c) .138
15. (a) $.49964 < \mu < .50116$,   (b) $.49901 < \mu < .50179$,   (c) 1.681,
    approximately .20, accept H_o, ±3.182,   (d) 1.681, approximately
    .10, accept H_o, 4.541
17. 1.373, 2.132, accept H_o, between .10 and .25
19. (a) $\mu_Z - \mu_X < 1$,   (b) $\mu_Z - \mu_X \geq 1$,   (c) difference-score t test,
    (d) 8.66, 6.965, between .005 and .01, reject H_o
21. .0051
23. 8.53, 11.0705, between .10 and .25, accept H_o
25. 15.13, 13.816, reject H_o
27. 8.33, 5.99147, between .01 and .025, reject H_o
29. (a) Approximately .025,   (b) 5.104,   (c) 6.6349,   (d) accept H_o
31. 5.15, 9.48773, approximately .25, accept H_o
33. (a) No. Assumption of independence is violated since 2000 > 800
    (b) No. Same reason

index

(Boldface italic or italicized numbers signify pages on which term appears in boldface italic or italics, respectively. Only the more important occurrences of a term are listed.)

A

Acceptance region, *234, 246*-247, ***261***, 263
Allocations, 33-39
Alternative hypothesis, *246*-247, ***260***, *262*-263
Assumptions, (*see* text on specific test)
Atomic outcomes, 55-*56*, 75-76

B

Bayes' formula, ***104***-109, 188-190, 193-195, 197-199, 217-221
Bernoullis theorem, **378**-383 (*see also* 51-55)

Biased sample, *57*-64
Biased selection, *56*, 58-60
Binomial probability law, 130-133, 142-146, 149-150, 153, 155, 350-352, 479-481, Table B-3 (*see also* 133-142)
Binomial tests, 241-264, 272-278, 307-311 (*see also* 228-234)

C

Central Limit Effect, *388*-433
  effect of skewness, 398-432
  factors influencing, 404, 430-432
  importance of tails, 392-432
  for *L*-shaped population, 390-432
  methods of investigating, 388-390
  for rectangular population, 388-389

Central Limit Theorem, *386*-388,
 504-512

Certainty, decision making under,
 182-185

Chance, 68-69, *73*

Chi-square, 297, 300, 307, 483-495,
 Table B-13

Combinations, *29*-33

Composite hypothesis, *259*

Concentration, test for, 277-278

Conditionally most probable state
 estimate, *215*-221

Conditional probabilities, 102-109

Confidence interval:
 defined, 312-*313*
 for difference between means, 445,
  448-449, 459-461, 467-468,
  473-477
 for mean, 438-440, 457-458, 463,
  465-466
 for median, 315
 for percentile, 313-316

Confidence level, *313*

Confidence limits, *313*

Consistent estimator, 316-*317*, 364,
 378-379

Containers, allocation to, 33-39

Correction for continuity, 480-481

Correlation, *284*-286, *334*
 defined, *284, 334*
 negative, *284*
 positive, *284*
 test for, 284-286 (*see also* 278-281)

Cost of sampling, 192-193

Counterbalancing sequential effects,
 27-29

Covariance, *340*, 348

Cox and Stuart's test, 276-277

Critical ties, 309-310

Critical value, 232, 234, *248*, 261

Cumulative probability, *148-149*,
 151-154

Cumulative proportion graph, *9*-10

D

Decision making, 161-270

Decision rules, *226*-227, 230-232, 244

Decision theory, 161-270

Density function, *9*-10

Dependence, test for, 278-281

Descriptive statistics, 3-18

Dice experiment, 85-89, 382-383

Dispersion, indices of, *12*-16

Distinguishable permutations, *20-21*,
 24-27, 32-33

Distribution-free tests and estimates,
 271-318

Distribution:
 chi square, 483-484
 of frequencies, 123-139, 349-352,
  378-383
 normal, 382-386
 of occurrences, 146-148, 349-353
 of sample mean, 360-433, 342-346
 of *t*, 461-462
 of test statistic (*see* specific test)
 of trials to criterion, 139-142, 349-
  354

Divisions of things into groups, 36-38

Dominated actions, *171*-172, 175-176

E

Equal likelihood, 50-56, 75-81, 83-84
 as an ideal, 79
 as a premise, 75-77
 inferred from ignorance, 79-81
 inferred from symmetry, 76-79
 insured by randomness, 76-77, 83-84

Equivalent decision rule, *226*-227,
 230-232, 244

Estimated objective probability, 74-75, 194-195

Estimation:
estimate, *215*
vs. hypothesis testing, 311-312
interval estimate, *215*, 311-316
point estimate, *215*-228, 311, *316*-318
of population indices, 316-318, 354-356, 364, 378
qualities of estimators, 316-318, 354-356
of states of nature, 215-228

Expectation, laws of, 329-334

Expected utility, 167-*168*, *185*

Expected value:
defined, *162, 329*
of perfect information, *192*
of sample information, 190-*192*
of sample statistics, 342-354
of utility, 167-*168*, *185*

Expedient announcements, 213-214

F

Fisher's test, 278-281

Fractionation method, 167

Frequency distribution, *4*-7, 8, 10, 16
cumulative, *4*-6, 8
cumulative relative, *4*-6, 8, 10
relative, *4*-7

Frequency polygon, 9-11

Friedman test, 293-298, 302-307

G

Games, 161-176

Geometric probability law, 141-142, 150, 153, 155, 350

Goodness-of-fit test, 493-494

Groupings, 33-39

H

Histogram, *7*-8

Hotelling and Pabst's test, 284-286

Hypergeometric probability law, 123-126, 129, 138-139, 149, 153, 155, 349-351, 482-483, Table B-6

Hypergeometric tests, 278-281

Hypotheses:
choosing between, 222-234 (*see also* 240-264)
composite, *259*
simple, *259*
varieties, 259-260

Hypothesis testing, 228-234, 240-264, 311-312

I

Ignorance, decision making under, 182-*183*

Independence:
as an assumption, 113, *487, 491*, 493
mutual (general case), *112*-116
tests of, 278-281, 284-286, 489-493
between two characteristics, *109*-111

Interaction, 302-307
defined, *303*-305
test for, 302-307

Interpercentile range, *14*-16
Interquartile range, *14*-16
Interval estimate, *215*, 311-316

K

Kruskal-Wallis test, 298-307
Kurtosis, *338*

L

Law of Large Numbers, 376-383 (*see
    also* 51-55, 312, 316)
  for sample mean, 376-378
  for sample proportion, *378*-383
Left-tail test, 248-*252*, *261*
Levels, *302*-303
Likelihood:
  equal, 50-58
  under random sampling, 57-58, 77
  under random selection, 55-57, 77
  relationship to probability, 78
Likelihood Ratio Test, 224, *228*-234,
    240-241, 244
Literary Digest poll, *62*-64
Location indices, *12*-13, 15-16
Lower-tail test, 248-*252*, *261*

M

Main effect, 302-307
  defined, *303*-305
  test for, 302-307
Mann-Whitney U-Test, 292-293
Maximin strategy, 169-*170*

Maximum likelihood estimate, *215*,
    221-222, 224-228
Mean:
  confidence intervals and tests for,
    437-479
  as estimator, 317, 354, 364, 376-378
  index of location, *13*, 15-16, 18,
    335-336
  of probability distributions, 342-354
  for sample statistics, 342-349
  for specific probability laws, 349-
    354
Mean-unbiased estimators, 316-*317*,
    354-356
Median:
  confidence interval for, 313-316
  as estimator, 317-318
  index of location, *12*-13, 15-16
  test for, 272-274
Median difference, test for, 274-276
Median-unbiased estimator, 316-318
Midrank method, 309-*310*
Minimax strategy, 169-*170*
Mixed strategy, *173*-175
Mode, *12*, 15-16
Molecular outcome, 55-*56*, 75-76
Moments, *335*-339
Most probable state estimate, 215-216
Multiaction games, 175
Multinomial probability law, 133-139,
    149-150, 155, 484-487
Multivariate hypergeometric probability
    law, 126-129, 136-139, 149-150,
    155
Mutually exclusive characteristics, *98*-
    101

N

Negative binomial probability law,
    139-142, 150, 153, 155, 350-354

Neyman-Pearson procedure, 234, 240-264

Normal approximation:
 to binomial, 479-481
 to hypergeometric, 482-483
 to Poisson, 482

Normal distribution, 382-386, Table B-11

Null hypothesis, *246, 260*

O

Objective probability, *74* (*see also* 73)

Observation, *3*, 17

Optimum sample size, 192

P

Parallel Axis Theorem, *336*

Parameter, *18*

Pascal's Rule, *32*

Pascal's Triangle, *32*-33

Payoffs, *168*-169

Percentile:
 confidence interval for, 313-316
 index of location, *13*, 15-16
 test for, 274

Permutations, *20*-29, 32-33

Point estimate, *215*, 311, *316*-318

Point probability, *148*-150

Poisson approximation to binomial, 142-146, 153, 155

Poisson probability law, 146-148, 150-153, 155, 350-353, 482, Table B-4

Population, *16*-18

Power, *246*-247, 250-264, 282-283, 441-443, 447, 451-452, 458-461, 465-468, 476-477

Power function, *250*-257, 262

Prior probability, *190*

Probability:
 of atomic outcome, 55-*56*, 76
 concept, 2, 69-73
  lay meaning, 69-70
  technical meaning, 70-73
 definitions or types, 73-75, 78-82, 125
  a posteriori, 81-*82*
  a priori, *78*-81, 125
  classical, 78
  empirical, 81-*82*
  mathematical, *78*-81
  objective, *74*
  radically objective, *73*
  subjective, *74*-75
 laws, 91-116, 121-155, Appendix B
 methods of obtaining or estimating, 75-89
  deductive, 75-81, 83-89
  frequency, 75, 81-82, 84-89
  subjective, 75, 82-89
 of molecular outcome, 55-*56*, 76
 under random sampling, 57-*58*, 77
 under random selection, 55-57, 77
 for a random variable, 77-78
 relationship to likelihood, 78
 under symmetry, 78-81

Probability distribution, *122*
 defined, *122*
 relationship to sampling distribution, 122

Probability level, *248*, 252, 255, *262*, 274, 276-278, 286, 443, 447, 489, 493, 494

Proportions, laws of, 91-116

Proportions, tests of, 240-260, 277-278, 487-489

Q

Quartiles, *13*, 15

R

Radically objective probability, *73*
RAND table, 58-59, Table B-1
Random numbers, 58-59, Table B-1
Randomness, 56-66
  attaining it, 58-60
  consequences of not attaining it,
    61-64
  as a prototype, 65-66
Random sample, *57* (*see also* 56-66,
    77)
Random sampling, 57-60, 66, 77
Random selection, 55-60, 77
Random variable, 64-*65*, 77-78
Range, *13*, 15
Rank tests, 281-310
Recursion formulas, *150*-151, 153
Reference class of similar incidents,
    70-73, 76
Rejection region, *234, 246*-247, *261*,
    263 (*see also* 240-264, 271-307,
    437-495)
Relative frequency polygon, 9-*10*
Revised probabilities, 188-190, 193-
    195, 217-221
Right-tail test, 252-254, *261*
Risk, decision making under, 182-*183*

S

Saddlepoint, *175*-176
St. Petersburg Paradox, *163*

Sample, *16*-18
Sample information:
  use in revising probabilities, 188-190,
    197
  value of, 190-192
Sample size, effect of, 51-55, 227, 232,
    250, 254, 257, 364, 370-383,
    388-433
Sampling distribution, 39-47, 122
  defined, *42*
  illustrated, 40-41
  relationship to probability
    distribution, 122
  shortcut methods, 42-47
Sampling without replacements, *17*-18
Sampling with replacements, 17-*18*
Scatter diagram, 10-*12*
Semi-interquartile range, *14*-15
Sensitivity of a test, *283*
Significance level, *248, 262*
Simple hypothesis, *259*
Skewness, *337*-339
Standard deviation, *14*-15
Standardized value, *336*-339
States of nature, 181
Statistic, *18*
Stirling's approximation, *26*
Strategy, 201
Subjective probability:
  defined, *74*-75
  as estimate of objective probability,
    82-83
  revised by data outcomes, 193-195
  convergence upon objective
    probability, *194*-195
Symmetry, probability under, 76-81

T

Tchebycheff inequalities, *365*-383

Tchebycheff inequalities (cont.)
general case, *365*-370
for sample mean, *370*-372, 375-378
for other sample statistics, 372-375, 378-383
Test, 224
Tests of:
concentration, 277-278
correlation, 284-286
dependence, 278-281, 489-493
difference between means, 443-452, 459-461, 466-477
goodness-of-fit, 493-494
independence, 278-281, 489-493
interaction, 302-307
main effects, 302-307
mean, 441-447, 457-460, 462-468
median, 272-274
median difference, 274-276
nonidentical distributions, 289-293, 298-302
percentile, 274
probabilities or proportions, 228-234, 241-259, 277-278, 487-489
treatment effects, 286-307
trend, 276-277, 286
Test statistics, 241, *246*, *260*-264
Tied observations, treatment of, 271, 307-310
Treatment effects, test for, 281-283, 286-310
Treatment populations, 275, 281-282
Trend, test for, 276-277, 286
t statistics, estimates and tests, 461-479
Two-action games, 170-175
Two-person zero-sum games, *168*-176
Two-tail test, 254-259, *261*
Type I Error, 234, *246*-248, *261*-263
(*see also* 241-264)
Type II Error, 234, *246*-247, *261*-263
(*see also* 241-264)

U

Uncertainty, decision making under, 182-*183*
Upper-tail test, 252-*254*, *261*
Utility, *165*-168

V

Value:
defined, *3*, 15, 17 (*see also* 161-168)
expected, *162*-165
objective, 161-162
of a sample, 190-193
subjective, 161-162, 165-168
Variability, 1-2
Variance:
of an average, 342
as estimator, 355-356, 453-456
index of dispersion, *14*-16, 18, *335*-336
laws of, 339-342
of probability distributions, 342-354
for sample statistics, 342-349
for specific probability laws, 349-354
Venn diagram, *100*-101

W

Wilcoxon tests, 286-293, 298, 305, 309-310

Z

Zero differences, treatment of, 271,
    307-311
Zero-sum games, 168-176

Z-scores, 336-339
Z statistics, estimates and tests, 437-
    453
Z statistics, estimates and tests,
    453-461